Applied Time Series Analysis

STATISTICS: Textbooks and Monographs

D. B. Owen

Founding Editor, 1972–1991

STATISTICS: Textbooks and Monographs

Recent Titles

Multivariate Statistical Analysis, Second Edition, Revised and Expanded, *Narayan C. Giri*

Computational Methods in Statistics and Econometrics, *Hisashi Tanizaki*

Applied Sequential Methodologies: Real-World Examples with Data Analysis, *edited by Nitis Mukhopadhyay, Sujay Datta, and Saibal Chattopadhyay*

Handbook of Beta Distribution and Its Applications, *edited by Arjun K. Gupta and Saralees Nadarajah*

Item Response Theory: Parameter Estimation Techniques, Second Edition, *edited by Frank B. Baker and Seock-Ho Kim*

Statistical Methods in Computer Security, *edited by William W. S. Chen*

Elementary Statistical Quality Control, Second Edition, *John T. Burr*

Data Analysis of Asymmetric Structures, *Takayuki Saito and Hiroshi Yadohisa*

Mathematical Statistics with Applications, *Asha Seth Kapadia, Wenyaw Chan, and Lemuel Moyé*

Advances on Models, Characterizations and Applications, *N. Balakrishnan, I. G. Bairamov, and O. L. Gebizlioglu*

Survey Sampling: Theory and Methods, Second Edition, *Arijit Chaudhuri and Horst Stenger*

Statistical Design of Experiments with Engineering Applications, *Kamel Rekab and Muzaffar Shaikh*

Quality by Experimental Design, Third Edition, *Thomas B. Barker*

Handbook of Parallel Computing and Statistics, *Erricos John Kontoghiorghes*

Statistical Inference Based on Divergence Measures, *Leandro Pardo*

A Kalman Filter Primer, *Randy Eubank*

Introductory Statistical Inference, *Nitis Mukhopadhyay*

Handbook of Statistical Distributions with Applications, *K. Krishnamoorthy*

A Course on Queueing Models, *Joti Lal Jain, Sri Gopal Mohanty, and Walter Böhm*

Univariate and Multivariate General Linear Models: Theory and Applications with SAS, Second Edition, *Kevin Kim and Neil Timm*

Randomization Tests, Fourth Edition, *Eugene S. Edgington and Patrick Onghena*

Design and Analysis of Experiments: Classical and Regression Approaches with SAS, *Leonard C. Onyiah*

Analytical Methods for Risk Management: A Systems Engineering Perspective, *Paul R. Garvey*

Confidence Intervals in Generalized Regression Models, *Esa Uusipaikka*

Introduction to Spatial Econometrics, *James LeSage and R. Kelley Pace*

Acceptance Sampling in Quality Control, *Edward G. Schilling and Dean V. Neubauer*

Applied Statistical Inference with MINITAB®, *Sally A. Lesik*

Nonparametric Statistical Inference, Fifth Edition, *Jean Dickinson Gibbons and Subhabrata Chakraborti*

Bayesian Model Selection and Statistical Modeling, *Tomohiro Ando*

Handbook of Empirical Economics and Finance, *Aman Ullah and David E. A. Giles*

Randomized Response and Indirect Questioning Techniques in Surveys, *Arijit Chaudhuri*

Applied Time Series Analysis, *Wayne A. Woodward, Henry L. Gray, and Alan C. Elliott*

Applied Time Series Analysis

Wayne A. Woodward
Southern Methodist University
Dallas, Texas, USA

Henry L. Gray
Southern Methodist University
Dallas, Texas, USA

Alan C. Elliott
University of Texas Southwestern Medical Center at Dallas
Dallas, Texas, USA

CRC Press is an imprint of the
Taylor & Francis Group, an **informa** business
A CHAPMAN & HALL BOOK

CRC Press
Taylor & Francis Group
6000 Broken Sound Parkway NW, Suite 300
Boca Raton, FL 33487-2742

Printed in the United States of America on acid-free paper
Version Date: 20110629

International Standard Book Number: 978-1-4398-1837-4 (Hardback)

Library of Congress Cataloging-in-Publication Data

Woodward, Wayne A.
Applied time series analysis / Wayne A Woodward, Henry L. Gray, and Alan C. Elliott.
p. cm. -- (Statistics: textbooks and monographs)
Includes bibliographical references and index.
ISBN 978-1-4398-1837-4 (alk. paper)
1. Time-series analysis. I. Gray, Henry L. II. Elliott, Alan C. III. Title.

QA280.W66 201
519.5'5--dc23 2011025090

Visit the Taylor & Francis Web site at
http://www.taylorandfrancis.com

and the CRC Press Web site at
http://www.crcpress.com

To Beverly, Becky, and E'Lynne

Contents

Preface

Why another time series book? That is a good question since there are so many excellent books available for a first and/or second course in time series analysis. *Applied Time Series Analysis* is a product of more than 25 years of teaching a course in introductory time series at Southern Methodist University (SMU). This course has always been substantially integrated with the use of an in-house time series package designed not only to do data analysis but also to educate users about properties of models. These time series analysis capabilities are now integrated into the software package `GW-WINKS` that can be downloaded from the website given later in the Preface.

Please keep in mind that this book is written to accomplish several distinct purposes. Among these are (1) provide a teachable book with innovative class-proven material that can be used in first and second semester time series courses, (2) cover areas such as long-memory models and data with time-varying frequencies/autocorrelations that are not typically dealt with in other such books, and (3) provide the student with the tools to solve *real* problems. Upon completion of this book, the student should have the understanding necessary to model, forecast, and identify the underlying components of a time series.

We believe that the writing style and the accompanying software will make this book a functional text for experts in the field of time series analysis along with being an invaluable resource for faculty members who have been "assigned" to teach the time series course but have no real experience in this area.

Features of the Book

Chapter 1 provides a treatment of stationarity and an analysis of stationary time series from both the time and frequency domains. We believe that it is important for students to become comfortable with the frequency domain from the get-go. The chapter contains a readable coverage of the frequency domain that does not assume a prior familiarity with it. Frequency domain concepts covered in this chapter include (1) frequency and period, (2) Nyquist frequency, (3) spectrum and spectral density, and (4) periodogram and window-based estimation of the spectral density.

Chapter 2 discusses basic filter material. In addition to the typical discussion of the general linear process and its relationship to stationarity, we provide a section containing practical filtering applications, which includes coverage of the frequency response function of a filter and how it determines the high pass, low pass, band pass, etc. properties of the filter. Unique to this statistical time series text is a brief discussion of the Butterworth filter, a widely used filter in science and engineering but one not usually covered in time series texts.

Chapter 3 gives an extensive coverage of ARMA(p,q) models. Learning the material in this chapter is greatly enhanced by using the `GW-WINKS` software, which provides unique capabilities for generating and plotting realizations from specified models along with their theoretical autocorrelations and spectral densities. Numerous simulation-based examples are given in this chapter to show properties of models. Although basic theorems are given and their proofs either included in the chapter's appendix or referenced, our coverage of these models focuses on applications over theory.

Factor tables: Unique to our treatment of ARMA processes is the use of factor tables that are based on a factoring of the pth and qth order characteristic polynomials of the autoregressive and moving average portions of an ARMA (p,q) model into their first and second order factors (see Woodward et al., 2009). We have used factor tables extensively in our introductory time series course at SMU for many years, and our experience is that these are indispensable tools for understanding ARMA models. The factor table and examples in the book are used to aid in understanding

- The frequency behavior of models
- How the proximity of a root (of the characteristic equation) affects the contribution of that root to the model characteristics
- The effect of near cancellation of factors in ARMA(p,q) models

Chapter 4 goes beyond the ARMA(p,q) models and discusses other important stationary models. We discuss the harmonic component model and its relationship to ARMA(p,q) models. We also give some basic material on ARCH/GARCH models that is designed to give readers an intuitive grasp of these models, which have proven to be useful in the analysis of economic time series data.

Chapter 5 contains a treatment of nonstationary models. These include signal-plus-noise models along with ARMA-type models with roots on the unit circle. Our treatment of nonstationary ARMA-type models has the following unique features:

- Use of results due to Findley (1978) and Quinn (1980) that illustrate the behavior of autocorrelations associated with roots on or close to the unit circle.
- An extension of the ARIMA(p,d,q) models to ARUMA(p,d,q) models that include the ARIMA models but also allow for complex conjugate roots on the unit circle. That is, these models allow for cyclic nonstationary behavior.
- A discussion of the seasonal model from the perspective of the associated factor table.

Chapter 6 covers forecasting, focusing on the Box–Jenkins approach. We discuss forecasting using both stationary and nonstationary (ARUMA) models. Several examples are given.

Chapter 7 provides basic material on parameter estimation. While the main focus is on maximum likelihood estimation, we discuss computationally efficient methods and discuss their advantages and disadvantages.

Chapter 8 discusses the topic of ARMA model identification. Our focus is on the use of AIC and its variations as the primary tool for ARMA model identification. Other model identification techniques such as the Box–Jenkins and array methods are discussed in an appendix. One unique feature is our use of "overfitting" to assist in model identification when some roots are on (i.e., nonstationary models) or close to the unit circle. We use factor tables to identify nonstationary ARIMA/ARUMA and seasonal time series models.

Chapter 9 is a relatively unique wrap-up of previous material that discusses issues related to selecting a final model for a set of data. Through examples (real and simulated), we give practical advice concerning checking the whiteness of residuals, deciding whether to use a stationary or nonstationary model, and whether to use signal-plus-noise or ARMA/ARUMA models. We also discuss checking the characteristics of realizations simulated from a fitted model to ascertain whether the characteristics of generated realizations are consistent with those of the actual data. Students have told us they like this chapter because it helps them put together everything they have learned to this point.

Chapter 10 covers multivariate time series and vector autoregressive models. We also provide an introduction to state-space models that will provide the reader with an overview of this important topic.

Chapter 11 discusses long-memory models that have received a considerable amount of recent attention in the literature. This chapter discusses not only the fractional (and FARMA) models typically covered in texts such as these, but also the Gegenbauer and GARMA models that allow for long-memory behavior associated with any frequency between 0 and 0.5 inclusive.

Chapter 12 covers wavelets and short-term Fourier transforms for analyzing data having behavior patterns that change with time and cannot successfully be modeled using ARMA/ARUMA models. We believe that this material will form the basis for a useful introduction to these ideas for the instructor who wants to spend a few weeks on wavelets.

Chapter 13 introduces innovative techniques to deal with data with time-varying frequencies (TVF) such as Doppler, bat echolocation, seismic, and chirp signals. The fundamental idea is that for many data sets with TVF behavior there is a transformation of the time axis on which the data are stationary. This is a topic that is based on recent developments in the literature and is totally unique to this book. GWS software, which runs on the `S-Plus` system, is available on the website.

Software Available with This Book

One of the unique features of this book is that we provide the (Windows®-based) software, GW-WINKS, which is easy to learn and use and is designed to help students understand time series models as well as analyze data. As mentioned, *we highly recommend using the GW-WINKS software with the book.* Sections entitled "Using GW-WINKS" are included throughout the book to discuss the use of GW-WINKS for performing analyses. GW-WINKS is a special purpose version of the WINKS SDA™ software package from TexaSoft (http://www.texasoft.com). A license for GW-WINKS is included with the book. While other programs are available for performing many of the procedures discussed in this book, GW-WINKS provides a "learning environment" that enhances the understanding of the material.

The website associated with this book is http://www.texasoft.com/atsa. This website will be referred to as the ATSA (*Applied Time Series Analysis*) website. This site can also be accessed by visiting http://faculty.smu.edu/waynew and selecting the link to Applied Time Series Analysis.

In addition to providing information on installing and running GW-WINKS, the website also contains R programs for performing computations related to some of the topics, primarily those covered in Chapters 11 and 12. Some of these programs can be run directly through GW-WINKS using data in the GW-WINKS data sheet. We also provide stand-alone R programs on the website. The R package can be downloaded free from the Comprehensive R Archive Network (CRAN) website at http://cran.r-project.org.

The GWS package for using the techniques discussed in Chapter 13 to analyze TVF data is also available on the ATSA website. GWS requires the S-Plus® package. Instructions for the use of the GWS software are provided on the ATSA website. Additionally, step-by-step directions for use of GWS are included in the Chapter 13 exercises. In the text, we also discuss the use of the Wavelets module in S-Plus®, along with MATLAB® (Chapter 2) and SAS® (Chapter 10) for performing certain analyses. Related code is provided on the ATSA website.

The ATSA website contains the following:

1. A *Getting Started Guide* for GW-WINKS that includes
   a. Instructions for downloading, installing, and activating the GW-WINKS program
   b. Step-by-step tutorials showing how to manage data in GW-WINKS and how to run a selection of GW-WINKS Time Series procedures
2. R programs for some procedures discussed in the book
3. Instructions on how to write, modify, and run WINKS-R programs (programs that require R software and are run from the GW-WINKS menus)

4. Instructions for downloading and running `GWS` software
5. Datasets used in examples and problems in the book
6. Updated information on software and a list of corrections for typos and other errors in the book

For MATLAB® and Simulink® product information, please contact:

The MathWorks, Inc.
3 Apple Hill Drive
Natick, MA, 01760-2098 USA
Tel: 508-647-7000
Fax: 508-647-7001
E-mail: info@mathworks.com
Web: www.mathworks.com

Acknowledgments

We would be remiss to not mention some of the people who have helped with this project, which actually started in the 1980s but got sidetracked for a variety of reasons. They include Nien Fan Zhang, who was involved in the early stages of the book, Tom Fomby, who has worked hard to keep us informed about econometric techniques, and countless graduate students who have taken the time series course and helped us to refine our pedagogical methods. The first author (Woodward) taught the graduate level time series course in the Spring 2010 semester using a preliminary set of notes for the first nine chapters. The following students did a great job of beta testing the software and copyediting the text, and they deserve to be thanked individually: Aymen Al-Rawashdeh, Fang Duan, Xiaowen Hu, Yalan Hu, Dong Lin, Runqi Lin, Long Luo, Katherine Sisk, Holly Stovall, Miao Zang, Peiqi Xhai, and Josh Zorsky. Special thanks go to Joel O'Hair who helped with the graphics and the wavelets material, Wenkai Bao who was a student in the time series class and also helped develop R code related to the long-memory models, and James Haney who played an integral role related to the material on the use of state-space methodology for computing additive components of an autoregressive model. We would also like to thank Giovanni Petris who has given the manuscript a careful reading and provided several very useful suggestions. Sheila Crain deserves a great deal of credit for her excellent job of inputting the manuscript and helping in all phases of the manuscript preparation. To all our colleagues with whom we have discussed these topics over the years, we thank you for your positive influence on this work.

1

Stationary Time Series

In basic statistical analysis, attention is usually focused on data samples, $X_1, X_2, \ldots, X_n$, where the X_is are independent and identically distributed random variables. In a typical introductory course in univariate mathematical statistics, the case in which samples are not independent, but are in fact correlated, is not generally covered. However, when data are sampled at neighboring points in time, it is very likely that such observations will be correlated. Such time-dependent sampling schemes are very common. Examples include the following:

- Daily Dow Jones stock market closes over a given period
- Monthly unemployment data for the United States
- Annual global temperature data for the past 100 years
- Monthly incidence rate of influenza
- Average number of sunspots observed each year since 1749
- West Texas monthly intermediate crude oil prices
- Average monthly temperatures for Pennsylvania

Note that in each of these cases, an observed data value is (probably) not independent of nearby observations. That is, the data are correlated and are therefore not appropriately analyzed using univariate statistical methods based on independence. Nevertheless, this type of data is abundant in fields such as economics, biology, and medicine, and the physical and engineering sciences, where there is interest in understanding the mechanisms underlying these data, producing forecasts of future behavior, and drawing conclusions from the data. Time series analysis is the study of this type of data, and in this book we will introduce you to the extensive collection of tools and models for using the inherent correlation structure in such data sets to assist in their analysis and interpretation.

As examples, in Figure 1.1a, we show monthly West Texas intermediate crude oil prices from January 2000 through October 2009, and in Figure 1.1b the average monthly temperatures in degrees Fahrenheit for Pennsylvania from January 1990 through December 2004. In both cases, the monthly data are certainly correlated. In the case of the oil process, it seems that prices for a given month are positively correlated with the prices for nearby (past and future) months. In the case of Pennsylvania temperatures, there is a clear

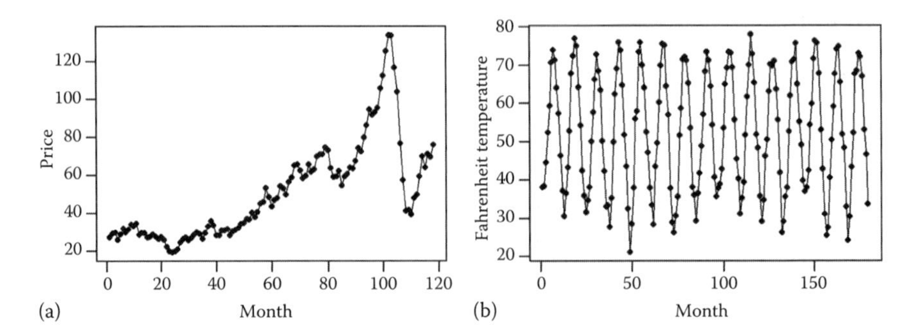

FIGURE 1.1
Two time series data sets. (a) West Texas intermediate crude. (b) PA average monthly temperatures.

12 month (annual) pattern as would be expected because of the natural seasonal weather cycles.

Time series analysis techniques are often classified into two major categories: time domain and frequency domain techniques. Time domain techniques include the analysis of the correlation structure, development of models that describe the manner in which such data evolve in time, and forecasting future behavior. Frequency domain approaches are designed to develop an understanding of time series data by examining the data from the perspective of their underlying cyclic (or frequency) content. The observation that the Pennsylvania temperature data tend to contain 12 month cycles is an example of examination of the frequency domain content of that data set. The basic frequency domain analysis tool is the power spectrum.

While frequency domain analysis is commonly used in the physical and engineering sciences, students with a statistics, mathematics, economics, or finance background may not be familiar with these methods. We do not assume a prior familiarity with frequency domain methods, and throughout the book we will introduce and discuss both time domain and frequency domain procedures for analyzing time series. In Sections 1.1 through 1.4, we discuss time domain analysis of time series data while in Sections 1.5 and 1.6 we present a basic introduction to frequency domain analysis and tools. In Section 1.7, we discuss several simulated and real-time series data sets from both a time and frequency domain perspective.

1.1 Time Series

Loosely speaking, a time series can be thought of as a collection of observations made sequentially in time. Our interest will not be in such series that are deterministic but rather in those whose values behave according to the laws of probability. In this chapter, we will discuss the fundamentals involved in the statistical analysis of time series. To begin with, we must be more careful in our definition of a time series. Actually, a time series is a special type of *stochastic process.*

Definition 1.1 A stochastic process $\{X(t);\ t \in T\}$ is a collection of random variables, where T is an index set for which all of the random variables, $X(t)$, $t \in T$, are defined on the same sample space. When T represents time, we refer to the stochastic process as a *time series.*

If T takes on a continuous range of values [e.g., $T=(-\infty,\infty)$ or $T=(0,\infty)$], the process is said to be a *continuous parameter* process. If, on the other hand, T takes on a discrete set of values (e.g., $T=\{0,1,2,\ldots\}$ or $T=\{0,\pm1,\pm2,\ldots\}$), the process is said to be a *discrete parameter* process. Actually, it is typical to refer to these as continuous and discrete processes, respectively.

We will use the subscript notation, X_t, when we are dealing specifically with a discrete parameter process. However, when the process involved is either continuous parameter or of unspecified type, we will use the function notation, $X(t)$. Also, when no confusion will arise, we often use the notation $\{X(t)\}$ or simply $X(t)$ to denote a time series. Similarly, we will usually shorten $\{X_t;\ t=0,\pm1,\ldots\}$ to X_t, $t=0,\ \pm1,\ldots$ or simply to X_t.

Recall that a random variable, Y, is a function defined on a sample space Ω whose range is the real numbers. An observed value of the random variable Y is a real number $y=Y(\omega)$ for some $\omega \in \Omega$. For a time series $\{X(t)\}$, its "value," $\{X(t,\omega);\ t \in T\}$ for some fixed $\omega \in \Omega$, is a collection of real numbers. This leads to the following definition.

Definition 1.2 A *realization* of the time series $\{X(t);\ t \in T\}$ is the set of real-valued outcomes, $\{X(t,\omega);\ t \in T\}$ for a fixed value of $\omega \in \Omega$.

That is, a realization of a time series is simply a set of values of $\{X(t)\}$, that result from the occurrence of some observed event. A realization of the time series $\{X(t);\ t \in T\}$ will be denoted $\{x(t);\ t \in T\}$. As before, we will sometimes use the notation $\{x(t)\}$ or simply $x(t)$ in the continuous parameter case and $\{x_t\}$ or x_t in the discrete parameter case when these are clear. The collection of all possible realizations is called an *ensemble*, and, for a given t, the expectation of the random variable $X(t)$, is called the ensemble mean and will be denoted $E[X(t)]=\mu(t)$. The variance of $X(t)$ is given by $Var[X(t)]=E\{[X(t)-\mu(t)]^2\}$ and is often denoted by $\sigma^2(t)$ since it also can depend on t.

Example 1.1 A Time Series with Two Possible Realizations

Consider the stochastic process $Y(t)$ for $t \in (-\infty,\infty)$ defined by $Y(t)=\sin(t+\varphi)$ where $P[\varphi=0]=0.5$ and $P[\varphi=\pi/2]=0.5$ and P denotes probability.

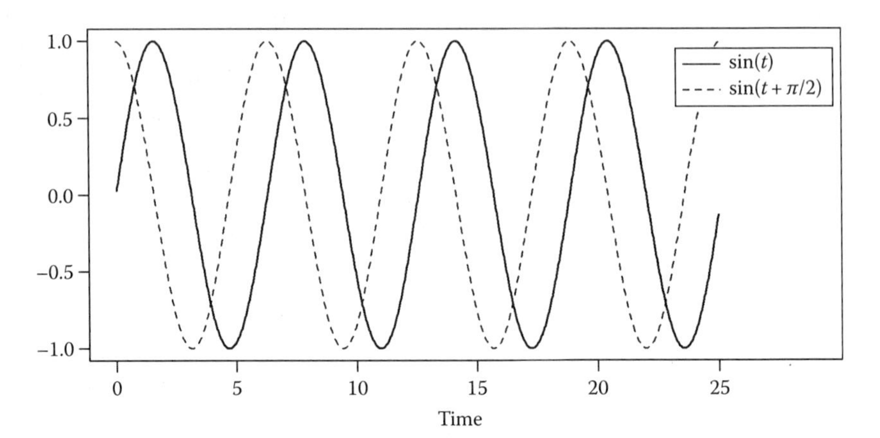

FIGURE 1.2
The two distinct realizations for $Y(t)$ in Example 1.1.

This process has only two possible realizations or sample functions, and these are shown in Figure 1.2 for $t \in [0, 25]$.

The individual curves are the realizations while the collection of the two possible curves is the *ensemble*. For this process, $E[Y(t)] = 0.5\sin(t+0) + 0.5\sin\left(t+\frac{\pi}{2}\right)$ So, for example,

$$E[Y(0)] = 0.5\sin(0) + 0.5\sin\left(\frac{\pi}{2}\right) = 0.5,$$

and

$$E\left[Y\left(\frac{\pi}{4}\right)\right] = \frac{\sqrt{2}}{2}.$$

Thus, $E[Y(t)] = \mu(t)$, that is, the expectation depends on t. Note that this expectation is an average "vertically" across the ensemble and not "horizontally" down the time axis. In Section 1.4, we will see how these "different ways of averaging" can be related. Of particular interest in the analysis of a time series is the covariance between $X(t_1)$ and $X(t_2)$, $t_1, t_2 \in T$. Since this is covariance within the same time series, we refer to it as the *autocovariance*.

Definition 1.3 If $\{X(t); t \in T\}$ is a time series, then for any $t_1, t_2 \in T$, we define

(a) The autocovariance function, $\gamma(\cdot)$, by

$$\gamma(t_1, t_2) = E\{[X(t_1) - \mu(t_1)][X(t_2) - \mu(t_2)]\}$$

(b) The autocorrelation function, $\rho(\cdot)$, by

$$\rho(t_1, t_2) = \frac{\gamma(t_1, t_2)}{\sigma(t_1)\sigma(t_2)}$$

1.2 Stationary Time Series

In the study of a time series, it is common that only a single realization from the series is available. Analysis of a time series on the basis of only one realization is analogous to analyzing the properties of a random variable on the basis of a single observation. The concepts of stationarity and ergodicity will play an important role in enhancing our ability to analyze a time series on the basis of a single realization in an effective manner. A process is said to be *stationary* if it is in a state of "statistical equilibrium." The basic behavior of such a time series does not change in time. As an example, for such a process, $\mu(t)$ would not depend on time and thus could be denoted μ for all t. It would seem that, since $x(t)$ for each $t \in T$ provides information about the ensemble mean, μ, it may be possible to estimate μ on the basis of a single realization. An *ergodic* process is one for which ensemble averages such as μ can be consistently estimated from a single realization. In this section, we will present more formal definitions of stationarity, but we will delay further discussion of ergodicity until Section 1.4.

The most restrictive notion of stationarity is that of *strict stationarity*, which we define as follows.

Definition 1.4 A process $\{X(t); t \in T\}$ is said to be strictly stationary if for any $t_1, t_2, \ldots, t_k \in T$ and any $h \in T$, the joint distribution of $\{X(t_1), X(t_2), \ldots, X(t_k)\}$ is identical to that of $\{X(t_1+h), X(t_2+h), \ldots, X(t_k+h)\}$.

NOTE: We have tacitly assumed that T is closed under addition, and we will continue to do so.

Strict stationarity requires, among other things, that for any $t_1, t_2 \in T$, the distributions of $X(t_1)$ and $X(t_2)$ must be the same, and further that all bivariate distributions of pairs $\{X(t), X(t+h)\}$ are the same for all h, etc. The requirement of strict stationarity is a severe one and usually is difficult to establish mathematically. In fact, for most applications, the distributions involved are not known. For this reason, less restrictive notions of stationarity have been developed. The most common of these is *covariance stationarity*.

Definition 1.5 (Covariance Stationarity) The time series $\{X(t); t \in T\}$ is said to be *covariance stationary* if

1. $E[X(t)] = \mu$ (constant for all t)
2. $\text{Var}[X(t)] = \sigma^2 < \infty$ (i.e., a finite constant for all t)
3. $\gamma(t_1, t_2)$ depends only on $t_2 - t_1$

Covariance stationarity is also called *weak stationarity*, *stationarity in the wide sense*, and *second-order stationarity*. In the remainder of this book, unless specified otherwise, the term *stationarity* will refer to covariance stationarity.

In time series, as in most other areas of statistics, uncorrelated data play an important role. There is no difficulty in defining such a process in the case of a discrete parameter time series. That is, the time series $\{X_t; t=0, \pm 1, \pm 2, \ldots\}$ is called a "purely random process" if the X_t are uncorrelated random variables. When considering purely random processes, we will only be interested in the case in which the X_t's are also identically distributed. In this situation, it is more common to refer to the time series as *white noise*. The following definition summarizes these remarks.

Definition 1.6 (Discrete white noise) The time series is called *discrete white noise* if

(i) The X_t's are identically distributed
(ii) $\gamma(t_1, t_2) = 0$ when $t_1 \neq t_2$
(iii) $\gamma(t, t) = \sigma^2$, where $0 < \sigma^2 < \infty$

We will also find the two following definitions to be useful.

Definition 1.7 (Gaussian Process) A time series is said to be Gaussian (normal) if for any positive integer k and any $t_1, t_2, \ldots, t_k \in T$, the joint distribution of $\{X(t_1), X(t_2), \ldots, X(t_k)\}$ is multivariate normal.

Note that for Gaussian processes, the concepts of strict stationarity and covariance stationarity are equivalent. This can be seen by noting that if the Gaussian process $X(t)$ is covariance stationary, then for any $t_1, t_2, \ldots, t_k \in T$ and any $h \in T$, the multivariate normal distributions of $\{X(t_1), X(t_2), \ldots, X(t_k)\}$ and $\{X(t_1+h), X(t_2+h), \ldots, X(t_k+h)\}$ have the same means and covariance matrices and thus the same distributions.

Definition 1.8 (Complex Time Series) A complex time series is a sequence of complex random variables $Z(t)$, such that

$$Z(t) = X(t) + iY(t),$$

where $X(t)$ and $Y(t)$ are real-valued random variables for each t.

It is easy to see that for a complex time series, $Z(t)$, the mean function, $\mu_Z(t)$, is given by

$$\begin{aligned}\mu_Z(t) &= E[Z(t)] \\ &= E[X(t)] + iE[Y(t)] \\ &= \mu_X(t) + i\mu_Y(t).\end{aligned}$$

A time series will be assumed to be real-valued in this book unless it is specifically indicated to be complex.

1.3 Autocovariance and Autocorrelation Functions for Stationary Time Series

In this section, we will examine the autocovariance and autocorrelation functions for stationary time series. If a time series is covariance stationary, then the autocovariance function $\gamma(t, t+h)$ only depends on h. Thus, for stationary processes, we denote this autocovariance function by $\gamma(h)$. Similarly, the autocorrelation function for a stationary process is given by $\rho(h) = \gamma(h)/\sigma^2$. Consistent with our previous notation, when dealing with a discrete parameter time series, we will use the subscript notation γ_k and ρ_k. The autocovariance function of a stationary time series satisfies the following properties:

(1) $\gamma(0) = \sigma^2$

(2) $|\gamma(h)| \leq \gamma(0)$ for all h

The inequality in (2) can be shown by noting that for any random variables X and Y it follows that

$$E[|(X - \mu_X)(Y - \mu_Y)|] \leq \{E[(X - \mu_X)^2]E[(Y - \mu_Y)^2]\}^{1/2}$$

by the Cauchy–Schwarz inequality. Now letting $X = X(t)$ and $Y = X(t+h)$, we see that

$$|\gamma(h)| \leq E|(X(t) - \mu)(X(t+h) - \mu)| \leq \sigma^2 = \gamma(0).$$

(3) $\gamma(h) = \gamma(-h)$

This result follows by noting that

$$\begin{aligned}\gamma(-h) &= E[(X(t)-\mu)(X(t-h)-\mu)] \\ &= E[(X(t-h)-\mu)(X(t)-\mu)] \\ &= E[(X(t_1)-\mu)(X(t_1+h)-\mu)],\end{aligned}$$

where $t_1 = t - h$. However, since the autocovariance does not depend on time t, this last expectation is equal to $\gamma(h)$.

(4) The function $\gamma(h)$ is positive semidefinite. That is, for any set of time points $t_1, t_2, \ldots, t_k \in T$ and all real $b_1, b_2, \ldots, b_k$, we have

$$\sum_{i=1}^{k}\sum_{j=1}^{k}\gamma(t_i - t_j)b_i b_j \geq 0. \tag{1.1}$$

To show this, let $W = \sum_{j=1}^{k} b_j X(t_j)$. Now, $\mathrm{Var}(W) \geq 0$ and

$$\begin{aligned}\mathrm{Var}(W) &= E\left[\left\{\sum_{j=1}^{k} b_j\left(X(t_j)-\mu\right)\right\}^2\right] \\ &= \sum_{i=1}^{k}\sum_{j=1}^{k} b_i b_j \gamma(t_i - t_j),\end{aligned}$$

and the result follows. Note that in the case of a discrete time series defined on $t = 0, \pm 1, \pm 2, \ldots$, then (1.1) is equivalent to the matrix

$$\boldsymbol{\Gamma}_k = \begin{pmatrix} \gamma_0 & \gamma_1 & \cdots & \gamma_k \\ \gamma_1 & \gamma_0 & \cdots & \gamma_{k-1} \\ \vdots & \vdots & & \vdots \\ \gamma_k & \gamma_{k-1} & \cdots & \gamma_0 \end{pmatrix} \tag{1.2}$$

being positive semi-definite for each k.

The autocorrelation function satisfies the following analogous properties:

(1) $\rho(0) = 1$

(2) $|\rho(h)| \leq 1$ for all h

(3) $\rho(h) = \rho(-h)$

(4) The function $\rho(h)$ is positive semidefinite, and for discrete time series defined on $t = 0, \pm 1, \pm 2, \ldots$, the matrix

$$\boldsymbol{\rho}_k = \begin{pmatrix} 1 & \rho_1 & \cdots & \rho_k \\ \rho_1 & 1 & \cdots & \rho_{k-1} \\ \vdots & \vdots & & \vdots \\ \rho_k & \rho_{k-1} & \cdots & 1 \end{pmatrix} \quad (1.3)$$

is positive semidefinite for each k.

Theorem 1.1 gives conditions that guarantee the stronger conclusion that $\boldsymbol{\Gamma}_k$ and equivalently $\boldsymbol{\rho}_k$ are positive definite for each k. The proof of this result can be found in Brockwell and Davis (1991), Proposition 5.1.1.

Theorem 1.1

If $\gamma_0 > 0$ and $\lim_{j\to\infty} \gamma_j = 0$, then $\boldsymbol{\Gamma}_k$ and $\boldsymbol{\rho}_k$ are positive definite for each integer $k > 0$.

It is important to understand the extent to which a plot of the autocorrelations for a given model describes the behavior in time series realizations from that model. In Figure 1.3, we show realizations from four stationary time

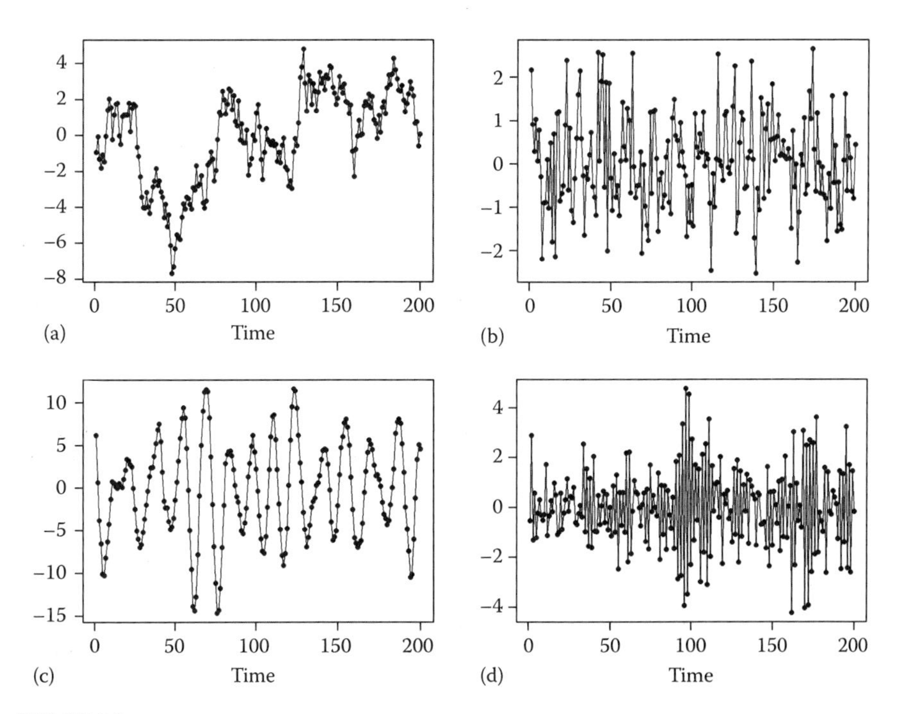

FIGURE 1.3
Four realizations from stationary processes. (a) Realization 1. (b) Realization 2. (c) Realization 3. (d) Realization 4.

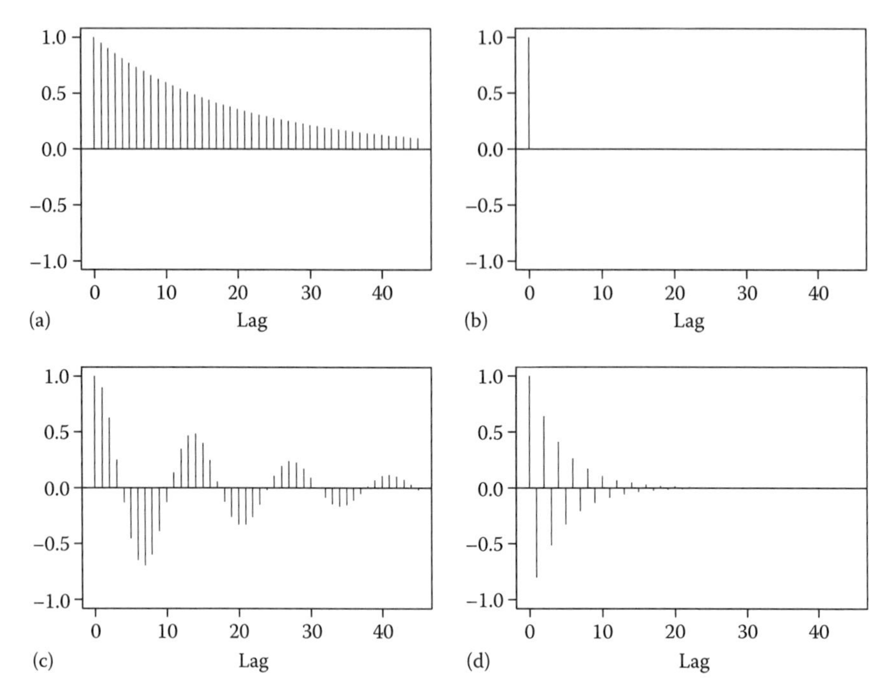

FIGURE 1.4
True autocorrelations from models associated with realizations in Figure 1.3. (a) Realization 1. (b) Realization 2. (c) Realization 3. (d) Realization 4.

series while Figure 1.4 shows the corresponding true autocorrelations for lags 0–45. Realization 1 in Figure 1.3a displays a wandering or piecewise trending behavior. Note that it is typical for x_t and x_{t+1} to be relatively close to each other, that is, the value of X_{t+1} is usually not very far from the value of X_t, and, as a consequence, there is a rather strong positive correlation between the random variables X_t and say X_{t+1}. Note also that, for large lags, there seems to be less correlation between X_t and X_{t+k} to the extent that there is very little correlation between X_t and say X_{t+40}. We see this behavior manifested in Figure 1.4a, which displays the true autocorrelations associated with the model from which realization 1 was generated. In this plot $\rho_1 \approx 0.95$ while as the lag increases, the autocorrelations decrease, and by lag 40 the autocorrelation has decreased to $\rho_{40} \approx 0.1$.

Realization 2 in Figure 1.3b shows an absence of pattern. That is, there appears to be no relationship between X_t and X_{t+1} or between X_t and X_{t+k} for any $k \neq 0$ for that matter. In fact, this is a realization from a white noise model, and, for this model, $\rho_k = 0$ whenever $k \neq 0$ as can be seen in Figure 1.4b. Notice that in all autocorrelation plots, $\rho_0 = 1$.

Realization 3 in Figure 1.3c is characterized by pseudo-periodic behavior with a cycle-length of about 14 time points. The corresponding autocorrelations in Figure 1.4c show a damped sinusoidal behavior. Note that, not surprisingly, there is a positive correlation between X_t and X_{t+14}. Also, there is a negative correlation at lag 7, since within the sinusoidal cycle, if x_t is above the average, then x_{t+7} would be expected to be below average and vice versa. Also note that due to the fact that there is only a pseudo-cyclic behavior in the realization, with some cycles a little longer than others, by lag 28 (i.e., two cycles), the autocorrelations are quite small.

Finally, realization 4 in Figure 1.3d shows a highly oscillatory behavior. In fact, if x_t is above average, then x_{t+1} tends to be below average, x_{t+2} tends to be above average, etc. Again, the autocorrelations in Figure 1.4d describe this behavior where we see that $\rho_1 \approx -0.8$ while $\rho_2 \approx 0.6$. Note also that the up-and-down pattern is sufficiently imperfect that the autocorrelations damp to near zero by about lag 15.

1.4 Estimation of the Mean, Autocovariance, and Autocorrelation for Stationary Time Series

As discussed previously, a common goal in time series analysis is that of obtaining information concerning a time series on the basis of a single realization. In this section, we discuss the estimation of μ, $\gamma(h)$, and $\rho(h)$ from a single realization. Our focus will be discrete parameter time series where $T = \{\ldots, -2, -1, 0, 1, 2, \ldots\}$, in which case a finite length realization $t = 1, 2, \ldots, n$ is typically observed. Results analogous to the ones given here are available for continuous parameter time series but will only be briefly mentioned in examples. The reader is referred to Parzen (1962) for the more general results.

1.4.1 Estimation of μ

Given the realization $\{x_t, t = 1, 2, \ldots, n\}$ from a stationary time series, the natural estimate of the common mean μ is the sample mean

$$\bar{x} = \frac{1}{n}\sum_{t=1}^{n} x_t. \tag{1.4}$$

It is obvious that the estimator $\bar{X}$ is unbiased for μ.

1.4.1.1 Ergodicity of $\bar{X}$

We will say that $\bar{X}$ is ergodic for μ if $\bar{X}$ converges in the mean square sense to μ as n increases, that is, $\lim_{n\to\infty} E\left[(\bar{X} - \mu)^2\right] = 0$.

Theorem 1.2

Suppose $\{X_t; t=0, \pm1, \pm2, \ldots\}$ is a stationary time series. Then, $\overline{X} = \frac{1}{n}\sum_{t=1}^{n} X_t$ is ergodic for μ if and only if

$$\lim_{k\to\infty} \frac{1}{k}\sum_{j=0}^{k-1} \gamma_j = 0. \tag{1.5}$$

See Yaglom (1962) for a proof of Theorem 1.2. We also have the following useful corollary.

Corollary 1.1

Let X_t be a discrete parameter stationary time series as in Theorem 1.2. Then, $\overline{X}$ is ergodic for μ if

$$\lim_{k\to\infty} \gamma_k = 0, \tag{1.6}$$

or equivalently if

$$\lim_{k\to\infty} \rho_k = 0. \tag{1.7}$$

The sufficient condition for the ergodicity of $\overline{X}$ given in Corollary 1.1 is quite useful and is one that holds for the broad class of stationary ARMA(p,q) time series that we will discuss in Chapter 3. Even though X_t's "close" to each other in time may have substantial correlation, the condition in Corollary 1.1 assures that for "large" separation, they are nearly uncorrelated.

Example 1.2 An Ergodic Process

Consider the process X_t, $t=0, \pm1, \pm2, \ldots$ defined by

$$X_t = 0.8X_{t-1} + a_t, \tag{1.8}$$

where a_t is discrete white noise with zero mean and variance σ_a^2. In other words, (1.8) is a "regression-type" model in which X_t (the dependent variable) is 0.8 times X_{t-1} (the independent variable) plus a random uncorrelated "noise." Notice that X_{t-k} and a_t are uncorrelated for $k > 0$. The process in (1.8) is an example of a first-order autoregressive process, denoted AR(1), which will be examined in detail in Chapter 3. We will show in that chapter that X_t as defined in (1.8) is a stationary process. By taking expectations of both sides of (1.8) and recalling that a_t has zero mean, it follows that

$$E[X_t] = 0.8E[X_{t-1}]. \tag{1.9}$$

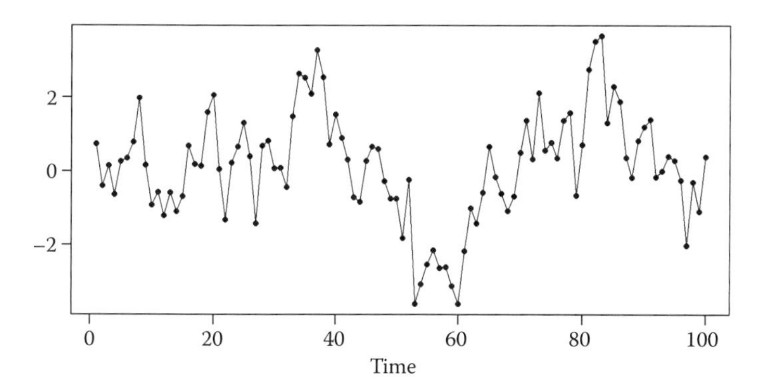

FIGURE 1.5
Realization from X_t in Example 1.2.

Since X_t is stationary, it follows that $E[X_t] = E[X_{t-1}] = \mu_X$. Now, (1.9) implies that, $\mu = 0.8\mu$, that is, $\mu = 0$. Letting $k > 0$, we next obtain an expression for $\gamma_k = E[X_t X_{t+k}] = E[X_{t-k} X_t]$ by premultiplying each term in (1.8) by X_{t-k} and taking expectations. This yields

$$E[X_{t-k}X_t] = 0.8E[X_{t-k}X_{t-1}] + E[X_{t-k}a_t], \tag{1.10}$$

and since $E[X_{t-k}a_t] = 0$, (1.10) simplifies to $\gamma_k = 0.8\gamma_{k-1}$, which implies $\gamma_k = \gamma_0(0.8)^k$ and $\rho_k = 0.8^k$. Thus, according to Corollary 1.1, $\overline{X}$ is ergodic for μ_X for model (1.8). See Section 3.2.4.

In order to see the impact of condition (1.7) on the ability to estimate μ based on a single realization, note that $\rho_1 = 0.8$, so that X_t and X_{t+1} are providing correlated (and somewhat redundant) information about μ, but, for example, $\rho_{20} = 0.01$, so that X_t and X_{t+20} are providing nearly uncorrelated information about μ. Figure 1.5 shows a realization of length 100 from the model in (1.8). Notice that the realization seems to "wander" back and forth around the mean $\mu = 0$ suggesting the possibility of a stationary process.

Example 1.3 A Stationary Cosine Process

Define $X(t)$ for $t \in (-\infty, \infty)$ by

$$X(t) = C\cos(\lambda t + Y), \tag{1.11}$$

where
 C and λ are real-valued constants
 Y is a random variable

Then, as we will show in the following, $X(t)$ is stationary if and only if $\varphi(1) = \varphi(2) = 0$, where $\varphi(u)$ is the characteristic function of Y, that is,

$$\varphi(u) = E[e^{iuY}] \quad \text{for } i = \sqrt{-1}.$$

Before verifying this result we first note that $\varphi(u)$ can be written as

$$\varphi(u) = E[\cos(uY)] + iE[\sin(uY)]. \tag{1.12}$$

We will also make use of the trigonometric identities

(i) $\cos(A + B) = \cos A \cos B - \sin A \sin B$

(ii) $\cos A \cos B = \frac{1}{2}\cos(A + B) + \frac{1}{2}\cos(A - B)$

Using identity (i), we see that $X(t)$ in (1.11) can be expressed as

$$X(t) = C[\cos \lambda t \cos Y - \sin \lambda t \sin Y], \tag{1.13}$$

while the product $X(t)X(t+h)$ can be rewritten using first (i) and then (ii) as

$$\begin{aligned} X(t)X(t+h) &= C\cos(\lambda t + Y)C\cos(\lambda(t+h) + Y) \\ &= \frac{C^2}{2}\cos(2\lambda t + \lambda h + 2Y) + \frac{C^2}{2}\cos(\lambda h) \\ &= \frac{C^2}{2}(\cos(2\lambda t + \lambda h)\cos(2Y) - \sin(2\lambda t + \lambda h)\sin(2Y) + \cos(\lambda h)) \end{aligned} \tag{1.14}$$

($\Leftarrow$) Suppose $\varphi(1) = \varphi(2) = 0$. Then, it is clear from (1.12) that

$$\begin{aligned} &\text{(a)} \quad E[\cos Y] = E[\sin Y] = 0 \\ &\text{(b)} \quad E[\cos 2Y] = E[\sin 2Y] = 0 \end{aligned}$$

From (1.13), we see that (a) implies $E[X(t)] = 0$. Also, upon taking the expectation of $X(t)X(t+h)$ in (1.14), it is seen that (b) implies

$$\gamma(h) = E[X(t)X(t+h)] = \frac{C^2}{2}\cos(\lambda h),$$

which depends on h but not on t. Moreover, $\sigma_X^2 = \gamma(0) = \frac{C^2}{2} < \infty$. Thus, $X(t)$ is covariance stationary.

($\Rightarrow$) Suppose $X(t)$ is covariance stationary. Then, since $E[X(t)]$ does not depend on t, it follows from (1.13) that $E[\cos Y] = E[\sin Y] = 0$, and so that $\varphi(1) = 0$. Likewise, since $E[X(t)X(t+h)]$ depends on h but not on t, it follows from the last equation in (1.14) that $E[\cos 2Y] = E[\sin 2Y] = 0$, that is, that $\varphi(2) = 0$ and the result follows.

We note that if $Y \sim \text{Uniform}[0, 2\pi]$, then

$$\begin{aligned}\varphi(u) &= \int_0^{2\pi} e^{iuy}\left(\frac{1}{2\pi}\right)dy \\ &= \frac{e^{2\pi iu} - 1}{2\pi iu},\end{aligned}$$

so that $\varphi(1) = \varphi(2) = 0$, and, thus, the uniform $[0, 2\pi]$ distribution satisfies the condition on Y, and the resulting process $X(t) = C\cos(\lambda t + Y)$ is stationary.

Figure 1.6 shows three realizations from the process described in (1.11) with $\lambda = 0.6$ and $Y \sim U[0, 2\pi]$. The realization in Figure 1.6a is for the case in which $Y = 1.5$. Unlike the realizations in Figures 1.3 and 1.5, the behavior in Figure 1.6a is very deterministic. In fact, inspection of the plot would lead one to believe that the mean of this process depends on t. However, this is not the case since Y is random. For example, Figure 1.6b and c show two other realizations for different values for the random variable Y. Note that the effect of the random variable Y is to randomly shift the "starting point" of the cosine wave in each realization. Y is called the *phase shift*. These realizations are simply three realizations from the ensemble of possible realizations. Note that for these three realizations, $X(10)$ equals 0.35, 0.98, and -0.48, respectively, which is consistent with our finding that $E[X(t)]$ does not depend on t.

The important point is that the property of stationarity cannot be determined by simple inspection of the data. Instead, it must be inferred from physical considerations, knowledge of the correlation structure, or more sophisticated techniques that we will consider in future chapters. Since Theorem 1.2 and its corollary relate to discrete parameter processes, they do not apply to $X(t)$ in the current example. However, results analogous to both Theorem 1.2 and Corollary 1.1 do hold for continuous parameter processes (see Parzen, 1962). Notice that, in this example, $\gamma(h)$ does not go to zero as $h \to \infty$, but $\overline{X}_t$, defined in this continuous parameter example as

$$\overline{X}_t = \frac{1}{t}\int_0^t X(s)ds, \tag{1.15}$$

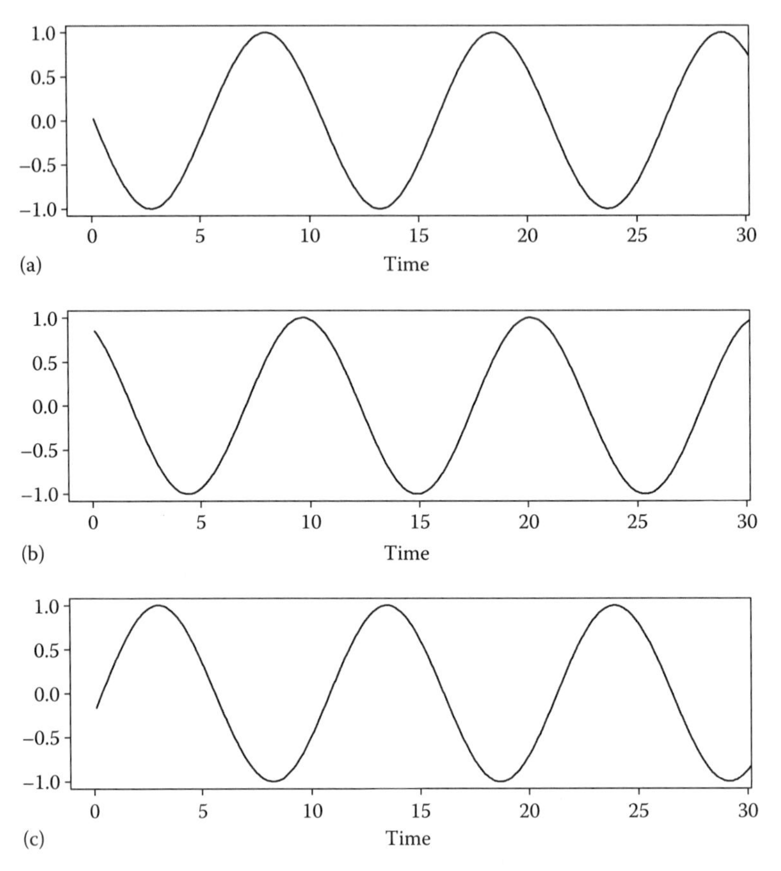

FIGURE 1.6
Three realizations from $X(t)$ of Example 1.3. (a) $Y = 1.5$. (b) $Y = 0.5$. (c) $Y = 4.5$.

can be shown to be ergodic for μ using the continuous analog to Theorem 1.2. The ergodicity of $\overline{X}_t$ seems reasonable in this example in that as t in (1.15) increases, it seems intuitive that $\overline{X}_t \to 0(=\mu)$.

Example 1.4 A Non-Ergodic Process

As a hypothetical example of a nonergodic process, consider a thread manufacturing machine that produces spools of thread. The machine is cleaned and adjusted at the end of each day after which the diameter setting remains unchanged until the end of the next day at which time it is recleaned, etc. The diameter setting could be thought of as a stochastic process with each day being a realization. In Figure 1.7, we show several realizations from this hypothetical process $X(t)$.

If μ is the average diameter setting across days, then μ does not depend on time t within a day. It is also clear that $\gamma(t, t+h)$ does not depend on t since

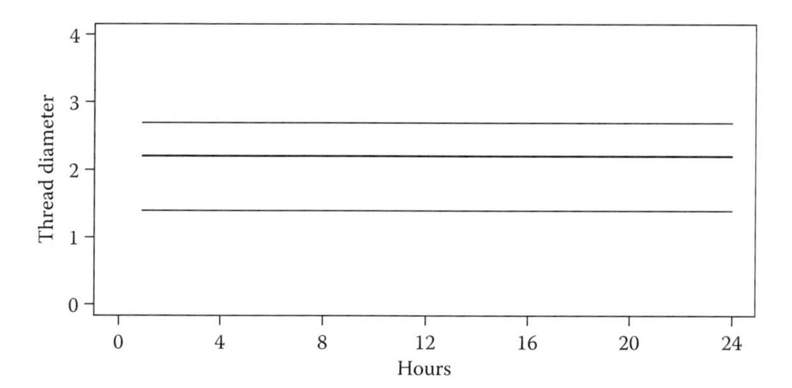

FIGURE 1.7
Three realizations from the process in Example 1.3.

$X(t) = X(t+h)$ and therefore $\gamma(h) = \gamma(0) < \infty$, for all h. Thus, the process is stationary. However, it obviously makes no sense to try to estimate μ by $\overline{X}$, that is, by averaging down the time axis for a single realization. Heuristically, we see that μ can be estimated by averaging across several realizations, but using *one* realization in essence provides only one piece of information. Clearly, in this case $\overline{X}_t$ is not ergodic for μ.

1.4.1.2 Variance of $\overline{X}$

Before leaving the topic of estimating μ from a realization from a stationary process, in Theorem 1.3, we give a useful formula for $\text{Var}(\overline{X})$. We leave the proof of the theorem as an exercise.

Theorem 1.3

If X_t is a discrete stationary time series, then the variance of $\overline{X}$ based on a realization of length n is given by

$$\text{Var}(\overline{X}) = \frac{\sigma^2}{n} \sum_{k=-(n-1)}^{n-1} \left(1 - \frac{|k|}{n}\right) \rho_k. \tag{1.16}$$

The result in (1.16) shows the effect of autocorrelation on the variance of $\overline{X}$, and if X_t is white noise, that is, $\gamma_k = 0$ if $k \neq 0$, then (1.16) becomes the well-known result $\text{Var}(\overline{X}) = \sigma^2/n$. In the following section, we discuss the estimation of γ_k and ρ_k for a discrete stationary time series. Using the notation $\hat{\rho}_k$ to denote the estimated (sample) autocorrelations [discussed in Section 1.4.2] and $\hat{\sigma}^2 = \hat{\gamma}_0$ to denote the sample variance, it is common practice to obtain approximate confidence 95% intervals for μ using

$$\left(\overline{X} - 1.96\sqrt{\frac{\hat{\sigma}^2}{n}\sum_{k=-(n-1)}^{n-1}\left(1-\frac{|k|}{n}\right)\hat{\rho}_k},\ \overline{X} + 1.96\sqrt{\frac{\hat{\sigma}^2}{n}\sum_{k=-(n-1)}^{n-1}\left(1-\frac{|k|}{n}\right)\hat{\rho}_k}\right). \tag{1.17}$$

1.4.2 Estimation of γ_k

Because of stationarity, $E[(X_t - \mu)(X_{t+k} - \mu)] = \gamma_k$ does not depend on t. As a consequence, it seems reasonable to estimate γ_k, from a single realization, by

$$\begin{aligned}\tilde{\gamma}_k &= \frac{1}{n-|k|}\sum_{t=1}^{n-|k|}(X_t - \overline{X})(X_{t+|k|} - \overline{X}), \quad k = 0, \pm 1, \pm 2, \ldots \pm (n-1)\\ &= 0, \quad |k| \geq n,\end{aligned} \tag{1.18}$$

that is, by moving down the time axis and finding the average of all products in the realization separated by k time units. Notice that if we replace $\overline{X}$ by μ in (1.18), then the resulting estimator is unbiased. However, the estimator $\tilde{\gamma}_k$ in (1.18) is only asymptotically unbiased.

Despite the intuitive appeal of $\tilde{\gamma}_k$, most time series analysts use an alternative estimator, $\hat{\gamma}_k$, defined by

$$\begin{aligned}\hat{\gamma}_k &= \frac{1}{n}\sum_{t=1}^{n-|k|}(X_t - \overline{X})(X_{t+|k|} - \overline{X}), \quad k = 0, \pm 1, \pm 2, \ldots, \pm (n-1)\\ &= 0, \quad |k| \geq n.\end{aligned} \tag{1.19}$$

From examination of (1.19), it is clear that $|\hat{\gamma}_k|$ values will tend to be "small" when $|k|$ is large relative to n. The estimator, $\hat{\gamma}_k$, as given in (1.19), has a larger bias than $\tilde{\gamma}_k$, but, in most cases, has smaller mean square error. The difference between the two estimators is, of course, most dramatic when $|k|$ is large with respect to n. The comparison can be made most easily when $\overline{X}$ is replaced by μ in (1.18) and (1.19), which we will assume for the discussion in the remainder of the current paragraph. The bias of $\hat{\gamma}_k$ in this case can be seen to be $(|k|/n)\gamma_k$. As $|k|$ increases toward n, the factor $|k|/n$ increases toward one, so that the bias tends to γ_k. As $|k|$ increases toward $n-1$, $\text{Var}(\tilde{\gamma}_k)$ will increase, and $\tilde{\gamma}_k$ values will tend to have a quite erratic behavior for the larger values of k. This will be illustrated in Example 1.5 that follows. The overall pattern of γ_k is usually better approximated by $\hat{\gamma}_k$ than it is by $\tilde{\gamma}_k$, and we will see later that it is often the *pattern* of the autocorrelation function that is of importance. The behavior of γ_k is also better approximated by $\hat{\gamma}_k$ in that $\hat{\gamma}_k$ is also positive semidefinite in the sense of property (4) of the autocovariance function given in Section 1.3, while $\tilde{\gamma}_k$ is not necessarily positive semidefinite. For further discussion of these topics, the reader is referred to Priestley (1981).

Example 1.5 Behavior of Sample Autocovariances

In Figure 1.8, we display theoretical autocovariances for the AR(1) realization of Example 1.2 along with sample autocovariances $\hat{\gamma}_k$ and $\tilde{\gamma}_k$. There it can be seen that the damped exponential nature of γ_k is better estimated with $\hat{\gamma}_k$ than $\tilde{\gamma}_k$. Ergodic results for $\hat{\gamma}_k$ (and for $\tilde{\gamma}_k$ because of their asymptotic equivalence) can also be obtained, but they are in general not of the straightforward nature of Theorem 1.2. Most of the results for $\hat{\gamma}_k$ are proved for the case in which μ is assumed to be known and, without loss of generality, equal to zero. For a discussion of ergodic theorems for $\hat{\gamma}_k$ and for some results concerning $\text{Var}(\hat{\gamma}_k)$and $\text{Cov}(\hat{\gamma}_k, \hat{\gamma}_j)$, see Priestley (1981). In the remainder of this book, we assume that the necessary conditions hold for the sample autocovariances to be ergodic for γ_k.

1.4.3 Estimation of ρ_k

The natural estimator of $\rho_k = \gamma_k/\gamma_0$ is $\hat{\rho}_k = \hat{\gamma}_k/\hat{\gamma}_0$, which we will henceforth refer to as the *sample autocorrelation function*. The estimator $\hat{\rho}_k$ has the desirable property that $|\hat{\rho}_k| \leq 1$, a property that $\tilde{\rho}_k = \tilde{\gamma}_k/\tilde{\gamma}_0$ does not possess. This property can be shown by relating $\hat{\rho}_k$ to the Pearson product-moment correlation coefficient between the two sets of $n+k$ observations

$$0, 0, \ldots, 0, X_1 - \overline{X}, X_2 - \overline{X}, \ldots, X_n - \overline{X}$$
$$X_1 - \overline{X}, X_2 - \overline{X}, \ldots, X_n - \overline{X}, 0, 0, \ldots, 0$$

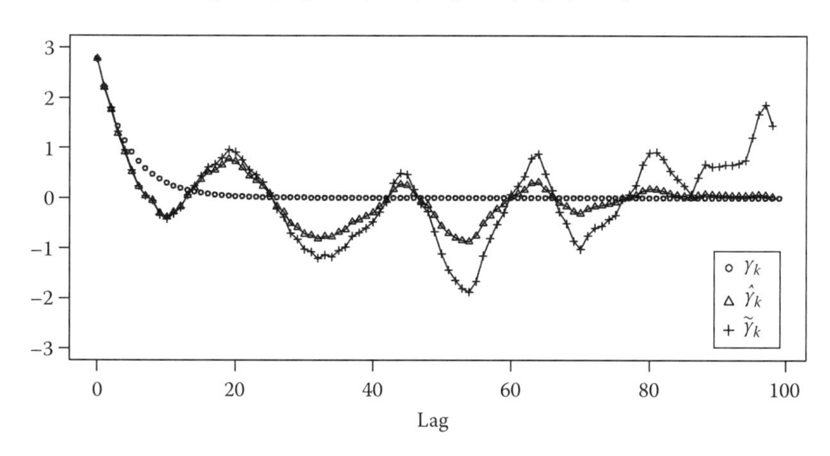

FIGURE 1.8
Theoretical and sample autocovariances based on the realization in Figure 1.5.

where, in each case, there are k zeros. For notational convenience, we will denote the first row by V_i, $i=1,\ldots,n+k$ and the second row by W_i, $i=1,\ldots,$ $n+k$. Thus, $\overline{V}=\overline{W}=0$ and the product-moment correlation between V and W is given by

$$r=\frac{\sum_{i=1}^{n+k}V_iW_i}{\sqrt{\sum_{i=1}^{n+k}V_i^2}\sqrt{\sum_{i=1}^{n+k}W_i^2}}.$$

The denominator can be rewritten as $\sum_{t=1}^{n}(X_t-\overline{X})^2$, and the numerator can be written as $\sum_{i=1}^{k}V_iW_i+\sum_{i=k+1}^{n}V_iW_i+\sum_{i=n+1}^{n+k}V_iW_i$. The first and last sums are zero, so that the numerator can be written as

$$\begin{aligned}\sum_{i=1}^{n+k}V_iW_i&=\sum_{i=k+1}^{n}V_iW_i\\&=\sum_{t=1}^{n-k}(X_t-\overline{X})(X_{t+k}-\overline{X}).\end{aligned}$$

Thus $r=\hat{\rho}_k$, and so $|\hat{\rho}_k|\le 1$.

Bartlett (1946) has obtained approximate expressions for $\mathrm{Cov}(\hat{\rho}_k,\hat{\rho}_j)$ when X_t is a stationary Gaussian process. In particular, he showed that

$$\begin{aligned}\mathrm{Cov}(\hat{\rho}_k,\hat{\rho}_{k+r})\approx&\frac{1}{n}\sum_{v=-\infty}^{\infty}\\&\times\{\rho_v\rho_{v+r}+\rho_{v+k+r}\rho_{v-k}-2\rho_k\rho_v\rho_{v-k-r}-2\rho_{k+r}\rho_v\rho_{v-k}+2\rho_k\rho_{k+r}\rho_v^2\},\end{aligned}$$

from which it follows, with $r=0$, that

$$\mathrm{Var}(\hat{\rho}_k)\approx\frac{1}{n}\sum_{v=-\infty}^{\infty}\{\rho_v^2+\rho_{v+k}\rho_{v-k}-4\rho_k\rho_v\rho_{v-k}+2\rho_k^2\rho_v^2\}. \quad (1.20)$$

When the time series has the property that $\rho_k\to 0$ as $k\to\infty$, we have for "large" k

$$\mathrm{Cov}(\hat{\rho}_k,\hat{\rho}_{k+r})\approx\frac{1}{n}\sum_{v=-\infty}^{\infty}\rho_v\rho_{v+r} \quad (1.21)$$

and

$$\mathrm{Var}(\hat{\rho}_k)\approx\frac{1}{n}\sum_{v=-\infty}^{\infty}\rho_v^2. \quad (1.22)$$

Bartlett's approximations will turn out to be quite important in our analysis of time series data. The results for approximating the variance of $\hat{\rho}_k$ are useful in determining when sample autocorrelations are significantly different from zero. Also, (1.21) shows that there can be correlation between neighboring sample autocorrelations. This is an important result to keep in mind when examining plots of $\hat{\rho}_k$ as we will be doing in our analyses of time series.

In the discussion in Section 1.3 relating to Figures 1.3 and 1.4, we described how patterns in the autocorrelations provide information about the basic nature of realizations from time series models. In Figure 1.9, we show sample autocorrelations, estimated using $\hat{\rho}_k$, associated with the realizations in Figure 1.3. These sample autocorrelations are estimates of the true autocorrelations, shown in Figure 1.4. It can be seen that these sample autocorrelations approximate the behavior of the true autocorrelations, but the correlation among $\hat{\rho}_k$ values should be noted. For example, notice that when a value of $\hat{\rho}_k$ tends to underestimate ρ_k, values of $\hat{\rho}_j$ for j "near" k also tend to underestimate the true value. This can produce a cyclic behavior in plots of $\hat{\rho}_k$ when no such behavior is present in the plot of ρ_k. This effect can be seen in Figure 1.9a and d. Also, it should be noted that sample autocorrelations for a white noise realization will of course not be exactly equal to zero. In Section 8.1, we will discuss tests for white noise that will help you decide whether sample

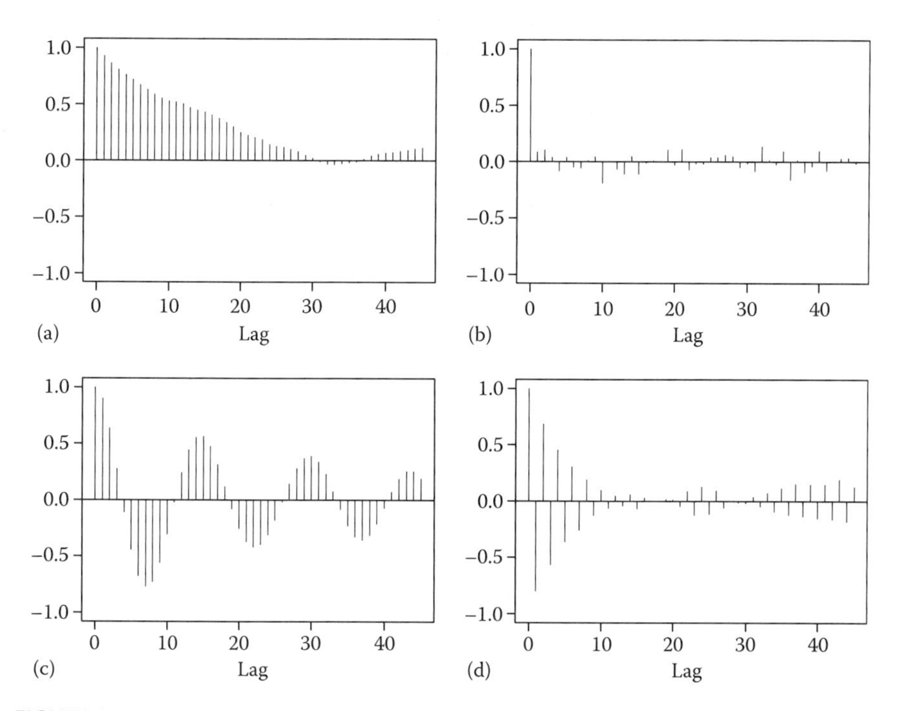

FIGURE 1.9
Sample autocorrelations associated with the four realizations in Figure 1.3. (a) Realization 1. (b) Realization 2. (c) Realization 3. (d) Realization 4.

autocorrelations such as those shown in Figure 1.9b are larger in magnitude than would be expected for a realization from a white noise process.

Using `GW-WINKS`: The sample mean for a column of data in the GW-WINKS data editor can be found using the `Analyze` menu by selecting `Descriptives`, followed by `Detail Statistics/One Variable` and then specifying the variable name assigned to the realization. The sample autocorrelations, $\hat{\rho}_k$, $k = 0, 1, \ldots, n - 1$, can be obtained using the `Time Series Analysis` menu, selecting `Sample Autocorrelations` and then specifying the variable name for the realization. The sample autocorrelations will be stored in a data editor column with a variable name beginning with `AUTOCORR`. Note that row 1 of the data editor contains $\hat{\rho}_0 = 1$, row 2 contains $\hat{\rho}_1$, etc. A data file containing the name `AUTOCORR` will be plotted in the style of Figure 1.9.

1.5 Power Spectrum

In Sections 1.1 through 1.4, we have discussed and described time series in terms of their realizations, mean, variance, and autocorrelations. These are all aspects of a time domain analysis of time series. It is often important to examine time series realizations and models from the perspective of their cyclic (of frequency) content. This type of analysis is commonplace in engineering curriculum, but, for statistics students taking their first time series course, this will likely be the first introduction to the frequency domain.

In 1807, Joseph Fourier astounded his contemporaries by showing that a very broad class of functions $f(x)$ could be written as a linear combination of sines and cosines. This finding led to the incredibly useful field of Fourier analysis, in which mathematicians and scientists have worked to develop an understanding of the fundamental properties of signals by breaking them into sums of trigonometric functions. This will be illustrated in Example 1.6.

Example 1.6 A Signal Containing Several Frequencies

Consider the upper left realization in Figure 1.10. Close examination reveals a repetitive nature to this signal, but its general appearance is similar to several realizations that we have seen in this chapter. Actually, this signal is the sum of sine waves with different cycle lengths and phase shifts that are shown in the other three plots in Figure 1.10. We will return to this data set in Example 1.7.

Actual realizations from time series models will not typically have such a simple mathematical form as the upper left signal in Figure 1.10. However, using the techniques of Fourier analysis applied to time series data via the spectrum, we will be able to analyze more complicated realizations in order to discover fundamental cyclical behavior that provides information about

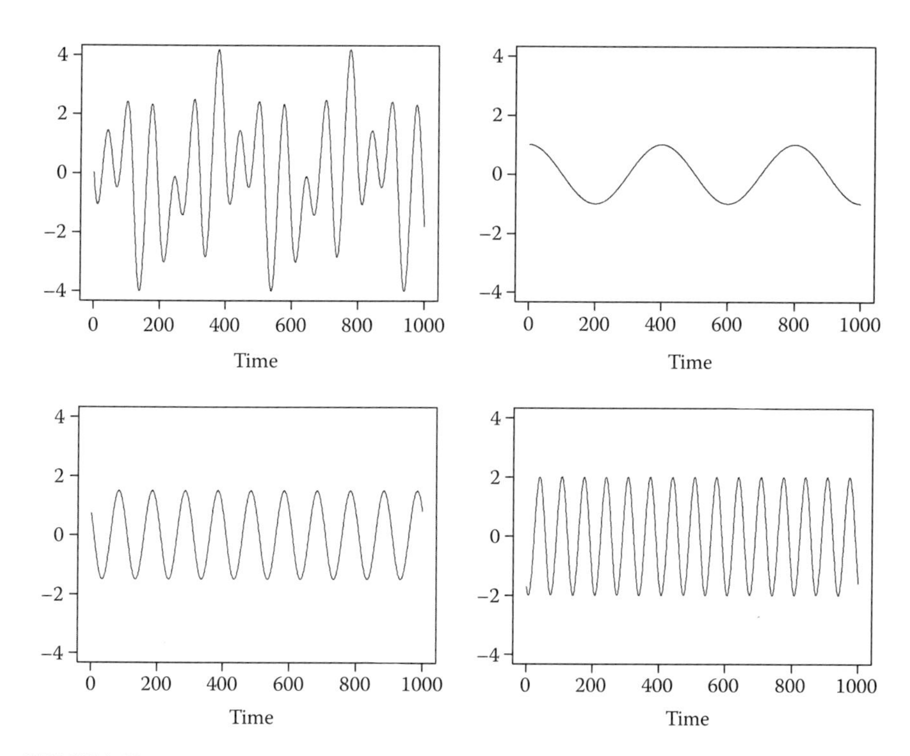

FIGURE 1.10
Signal (upper left) which is the sum of the three sine waves given in three following panels.

the underlying model generating the data. The examination of underlying periodicities in a time series is called *spectral analysis*, and the tools that we will use to measure the frequency content are the *power spectrum* and the *spectral density*. We first introduce needed terminology and definitions.

Definition 1.9 A function $g(t)$ is *periodic* with *period* (or cycle-length) $p > 0$ if p is the smallest value such that $g(t) = g(t + kp)$ for all t and integers k. A function $g(t)$ is said to be *aperiodic* if no such p exists.

Clearly, $g(t) = \cos(t)$ is periodic with period (cycle length) 2π, $\cos(2\pi t)$ is periodic with period 1, and $\cos(2\pi ft)$ has period $1/f$. In our study, we will often discuss cyclic behavior in terms of *frequency*. The definition we will use for frequency is specified in the following.

Definition 1.10 Frequency (f) measures the number of periods or cycles per sampling unit, that is, $f = 1/\text{period}$.

If the data are collected annually, monthly, or hourly, the sampling units are 1 year, 1 month, and 1 h, respectively, and the corresponding frequencies measure the number of cycles per year, month, and hour. In the scientific and

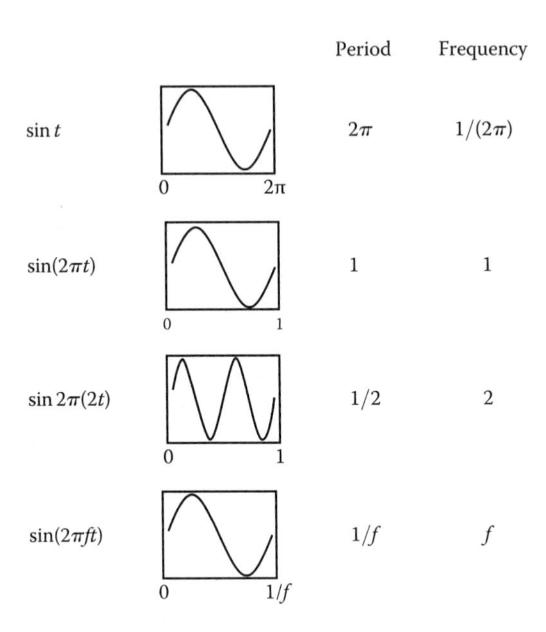

FIGURE 1.11
Examples showing the relationship between period and frequency.

engineering literature, it is common to measure frequency in Hertz (Hz), which is the number of cycles per second.

From the discussion following Definition 1.9, it follows that $\sin(t)$ has frequency $1/(2\pi)$, $\sin(2\pi t)$ has frequency 1, and $\sin(2\pi ft)$ has frequency f. These facts are illustrated in Figure 1.11.

Figure 1.12 shows the digitized record taken at 8000 Hz (samples per second) of voltage readings obtained from the acoustical energy generated by the first author speaking the words "King Kong Eats Grass" while a fan was blowing in the background. During the silent periods between words, it can be seen that the fan noise is exhibited by low-amplitude cyclic behavior with a frequency of about 136 Hz (cycles per second). Also, note the similarity of the frequency content in the words "King" and "Kong." The "K" at the beginning and the "g" at the end of the words "look" similar in frequency behavior, while the word "eats" is quite different in both its frequency and amplitude while the "s" sounds at the ends of "eats" and "grass" are similar. Thus, inspection of Figures 1.10 and 1.12 should convince us that there can be something to be gained by understanding the frequency content of a time series. This conclusion is valid and, in fact, as we will see, the frequency interpretation of a time series may often furnish us with the most relevant information about the underlying process.

As we have indicated, the role of the autocorrelation in time series analysis is a central one. However, when it is desirable to determine the frequency or cyclic content of data, the relevant information in the autocorrelation is more readily conveyed through the Fourier transform of the autocorrelation, called

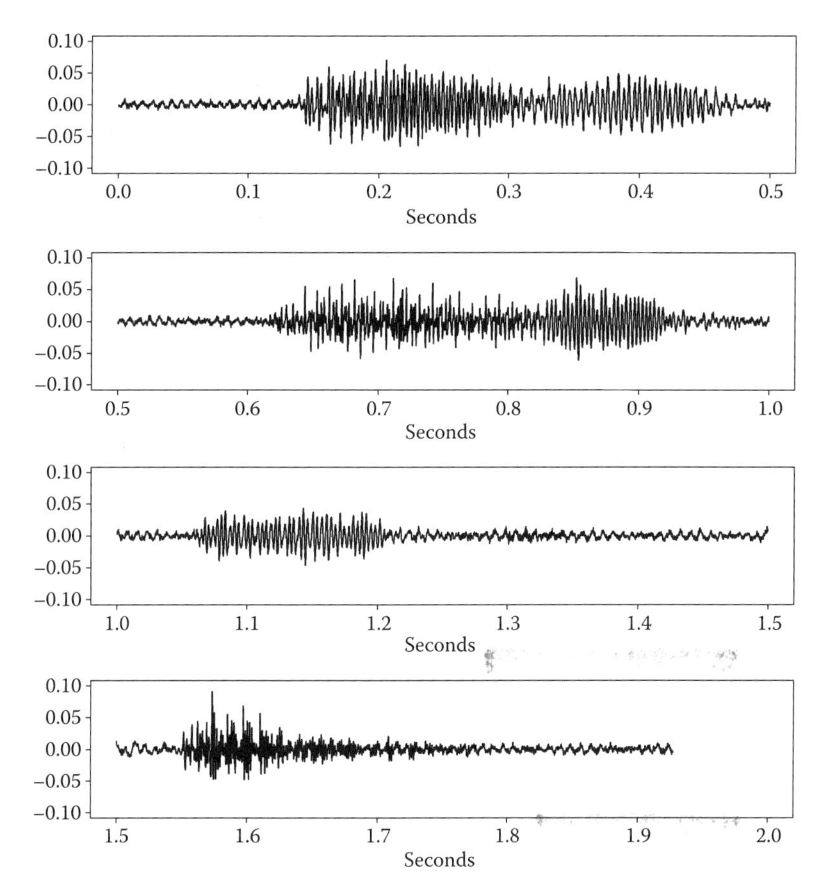

FIGURE 1.12
Digitized record of the phrase *King Kong Eats Grass.*

the *spectral density*, rather than through the autocorrelation itself. The power spectrum and spectral density, the primary tools for examining frequency content in time series data, are defined in the following.

Definition 1.11 Let $\{X(t); t \in (-\infty, \infty)\}$ be a stationary time series.

(a) The power spectrum, $P_X(f)$, is defined by

$$P_X(f) = \int_{-\infty}^{\infty} e^{-2\pi i f h} \gamma(h) dh \tag{1.23}$$

where $-\infty < f < \infty$, $i = \sqrt{-1}$, and $\gamma(h)$ is the autocovariance function.

(b) The spectral density is a normalized version of $P_X(f)$. Specifically, we define the spectral density, $S_X(f)$, by

$$S_X(f) = \frac{P_X(f)}{\sigma_X^2} = \int_{-\infty}^{\infty} e^{-2\pi i f h} \rho(h) dh. \tag{1.24}$$

The power spectrum will often be referred to as simply the spectrum. Definition 1.11 requires the index set T to be continuous. It is possible to give a more general definition that also includes the discrete case. However, to do so would require us to use the Stieltjes integral in the definition. The reader should refer to Priestley (1981) and Koopmans (1974) for this approach. We have chosen instead to consider the discrete parameter case separately.

Definition 1.12 Let $\{X_t;\, t = 0, \pm 1, \pm 2, \ldots\}$ be a stationary time series and let $|f| \leq 0.5$.

(a) The power spectrum of X_t is defined by

$$P_X(f) = \sum_{k=-\infty}^{\infty} e^{-2\pi i k f} \gamma_k. \tag{1.25}$$

(b) The spectral density is defined by

$$S_X(f) = \frac{P_X(f)}{\sigma_X^2} = \sum_{k=-\infty}^{\infty} e^{-2\pi i k f} \rho_k. \tag{1.26}$$

If

$$\sum_{k=-\infty}^{\infty} |\gamma_k| < \infty, \tag{1.27}$$

then it can be shown that the series in (1.25) and (1.26) converge and

$$\gamma_k = \int_{-0.5}^{0.5} e^{2\pi i f k} P_X(f) df, \tag{1.28}$$

that is, the spectrum and the autocovariance function are Fourier transform pairs. Consequently, given the spectrum of a stationary process satisfying (1.27), the autocovariance function can be obtained mathematically from it and vice versa. The results in (1.25) and (1.28) show that the autocovariance and the spectrum contain equivalent of information about the process. The autocovariance conveys this information in the time domain while the spectrum expresses the equivalent information in the frequency domain. We will see that for a particular application, one of these may provide more understandable information about the process than does the other. A similar relationship exists between the autocorrelation and the spectral density.

We leave it as an exercise to show that $P_X(f)$ and $S_X(f)$ in Definition 1.12 can be written as

$$P_X(f) = \sigma_X^2 + 2\sum_{k=1}^{\infty} \gamma_k \cos(2\pi f k), \tag{1.29}$$

and

$$S_X(f) = 1 + 2\sum_{k=1}^{\infty} \rho_k \cos(2\pi f k), \tag{1.30}$$

again with the restriction that $|f| \le 0.5$.

Nyquist frequency

Definition 1.12 seems the rather natural discrete version of Definition 1.11 except for the restriction on the range of the parameter f. For discrete time series data sampled at unit increments, the shortest period that can be observed is a period of 2, that is, the highest observable frequency is 0.5. This is called the *Nyquist frequency*. For example, consider the two signals $Y_t = \cos(2\pi(0.3)t)$ and $Z_t = \cos(2\pi(1.3)t)$, which have frequencies 0.3 and 1.3, respectively. These two signals are defined along a continuum, but we consider the case in which they are sampled at the integers. In Figure 1.13, we show an overlay of these two signals, and it is interesting to note that the two signals are equal to each other at integer values of t. These common values are indicated by the circles on the plot. Consequently, the two curves are indistinguishable when sampled at the integers. This phenomenon is called *aliasing*, since the frequency $f = 1.3$ "appears" to be $f = 0.3$. For discrete data, because it would be impossible to detect a period less than 2 sample units, it should not be surprising that we must restrict ourselves to $|f| \le 0.5$ in the discrete case. It should be noted that if higher frequencies are present, then in order to be able to detect them we would need to sample more rapidly. This motivates the following definition.

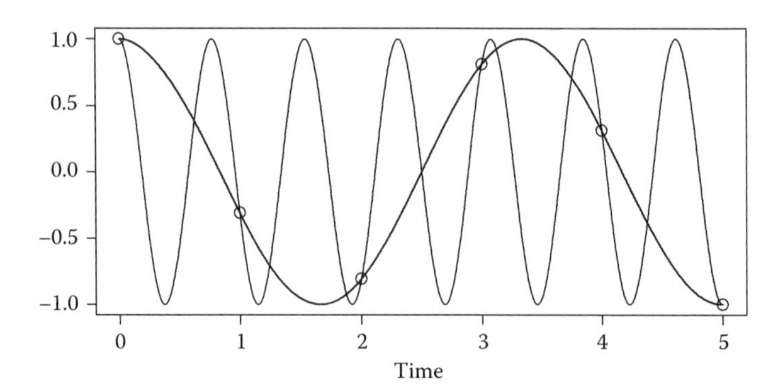

FIGURE 1.13
Plots of $\cos(2\pi(0.3)t)$ and $\cos(2\pi(1.3)t)$ (higher frequency) illustrating that these two signals agree at the integers.

Definition 1.13 If data are sampled at increments of Δ, then the Nyquist frequency is $1/(2\Delta)$, and the shortest observable period is 2Δ.

Example 1.7 Spectra for Examples 1.3 and 1.6

In this example, we evaluate the spectrum for the processes of Examples 1.3 and 1.6.

(a) In Example 1.3

$$X(t) = C\cos(\lambda t + Y), \tag{1.31}$$

where, as before, C and λ are real-valued constants, and we examine the special case in which $Y \sim$ Uniform $[0, 2\pi]$. We recall from Example 1.3 that $\mu = 0$ and $\gamma(h) = (C^2/2)\cos(\lambda h)$. Thus, $\gamma(h)$ is a periodic function of period $2\pi/\lambda$ or said another way, $\gamma(h)$ has a frequency $\lambda/(2\pi)$. The spectrum as defined in (1.23) is

$$P_X(f) = \int_{-\infty}^{\infty} e^{-2\pi i f h}\gamma(h)dh$$

$$= \frac{C^2}{2}\int_{-\infty}^{\infty} e^{-2\pi i f h}\cos(\lambda h)dh \tag{1.32}$$

$$= \frac{C^2}{4}\left\{\delta\left(f - \frac{\lambda}{2\pi}\right) + \delta\left(f + \frac{\lambda}{2\pi}\right)\right\}, \tag{1.33}$$

where δ denotes the *Dirac delta function*.

Remark: The integral in (1.32) is not convergent if one considers it as an improper Riemann integral. Moreover, the δ-function in (1.33) is not really a function but is instead a functional. Nevertheless, the results of (1.32) and (1.33) can be made rigorous by interpreting the integral in (1.32) as the limit of a sequence of integrals and by interpreting δ as a "generalized function." For discussions of generalized functions, see Kanwal (1998) and Lighthill (1959). Properties of the Dirac delta function that we will need in this book are the following:

$$\int_{-\infty}^{\infty} \delta(t - t_0)F(t)dt = F(t_0)$$
$$\int_{-\infty}^{\infty} e^{-2\pi i t u} du = \delta(t). \tag{1.34}$$

Figure 1.14 shows a plot of $P_X(f)$ given in (1.33) for $f \geq 0$ and for the special case $\lambda = 0.2\pi$. The fact that the "frequency content" of the data is concentrated at the single frequency $\lambda/(2\pi) = 0.1$ is indicated by the infinite spike at 0.1.

(b) Now, we consider again the data set plotted at the top left of Figure 1.10, which is a realization from the continuous parameter process

$$X(t) = \cos(2\pi(0.025)t + \psi_1) + 1.5\cos(2\pi(0.1)t + \psi_2) + 2\cos(2\pi(0.15)t + \psi_3) \tag{1.35}$$

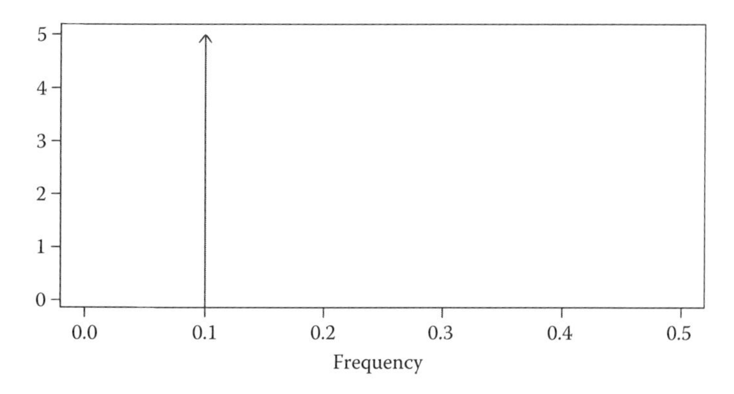

FIGURE 1.14
Power spectrum for $X(t)$ in Example 1.3 with $c = 1$, $\lambda = 0.6$.

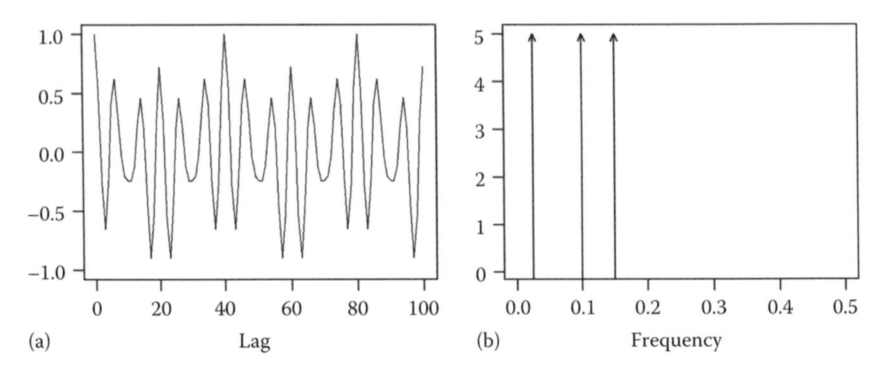

FIGURE 1.15
(a) Autocorrelations and (b) spectrum for $X(t)$ in (1.35).

where ψ_i, $i = 1$, 2, and 3 are independent random variables distributed Uniform $[0, 2\pi]$. In the particular realization plotted in Figure 1.10, $\psi_1 = 0$, $\psi_2 = 1$, and $\psi_3 = 2.5$. The first component (upper right figure) has a frequency $f = 0.025$ (period or cycle length of $1/0.025 = 40$), the second component (lower left figure) has frequency $f = 0.1$ (period = 10), while the third component (lower right figure) has frequency $f = 0.15$ (period = 6.67). The autocorrelations (see Exercise 1.11) and spectrum for $X(t)$ in (1.35) are given in Figure 1.15. It is crystal clear from the spectrum in Figure 1.15b that the frequency content is entirely concentrated at the frequencies 0.025, 0.1, and 0.15. Certainly, the spectrum does a much better job of displaying the underlying frequency content of the data than can be obtained by simple examination of the autocorrelations in Figure 1.15a or the realization in the top left of Figure 1.10. We usually plot the spectrum or the spectral density as a function of f in order to examine their characteristics. A peak in the spectrum or spectral density is indicative of the presence of cyclic behavior in the data at the associated frequency. In discrete parameter data, which we will typically analyze in this book, the peaks will not be of the "spiked" nature of Figures 1.14 and 1.15, but will be broader, indicating additional frequency content near the peak. This is illustrated in the following example.

Example 1.8 A Signal Containing Two Dominant Frequency Components

In Figure 1.16a, we show a realization of length 250 where it is clear that the data have a pseudo-periodic behavior with a period of about 25 (there are about 10 cycles in the realization of 250 time points). Also apparent is a higher frequency behavior indicated by the "jagged" behavior. Figure 1.16a is a realization from an autoregressive model of order 4 (we will discuss these

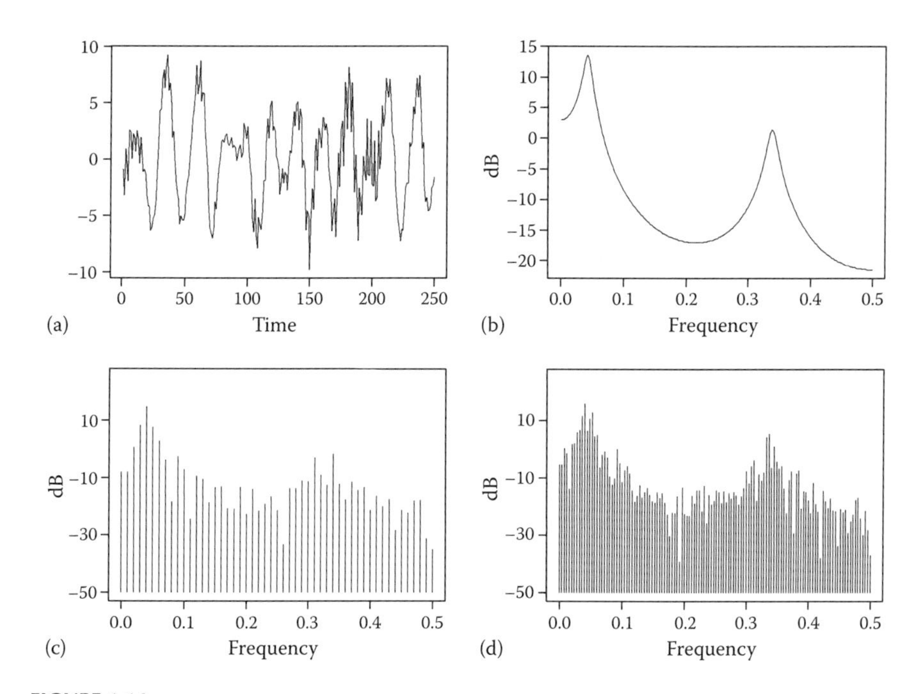

FIGURE 1.16
(a) Realization, (b) true spectral density, (c) periodogram based on first 100 points, and (d) periodogram based on all 250 points associated with Example 1.8.

in Chapter 3), and in Figure 1.16b, we show the true spectral density associated with this model. The spectral density has a peak at about $f = 0.04 = 1/25$ associated with cyclic behavior in the data with period about 25, and it also has a peak at about $f = 0.33$ associated with the higher frequency content in the data with period length of about 3. Also notice that the peaks in the spectral density in Figure 1.16b are not spikes as in Figures 1.14 and 1.15 but instead indicate the presence of periodic behavior for frequencies close to 0.04 and 0.33.

We note two points before proceeding.

(1) For real-time series $P(f) = P(-f)$ and likewise $S(f) = S(-f)$ (see Problem 1.19(a)). For this reason, plots of the spectrum or spectral density of real data are usually only shown for $f \geq 0$ as was done in Figures 1.14 through 1.16.

(2) In Figure 1.16b, we actually plot the spectrum on the log scale. There are two reasons for this. Secondary peaks in the spectrum are more discernible in such a plot, and $10 \log_{10} S_X(f)$ puts the ordinate axis in units of decibels (dB), which is the standard unit for measuring acoustical energy and is thus widely used in engineering.

1.6 Estimating the Power Spectrum and Spectral Density for Discrete Time Series

Estimation of the power spectrum and spectral density from data can be a challenging problem, and there are a variety of estimation techniques in the literature. We will briefly introduce the topic here but will come back to it in later chapters. Given the spectrum in (1.29) for a discrete parameter time series, it seems that a natural estimator for the spectrum could be obtained by simply substituting the sample autocovariances discussed in Section 1.4 into (1.29). The resulting estimator is called the *sample spectrum* and is defined in Definition 1.14.

Definition 1.14 Let X_t be a discrete stationary time series defined on the integers. The sample spectrum based on a realization of length n from this time series is given by

$$\hat{P}_X(f) = \hat{\sigma}_X^2 + 2\sum_{k=1}^{n-1} \hat{\gamma}_k \cos(2\pi f k), \tag{1.36}$$

for $|f| \leq 0.5$. The sample spectral density is given by

$$\hat{S}_X(f) = 1 + 2\sum_{k=1}^{n-1} \hat{\rho}_k \cos(2\pi f k). \tag{1.37}$$

The limit, $n-1$, on the sum is reasonable, since sample autocovariances and autocorrelations for $k \geq n$ are defined to be zero.

A related estimator is the *periodogram* that is simply the sample spectrum evaluated at the harmonics of the *fundamental frequency* $f_1 = 1/n$. Specifically, the *harmonics*, f_j, are given by

$$\begin{aligned} f_j &= \frac{j}{n}, \quad j = 1, \ldots, \frac{n}{2} \quad \text{(if } n \text{ is even)} \\ &= \frac{j}{n}, \quad j = 1, \ldots, m \quad \text{(if } n \text{ is odd and } n = 2m+1\text{)}. \end{aligned}$$

Definition 1.15 Let X_t be a discrete time series defined on the integers. The periodogram based on a realization of length n from this time series is given by

$$I(f_j) = \hat{\sigma}_X^2 + 2\sum_{k=1}^{n-1} \hat{\gamma}_k \cos 2\pi f_j k, \tag{1.38}$$

where the f_j are the harmonics as defined earlier.

The periodogram can alternatively be defined in terms of the finite Fourier series representation of the signal and as the modulus squared of the finite Fourier transform of the data. See Appendix 1.A. While the sample spectrum and periodogram are obvious estimators of the spectrum, they are not well-behaved as estimators. Theorem 1.4 that follows outlines the properties of the sample spectrum.

Theorem 1.4

Let X_t be a discrete stationary time series defined on the integers, and let $\hat{P}_X(f)$ denote the sample spectrum given in Definition 1.14. $\hat{P}_X(f)$ has the following properties:

(i) $\hat{P}_X(f)$ is asymptotically unbiased, that is, $\lim_{n\to\infty} E\left[\hat{P}_X(f)\right] = P(f)$.
(ii) $\hat{P}_X(f)$ is not a consistent estimator of $P_X(f)$, and $\mathrm{Var}\left[\hat{P}_X(f)\right]$ does not decrease as $n \to \infty$.

For more details, see Parzen (1962). In Figure 1.16c and d, we show the periodograms (plotted as vertical bars in dB at the harmonics of the fundamental frequency) based on the first 100 points and the entire realization, respectively. The periodograms are somewhat visually different, since (c) is plotted at the points $f_j = j/100$, $j = 1, \ldots, 50$ while the periodogram in (d) is based on a finer partition (i.e., it is plotted at $f_j = j/250$, $j = 1, \ldots, 125$). The lack of consistency is illustrated by the fact that the periodogram based on $n = 250$ is just as variable and erratic as the one for $n = 100$, that is, we have gained very little with the increased n. It should be noted that although the periodogram as defined in (1.38) and its alternative forms are commonly studied as estimators of the spectrum, in this book, we plot standardized periodograms (in dB) that estimate the spectral density [i.e., we use $\hat{\rho}_k$ instead $\hat{\gamma}_k$ in (1.38)].

The basic problem with the sample spectrum and the periodogram is that the estimates for γ_j whenever j is close to n are extremely variable regardless of the magnitude of n. Much research has gone into the problem of smoothing the sample spectrum in order to obtain better behaved spectral estimators. A common type of smoothed spectral estimator takes the general form

$$\hat{P}_S(f) = \lambda_0 \hat{\sigma}_X^2 + 2\sum_{k=1}^{M} \lambda_k \hat{\gamma}_k \cos(2\pi f k), \tag{1.39}$$

where the truncation point M is chosen to be substantially smaller than n, and λ_k, $k=0, 1, \ldots, M$ is a window function that damps as k increases. The weights for three classic window functions are as follows:

(a) *Bartlett (triangular) window*

$$\lambda_k = 1 - k/M, \quad k = 0, 1, \ldots, M$$
$$= 0, \quad k > M$$

(b) *Tukey window*

$$\lambda_k = \frac{1}{2}[1 + \cos(k\pi/M)], \quad k = 0, 1, \ldots, M$$
$$= 0, \quad k > M$$

(c) *Parzen window*

$$\lambda_k = 1 - 6(k/M)^2 + 6(|k|/m)^3, \quad k| \leq M/2$$
$$= 2(1 - |k|/M)^3, \quad M/2 < |k| \leq M$$
$$= 0, \quad k| > M.$$

In Figure 1.17, we show smoothed versions of the periodogram for the realization of length $n=250$ in Figure 1.16a using the Parzen window

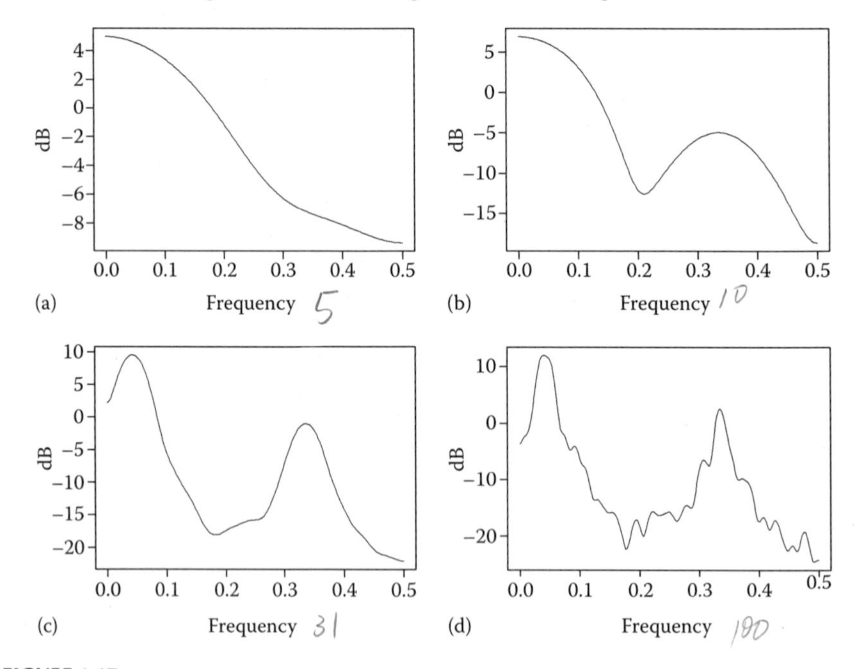

FIGURE 1.17
Windowed spectral estimators for data in Figure 1.16 using Parzen's window with various values of M. Truncation points: (a) $M=5$, (b) $M=10$, (c) $M=31$, and (d) $M=100$.

with four different values for M. The choice of M is somewhat subjective, and the goal is to balance resolution against variability. It can be seen that as M decreases the variability of $\hat{P}_S(f)$ decreases while the bias increases. Clearly, in Figure 1.17a, it can be seen that $M=5$ produces a very smooth spectral estimator that essentially looses the key spectral feature of two peaks. Figure 1.17b shows that $M=10$ does a better job of separating the peaks, but the low frequency peak is sufficiently smeared that the actual peak in the smoothed spectral estimator is $f=0$. Moreover, $M=100$ seems too large, because the associated estimator in Figure 1.17d is not much smoother than the sample spectrum (periodogram) shown in Figure 1.16d. A common choice for the truncation point is $M = 2\sqrt{n}$, where n is the realization length (in this case $M=31$) which corresponds to Figure 1.17c where it can be seen that the resulting estimator is fairly smooth yet clearly shows the two peaks. The selection of window types and truncation points is referred to as window carpentry. See Percival and Walden (1993) and Jenkins and Watts (1968).

Other types of spectral estimators include techniques involving data tapering, which include popular multitapering techniques. See Percival and Walden (1993) and Thomson (1982). Autoregressive and autoregressive-moving average spectral estimators are discussed in Section 7.7.

Using `GW-WINKS`: The periodogram and window-based spectral density estimates can be obtained by selecting `Spectrum` on the `Time Series Analysis` menu. The window-based options are based on the Bartlett, Tukey, and Parzen windows. You can choose whether or not to plot the log of the spectral density and you can specify the truncation point (default $M = 2\sqrt{n}$). The spectral estimates calculated are each placed in columns in the `GW-WINKS` data editor. A data editor column containing a periodogram has m values where $m=n/2$ if n is even and $n=2m+1$ if n is odd. The kth row contains the periodogram value calculated at $f=k/n$. The columns containing windowed spectral estimates have 251 data values that should be considered to be the 251 values of $\hat{S}_X(f), f=0, 0.002, 0.004, \ldots, 0.5$. `GW-WINKS` assigns default names beginning with `PERIODOGRAM`, `PARZEN`, etc., for these columns. These files can be plotted using the `Graphs/Chart` menu item and then selecting `Line/Time Series Plots` followed by `Line plots one or multiple`. You may select up to four data fields to be plotted. The data file containing the name `PERIODOGRAM` will be plotted in the style of Figure 1.16c and d. A variable name containing `BARTLETT`, `PARZEN`, or `TUKEY` causes a plot of the style in Figure 1.17. Plot styles can be modified in graphics edit mode, which is activated with a right click on the graph screen. You can also overlay more than one window-based spectral plot using the `Overlay line plot` option instead of requesting multiple plots.

1.7 Time Series Examples

1.7.1 Simulated Data

In this section, we examine time series realizations simulated from known time series models. The autocorrelations and spectral densities will be studied for each realization in order to provide a better understanding of the information that each conveys. The autoregressive models of order p (AR(p)) used to generate these realizations will be examined in detail in Chapter 3, and we only consider these examples at this point for illustrative purposes. For these models, ρ_k and $S(f)$ can be obtained using results to be given in Chapter 3.

Example 1.9 Two AR(1) Models

Figures 1.18a and 1.19a shows realizations of length 250 from the AR(1) models

$$X_t = 0.9X_{t-1} + a_t \tag{1.40}$$

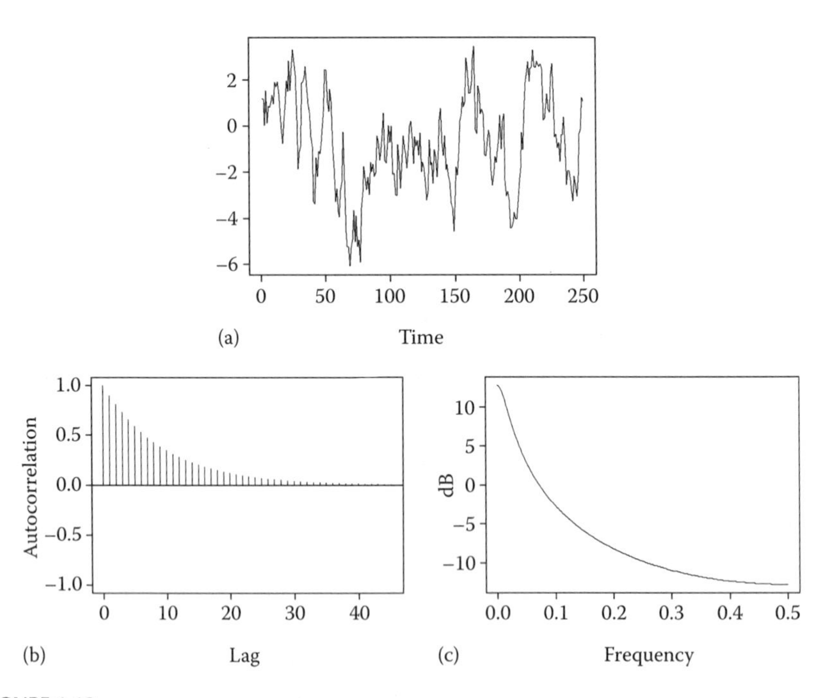

FIGURE 1.18
Plots associated with the model (1.40) of Example 1.9. (a) Realization. (b) Autocorrelation plot. (c) Spectral density plot.

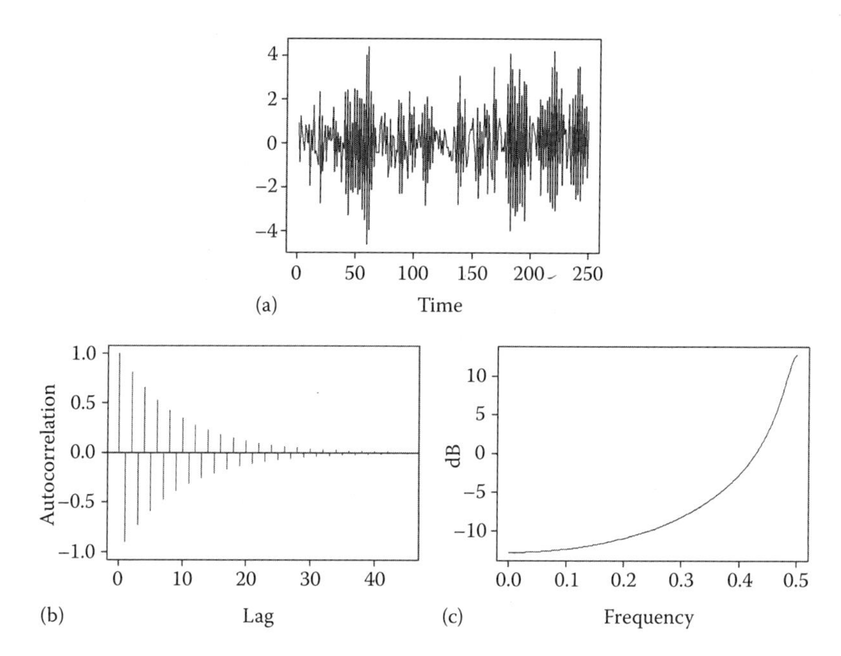

FIGURE 1.19
Plots related to model (1.41) of Example 1.9. (a) Realization. (b) Autocorrelation plot. (c) Spectral density plot.

and by

$$X_t = -0.9X_{t-1} + a_t, \tag{1.41}$$

respectively, where a_t is Gaussian white noise with zero mean and unit variance. We notice that the series in Figure 1.18a has a wandering type behavior with no distinct periodic tendency. Also note that there seems to be a rather high correlation between X_t and X_{t+k} for small k. On the other hand, note that the realization in Figure 1.19a shows rapid oscillation.

The true autocorrelations, ρ_k, for models (1.40) and (1.41) are displayed in Figures 1.18b and 1.19b, respectively. In Figure 1.18b, it can be seen that the true autocorrelations damp exponentially to zero, but that there is relatively high positive correlation between X_t and X_{t+k} for small k, which is consistent with our observation from the data themselves. The autocorrelations shown in Figure 1.19b appear to be a damped oscillating exponential. In Figure 1.18c, the peak at $f=0$ indicates that associated realizations such as the one in Figure 1.18a show little or no periodic behavior, that is, the data are primarily low frequency with no real periodic tendency. On the other hand, the spectral plot shown in Figure 1.19c is close to zero near $f=0$ and has its maximum at $f=0.5$ indicating that the data have a highly oscillatory nature, that is, it are predominantly high frequency with a period of about 2.

Example 1.10 An AR(2) Model

Figure 1.20 shows plots related to the AR(2) model

$$X_t = 1.6X_{t-1} - 0.95X_{t-2} + a_t, \tag{1.42}$$

where again a_t is Gaussian white noise with zero mean and unit variance. In Figure 1.20a, we display a realization of length 250 from this model. The realization shows a pseudo-periodic tendency with an indication of a period of about 10 time units, since the data go through about 25 cycles in the 250 time points. In Figure 1.20b, we show the true autocorrelations, ρ_k, based on the AR(2) model, and these display a damped sinusoidal behavior with a period of about 11 units. However, the precise periodic content of the data is still vague, since the apparent cycles in the data are of differing lengths. Figure 1.20c displays the spectral density in dB for this time series model where we see a rather sharp peak at about $f = 0.09$, indicating a pseudo-periodic behavior in the data with dominant period $1/0.09 = 11.1$. Notice that frequencies very near $f = 0.09$ also have rather large values in the true spectral density, as evidenced in the data by varying cycle lengths near 11. This is in contrast to the spectrum in Figure 1.14 in which case the data have a "pure" periodic component, and the true spectral density contains a spike with no indication of contribution from nearby frequencies.

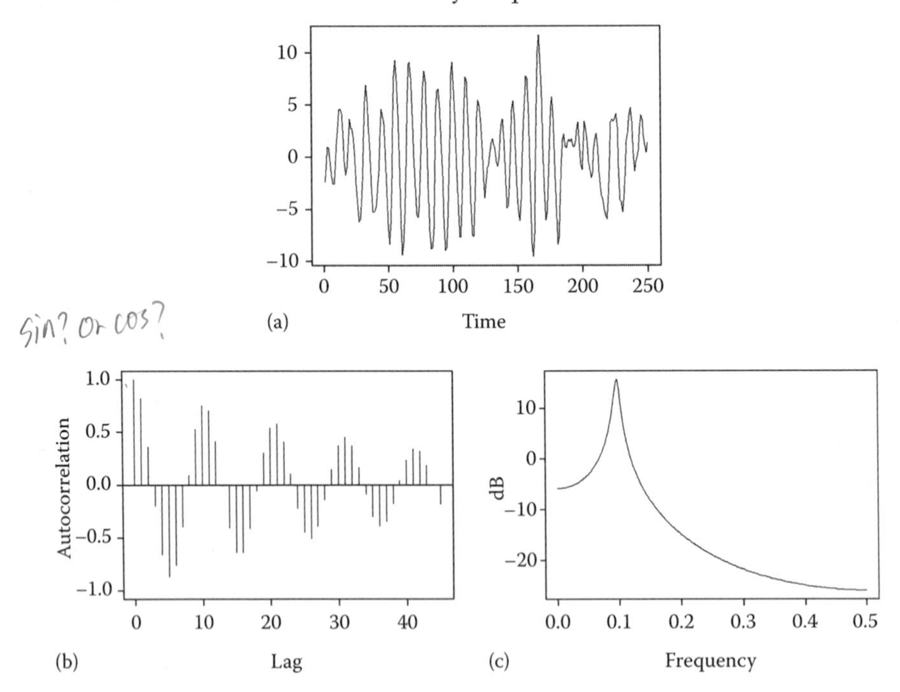

FIGURE 1.20
Plots related to the model of Example 1.10. (a) Realization. (b) Autocorrelation plot. (c) Spectral density plot.

Example 1.11 An AR(4) Process with More Than One Dominant Frequency Component

In Figure 1.21, we show plots based on the model

$$X_t = 0.85X_{t-1} + 0.03X_{t-2} + 0.745X_{t-3} - 0.828X_{t-4} + a_t, \quad (1.43)$$

where a_t is normal zero mean, unit variance white noise. In Figure 1.21a and c, we display the realization and true spectral density, which are somewhat similar to those given in Figure 1.16a and b. In the current data set, we notice a low frequency component associated with a period of about 30, while there is also indication of a higher frequency component in the data. The spectral density plot in Figure 1.21c clearly indicates the presence of two frequencies in the data, one at about $f=0.04$ and another at about $f=0.34$. The autocorrelations in Figure 1.21b exhibit a sinusoidal nature that is similar to that in Figure 1.20b. However, notice that the autocorrelations in Figure 1.21b show no real indication of the presence of a higher frequency component. This illustrates the ability of the spectrum to detect the presence of cycles in the data when the autocorrelation does not, even though, as discussed previously, they contain equivalent mathematical information.

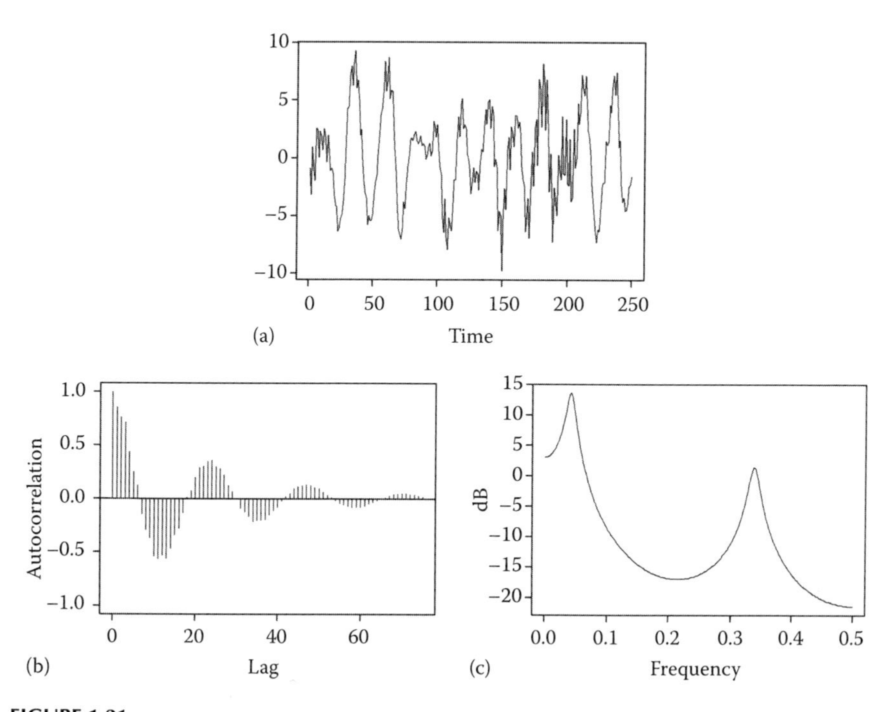

FIGURE 1.21
Plots associated with the model of Example 1.11. (a) Realization. (b) Autocorrelation plot. (c) Spectral density plot.

Example 1.12 White Noise

Figure 1.22a shows a realization from the white noise series $X_t = a_t$ where as before, a_t is normal with zero mean and unit variance, and, in Figure 1.22b, we show the true autocorrelations. This is the same realization and theoretical autocorrelations that were shown in Figures 1.3b and 1.4b. As before, notice that the data appear to be completely random with no discernible cyclic or trending pattern. The true autocorrelations, ρ_k, are zero except for $k = 0$. In Figure 1.22c, we show the true spectral density. For discrete white noise, (1.30) becomes $S(f) = 1$, $|f| \leq 0.5$. Thus, discrete white noise has a constant spectral density, and a plot of the spectral density is simply that of a straight horizontal line. Since we plot dB in Figure 1.22c, the spectral density plot is a horizontal line at dB = 0.

1.7.2 Real Data

The following examples contain realizations from four well-studied time series. In these examples, for which of course we do not know the theoretical model, we will show plots of the realizations, the sample autocorrelations, and estimates of the spectral density.

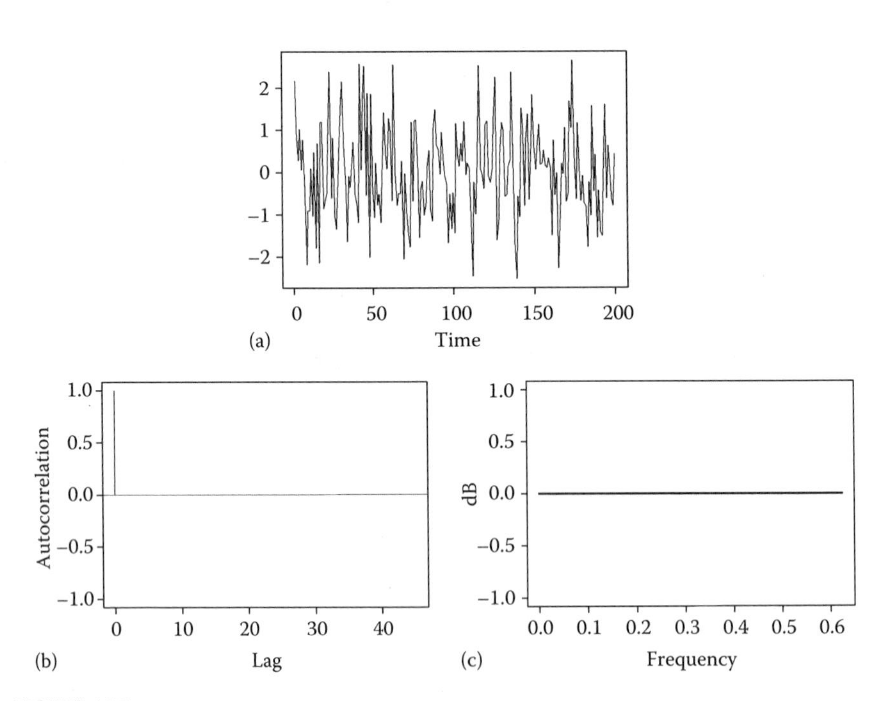

FIGURE 1.22
(a) White noise realization, (b) autocorrelations, and (c) spectral density plot.

Example 1.13 Sunspot Data

One of the most intriguing time series data sets is that of the sunspot data. Sunspots are areas of solar explosions or extreme atmospheric disturbance on the sun. The sunspots have created much interest among the scientific community partially because of the fact that they seem to have some effect here on the earth. For example, high sunspot activity tends to cause interference with radio communication and to be associated with higher ultraviolet light intensity and northern light activity. Other studies have attempted to relate sunspot activity to precipitation, global warming, etc., with less than conclusive results. In 1848, the Swiss astronomer Rudolf Wolf introduced a method of enumerating the daily sunspot activity, and monthly data using his method are available since 1749. See Waldmeier (1961) and various internet sites.

Our interest is in annual data, and in Figure 1.23a we plot these yearly averages for the years 1749–2008. The problem of modeling these data has been studied by many investigators in the past half century. In particular, Yule (1927) proposed the autoregressive model as a tool for describing the disturbed periodic nature of the data.

The sunspot data in Figure 1.23a reveal a pseudo-periodic nature, where there is an indication of a period somewhere between 9 and 11 years. It is

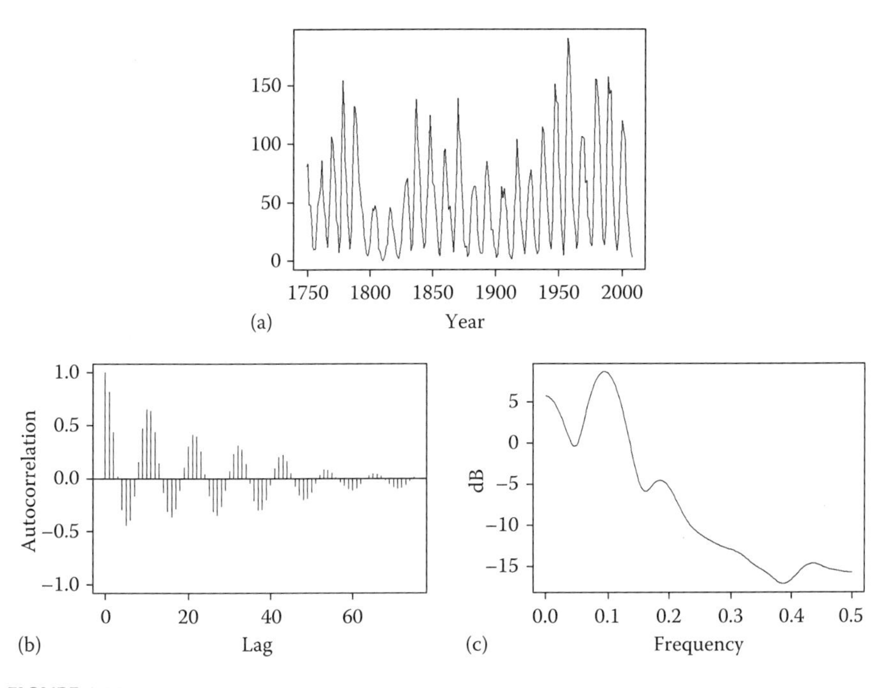

FIGURE 1.23
Plots related to sunspot data. (a) Annual sunspot numbers. (b) Sample autocorrelations. (c) Parzen window spectral estimator.

worth noting that the sunspot data are not sinusoidal in nature, and, in fact, there is much more variability in the peaks than there is in the troughs. Nonlinear models have been proposed for these data (see Tong, 1990). In Figure 1.23b, we show the sample autocorrelations for the sunspot data. Figure 1.23c displays the spectral density where we have estimated $S(f)$ using the Parzen window with $M = 2\sqrt{n} = 32$. In Figure 1.23c, we notice a reasonably sharp peak at approximately $f = 0.09$ indicating a predominant period of about 11 years in the data. It is also interesting to note that there is also a peak at $f = 0$ indicating possibly a longer period or aperiodic behavior.

Example 1.14 Global Temperature Data

The concern over global warming has attracted a great deal of attention over the past 25 years. Of interest are data sets that provide a measure of the average annual temperature. One such data set was collected in England by the Climatic Research at the University of East Anglia in conjunction with the Met Office Hadley Centre. The data given in Figure 1.24a are annual temperature information (based on temperature stations positioned around the

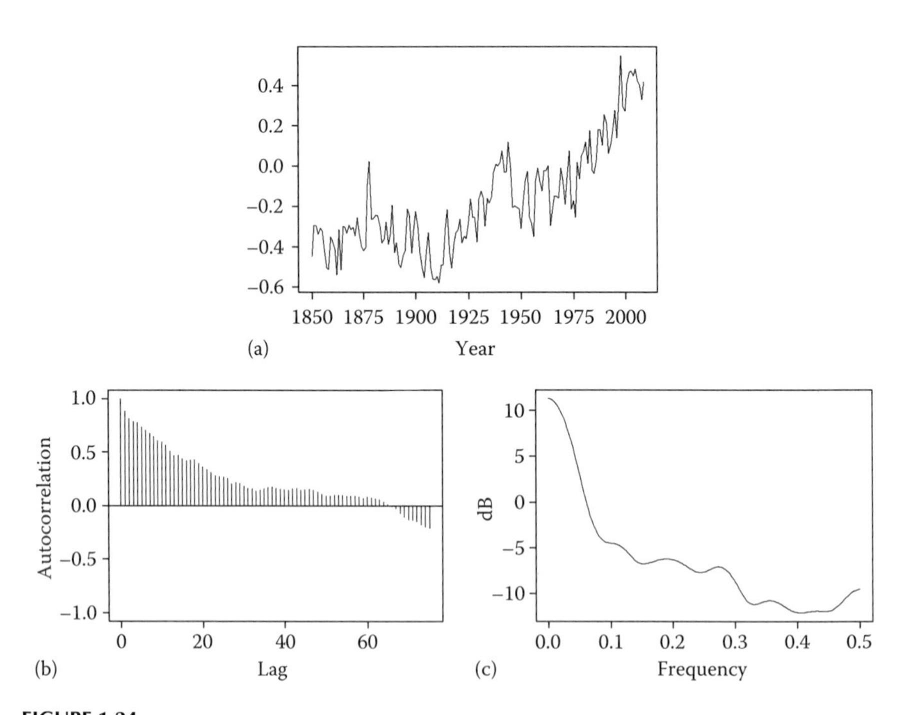

FIGURE 1.24
Plots related to global temperature data. (a) Global temperature data. (b) Sample autocorrelations. (c) Parzen window spectral estimator.

globe) for the years 1850–2009. The data plotted are the temperature anomalies (i.e., departures from the average for the period 1961–1990) based on both land and marine temperature stations. In the data set, it is clear that there has been a rise in temperatures over this period. There is also considerable year to year variability but no apparent cyclic behavior. In Figure 1.24b, we see that the autocorrelations for the lower lags are positive indicating that, for example, there is substantial positive correlation between this year's and next year's annual temperatures. The spectral estimator in Figure 1.24c (smoothed sample spectrum using Parzen's window with $M = 2\sqrt{n} = 25$) also suggests the lack of periodic behavior. The peak at $f = 0$ suggests either a very long period or aperiodic behavior. The smaller peaks in the spectral estimator may or may not be meaningful.

Example 1.15 Airline Data

Figure 1.25a is a graph of the natural logarithms of monthly total numbers (in thousands) of passengers in international air travel for 1949–1960. This series is examined by Box et al. (2008) and in earlier editions of that text and is often analyzed as a classical example of a *seasonal* time series. A strong seasonal pattern is expected due to the seasonal nature of air travel, for example, heavier travel in the summer months and in November and

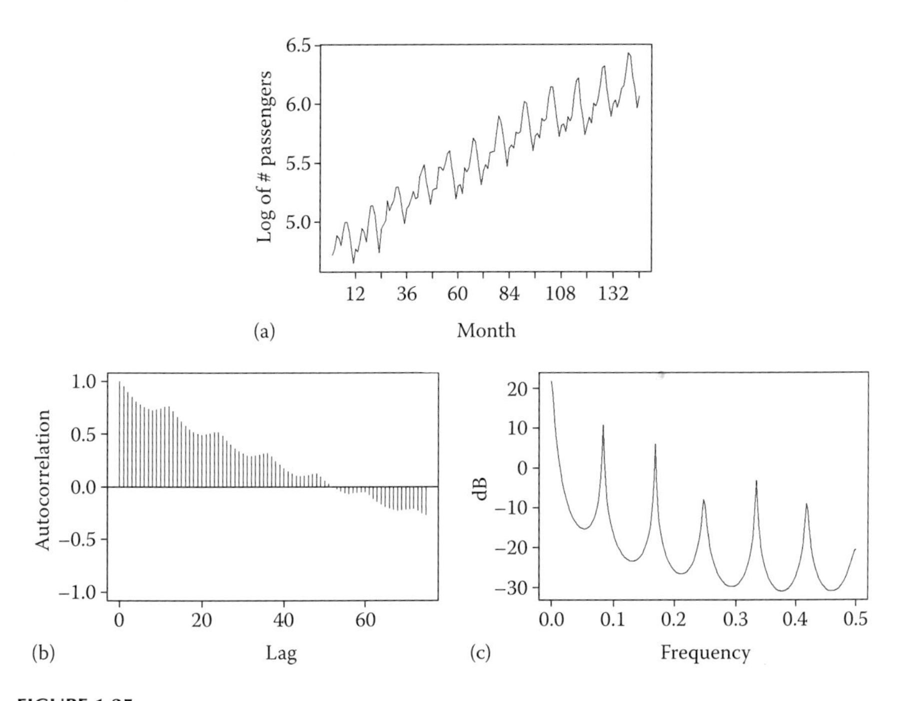

FIGURE 1.25
Plots related to the airline data. (a) International airline passengers. (b) Sample autocorrelations. (c) AR spectral estimator.

December. An inspection of the data shows the heavier summer travel, a 12 month annual pattern, and an apparent upward trend. Note that the 12 month cycle does not appear to be sinusoidal but rather tends to "copy" the pattern of the previous year. We will discuss seasonal models in Chapters 3 and 5. Figure 1.25b shows the sample autocorrelations. The analysis of the seasonal and trend behavior in the data is difficult here and simple inspection of the plot of these sample autocorrelations is inadequate for this analysis. Figure 1.25c shows the autoregressive spectral density estimator (see Section 7.7). Note the high power at zero indicating the strong trend (aperiodic) component in the data. In addition, the high power at $f=1/12$ and at the harmonics of 1/12, that is, $f=i/12$, $i=2, 3,\ldots,6$ are quite apparent. That is, the 12 month, 6 month, ..., 2 month periods in the data are vividly displayed here.

Example 1.16 Bat Echolocation Signal

In Figure 1.26a, we show 96 observations taken from a Nyctalus noctula hunting bat echolocation signal sampled at 4×10^{-5} s intervals. At first glance, the signal in Figure 1.26a appears to be periodic with behavior similar to that seen in Figure 1.20a and somewhat similar to the sunspot

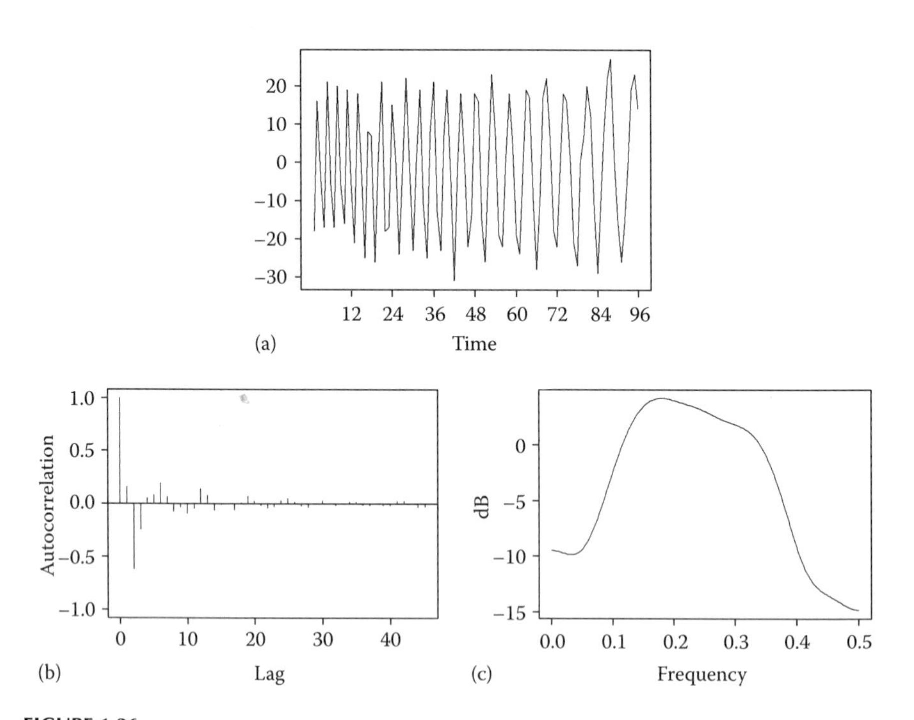

FIGURE 1.26
Plots related to Nyctalus noctula hunting bat echolocation signal. (a) Echolocation signal. (b) Sample autocorrelations. (c) Parzen window spectral estimator.

data in Figure 1.23a. However, the sample autocorrelations seem unusual, since there is no strong cyclic behavior as there was in Figures 1.20b and 1.23b. Further, examination of the spectral estimate in Figure 1.26c shows no sharp peak, and, in fact, there is a wide band of frequencies between $f = 0.15$ and $f = 0.35$ associated with a broad "peak." Spectra such as the one shown in Figure 1.26c are referred to as *spread spectra* and indicate that the behavior in the signal is not concentrated at and around a single frequency, but instead there is a broad range of frequency behavior. Closer examination of Figure 1.26a shows that the frequency behavior is changing with time. More specifically, the cycle lengths tend to lengthen (frequencies decrease) in time. Consequently, it appears that the data may not be stationary and therefore the question arises, "What does the spectral density or autocorrelation function mean if periods are changing with time?" Analysis of this type of data with time-varying frequencies will be discussed in Chapters 12 and 13.

These and other time series will be investigated in more detail in future chapters. The following chapters discuss in detail the use of tools such as the autocorrelation and spectral density, and models such as the autoregressive model for properly analyzing stationary and nonstationary time series data. For now, we have simply introduced these ideas briefly to help the reader begin to understand the concepts.

1.A Appendix

In Definition 1.15, we define the periodogram of a discrete time series X_t to be

$$\text{I. } I(f_j) = \hat{\sigma}_X^2 + 2\sum_{k=1}^{n-1} \hat{\gamma}_k \cos 2\pi f_j k, \tag{1.A.1}$$

where f_j are the harmonics of the fundamental frequency $1/n$. Specifically,

$$\begin{aligned} f_j &= \frac{j}{n}, \quad j = 1, \ldots, \frac{n}{2} \quad \text{(if } n \text{ is even)} \\ &= \frac{j}{n}, \quad j = 1, \ldots, m \quad \text{(if } n \text{ is odd and } n = 2m + 1\text{)}. \end{aligned}$$

It is interesting to note that the periodogram has two useful alternate representations. The periodogram in (1.A.1) can be obtained as $1/n$ times the modulus squared of the Fourier transform of the data as given by (1.A.2).

$$\text{II. } I(f_j) = \frac{1}{n}\left|\sum_{t=1}^{n} x_t e^{-2\pi ijt/n}\right|^2. \tag{1.A.2}$$

The proof of the equality of forms (1.A.1) and (1.A.2) is left as Problem 1.20.

The third form involves the Fourier series. In this appendix, we discuss the topic of fitting a Fourier series to a discrete time series realization x_t, $t = 1, \ldots, n$, and we assume that $n = 2m + 1$ is odd. That is, we approximate x_t using a finite Fourier series $\sum_{j=1}^{m}\left[a_j \cos\left(\frac{2\pi jt}{n}\right) + b_j \sin\left(\frac{2\pi jt}{n}\right)\right]$. The coefficients a_j and b_j are estimated using least squares, that is, we find the a_j and b_j that minimize

$$Q = \sum_{t=1}^{n}\left\{x_t - \sum_{j=1}^{m}\left(a_j \cos\left(\frac{2\pi jt}{n}\right) + b_j \sin\left(\frac{2\pi jt}{n}\right)\right)\right\}^2. \tag{1.A.3}$$

The a_j and b_j, $j = 1, \ldots, m$ that minimize Q in (1.A.3) are

$$a_j = \frac{2}{n}\sum_{t=1}^{n} x_t \cos\left(\frac{2\pi jt}{n}\right) = \frac{2}{n}\sum_{t=1}^{n}(x_t - \bar{x})\cos\left(\frac{2\pi jt}{n}\right)$$

$$b_j = \frac{2}{n}\sum_{t=1}^{n} x_t \sin\left(\frac{2\pi jt}{n}\right) = \frac{2}{n}\sum_{t=1}^{n}(x_t - \bar{x})\sin\left(\frac{2\pi jt}{n}\right).$$

The second equality in each expression follows from the fact that

$$\sum_{t=1}^{n}\cos\left(\frac{2\pi jt}{n}\right) = \sum_{t=1}^{n}\sin\left(\frac{2\pi jt}{n}\right) = 0$$

for each j. Using this terminology, the third form of the periodogram is given by

$$\text{III. } I(f_j) = \frac{n}{4}\left(a_j^2 + b_j^2\right), \quad j = 1, \ldots, m. \tag{1.A.4}$$

Using Parseval's equality (see Jenkins and Watts, 1968), we obtain

$$\frac{1}{n}\sum_{t=1}^{n}(x_t - \bar{x})^2 = \frac{1}{2}\sum_{j=1}^{m}\left(a_j^2 + b_j^2\right), \quad j = 1, \ldots, m. \tag{1.A.5}$$

The result in (1.A.5) gives an analysis of variance breakdown showing the contribution of each harmonic to the sample variance. That is, large values of $a_j^2 + b_j^2$ indicate the presence of frequency behavior at or near $f_j = j/n$ in the data. The proof of the equality of (1.A.4) with the other two forms is left as Problem 1.20.

When $n = 2m$ is even, the formulas discussed earlier vary slightly. Specifically, the formulas for a_j and b_j hold for $j = 1, \ldots, m-1$. However, $a_m = \frac{1}{n}\sum_{t=1}^{n}(-1)^t x_t$, $b_m = 0$, and $I(f_m) = na_m^2$.

Exercises

Applied Problems

If you use **GW-WINKS** to work following problems, you will find the `Generate Series`, `Sample Autocorrelations`, and `Spectrum` options of the `Time Series Analysis` menu to be useful. Time series data sets mentioned here are available on the website www.texasoft.com/ATSA.

1.1 The following data are annual sales of a hypothetical company in millions of dollars:

Period	Sales
1	76
2	70
3	66
4	60
5	70
6	72
7	76
8	80

Compute by hand (i.e., calculator) the estimates $\hat{\gamma}_0$, $\hat{\gamma}_1$, $\hat{\rho}_0$, and $\hat{\rho}_1$.

1.2 Figure 1.1a shows plots of West Texas intermediate crude oil prices from January 2000 through October, 2009, and Figure 1.19b shows monthly average temperatures for Pennsylvania for January, 1990 through December, 2009. These two data sets are `WTCRUDE.XLS` and `PATEMP.XLS`, respectively. For each of these realizations, plot the sample autocorrelations, periodogram, and a window-based spectral density estimate. Explain how these plots describe (or fail to describe) the behavior in the data.

1.3 Figure 1.25c shows an autoregressive spectral estimator (in dB) for the log airline data given in AIRLOG.XLS. Using the GW-WINKS software, find the windowed spectral density estimators for these data using Tukey's window with truncation points $M = 10$, 24, and 75. Also, plot the periodogram (all of these in dB). Compare and contrast the four spectral density estimates you obtain and also compare them with the autoregressive spectral density estimator in Figure 1.25 regarding the ability to see the less dominant spectral peak associated with the higher frequency?

1.4 Following are displayed three sets of figures each containing four plots. The first set shows four realizations of length $n = 100$ each generated from a time series model. The second set contains four autocorrelation functions based on the models used to generate realizations (in random order), and the third set displays the four spectral densities. Match each realization with the corresponding autocorrelations and spectral density. Explain your answers.

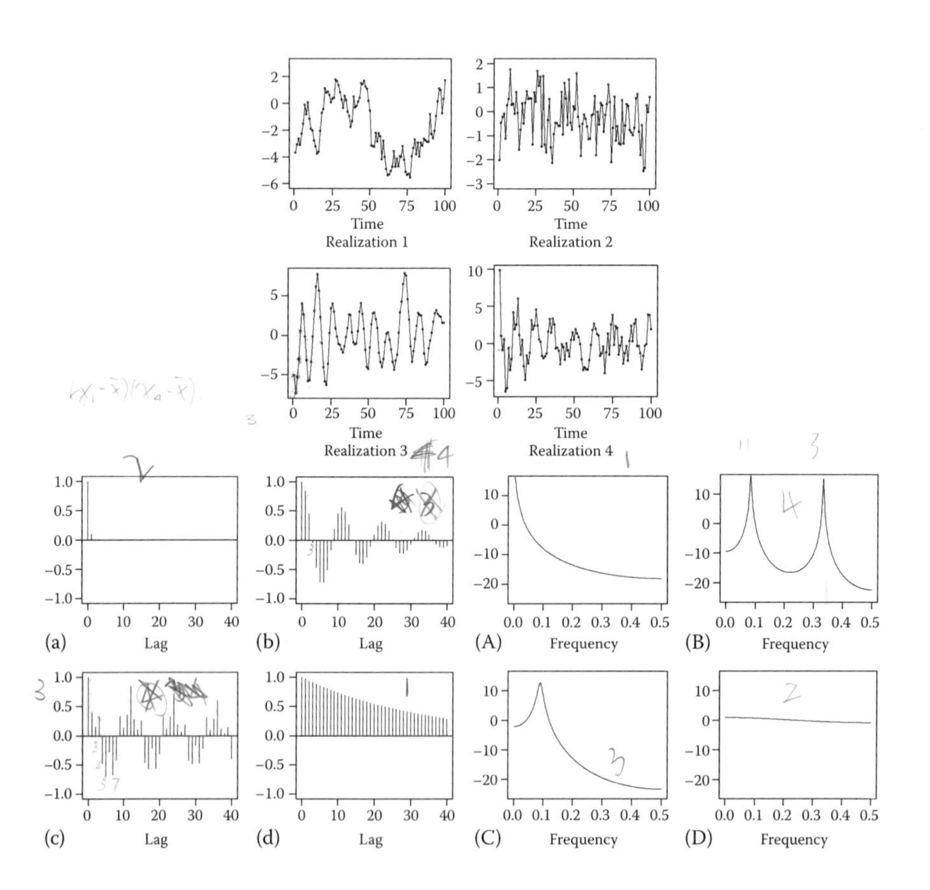

1.5 The file `FIG1.21A.XLS` contains the realization of length $n = 250$ shown in Figure 1.21a. Notice that the data and the spectral density (but not the autocorrelations) show evidence of two frequencies, a lower one at about $f = 0.05$ and a higher frequency of about $f = 0.33$. Find the Tukey spectral density estimate for this realization with $M = 31$. Plot this spectral density estimate in dB (10 log 10) and again without taking the logarithms. Comment on the information visible in the spectral density estimates concerning the two dominant frequencies. What impact has plotting the logarithm (as dB) had?

1.6 Generate a realization of length $n = 100$ from the signal-plus-noise model

$$X_t = 3\cos(2\pi(0.05)t) + 1.5\cos(2\pi(0.35)t + 2) + a(t)$$

where a_t is N(0,1) white noise. For these data, plot the

(a) Realization
(b) Sample autocorrelations
(c) Periodogram
(d) Parzen window spectral density estimator with default truncation point

Discuss cyclic features of the data that can be seen from each of the plots. Which of these plots provides the best indication of the two dominant frequencies (i.e., 0.05 and 0.35) in the data?

NOTE: A realization from the model

$$X_t = c_1\cos(2\pi f_1 t) + c_2\cos(2\pi f_2 t) + N_t \tag{4.13}$$

where N_t is normal white noise can be obtained using the `Generate Series` option on the `Time Series Analysis` menu. Select the `Signal + Noise` button and then you will be asked to enter c_1, f_1, c_2, f_2, and white noise variance σ_a^2 along with the series length and seed.

1.7 Find a time series data set (on the internet, etc.) not discussed in the book or in class. For this time series:

(a) Plot the time series realization.
(b) Plot the sample autocorrelations. Write code (in R or otherwise) to calculate the sample autocorrelations (i.e., do not use a prewritten function for this.)
(c) Describe the behavior (wandering, periodic, etc.) in the realization and how that behavior manifests itself in the sample autocorrelations.
(d) Plot the periodogram.
(e) Plot smoothed versions of the periodogram using at least two different window truncation points (M). Describe the behavior (wandering, periodic, etc.) in the realization and how that behavior manifests itself in the spectral estimators.

Theoretical Problems

1.8 Using the definition of autocovariance of a stationary process, show that it does not matter whether we look backward or forward in time; that is

$$\mathrm{Cov}(Z_t, Z_{t+k}) = \mathrm{Cov}(Z_t, Z_{t-k}) = \gamma_k.$$

1.9 Consider the stochastic process $Z_t = a_t$, where a_t is drawn independently for any t and where the distribution of a_t is given by

$$a_t = \begin{cases} c \text{ with probability } 1/2 \\ -c \text{ with probability } 1/2 \end{cases}$$

for $c > 0$.

(a) Is the process Z_t stationary? Why or why not?

(b) What are the values of the parameters μ, γ_0, γ_1, and γ_j, with $j > 1$ for this process?

1.10 Show that the sum of k uncorrelated covariance stationary processes is also covariance stationary.

1.11 Let $X(t) = \cos(2\pi(0.025)t + \psi_1) + 1.5\cos(2\pi(0.1)t + \psi_2) + 2\cos(2\pi(0.15)t + \psi_3)$ be the time series given in Example 1.7(b), Equation 1.31. Show that $X(t)$ is covariance stationary and (using the results in Example 1.3) find expressions for

(a) $\gamma(h)$

(b) σ_X^2

(c) $\rho(h)$

1.12 Let Y_t be given by

$$\begin{aligned} Y_t &= C_1 \cos(2\pi f_1 t + \psi_1) + N_t, \quad t = 0, \pm 1, \ldots, \\ &= H_t + N_t, \end{aligned}$$

where

N_t is discrete white noise

ψ_1 is distributed uniform$[0, 2\pi]$

N_t and ψ_1 are uncorrelated for each t

Show that H_t is covariance stationary and that the autocovariance of H_t is given by

$$\begin{aligned} \gamma_k &= \frac{C_1^2}{2}\cos 2\pi f_1 k, \quad k \neq 0 \\ &= \frac{C_1^2}{2} + \sigma_N^2, \quad k = 0. \end{aligned}$$

1.13 Let Z_t, $t=0, \pm 1, \ldots$ be independent normal random variables each with mean 0 and variance σ^2, and let a, b, and c be constants. Which, if any, of the following processes are stationary? For each stationary process, specify the mean and covariance function.
(a) $X_t = a + bZ_t + cZ_{t-1}$
(b) $X_t = a + bZ_0$
(c) $X_t = Z_t \cos(ct) + Z_{t-1}(\sin ct)$
(d) $X_t = Z_0 \cos(ct)$

1.14 Let $\{X_t;\ t=0,\ 1, \ldots\}$ be a covariance stationary process and define the processes

$$\left\{X_t^{(i)}, t = 0, 1, \ldots\right\} \quad i = 1, 2$$

as the difference processes

$$X_t^{(1)} = X_t - 0.5X_{t-1}$$
$$X_t^{(2)} = X_t^{(1)} - X_{t-1}^{(1)}.$$

Show that $X_t^{(1)}$ and $X_t^{(2)}$ are also covariance stationary.

1.15 Let X be a random variable with mean μ and variance $\sigma^2 < \infty$. Define $Z_t = X$ for all t where $t \in I$ where $I = 1, 2, \ldots$
(a) Show that $Z_t = X$ is strictly and weakly stationary.
(b) Find the autocorrelation function.
(c) Draw a "typical" realization.
(d) Determine whether or not $\overline{Z}$ is ergodic for μ.

1.16 Let $Y_t = a + bt + X_t$ where X_t is a covariance stationary process.
(a) Is Y_t covariance stationary? Why or why not?
(b) Let $Z_t = Y_t - Y_{t-1}$. Is Z_t covariance stationary? Why or why not?

1.17 Consider the two processes

$$X_t^{(1)} = a_t - \theta a_{t-1}$$

$$X_t^{(2)} = a_t - \frac{1}{\theta} a_{t-1},$$

where a_t is a white noise process with mean zero and variance $\sigma_a^2 < \infty$. Show that $X_t^{(1)}$ and $X_t^{(2)}$ have the same autocorrelation functions.

1.18 Show that if a stationary time series, Y_t, satisfies the difference equation $Y_t - Y_{t-1} = e_t$, then $E[e_t^2] = 0$.

1.19 Show that for a covariance stationary process $\{X_t, t = 0, \pm 1, \pm 2, \ldots\}$ the following hold:

(a) $P_X(f) = P_X(-f), \quad |f| \le 0.5$
$S_X(f) = S_X(-f), \quad |f| \le 0.5$

(b) $P_X(f) = \sigma_X^2 + 2\sum_{k=1}^{\infty} \gamma_k \cos(2\pi f k), \quad |f| \le 0.5$

$$S_X(f) = 1 + 2\sum_{k=1}^{\infty} \rho_k \cos(2\pi f k), \quad |f| \le 0.5$$

(c) $\gamma_k = \int_{-0.5}^{0.5} P_X(f) \exp(2\pi i f k) df$

$$\rho_k = \int_{-0.5}^{0.5} S_X(f) \exp(2\pi i f k) df$$

(d) $\sigma_X^2 = \int_{-0.5}^{0.5} P_X(f) df$

$$1 = \int_{-0.5}^{0.5} S_X(f) df$$

(e) $\gamma_k = 2\int_{0}^{0.5} P_X(f) \cos(2\pi f k) df$

$$\rho_k = 2\int_{0}^{0.5} S_X(f) \cos(2\pi f k) df$$

1.20 Show the equivalence of the three forms for the periodogram shown in Appendix 1.A. Specifically, show that for $k = 1, 2, \ldots, m$ (where $n = 2m + 1$) the following holds:

$$\frac{1}{n}\left|\sum_{t=1}^{n} x_t e^{-2\pi i j t/n}\right|^2 = \frac{n}{4}\left(a_j^2 + b_j^2\right) = \hat{\sigma}_X^2 + 2\sum_{k=1}^{n-1} \hat{\gamma}_k \cos(2\pi j k/n).$$

1.21 Prove Theorem 1.3.

1.22 Prove Theorem 1.4.

2

Linear Filters

Many of the time series that we will study can be viewed as the output of a linear operator or filter. In this chapter, we introduce the basic concepts and terminology of linear filter theory. We also discuss the general linear process, which is a linear filter with white noise input, that plays a fundamental role in the study of stationary processes. Finally, we discuss the application of linear filters for purposes of removing or retaining certain types of frequency behavior from a time series realization.

2.1 Introduction to Linear Filters

Consider the process $\{X_t - \mu;\ t = 0, \pm 1, \pm 2, \ldots\}$ given by

$$X_t - \mu = \sum_{j=-\infty}^{\infty} h_j Z_{t-j}, \tag{2.1}$$

where

h_j are constant
Z_t is a random variable with $E[Z_t] = 0$ for each t

The process X_t is referred to as the output of a linear filter, and Z_t is the input. If $h_j = 0$ for $j < 0$, then this is called a *causal linear filter*, since X_t only depends on present and past values of the input Z_t. In this book, we focus on causal filters, and, thus, although Theorems 2.1 and 2.2 and Definition 2.1, which follow, have noncausal analogs, we state them here only for causal filters. We will find it convenient to use the backshift operator B, defined by $B^k Z_t = Z_{t-k}$, to write (the causal version of) (2.1) as

$$\begin{aligned} X_t - \mu &= \sum_{j=0}^{\infty} h_j B^j Z_t \\ &= H(B) Z_t, \end{aligned} \tag{2.2}$$

where $H(B) = \sum_{j=0}^{\infty} h_j B^j$. The linear filter in (2.1) is symbolically described as follows:

$$Z_t = \text{Input} \rightarrow \underset{\text{FILTER}}{\boxed{H(B)}} \rightarrow \text{Output} = X_t$$

We will use the following notation:

X_t is the output
Z_t is the input
h_j are the filter weights
collectively, $\{h_j\}$ is called the *impulse response function*
$H(e^{-2\pi i f})$ is the *frequency response function* (defined in (2.4))

The sense in which an infinite series of random variables such as (2.1) converges will always be taken in this book to be in the mean square sense. Theorem 2.1 that follows gives sufficient conditions for the series on the right-hand side of (2.1) to converge in mean square to $X_t - \mu$.

Theorem 2.1

If the h_j, $j=0, 1, \ldots$ of a causal linear filter are such that $\sum_{j=0}^{\infty} |h_j| < \infty$, if for each t, $E\left[Z_t^2\right] \leq M$ for some finite M, and if μ is a real number, then there is a unique stochastic process, X_t, such that for each t,

$$\lim_{m \to \infty} E\left\{\left[X_t - \mu - \sum_{j=0}^{m} h_j Z_{t-j}\right]^2\right\} = 0$$

Proof:

See Appendix 2.A.

Thus, the equality in (2.1) indicates that the mean square convergence in Theorem 2.1 holds, and when infinite series of random variables such as those in (2.1) are given without comment in this book, the reader may assume that the mathematical structure is sufficiently rich to guarantee that under some set of assumptions, which are broad enough to be useful, the indicated infinite series converges in mean square. We have included a few problems dealing with these concepts at the end of the chapter. We will not undertake an extensive mathematical study of linear filters in this book. For more discussion of linear filters, see Koopmans (1974), Brockwell and Davis (1991), and Shumway and Stoffer (2006).

2.1.1 Relationship between the Spectra of the Input and Output of a Linear Filter

One special use of the spectrum and spectral density is to describe the effect of a linear filter. The following result is fundamental in this respect.

Theorem 2.2

If Z_t and X_t are the input and output of the linear filter in (2.1) where $\sum_{j=0}^{\infty} |h_j| < \infty$, then

$$P_X(f) = \left|H(e^{-2\pi if})\right|^2 P_Z(f), \tag{2.3}$$

where $H(e^{-2\pi if})$ is the frequency response function given by

$$H(e^{-2\pi if}) = \sum_{k=0}^{\infty} h_k e^{-2\pi ifk}. \tag{2.4}$$

Moreover, $S_X(f) = \left|H(e^{-2\pi if})\right|^2 S_Z(f)$.

Proof:

See Appendix 2.A

The importance of Theorem 2.2 is that it provides information about how the filter weights, h_k, impact the frequency behavior of the output signal. This can be useful in designing filters with specific properties, which will be discussed in Section 2.4. In the following, we will refer to $\left|H(e^{-2\pi if})\right|^2$ as the *squared frequency response*.

2.2 Stationary General Linear Processes

Many of the statistical time series that we will study can be viewed as the output, X_t, of a linear filter with white noise input. Such a process is called a *general linear process*, which is defined in the following.

Definition 2.1 (General Linear Process) The process X_t, $t = 0, \pm 1, \pm 2, \ldots$ given by

$$X_t - \mu = \sum_{j=0}^{\infty} \psi_j a_{t-j}, \tag{2.5}$$

is defined to be a *general linear process* if a_t is a white noise process with zero mean and finite variance and $\sum_{j=0}^{\infty} |\psi_j| < \infty$.

Using the backshift operator B (2.5) can be written as

$$X_t - \mu = \sum_{j=0}^{\infty} \psi_j B^j a_t$$
$$= \psi(B) a_t,$$

where

$$\psi(B) = \sum_{j=0}^{\infty} \psi_j B^j.$$

In this section, we will consider the problem of finding stationarity conditions for X_t in (2.5) along with expressions for the autocovariance function and power spectrum. In this connection the following theorem is helpful. We include its proof here because of its use of fundamental concepts.

Theorem 2.3

If X_t is a causal general linear process, then X_t is stationary.

Proof:

Without loss of generality assume that $\mu = 0$. Since $E[a_t] = 0$, it follows immediately that $E[X_t] = 0$, and thus the autocovariance function for X_t is given by, $E[X_t X_{t+k}]$. Now, noticing that

$$X_{t+k} = \sum_{m=0}^{\infty} \psi_m a_{t+k-m},$$

we have

$$E[X_t X_{t+k}] = E\left[\left(\sum_{j=0}^{\infty} \psi_j a_{t-j}\right)\left(\sum_{m=0}^{\infty} \psi_m a_{t+k-m}\right)\right]$$
$$= E\left[\sum_{j=0}^{\infty}\sum_{m=0}^{\infty} \psi_j \psi_m a_{t-j} a_{t+k-m}\right]$$
$$= \sum_{j=0}^{\infty}\sum_{m=0}^{\infty} \psi_j \psi_m E[a_{t-j} a_{t+k-m}]. \tag{2.6}$$

Since

$$E[a_{t-j}a_{t+k-m}] = \sigma_a^2 \quad \text{if } m = j+k$$
$$= 0 \quad \text{elsewhere,}$$

it follows from (2.6) that

$$E[X_t X_{t+k}] = \sigma_a^2 \sum_{j=0}^{\infty} \psi_j \psi_{j+k}. \tag{2.7}$$

That is, $E[X_t X_{t+k}]$ depends on k and not t. The following properties have been shown to hold:

1. $E[X_t] = 0$ (a constant for all t)
2. $\mathrm{Var}[X_t] = E[X_t X_t]$
$$= \sigma_a^2 \sum_{j=0}^{\infty} \psi_j^2 \text{ (a constant)}$$
$$< \infty \left(\text{since } \sum_{j=0}^{\infty} |\psi_j| \text{ converges}\right)$$
3. $E[X_t X_{t+k}] = \gamma_k$ (depends only on k from (2.7))

Thus, X_t is stationary, and the proof is complete.

2.2.1 Spectrum and Spectral Density for a General Linear Process

Using (2.3) and the fact that the spectrum for white noise is constant and given by $P_a(f) = \sigma_a^2$, the spectrum of the general linear process X_t is given by

$$P_X(f) = \left|\psi(e^{-2\pi i f})\right|^2 P_a(f)$$
$$= \sigma_a^2 \left|\psi(e^{-2\pi i f})\right|^2, \tag{2.8}$$

and the spectral density is given by

$$S_X(f) = \frac{\sigma_a^2}{\sigma_X^2} \left|\psi(e^{-2\pi i f})\right|^2. \tag{2.9}$$

2.3 Wold Decomposition Theorem

Before proceeding, we state the Wold Decomposition Theorem that emphasizes the central role of the general linear process in the study of covariance stationary processes. For a proof and discussion of this result, the reader is referred to Brockwell and Davis (1991).

Theorem 2.4

Let $\{X_t;\ t = 0,\ \pm 1,\ \pm 2, \ldots\}$ be a zero mean, covariance stationary process. Then, X_t can be expressed as the sum of two zero mean, uncorrelated, covariance stationary processes,

$$X_t = U_t + V_t,$$

such that

(i) U_t is a general linear process with $\psi_0 = 1$
(ii) V_t is completely determined by a linear function of its past values

The process V_t is usually referred to as the *deterministic component* of X_t while U_t is called the *nondeterministic component*. In Chapter 3, it will be shown that stationary autoregressive, moving average, and autoregressive moving average processes can all be expressed as general linear processes. Thus, the Wold decomposition provides justification for these models to be fundamentally well suited for the analysis of stationary processes.

2.4 Filtering Applications

A common application of linear filters in many fields of science and engineering is to "filter out" certain types of frequencies from a set of data. Before leaving the topic of linear filters, we provide some of the basics. *Low-pass filters* are designed to remove the higher frequency behavior in a set of data while *high pass filters* remove the lower frequencies. Other types of filters are *band-pass filters* that pass frequencies in a certain frequency band through the filter and *band-stop (or notch) filters* that pass frequencies except those in a certain frequency band or range. We will briefly discuss filters and provide a few examples.

Theorem 2.2 forms the basis for developing filters with desired properties. For example, suppose you want a low-pass filter that keeps frequencies in the

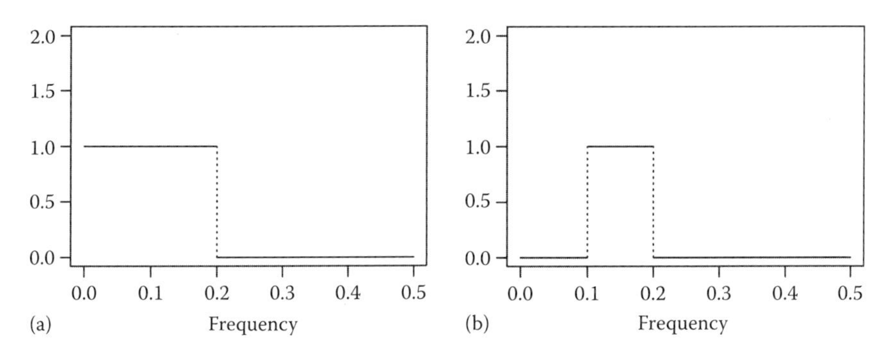

FIGURE 2.1
Squared frequency response function, $|H(e^{-2\pi if})|^2$, for ideal (a) low-pass and (b) band-pass filters.

data less than or equal to the *cutoff* frequency $f_c = 0.2$ and removes frequencies greater than $f_c = 0.2$. The result in (2.3) says that ideally, we would design our filter, so that

$$\begin{aligned} |H(e^{-2\pi if})|^2 &= 1, \quad f \leq 0.2 \\ &= 0, \quad f > 0.2, \end{aligned} \tag{2.10}$$

since according to Theorem 2.2, this would eliminate frequency behavior from the spectrum of the output for $f > 0.2$ and leave it unchanged for $f \leq 0.2$. Figure 2.1a shows the squared frequency response for this ideal (brick-wall) filter, and Figure 2.1b shows the ideal squared frequency response for a band-pass filter that passes through frequencies $0.1 \leq f \leq 0.2$.

Example 2.1 First Difference Filter

Consider a time series realization x_t, $t = 1, \ldots, n$, and suppose we compute the first difference $y_t = x_t - x_{t-1}$. In order to see the effect of this differencing operation, we note that the first difference is a linear filter of the form in (2.1) where $h_0 = 1$, $h_1 = -1$, and $h_j = 0$, $j \neq 0$ or 1. We can thus calculate $H(e^{-2\pi if}) = 1 - \cos 2\pi f + i \sin 2\pi f$ and $|H(e^{-2\pi if})|^2 = 2(1 - \cos 2\pi f)$. The plot of $|H(e^{-2\pi if})|^2$ is given in Figure 2.2b where it can be seen that a difference is a high-pass filter, but that it is not a sharp filter in that it allows some low frequency behavior to leak through. In Figure 2.2a, we show the realization previously shown in Figure 1.3a, which shows a wandering behavior associated with low frequency behavior, along with some higher frequency oscillation. The first difference data are shown in Figure 2.2c where it can be seen that the low-frequency wandering behavior has been removed with only the higher frequency behavior remaining. The first difference is widely used in time series analysis, but if the goal is to remove a specific range of low frequencies from the data, there are better approaches.

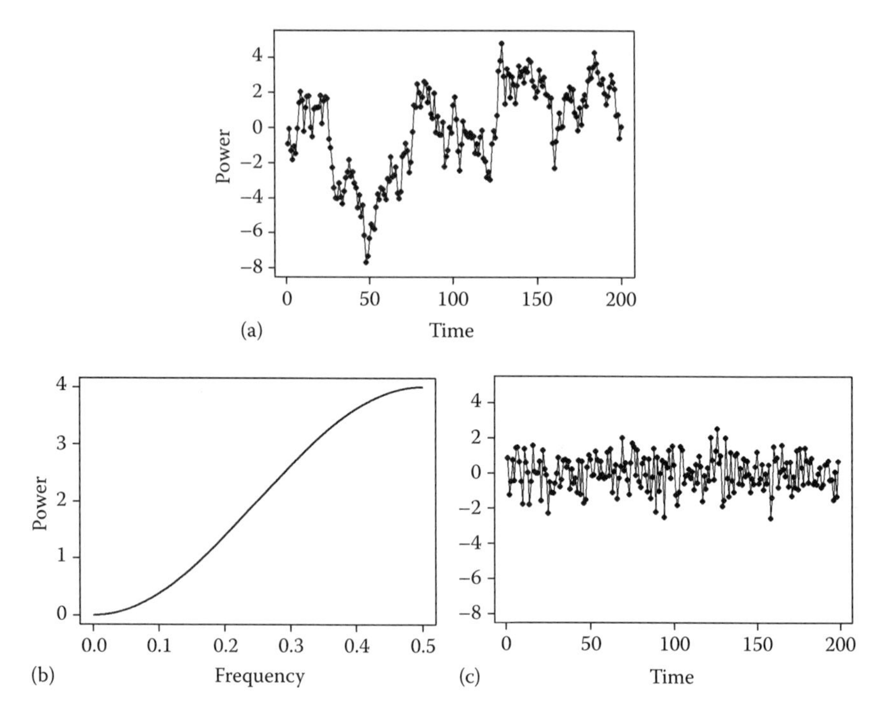

FIGURE 2.2
(a) Realization previously shown in Figure 1.3(a), (b) squared frequency response function for a first difference, and (c) first difference of the data in (a).

Example 2.2 Sum and Moving Average Filters

Consider again a time series realization x_t, $t=1,\ldots,n$, and suppose we now compute the sum $y_t=x_t+x_{t-1}$, which is also a linear filter where in this case $h_0=1$, $h_1=1$, and $h_j=0$, $j\neq 0$ or 1. Calculations show that

$$H(\mathrm{e}^{-2\pi if}) = 1+\cos 2\pi f - i\sin 2\pi f$$

and

$$\left|H(\mathrm{e}^{-2\pi if})\right|^2 = 2(1+\cos 2\pi f).$$

The plot of $\left|H(\mathrm{e}^{-2\pi if})\right|^2$ is given in Figure 2.3, where it can be seen that it is a mirror image of Figure 2.2b for the difference filter, and, as a consequence, low frequencies are passed and high frequencies tend to be removed. Again, this is not a sharp filter and can be improved.

Closer observation of Figures 2.2b and 2.3 shows that $\left|H(e^{-2\pi if})\right|^2=4$ when $f=0.5$ for the first difference filter and when $f=0$ for the sum filter. The goal

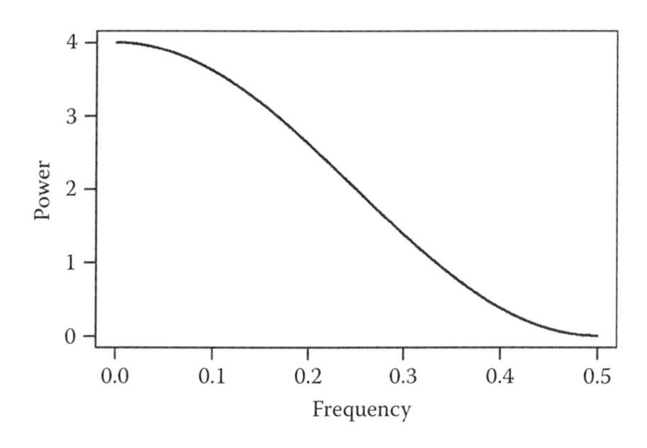

FIGURE 2.3
Squared frequency response functions for a 2-point sum filter.

of a filter is to leave some frequency content unchanged and remove others. Thus, according to Theorem 2.2, it is desirable that $|H(e^{-2\pi if})|^2 \approx 1$ for the range of frequencies that are to be retained. To this end, we consider the three moving average filters:

$$y_t = \frac{1}{2}(x_t + x_{t-1}) \quad \text{2-point moving average}$$
$$y_t = \frac{1}{3}(x_{t-1} + x_t + x_{t+1}) \quad \text{3-point moving average}$$
$$y_t = \frac{1}{7}(x_{t-3} + x_{t-2} + x_{t-1} + x_t + x_{t+1} + x_{t+2} + x_{t+3}) \quad \text{7-point moving average filter}$$

for which the corresponding frequency response functions are

$$\begin{aligned} H(e^{-2\pi if}) &= \frac{1}{2}(1 + \cos 2\pi f - i \sin 2\pi f) \quad \text{2-point} \\ &= \frac{1}{3}(1 + 2\cos 2\pi f) \quad \text{3-point} \\ &= \frac{1}{7}(1 + 2\cos 2\pi f + 2\cos 4\pi f + 2\cos 6\pi f) \quad \text{7-point.} \end{aligned}$$

The squared frequency response functions for these moving average filters are shown in Figure 2.4 where it can be seen that these are all low-pass filters (smoothers). Also, as the number of data values involved in the moving average filter increases, the "cutoff" frequency decreases and the filter becomes "sharper" in that the squared frequency response is closer to the ideal filter. Figure 2.5 shows the result of applying the three moving average filters to the data in Figure 2.2a. As would be expected from Figure 2.4, it can be seen that as the number of points in the moving average increases, the smoother the output series becomes. In all these cases, the low frequency "wandering behavior" in the original series is retained.

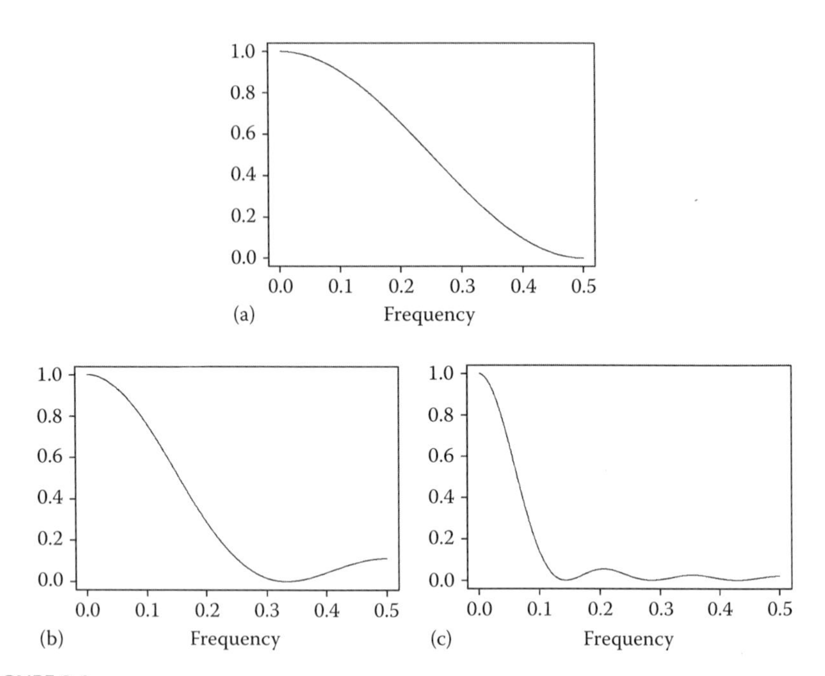

FIGURE 2.4
Squared frequency response functions for three moving average smoothers. (a) 2-point moving average. (b) 3-point moving average. (c) 7-point moving average.

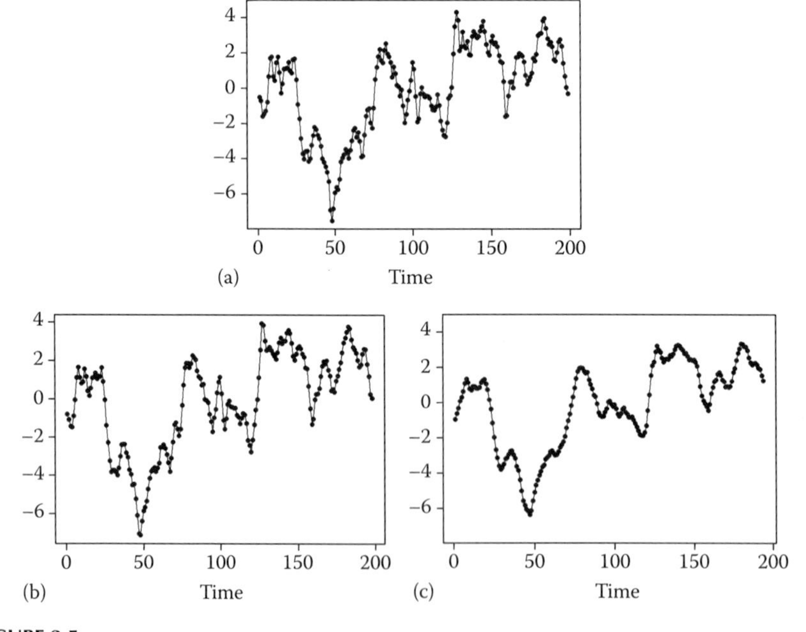

FIGURE 2.5
Output series from applying three moving average smoothers to the data in Figure 2.2a. (a) 2-point moving average. (b) 3-point moving average. (c) 7-point moving average.

2.4.1 Butterworth Filters

One of the more common filters used by scientists and engineers is the Butterworth filter (see Butterworth, 1930; Porat, 1997). This filter will be briefly introduced here. While the Butterworth filter can be of any type (low-pass, high-pass, etc.) our discussion will focus on an Nth order low-pass Butterworth filter, which has a squared frequency response function

$$\left|H(e^{-2\pi if})\right|^2 = \frac{1}{1+\left(\frac{f}{f_c}\right)^{2N}}. \tag{2.11}$$

Increasing the value of N sharpens the filter as shown in Figure 2.6 where we show the squared frequency response functions for low-pass Butterworth filters with $f_c = 0.2$ for various values of N. That is, for large values of N, the squared frequency response function of a low-pass Butterworth filter with cutoff, f_c, more closely satisfies the ideal filter

$$\begin{aligned}\left|H(e^{-2\pi if})\right|^2 &= 1, \quad f \le f_c \\ &= 0, \quad f > f_c,\end{aligned}$$

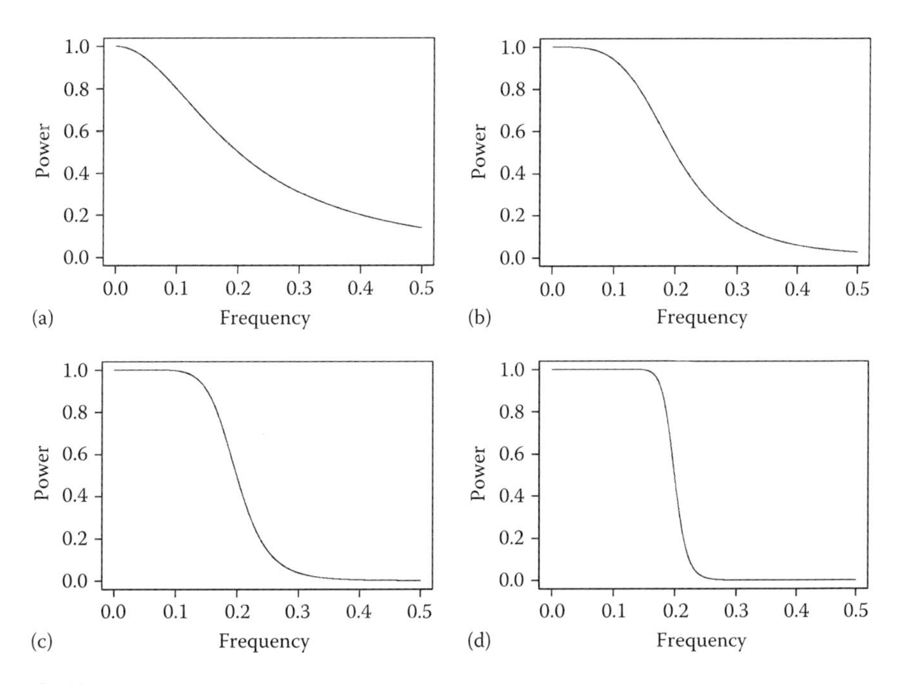

FIGURE 2.6
Squared frequency response functions, $\left|H(e^{-2\pi if})\right|^2$, for low-pass Butterworth filters with four values of N. (a) $N=1$. (b) $N=2$. (c) $N=4$. (d) $N=10$.

shown in Figure 2.1a. We will not discuss issues of filter design such as the choice of N or phase. For discussion of such issues, see Porat (1997).

Although the impulse response function associated with (2.11) is of infinite order, it can be well approximated by a ratio of two Nth order polynomials

$$\sum_{k=0}^{\infty} h_k B^k \approx \frac{\sum_{k=0}^{N} \alpha_k B^k}{\sum_{k=0}^{N} \beta_k B^k}, \tag{2.12}$$

which leads to the easy to implement recursive filter

$$y_t = \sum_{k=0}^{N} \alpha_k x_{t-k} - \sum_{k=0}^{N} \beta_k y_{t-k}.$$

Example 2.3 Butterworth Low-Pass Filter Example

In Figure 2.7a, we show a realization of length $n = 200$ from signal

$$X_t = \cos(2\pi(0.025)\,t) + 3\cos(2\pi(0.125)\,t) \tag{2.13}$$

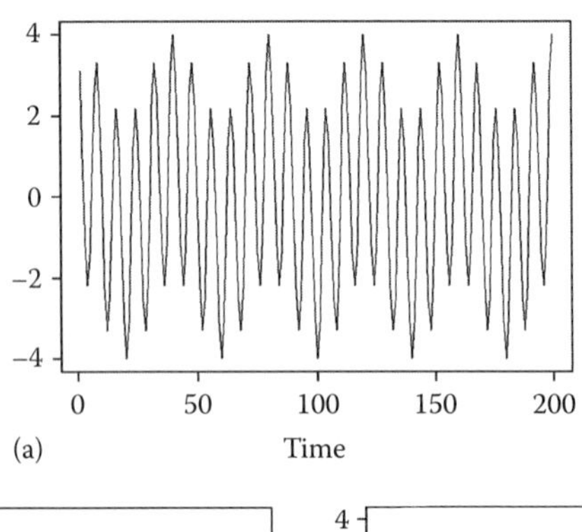

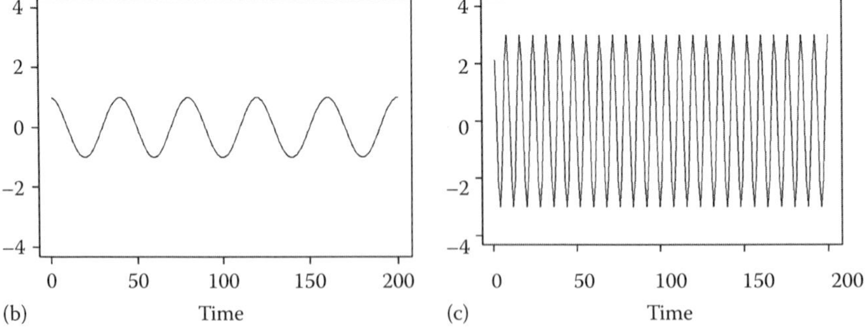

FIGURE 2.7
Signal in (2.13) along with its two additive components. (a) Original signal. (b) Low-frequency component. (c) High-frequency component.

which has two cycles, with frequencies $f = 0.025$ and $f = 0.125$. Figure 2.7b and c are plots of the two additive components that sum to give X_t. For illustration, we apply a low-pass Butterworth filter with cutoff frequency $f_c = 0.04$ in order to retain the cycle associated with $f = 0.025$ and filter out the higher frequency cycle associated with $f = 0.125$. That is, we want to recover the component shown Figure 2.7b. In Figure 2.8c and d, we show results using Butterworth filters with orders $N = 1$ and $N = 3$, and, in Figure 2.8a and b, we show the squared frequency response functions associated with these two filters. In Figure 2.8a, we see that the Butterworth filter with $N = 1$ would not be expected to perform well, because the squared frequency still has some power at $f = 0.125$, which is the frequency to be removed. On the other hand, Figure 2.8b shows that the squared frequency response for $N = 3$ has very little power beyond about $f = 0.08$, and thus the Butterworth filter with $N = 3$ should produce the desired results. In Figure 2.8c, we see that, as expected, some of the higher frequency behavior leaked through the filter, while Figure 2.8d shows that the filtered series resulting from a Butterworth filter with $N = 3$ looks very similar to Figure 2.7b.

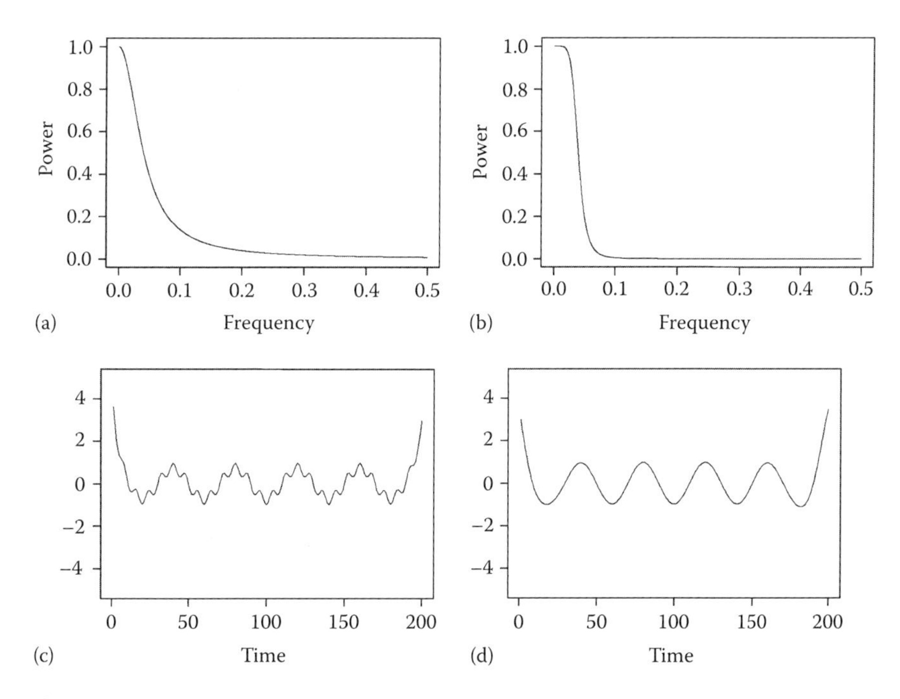

FIGURE 2.8
Squared frequency response functions, $|H(e^{-2\pi if})|^2$, and filtering results using low-pass Butterworth filters of orders $N = 1$ and $N = 3$ on the realization in Figure 2.7a. (a) $N = 1$. (b) $N = 3$. (c) Filtered data ($N = 1$). (d) Filtered data ($N = 3$).

One point that should be made is that the Butterworth filter has end-effect problems that are associated with all linear filters. This is a result of the fact that in order to obtain, for example, the filtered value at $t=1$, the filter ideally needs to use values prior to $t=1$, which do not exist. Several methods for dealing with end effects in filters have been recommended, but will not be discussed here.

Using `GW-WINKS` (*and* `MATLAB`®): `GW-WINKS` allows you to do low and high-pass filtering using a third-order `Butterworth` filter. The results in Figure 2.8d were obtained using the `Butterworth` option on the `Time Series Analysis` menu, selecting `Cutoff Frequency` = 0.04, `Sampling Interval` = 1, and choosing `Low Pass`.

For comparison, the results in Figure 2.8c were obtained using `MATLAB` with the `signal processing tool kit`. Specifically, letting `data` denote the file containing the original data set and `filtdata` the filtered data set, then the following MATLAB commands produce the results in Figure 2.8c:

```
[b,a] = butter(1,0.08]
filtdata = filtfilt(b,a,data)
```

It is very important to note that the frequency, f_M, used in MATLAB is given by $f_M = 2f$. That is, frequency in MATLAB goes from 0 to 1 instead of 0 to 0.5. The first command sets up the coefficients needed to perform a first-order low-pass Butterworth filter with cutoff frequency $f_M = 0.08$, which corresponds to the cutoff frequency $f_c = 0.04$ as discussed earlier. The second command performs the filter and places the results in file `filtdata`. Figure 2.8c is a plot of `filtdata`.

Example 2.4 Filtering the King Kong Eats Grass Data

In Figure 1.12, we showed the acoustical signal of the phrase "King Kong Eats Grass." It was noted that a fan was operating in the room, and the associated cyclic behavior (of about 136 Hz) was apparent in the "silent spaces" in the signal. Using the software package MATLAB, we applied a fourth-order zero-phase band-stop Butterworth filter to attempt to remove the fan noise. Our sampling rate was 8000 Hz (samples per second), so a frequency of 136 Hz corresponds to 136 cycles per 8000 units, or a frequency of $f = 136/8000 = 0.017$. We applied a fourth-order band-stop filter designed to filter out frequencies in the band $f \in [0.0075, 0.0275]$. The filtered signal is shown in Figure 2.9 where it can be seen that cyclic behavior due to the fan has been largely removed (with some ambient noise still remaining) while leaving the speech signal essentially unchanged.

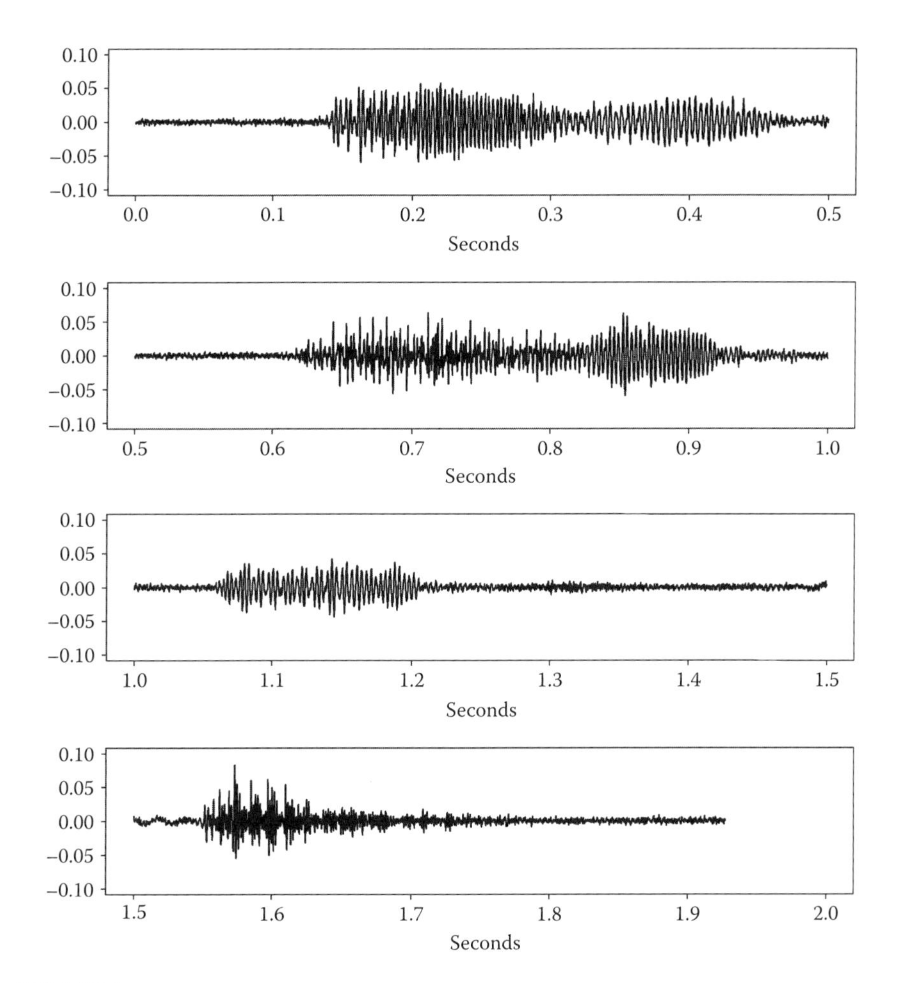

FIGURE 2.9
King Kong Eats Grass data from Figure 1.12 with fan noise removed.

Butterworth filtering using MATLAB: The commands below calculate the filtered data "King Kong Eats Grass" data where `kkeg` and `fkkeg` denote the original and filtered data:

```
[b,a] =butter(4,[.015,.055],'stop');
fkkeg=filtfilt(b,a,kkeg);
```

The first command sets up the coefficients needed to perform a fourth-order band-stop Butterworth filter with stop band [0.015, 0.055]. Notice the modification needed, because the frequency used by MATLAB, f_M, satisfies $f_M = 2f$.

2.A Appendix

2.A.1 Proof of Theorem 2.1

Without loss of generality, we assume $\mu = 0$. If $m > k$ then

$$
\begin{aligned}
E\left[\left(\sum_{j=0}^{m} h_j Z_{t-j} - \sum_{j=0}^{k} h_j Z_{t-j}\right)^2\right] &= E\left[\left(\sum_{j=k+1}^{m} h_j Z_{t-j}\right)^2\right] \\
&\le \left[\sum_{j=k+1}^{m} \left\{E|h_j Z_{t-j}|^2\right\}^{1/2}\right]^2 \quad \text{(Minkowski inequality)} \\
&= \left[\sum_{j=k+1}^{m} |h_j| \sqrt{\mathrm{Var}(Z_{t-j})}\right]^2 \\
&\le \left[\sqrt{M} \sum_{j=k+1}^{m} |h_j|\right]^2. \qquad (2.A.1)
\end{aligned}
$$

As $m \to \infty$ and $k \to \infty$, then the right side of (2.A.1) converges to zero. It follows then that $\sum_{j=0}^{m} h_j Z_{t-j}$ is a Cauchy sequence and hence converges, say to X_t, which will be denoted

$$
X_t = \sum_{j=0}^{\infty} h_j Z_{t-j}.
$$

The fact that each Cauchy sequence of random variables converges to a random variable is based on Hilbert space concepts. See Brockwell and Davis (1991, 2002) for a discussion.

To show the uniqueness of X_t, we assume that there exists another random variable Y_t, such that $\sum_{j=0}^{m} h_j Z_{t-j}$ also converges in mean square to Y_t. Then, again using the Minkowski inequality,

$$
\begin{aligned}
E\left(|X_t - Y_t|^2\right) &= E\left[\left|X_t - \sum_{j=0}^{m} h_j Z_{t-j} + \sum_{j=0}^{m} h_j Z_{t-j} - Y_t\right|^2\right] \\
&\le \left[\left(E\left|X_t - \sum_{j=0}^{m} h_j Z_{t-j}\right|^2\right)^{1/2} + \left(E\left|\sum_{j=0}^{m} h_j Z_{t-j} - Y_t\right|^2\right)^{1/2}\right]^2,
\end{aligned}
$$

and thus $E\left(|X_t - Y_t|^2\right) \to 0$ as $m \to \infty$.

2.A.2 Proof of Theorem 2.2

Letting $\gamma_X(k)$ and $\gamma_Z(k)$ denote the autocovariance functions of the processes X_t and Z_t, respectively, then

$$\begin{aligned}
\gamma_X(k) &= E[(X_t - \mu)(X_{t+k} - \mu)] \\
&= E\left[\left(\sum_{j=0}^{\infty} h_j Z_{t-j}\right)\left(\sum_{m=0}^{\infty} h_m Z_{t+k-m}\right)\right] \\
&= \sum_{j=0}^{\infty}\sum_{m=0}^{\infty} h_j h_m \gamma_Z(k-m+j) \\
&= \sum_{j=0}^{\infty}\sum_{m=0}^{\infty} h_j h_m \int_{-0.5}^{0.5} e^{2\pi i f(k-m+j)} P_Z(f)df \\
&= \int_{-0.5}^{0.5} \left(\sum_{j=0}^{\infty} h_j e^{2\pi i f j}\right)\left(\sum_{m=0}^{\infty} h_m e^{-2\pi i f m}\right) e^{2\pi i f k} P_Z(f)df \\
&= \int_{-0.5}^{0.5} e^{2\pi i f k} \left|\sum_{j=0}^{\infty} h_j e^{-2\pi i f j}\right|^2 P_Z(f)df.
\end{aligned}$$

It can be shown that if $P_V(f)$ and $P_W(f)$ are spectra, and if for each k

$$\int_{-0.5}^{0.5} e^{2\pi i f k} P_V(f)df = \int_{-0.5}^{0.5} e^{2\pi i f k} P_W(f)df,$$

then $P_V(f) = P_W(f)$. Using this fact, we see that $\gamma_X(k) = \int_{-0.5}^{0.5} e^{2\pi i f k} P_X(f)$ by definition, and $\gamma_X(k) = \int_{-0.5}^{0.5} e^{2\pi i f k} \left|\sum_{j=0}^{\infty} h_j e^{-2\pi i f j}\right|^2 P_Z(f)df$ from the preceding calculations. So, it follows that

$$P_X(f) = \left|\sum_{j=0}^{\infty} h_j e^{-2\pi i f j}\right|^2 P_Z(f).$$

Exercises

When using **GW-WINKS** to work problems below, you will find the `Generate Series`, `Spectrum`, and `Butterworth Filter` on the `Time series Analysis` menu to be useful.

Applied Problems

2.1 Generate a realization of length $n = 100$ from the signal-plus-noise model in Problem 1.6.

(a) Apply a third-order low-pass Butterworth filter to the original realization with cut-off point 0.2.

(b) Apply a third-order high-pass Butterworth filter to the original realization with cut-off point 0.2.

(c) Apply a third-order low-pass Butterworth filter with cut-off point 0.2 to the high-pass filtered realization in (b).

For the original data and the three filtered realizations obtained in (a)–(c) earlier, plot the following:

(i) The realization

(ii) The sample autocorrelations

(iii) The Parzen window-based spectral density estimate using the default truncation point

Discuss the cyclic behavior in the original data along and for (a)–(c) and discuss the effect of the filters applied.

2.2 Using the signal-plus-noise realization generated for Problem 2.1:

(a) Apply a third-order low-pass Butterworth filter with cut-off frequency 0.1.

(b) Apply a third-order low-pass Butterworth filter with cut-off frequency 0.075.

For the two filtered realizations obtained here, plot the following:

(i) The realization

(ii) The sample autocorrelations

(iii) The Parzen window-based spectral density estimate using the default truncation point

Compare these plots with the plots in Problem 2.1(a) associated with a low-pass Butterworth frequency with cut-off point 0.2. Discuss the effect of moving the cut-off point closer to the actual frequency in the data.

2.3 Using the signal-plus-noise realization generated for Problem 2.1, difference the data. That is, compute the realization of length 99 given by $y_t = x_t - x_{t-1}$, where x_t denotes the original realization. Plot the following:

(i) The differenced data (i.e., y_t)

(ii) The sample autocorrelations

(iii) The Parzen window-based spectral density estimate using the default truncation point

Discuss the effect of the differencing. What type of filter (high-pass or low-pass) is the difference? How does it compare with the low-pass Butterworth filter for filtering out the frequency 0.05?

NOTE: The **GW-WINKS** program can be used to difference a data set by using the `Edit` menu, then selecting `New variable, Common Transformations`, and then selecting `by Differencing series`.

Choose the original data set as the data to be differenced and then indicate that a first-order difference is desired.

2.4 Verify the frequency response functions given in the text for 2-point, 3-point, and 7-point moving averages. Find the frequency response function for a 5-point moving average. In each case, also derive $|H(e^{-2\pi if})|^2$. Plot the squared frequency response for the 5-point moving average using a program or language of your choice.

Theoretical Problems

2.5 Suppose that for each t, the sequence $\{X_t^{(m)}: m = 1, 2, \ldots\}$ of random variables converges to X_t in mean square for each t as $m \to \infty$. Similarly, let $Y_t^{(m)}$ converge to Y_t in mean square as $m \to \infty$. Show that $a_1 X_t^{(m)} + b_1 Y_t^{(m)} \to a_1 X_t + b_1 Y_t$ in mean square as $m \to \infty$, where a_1 and b_1 are real constants. You may assume that each of these random variables is in L^2, that is, that each random variable, X, is such that $E[X^2] < \infty$.

2.6 Let $\{X_t\}$ be a sequence of independent random variables with $E(X_t) = 0$ and $\text{Var}(X_t) = \sigma_t^2 < \infty$. Show that $\sum_{t=1}^{\infty} X_t = Y$ exists as a limit in the mean square sense with $Y \in L^2$ if and only if $\sum_{t=1}^{\infty} \sigma_t^2 < \infty$.

2.7 Let $X_t^{(k)}$, $k = 1, 2, \ldots, m$ be a stationary general linear processes given by

$$X_t^{(k)} = \sum_{j=0}^{\infty} h_j^{(k)} a_{t-j},$$

where a_t is zero mean white noise and $\sum_{j=0}^{\infty} |h_j^{(k)}| < \infty, k = 1, 2, \ldots, m$. Show that the process $X_t = \sum_{k=1}^{m} b_k X_t^{(k)}$ is stationary where the b_k are finite constants.

2.8 Let $\{h_k\}$ and $\{b_k\}$ be sequences of real numbers such that $\sum_{k=0}^{\infty} |h_k| < \infty$ and $\sum_{k=0}^{\infty} |b_k| < \infty$. Further, let $E[Z_t^2] \le M < \infty$ for $t = \pm 1, \pm 2, \ldots$ and define

$$X_t = \sum_{k=0}^{\infty} h_k Z_{t-k}$$

$$Y_t = \sum_{k=0}^{\infty} b_k Z_{t-k}.$$

Show the following:

(i) $E(X_t) = \lim_{n \to \infty} \sum_{k=0}^{n} h_k E(Z_{t-k})$

(ii) $E(X_t Y_t) = \lim_{n \to \infty} \sum_{k=0}^{n} \sum_{j=0}^{n} h_k b_j E(Z_{t-k} Z_{t-j})$.

2.9 For the following, determine whether the associated filter is high pass, low pass, band pass, band stop, etc. and explain your answer.

(a) A filter that has the squared frequency response function

$$\left|H(e^{-2\pi if})\right|^2 = \frac{1}{1+\left(\frac{1}{0.2f}\right)^{8}}$$

(b) A filter with frequency response function

$$H(e^{-2\pi if}) = 1 - 2(\cos 2\pi f - i\sin 2\pi f) + \cos 2\pi f - i\sin 4\pi f$$

3

ARMA Time Series Models

In this chapter we will discuss some of the important models that are used to describe the behavior of a time series. Of particular interest will be the autoregressive (AR), moving average (MA), and autoregressive–moving average (ARMA) models. These models were made popular by George Box and Gwilym Jenkins in their classic time series book, the current edition of which is Box et al. (2008). These models have become fundamental to the theory and applications of time series. It should be noted that we use simulated realizations in this chapter almost exclusively because we are focusing on the properties and developing an understanding of known models. Also, unless otherwise specified, the autocorrelations and spectral densities illustrated in this chapter are the theoretical quantities for the models being studied. In later chapters, we will use the lessons learned in this chapter as we analyze actual time series data.

3.1 Moving Average Processes

We begin by defining an MA process of order q, which will be denoted MA(q).

Definition 3.1 Let a_t be a white noise process with zero mean and finite variance σ_a^2, and let the process $X_t, t = 0, \pm 1, \pm 2, \ldots$ be defined by

$$X_t - \mu = a_t - \theta_1 a_{t-1} - \cdots - \theta_q a_{t-q}, \tag{3.1}$$

where the $\theta_j, j = 1, \ldots, q$ are real constants with $\theta_q \neq 0$. The process X_t defined in (3.1) is called an MA *process of order* q, which is usually shortened to MA(q).

We will often write (3.1) in the form

$$X_t - \mu = \theta(B) a_t,$$

where

$$\theta(B) = 1 - \theta_1 B - \theta_2 B^2 - \cdots - \theta_q B^q.$$

NOTE: Before proceeding, we point out that some authors and writers of software packages write (3.1) as $X_t - \mu = a_t + \theta_1 a_{t-1} + \cdots + \theta_q a_{t-q}$. Consider the MA(2) model $X_t = a_t - 1.2a_{t-1} + 0.8a_{t-2}$. Using the definition in (3.1), it follows that $\theta_1 = 1.2$ and $\theta_2 = -0.8$, while the coefficients would be $\theta_1 = -1.2$ and $\theta_2 = 0.8$ using the alternative representation. Before entering coefficients into a time series computer package or comparing formulas involving the θ's in different textbooks, it is very important to check the MA(q) definition.

Notice that an MA(q) process is a general linear process with a finite number of terms. In particular

$$\begin{aligned}\psi_0 &= 1\\ \psi_k &= -\theta_k, \quad k = 1, 2, \ldots, q\\ \psi_k &= 0, \quad k > q.\end{aligned}$$

Using the notation in (2.8), note that $\psi(z) = 1 - \sum_{j=1}^{q} \theta_j z^j$, where z is real or complex. For an MA(q) process, we denote $\psi(z)$ by $\theta(z)$. An MA(q) process can be shown to be stationary using Theorem 2.3, because there are only a finite number of ψ-weights. Since $\sum_{j=0}^{\infty} \psi_j^2 = 1 + \theta_1^2 + \cdots + \theta_q^2$, it follows that

$$\sigma_X^2 = \sigma_a^2\left(1 + \theta_1^2 + \cdots + \theta_q^2\right) < \infty. \tag{3.2}$$

Also, from (2.7)

$$\begin{aligned}\gamma_k &= \sigma_a^2\left\{-\theta_k + \sum_{j=1}^{q-k} \theta_j \theta_{j+k}\right\}, \quad k = 1, 2, \ldots, q\\ &= 0, \quad k > q.\end{aligned} \tag{3.3}$$

It follows then that

$$\begin{aligned}\rho_k &= \frac{-\theta_k + \sum_{j=1}^{q-k} \theta_j \theta_{j+k}}{1 + \sum_{j=1}^{q} \theta_j^2}, \quad k = 1, 2, \ldots, q\\ &= 0, \quad k > q.\end{aligned} \tag{3.4}$$

Here and throughout this text, we define the sum $\sum_{j=m}^{n} f(j)$ to be identically zero if $n < m$. The results in (3.3) and (3.4) show that autocovariances of an MA(q) process trivially satisfy the conditions of Theorem 1.1 (i.e., that $\gamma_k \to 0$ as $k \to \infty$) and so the covariance matrix, $\mathbf{\Gamma}_k$, defined in (1.2), is positive definite for each integer $k > 0$.

The spectrum can be easily obtained for an MA(q) process X_t. From (2.8)

$$\begin{aligned} P_X(f) &= \sigma_a^2 |\psi(e^{-2\pi if})|^2 \\ &= \sigma_a^2 |\theta(e^{-2\pi if})|^2 \\ &= \sigma_a^2 \left| 1 - \sum_{j=1}^{q} \theta_j e^{-2\pi ifj} \right|^2 \end{aligned} \tag{3.5}$$

and

$$S_X(f) = \frac{\left| 1 - \sum_{j=1}^{q} \theta_j e^{-2\pi ifj} \right|^2}{1 + \sum_{j=1}^{q} \theta_j^2}. \tag{3.6}$$

Recall from (1.29) that for a stationary process, X_t, the power spectrum is defined as

$$P_X(f) = \gamma_0 + 2\sum_{j=1}^{\infty} \gamma_j \cos 2\pi fj,$$

so that for an MA(q) process we have simplified expressions for $P_X(f)$ and $S_X(f)$ given by

$$P_X(f) = \gamma_0 + 2\sum_{j=1}^{q} \gamma_j \cos 2\pi fj \tag{3.7}$$

and

$$S_X(f) = 1 + 2\sum_{j=1}^{q} \rho_j \cos 2\pi fj, \tag{3.8}$$

where γ_j and $\rho_j, j = 0, 1, \ldots, q$ are given in (3.3) and (3.4).

3.1.1 MA(1) Model

Let X_t be the MA(1) process defined by

$$\begin{aligned} X_t - \mu &= a_t - \theta_1 a_{t-1} \\ &= (1 - \theta_1 B) a_t. \end{aligned} \tag{3.9}$$

From (3.2) and (3.3), we obtain

$$\begin{aligned}\gamma_0 &= \sigma_a^2[1+\theta_1^2] = \sigma_X^2 \\ \gamma_1 &= \sigma_a^2[-\theta_1] \\ \gamma_j &= 0, \quad j > 1.\end{aligned} \tag{3.10}$$

Thus, for an MA(1) process,

$$\begin{aligned}\rho_0 &= 1 \\ \rho_1 &= \frac{-\theta_1}{1+\theta_1^2} \\ \rho_j &= 0, \quad j > 1.\end{aligned} \tag{3.11}$$

It is easily shown that the function $f(u) = |u|/(1+u^2)$ has a maximum of 0.5 at $u = \pm 1$, from which it follows that for an MA(1) process, $\max|\rho_1| = 0.5$. In particular, $\rho_1 = 0.5$ when $\theta_1 = -1$ and $\rho_1 = -0.5$ when $\theta_1 = 1$. From (3.6), the spectral density of an MA(1) is given, for $|f| \le 0.5$, by

$$\begin{aligned}S_X(f) &= \frac{\left|1-\sum_{j=1}^{q}\theta_j e^{-2\pi i f j}\right|^2}{1+\sum_{j=1}^{q}\theta_j^2} \\ &= \frac{|1-\theta_1 e^{-2\pi i f}|^2}{1+\theta_1^2} \\ &= \frac{1}{1+\theta_1^2}\{1-\theta_1 e^{-2\pi i f}\}\{1-\theta_1 e^{2\pi i f}\} \\ &= \frac{1}{1+\theta_1^2}\{1+\theta_1^2-2\theta_1\cos 2\pi f\} \\ &= 1-\frac{2\theta_1\cos 2\pi f}{1+\theta_1^2}.\end{aligned} \tag{3.12}$$

As noted in Chapter 1, the spectral density provides information concerning the periodicities present in the data. From (3.12), $S_X(f)$ for an MA(1) model can be seen to be maximized, with respect to f, whenever the numerator of the right-hand side is minimized. If $\theta_1 \ge 0$, this numerator is minimized when $f = \pm 0.5$, while if $\theta_1 < 0$, the minimum occurs at $f = 0$. Consequently, data for which $\theta_1 > 0$ are higher frequency than data for which $\theta_1 < 0$.

Notice that as $\theta_1 \to 0$, the MA(1) process approaches white noise. This observation is consistent with the fact that as $\theta_1 \to 0$, ρ_1 as given in (3.11) approaches 0, and $S_X(f)$ as given in (3.12) approaches a constant spectral density.

Example 3.1 MA(1) Models

Consider the MA(1) models given by

(a) $X_t = a_t - 0.99a_{t-1}$
(b) $X_t = a_t + 0.99a_{t-1}$,

where a_t is white noise with $\sigma_a^2 = 1$. Note that $\theta_1 = 0.99$ for model (a) and $\theta_1 = -0.99$ for model (b). For both models, $\rho_0 = 1$ and $\rho_j = 0, j > 1$. Using (3.11), we see that for model (a), $\rho_1 = -0.49997$ while for model (b), $\rho_1 = 0.49997$. The spectral densities, from (3.12), are

$$S_X(f) = 1 - \frac{1.98}{1.9801}\cos 2\pi f$$

and

$$S_X(f) = 1 + \frac{1.98}{1.9801}\cos 2\pi f,$$

for models (a) and (b), respectively. Realizations of length 100 from each of these two models are plotted in Figure 3.1.

In Figure 3.2, the true autocorrelations are shown while in Figure 3.3 the true ($10 \log_{10}$) spectral densities are plotted. Notice that the realization from model (a) exhibits a high-frequency behavior, that is, there seems to be a very short period in the data. This behavior can be attributed to the negative correlation between successive values in the series due to the fact that $\rho_1 = -0.497$. However, notice that interestingly, the autocovariances are zero for $k \geq 2$. Additionally, this high-frequency behavior is illustrated in the spectral density via the maximum at $f = 0.5$. On the other hand, the data for

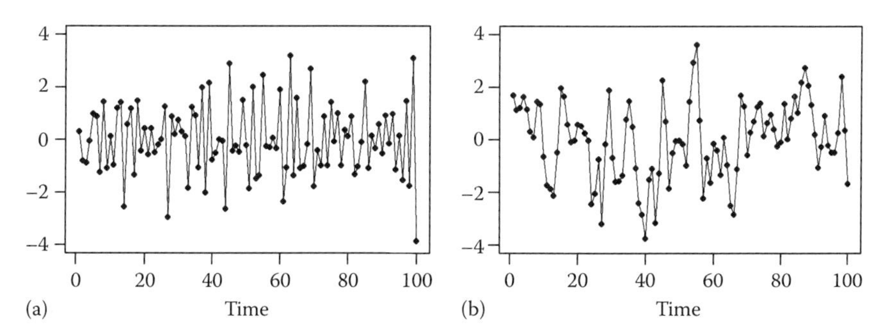

FIGURE 3.1
Realizations from MA(1) models. (a) $X_t = a_t - 0.99a_{t-1}$. (b) $X_t = a_t + 0.99a_{t-1}$.

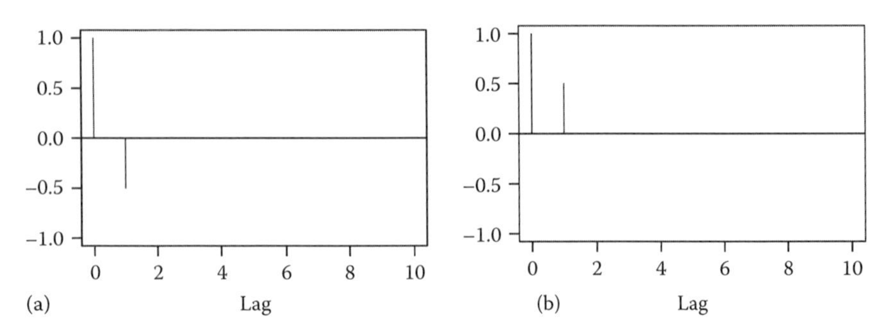

FIGURE 3.2
Autocorrelations from MA(1) models. (a) $X_t = a_t - 0.99a_{t-1}$. (b) $X_t = a_t + 0.99a_{t-1}$.

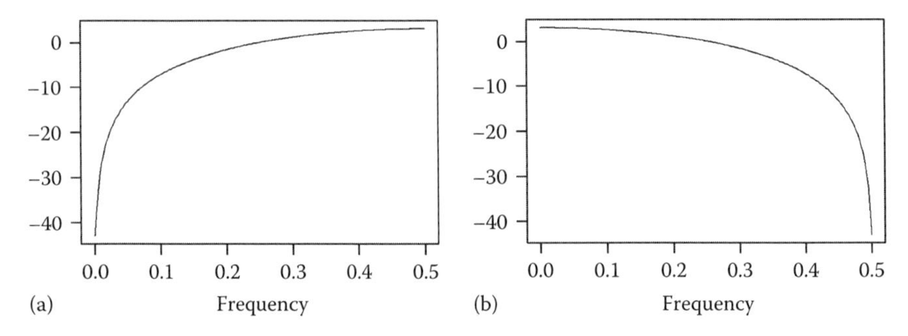

FIGURE 3.3
Spectral densities of MA(1) models. (a) $X_t = a_t - 0.99a_{t-1}$. (b) $X_t = a_t + 0.99a_{t-1}$.

model (b) clearly are not as highly oscillatory, and this is further illustrated by the maximum of the spectral density at $f = 0$ and its steady decline as f approaches 0.5.

Using `GW-WINKS`: Figures such as those shown in Figures 3.1 through 3.3 can be obtained with the `GW-WINKS` software using the `Generate Realization` option in the `Time Series Analysis` menu. You will be asked to specify the model (you can enter coefficients or factors—to be discussed later), the realization length, white noise variance, and a seed. Be careful with the signs as you enter the coefficients for the model. For example, the coefficient for model (a) in Example 3.1 is $\theta_1 = 0.99$ while for model (b) we have $\theta_1 = -0.99$. The `Generate Realization` option will place four columns of data in the spreadsheet. The `ARMA_GENERATED` column contains the generated realization, the `CORR` column contains the true autocorrelations for the model from which the data are generated, `SPECTRUM` contains the true spectral density associated with the model specified, and `SAMPCORR` contains the sample autocorrelations for the realization. These can be plotted using the `Graphs/Charts` menu by selecting `Line/Time Series Plots` and then `Line plots one or multiple`.

3.1.2 MA(2) Model

Let X_t denote the MA(2) model given by

$$X_t = a_t - \theta_1 a_{t-1} - \theta_2 a_{t-2}.$$

For this model, it follows that

$$\begin{aligned}
\gamma_0 &= \sigma_a^2(1 + \theta_1^2 + \theta_2^2) = \sigma_X^2 \\
\rho_1 &= \frac{-\theta_1 + \theta_1\theta_2}{1 + \theta_1^2 + \theta_2^2} \\
\rho_2 &= \frac{-\theta_2}{1 + \theta_1^2 + \theta_2^2} \\
\rho_k &= 0, \quad k \geq 3.
\end{aligned}$$

Using (3.6), the spectral density of the MA(2) process is given by

$$S_X(f) = \frac{\sigma_a^2}{\sigma_X^2}\left\{1 + \theta_1^2 + \theta_2^2 - 2\theta_1(1 - \theta_2)\cos 2\pi f - 2\theta_2 \cos 4\pi f\right\}.$$

3.2 Autoregressive Processes

Let a_t be a white noise process and let $\phi_1, \phi_2, \ldots, \phi_p$ be real constants with $\phi_p \neq 0$. Then, loosely speaking, if X_t, $t = 0, \pm 1, \pm 2, \ldots$ is specified by

$$X_t - \mu - \phi_1(X_{t-1} - \mu) - \phi_2(X_{t-2} - \mu) - \cdots - \phi_p(X_{t-p} - \mu) = a_t, \qquad (3.13)$$

then X_t is called an *AR process of order p*, which is usually shortened to AR(p). We will often write (3.13) in the operator form

$$\phi(B)(X_t - \mu) = a_t \qquad (3.14)$$

where

$$\phi(B) = 1 - \phi_1 B - \phi_2 B^2 - \cdots - \phi_p B^p.$$

The AR(p) model in (3.14) can be rewritten in the form $\phi(B)X_t = \phi(B)\mu + a_t$, which by noting that

$$\begin{aligned}
\phi(B)\mu &= \left(1 - \phi_1 B - \cdots - \phi_p B^p\right)\mu \\
&= (1 - \phi_1 - \cdots - \phi_p)\mu \\
&= \phi(1)\mu,
\end{aligned}$$

can be written as

$$\phi(B)X_t = \phi(1)\mu + a_t. \tag{3.15}$$

The quantity $\phi(1)\mu$ is referred to as the MA *constant*. The model in (3.13) is often simplified by letting $\tilde{X}_t = X_t - \mu$, which results in

$$\tilde{X}_t - \phi_1\tilde{X}_{t-1} - \phi_2\tilde{X}_{t-2} - \cdots - \phi_p\tilde{X}_{t-p} = a_t, \tag{3.16}$$

corresponding to (3.13). Henceforth, unless specified differently, we will without loss of generality assume $\mu = 0$ so $\tilde{X}_t \equiv X_t$, and we will specify the AR(p) model as

$$X_t - \phi_1X_{t-1} - \phi_2X_{t-2} - \cdots - \phi_pX_{t-p} = a_t \tag{3.17}$$

or

$$\phi(B)X_t = a_t. \tag{3.18}$$

Likewise, the MA(q) model in (3.1) will usually be written assuming $\mu = 0$.

NOTE: When using the AR(p) or MA(q) to model a given set of stationary time series data, it will be the standard procedure to first estimate μ by $\bar{X}$, subtract $\bar{X}$ from each observation, and model the remaining data using the "zero mean" model.

As noted previously, the MA(q) process is a special case of the general linear process, in which only a finite number of the ψ-weights are nonzero. Examination of (3.17), however, does not immediately reveal the relationship between the AR(p) process and the general linear process. Actually, the specification of an AR(p) process as given by (3.17) is inadequate since (3.17) has many distinct solutions and, to this point, we have failed to specify which solution we wish to consider. For example, if X_t is a solution of (3.17) and $\phi(B)y_t = 0$, then $X_t + cy_t$ is also a solution of (3.17) for any constant c. Consequently, (3.17) has an uncountable number of solutions. However, if $c \neq 0$, then all of these solutions are nonstationary, and since for now our interest is in stationary solutions, we need to be more specific. This leads us to the following definition.

Definition 3.2 Suppose $\{X_t: t = 0, \pm 1, \pm 2, \ldots\}$ is a causal, stationary process satisfying the difference equation

$$\phi(B)(X_t - \mu) = a_t. \tag{3.19}$$

Then, X_t will be called an AR process of order p and denoted AR(p). Again, we will take $\mu = 0$ unless indicated to the contrary.

To clarify Definition 3.2, consider the AR(1) model

$$\phi(B)X_t = (1 - \alpha B)X_t = a_t, \tag{3.20}$$

where $|\alpha| < 1$. We note that

$$(1 - \alpha B)X_t = \alpha^t(1 - B)\alpha^{-t}X_t,$$

so that the two operators $1 - \alpha B$ and $\alpha^t(1 - B)\alpha^{-t}$ are equivalent, that is,

$$1 - \alpha B \equiv \alpha^t(1 - B)\alpha^{-t}. \tag{3.21}$$

Therefore,

$$\alpha^t(1 - B)\alpha^{-t}X_t = a_t.$$

So

$$(1 - B)\alpha^{-t}X_t = \alpha^{-t}a_t,$$

and

$$\sum_{k=-\infty}^{t} (1 - B)\alpha^{-k}X_k = \sum_{k=-\infty}^{t} \alpha^{-k}a_k,$$

which gives

$$\alpha^{-t}X_t - \lim_{j\to-\infty} \alpha^{-j}X_j = \sum_{k=-\infty}^{t} \alpha^{-k}a_k. \tag{3.22}$$

Since $|\alpha| < 1$, it follows that $\lim_{j\to-\infty} \alpha^{-j}X_j = 0$ with probability 1. Consequently,

$$X_t = \sum_{k=-\infty}^{t} \alpha^{t-k}a_k$$
$$= \sum_{h=0}^{\infty} \alpha^h a_{t-h}. \tag{3.23}$$

Clearly, X_t in (3.23) satisfies (3.20), is causal, and, by Theorem 2.3, is stationary since $|\alpha| < 1$ implies $\sum_{h=0}^{\infty} |\alpha|^h < \infty$. Therefore, X_t is a (causal and stationary) AR(1) process. Because the polynomial operator $\phi(B)$ can always be factored as

$$
\begin{aligned}
X_t - \phi_1 X_{t-1} - \cdots - \phi_p X_{t-p} &= \phi(B)X_t \\
&= (1-\alpha_1 B)(1-\alpha_2 B)\cdots(1-\alpha_p B)X_t \\
&= \prod_{k=1}^{p} \alpha_k^t (1-B)\alpha_k^{-t} X_t, \qquad (3.24)
\end{aligned}
$$

where the α_k may be real or complex, the following theorem can be shown.

Theorem 3.1

Let X_t be causal and suppose

$$
\phi(B)X_t = (1-\alpha_1 B)(1-\alpha_2 B)\cdots(1-\alpha_p B)X_t = a_t, \qquad (3.25)
$$

where the α_k are real or complex numbers, $\alpha_k \neq \alpha_j, k \neq j, |\alpha_k| < 1, k=1,\ldots,p$, and a_t is zero mean white noise. Then, X_t is an AR(p) process and

$$
X_t = \sum_{k=0}^{\infty} \psi_k(p) a_{t-k}, \qquad (3.26)
$$

where

$$
\begin{aligned}
&\text{(a)} \quad \psi_\ell(1) = \alpha_1^\ell \quad \text{for } p=1 \\
&\text{(b)} \quad \psi_\ell(p) = \sum_{k=1}^{p-1} \frac{\alpha_{p-k}^{p-2}\left(\alpha_p^{\ell+1} - \alpha_{p-k}^{\ell+1}\right)}{(\alpha_p - \alpha_{p-k})\prod_{\substack{h=1 \\ h\neq p-k}}^{p-1} (\alpha_{p-k} - \alpha_h)}, \quad \text{for } p \geq 2. \qquad (3.27)
\end{aligned}
$$

Proof

see Appendix 3.A.

We note that Theorem 3.1 gives the ψ-weights for the case of no repeated roots. Problem 3.33 shows how (3.27) can be used in the case of a repeated root. The constants $\psi_k(p)$ will be referred to as the "ψ-weights" for an AR(p) process. Definition 3.3 introduces the characteristic equation, which is a useful tool in the analysis of AR(p) models.

Definition 3.3 If X_t is the AR(p) process defined by (3.17), then the algebraic equation $\phi(z) = 0$ is called the associated *characteristic equation* where

$$
\phi(z) = 1 - \phi_1 z - \cdots - \phi_p z^p,
$$

and z is a real or complex number. The polynomial $\phi(z)$ is called the *characteristic polynomial*.

Theorem 3.2 that follows is an important result for determining whether a given set of AR(p) coefficients yields a causal, stationary AR(p) process.

Theorem 3.2

The process X_t with ψ-weights as defined in Theorem 3.1 is a stationary and causal AR(p) process if and only if the roots of the characteristic equation $\phi(z) = 0$ all lie outside the unit circle.

Proof

See Appendix 3.A.

Thus, X_t in (3.25) is an AR(p) process if and only if $|\alpha_k| < 1$, $k = 1, \ldots, p$, since α_k is the reciprocal of the kth root of the characteristic equation. From (3.26), it is also clear that in this case, X_t is a general linear process.

3.2.1 Inverting the Operator

It is worth noting that properties of $\phi(B)$ can be found by examining the algebraic counterpart, $\phi(z)$, when z is real or complex. That is, it can be shown (see Brockwell and Davis, 1991) that for

$$\begin{aligned}\phi(z) &= 1 - \phi_1 z - \cdots - \phi_p z^p \\ &= \prod_{k=1}^{p} (1 - \alpha_k z),\end{aligned}$$

where $|\alpha_k| < 1$, $k = 1, \ldots, p$, that $\phi^{-1}(B) = \phi^{-1}(z)|_B$. Thus, for example, if $(1 - \phi_1 B)X_t = a_t$ and $|\phi_1| < 1$, then since $(1 - \phi_1 z)^{-1} = \sum_{h=0}^{\infty} \phi_1^h z^h$, it follows that

$$\begin{aligned}X_t &= (1 - \phi_1 B)^{-1} a_t \\ &= \sum_{k=0}^{\infty} \phi_1^k B^k a_t \\ &= \sum_{k=0}^{\infty} \phi_1^k a_{t-k}.\end{aligned}$$

In this book, we will often use this formal procedure for inverting operators because of its ease of use. Note that the proof of Theorem 3.1 is actually based on the use of $\phi(B)$ as an operator.

3.2.2 AR(1) Model

We first consider the properties of the AR(1) model given by

$$X_t - \phi_1 X_{t-1} = a_t, \tag{3.28}$$

where $|\phi_1| < 1$.

(a) *ψ-weights and stationarity conditions for an AR(1)*
From (3.27), we see that the general linear process form of an AR(1) process is

$$X_t = \sum_{k=0}^{\infty} \phi_1^k a_{t-k}.$$

That is, the ψ-weights are $\psi_k(1) = \phi_1^k$. The characteristic equation is given by $1 - \phi_1 r = 0$, which has root $r_1 = \phi_1^{-1}$ where $|r_1| = |\phi_1^{-1}| > 1$ since $|\phi_1| < 1$, from which it follows that the model in (3.28) is in fact stationary and causal. Note that we also obtained the ψ-weights of an AR(1) model in the preceding discussion about inverting the operator.

(b) *Autocorrelations of an AR(1) process*
In order to find γ_k and ρ_k for an AR(1) process, it is necessary to evaluate the expectation $E[X_t X_{t+k}] = E[X_{t-k} X_t]$. To obtain γ_k, multiply both sides of the equality in (3.28) by X_{t-k}, where $k > 0$, to yield

$$X_{t-k}\{X_t = \phi_1 X_{t-1} + a_t\}.$$

Taking the expectation of both sides we obtain

$$E[X_{t-k}X_t] = \phi_1 E[X_{t-k}X_{t-1}] + E[X_{t-k}a_t]. \tag{3.29}$$

Now, $X_{t-k} = \sum_{j=0}^{\infty} \phi_1^j a_{t-k-j}$ and thus X_{t-k} depends only on $a_{t-k-j}, j = 0, 1, \ldots$. It follows that X_{t-k} is uncorrelated with a_t, that is, $E[X_{t-k}a_t] = 0$ for $k > 0$. Thus, from (3.29), $\gamma_k = \phi_1 \gamma_{k-1}$ for $k > 0$, and

$$\gamma_1 = \phi_1 \gamma_0$$
$$\gamma_2 = \phi_1 \gamma_1 = \phi_1^2 \gamma_0,$$

so that in general

$$\gamma_k = \phi_1^k \gamma_0 \tag{3.30}$$

and

$$\rho_k = \phi_1^k. \tag{3.31}$$

Since $\gamma_k = \gamma_{-k}$ and $\rho_k = \rho_{-k}$, then for $k = 0, \pm 1, \pm 2, \ldots$, (3.30) and (3.31), respectively, become

$$\gamma_k = \phi_1^{|k|} \gamma_0$$

and

$$\rho_k = \phi_1^{|k|}. \tag{3.32}$$

Further, γ_0 can be obtained by premultiplying both sides of (3.28) by X_t and taking expectations as before, to obtain

$$E\left[X_t^2\right] = \phi_1 E[X_t X_{t-1}] + E[X_t a_t].$$

Thus

$$\gamma_0 = \phi_1 \gamma_1 + E[X_t a_t]. \tag{3.33}$$

The term $E[X_t a_t]$ is found using an analogous procedure, this time by postmultiplying both sides of (3.28) by a_t and taking expectations, which gives

$$E[X_t a_t] = \phi_1 E[X_{t-1} a_t] + E[a_t^2].$$

Previously, it was shown that $E[X_{t-1} a_t] = 0$, and therefore, from (3.28), $E[X_t a_t] = E[a_t^2] = \sigma_a^2$. Substituting into (3.33) yields

$$\gamma_0 = \phi_1 \gamma_1 + \sigma_a^2, \tag{3.34}$$

so that using (3.30), we have

$$\gamma_0 = \frac{\sigma_a^2}{1 - \phi_1^2}. \tag{3.35}$$

(c) *Spectrum and spectral density for an AR(1)*
The discussion in Section 3.2.1 regarding inverting the operator allows us to apply (2.8) to obtain the spectrum yielding

$$\begin{aligned} P_X(f) &= \sigma_a^2|\psi(e^{-2\pi if})|^2 \\ &= \sigma_a^2|\phi^{-1}(e^{-2\pi if})|^2 \\ &= \frac{\sigma_a^2}{|\phi(e^{-2\pi if})|^2} \\ &= \frac{\sigma_a^2}{|1-\phi_1 e^{-2\pi if}|^2} \end{aligned}$$

Using (3.35), the spectral density of an AR(1) process is thus given by

$$\begin{aligned} S_X(f) &= \frac{\sigma_a^2}{\gamma_0}\frac{1}{|1-\phi_1 e^{-2\pi if}|^2} \\ &= \frac{1-\phi_1^2}{|1-\phi_1 e^{-2\pi if}|^2}. \end{aligned}$$

Example 3.2 AR(1) Models

Figure 3.4 shows plots of autocorrelations for the AR(1) models with $\phi_1 = \pm 0.95$ and ± 0.7. The autocorrelations in Figure 3.4a and c associated with positive ϕ_1 show that there is strong positive correlation between adjacent observations. This suggests that in both cases, realizations should exhibit a tendency for x_{t+1} to be fairly close to x_t. This positive correlation associated with $\phi_1 = 0.95$ holds up much longer than it does for $\phi_1 = 0.7$, and, for example, the correlation between X_t and X_{t+10} is about 0.6 for $\phi_1 = 0.95$ and about 0.03 for $\phi_1 = 0.7$. In Figure 3.5, we show realizations of length $n = 100$ from the four AR(1) models associated with the autocorrelations in Figure 3.4, with $\sigma_a^2 = 1$, where $\mu = 0$ in all cases. In Figure 3.5a and c, we see the behavior suggested by the autocorrelations in Figure 3.4a and c. Specifically, in both these realizations, there is strong correlation between adjacent observations. Figure 3.5a shows a tendency for observations to remain positively correlated across several lags resulting in several lengthy runs or a wandering behavior not seen in Figure 3.5c. This is due to the fact that the positive autocorrelation between X_t and X_{t+k}, that is, $\rho_k = 0.95^k$, is fairly strong (or persistent) whenever k is small or even moderately large. We will refer to this persistence of the correlation when $\phi_1 \approx 1$ as *near nonstationary* behavior. In Figure 3.4b and d, we see that the

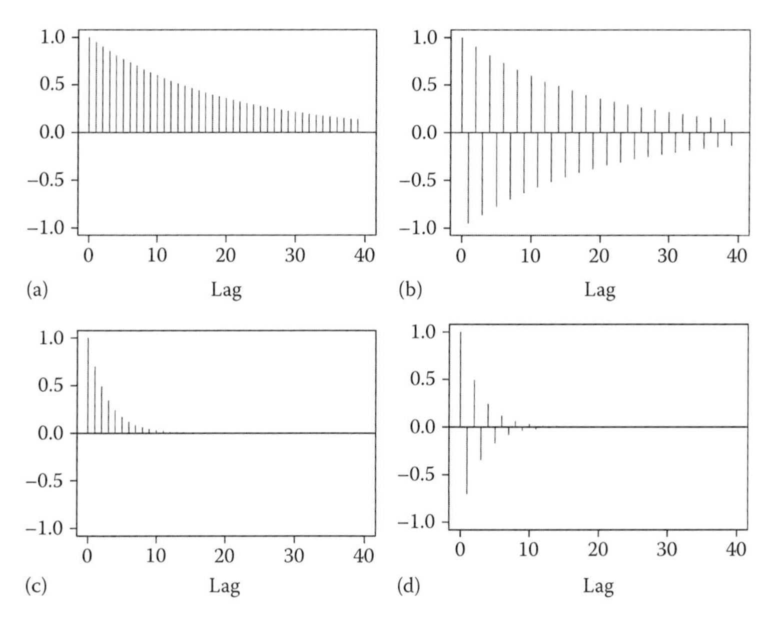

FIGURE 3.4
Autocorrelations for AR(1) models. (a) $X_t - 0.95X_{t-1} = a_t$. (b) $X_t + 0.95X_{t-1} = a_t$. (c) $X_t - 0.7X_{t-1} = a_t$. (d) $X_t + 0.7X_{t-1} = a_t$.

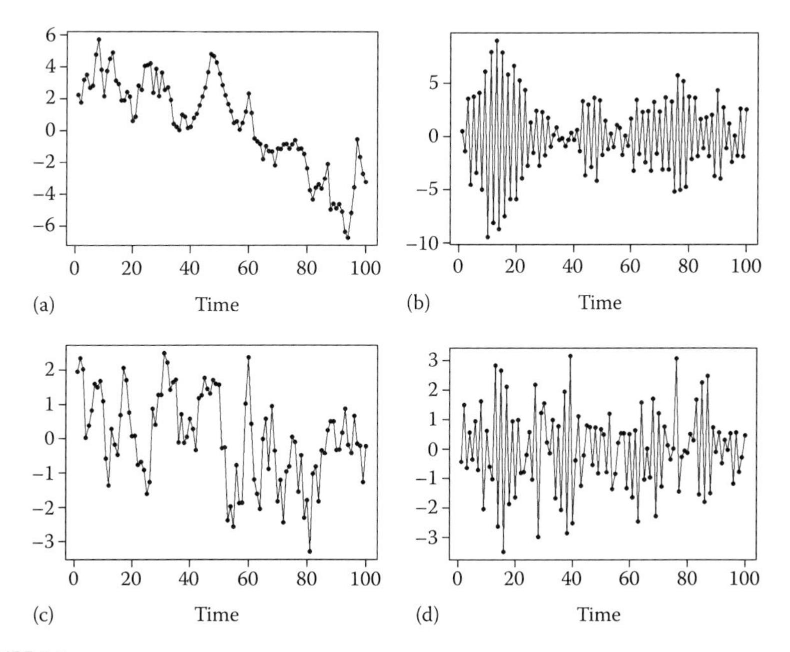

FIGURE 3.5
Realization from AR(1) models. (a) $X_t - 0.95X_{t-1} = a_t$. (b) $X_t + 0.95X_{t-1} = a_t$. (c) $X_t - 0.7X_{t-1} = a_t$. (d) $X_t + 0.7X_{t-1} = a_t$.

autocorrelations for negative values of ϕ_1 show an oscillating behavior. For example, for the $\phi_1 = -0.95$ case, $\rho_1 = -0.95$ indicating that if a value at time t is above the mean then the value at time $t+1$ will tend to be below the mean, and vice versa. Similarly, since $\rho_2 = (-0.95)^2 = 0.9025$, then the value at time $t+2$ will tend to be on the same side of the mean as the value at time t. This oscillating tendency in the realization for $\phi_1 = -0.95$ is stronger than it is for $\phi_1 = -0.7$ since the correlation structure is stronger in the former. As with Figure 3.4a and c, the strength of the correlations holds up longer for $\phi_1 = -0.95$ in Figure 3.4b than for $\phi_1 = -0.7$ in Figure 3.4d. As expected, the realizations in Figure 3.5b and d associated with negative ϕ_1 have an oscillatory nature. Also note that when $|\phi_1| = 0.95$, the variance of the realization appears to be larger. As $\phi_1 \to 0$, the AR(1) approaches white noise, so that realizations with $\phi_1 \approx 0$ can be expected to look much like white noise. In Example 1.9, we examined AR(1) processes with $\phi_1 = \pm 0.9$, and the observations made in that example are consistent those discussed here. In Figure 3.6, we show spectral densities for the four AR(1) models considered in Figures 3.4 and 3.5. From the figure, it can be seen that when $\phi_1 > 0$, the spectral densities have a peak at $f = 0$ (i.e., the denominator in the spectral density formula is a minimum at $f = 0$) indicating, as mentioned in

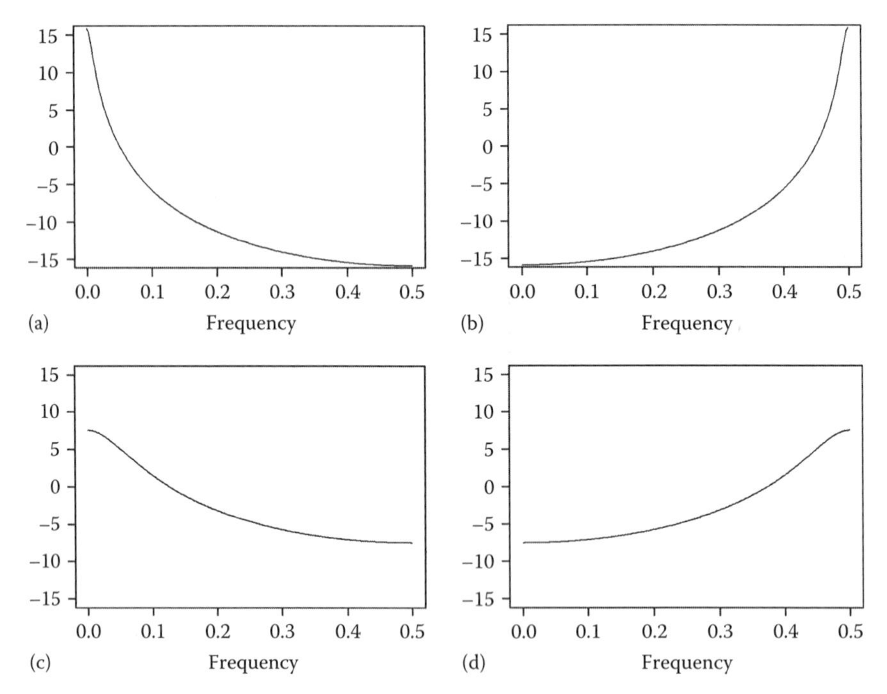

FIGURE 3.6
Spectral densities for AR(1) models. (a) $X_t - 0.95X_{t-1} = a_t$. (b) $X_t + 0.95X_{t-1} = a_t$. (c) $X_t - 0.7X_{t-1} = a_t$. (d) $X_t + 0.7X_{t-1} = a_t$.

Example 1.9, that the associated realizations will tend to wander about the mean with no distinct periodic behavior. That is, the data are primarily low frequency with no real periodic tendency. The spectral densities for $\phi_1 < 0$ have a maximum at $f = 0.5$, indicating that the data will tend to have a highly oscillatory nature with period of about 2. These behaviors are seen in the realizations in Figure 3.5. It should also be noticed that the spectral densities associated with $|\phi_1| = 0.95$ have a higher peak and less power away from the peak than those associated with $|\phi_1| = 0.7$, indicating that the associated behaviors in realizations will be more pronounced for $|\phi_1| = 0.95$.

Using `GW-WINKS`: Figures 3.4 through 3.6 can be obtained using the `Generate Realization` option of the `Time Series Analysis` menu discussed in Section 3.1. Again remember to be careful about the signs of the coefficients.

Example 3.3 Nonstationary Models

In Figure 3.7, realizations are shown from the model in (3.28) with $\phi_1 = 1$ and $\phi_1 = 1.1$ where $\sigma_a^2 = 1$. Since the roots of the two characteristic equations are $r = 1$ and $r = 0.9$, respectively, it is clear that these are nonstationary models. Examination of the realizations shows that for $\phi_1 = 1$, the "quasi-linear behavior" indicated in Figure 3.5 for $\phi_1 = 0.95$ and $\phi_1 = 0.7$ is even more pronounced. In fact for $\phi_1 = 1$, there is no attraction to the mean as in stationary processes. For $\phi_1 = 1$, the process X_t is given by $X_t = X_{t-1} + a_t$. The realizations shown in this book are only *finite* pieces of realizations that extend infinitely in both directions.

The realization associated with $\phi_1 = 1.1$ demonstrates what Box et al. (2008) refer to as "explosive" behavior. The random wandering seen in Figure 3.7 associated with $\phi_1 = 1$ is typical of certain real data series such as daily stock prices. However, the explosive realization associated with $\phi_1 = 1.1$ is not similar to that of any data sets that we will attempt to analyze, being better modeled as a deterministic process.

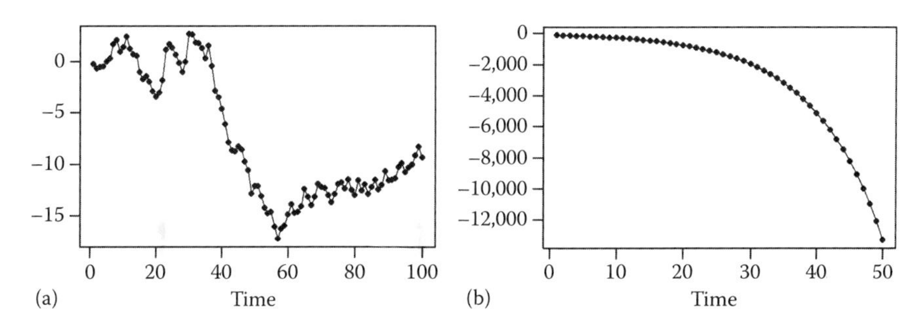

FIGURE 3.7
Realizations from first-order nonstationary models. (a) $X_t - X_{t-1} = a_t$. (b) $X_t - 1.1X_{t-1} = a_t$.

3.2.3 AR(p) Model for $p \geq 1$

Now that we have discussed the properties of the simplest of all AR models, the AR(1), we turn our attention to the more general case in which $p \geq 1$. Theorem 3.2 provides necessary and sufficient conditions for a model of the form (3.17) to be causal and stationary. We recommend finding the roots of the characteristic equation numerically and examining the resulting factored characteristic polynomial in a *factor table*. This comment will be clarified in Section 3.2.11.

3.2.4 Autocorrelations of an AR(p) Model

In order to find the autocorrelations of an AR(p) process, premultiply each side of (3.17) by X_{t-k} where $k > 0$, to obtain

$$X_{t-k}\{X_t - \phi_1 X_{t-1} - \cdots - \phi_p X_{t-p} = a_t\}. \tag{3.36}$$

Taking expectations in (3.36) yields

$$\gamma_k - \phi_1 \gamma_{k-1} - \cdots - \phi_p \gamma_{k-p} = 0,$$

where we have used the fact that, as in the AR(1) case, $E[X_{t-k}a_t] = 0$ for $k > 0$. It follows that for $k > 0$,

$$\rho_k - \phi_1 \rho_{k-1} - \cdots - \phi_p \rho_{k-p} = 0. \tag{3.37}$$

Writing (3.37) as

$$\rho_k = \phi_1 \rho_{k-1} + \cdots + \phi_p \rho_{k-p}, \tag{3.38}$$

the first p equations specified in (3.38) can be written as

$$\begin{aligned}
\rho_1 &= \phi_1 + \phi_2 \rho_1 + \cdots + \phi_p \rho_{p-1} \\
\rho_2 &= \phi_1 \rho_1 + \phi_2 + \cdots + \phi_p \rho_{p-2} \\
&\vdots \\
\rho_p &= \phi_1 \rho_{p-1} + \phi_2 \rho_{p-2} + \cdots + \phi_p.
\end{aligned} \tag{3.39}$$

The equations in (3.39) are called the *Yule–Walker* equations. When the coefficients $\phi_1, \ldots, \phi_p$ are known, the Yule–Walker equations are p equations that can be solved for the p unknowns $\rho_1, \ldots, \rho_p$. Note that after the first p autocorrelations are found by solving the Yule–Walker equations, then ρ_k, $k > p$ can iteratively be found using the relationship in (3.38).

Also note that for a given set of data assumed to follow an unknown AR(p) model, the model parameters $\phi_1, \ldots, \phi_p$ can be estimated by first replacing $\rho_k, k = 1, 2, \ldots, p$ with the estimates $\hat{\rho}_k$, $k = 1, 2, \ldots, p$ and solving the resulting p equations for $\hat{\phi}_1, \ldots, \hat{\phi}_p$. These estimates are called the Yule–Walker estimates and will be discussed in more detail in Chapter 7.

In order to find $\gamma_0 = \sigma_X^2$, multiply every term in (3.17) by X_t and take expectations to obtain

$$\gamma_0 - \phi_1\gamma_1 - \cdots - \phi_p\gamma_p = E[X_t a_t]. \tag{3.40}$$

The expectation $E[X_t a_t]$ can be found by postmultiplying each term in (3.17) by a_t and taking expectations, which gives

$$\begin{aligned} E[X_t a_t] &= \sum_{k=1}^{p} \phi_k E[a_t X_{t-k}] + \sigma_a^2 \\ &= \sigma_a^2, \end{aligned} \tag{3.41}$$

since $E[a_t X_{t-k}] = 0$ for $k > 0$. From (3.40) and (3.41), we find that

$$\gamma_0 - \phi_1\gamma_1 - \cdots - \phi_p\gamma_p = \sigma_a^2, \tag{3.42}$$

and hence

$$\sigma_X^2(1 - \phi_1\rho_1 - \cdots - \phi_p\rho_p) = \sigma_a^2.$$

Therefore,

$$\sigma_X^2 = \frac{\sigma_a^2}{1 - \phi_1\rho_1 - \cdots - \phi_p\rho_p}. \tag{3.43}$$

3.2.5 Linear Difference Equations

We have discussed the fact that autocorrelations of an AR(p) model satisfy (3.37) and can be found using (3.39). However, the behavior of these autocorrelations is not immediately apparent. The equation in (3.37) is an example of *a linear homogenous difference equation with constant coefficients,* which we briefly discuss here.

A function, g_k, defined on a set of consecutive integers, I, is said to satisfy (or is a solution of) a linear homogeneous difference equation of order m with constant coefficients, if g_k is such that

$$g_k - b_1 g_{k-1} - \cdots - b_m g_{k-m} = 0 \tag{3.44}$$

for $k \in I$, where $b_1, b_2, \ldots, b_m$ are constants. Thus, from (3.37), it follows that ρ_k satisfies a pth-order linear homogeneous difference equation with constant coefficients for $k > 0$. The general solution to the difference equation in (3.44) is well known (see Brockwell and Davis, 1991). Specifically, if g_k satisfies (3.44) for $k > 0$, then

$$g_k = \sum_{i=1}^{v} \sum_{j=0}^{m_i-1} C_{ij} k^j r_i^{-k}, \tag{3.45}$$

where the r_i's are the roots of the characteristic equation

$$1 - b_1 z - b_2 z^2 - \cdots - b_m z^m = 0,$$

v is the number of distinct roots, m_i is the multiplicity of r_i (i.e., if r_i is repeated once, $m_i = 2$, and if r_i is not repeated, then $m_i = 1$), $\sum_{i=1}^{v} m_i = m$, and C_{ij} are uniquely determined by the initial conditions

$$\begin{aligned} g_1 &= h_1 \\ g_2 &= h_2 \\ &\vdots \\ g_m &= h_m, \end{aligned}$$

where the h_j are constants. When there are no repeated roots, (3.45) simplifies to

$$g_k = C_1 r_1^{-k} + C_2 r_2^{-k} + \cdots + C_m r_m^{-k}. \tag{3.46}$$

From (3.38) and (3.45), it follows that for an AR(p) model,

$$\rho_k = \sum_{i=1}^{v} \sum_{j=0}^{m_i-1} C_{ij} k^j r_i^{-k}, \tag{3.47}$$

where

$$\sum_{i=1}^{v} m_i = p,$$

and the initial conditions are given by

$$\rho_k = \phi_1 \rho_{k-1} + \cdots + \phi_p \rho_{k-p}, \tag{3.48}$$

$k = 1, 2, \ldots, p$. Note that $\rho_k \to 0$ as $k \to \infty$ in (3.47) if and only if $|r_i| > 1, i = 1, \ldots, p$, and thus it follows from Theorem 3.2 and the comments following it that $\rho_k \to 0$ as $k \to \infty$ is a necessary and sufficient condition for the model in (3.17) to be a causal and stationary AR(p) process. In addition, note that if none of the roots are repeated, it follows from (3.46) that for $k > 0$

$$\rho_k = C_1 r_1^{-k} + C_2 r_2^{-k} + \cdots + C_p r_p^{-k}. \tag{3.49}$$

For an AR(p) process, it follows from (3.47) that $\rho_k \to 0$ (and equivalently $\gamma_k \to 0$) as $k \to \infty$. From Theorem 1.1, it then follows that $\mathbf{\Gamma}_k$ and $\boldsymbol{\rho}_k$ defined in (1.2) and (1.3), respectively, are positive definite for each integer $k > 0$ in the case of an AR process.

Example 3.4 Autocorrelations of an AR(1) Model

For the case of the AR(1) model $X_t - \phi_1 X_{t-1} = a_t$, the autocorrelation ρ_k satisfies the difference equation

$$\rho_k - \phi_1 \rho_{k-1} = 0, \quad k > 0 \tag{3.50}$$

Since there is only one root (and thus no repeated roots), we use (3.49), and ρ_k satisfies $\rho_k = C_1 r_1^{-k}, k > 0$. Also, for an AR(1) model, $r_1 = 1/\phi_1$, so this can be rewritten

$$\rho_k = C_1 \phi_1^k, \quad k > 0. \tag{3.51}$$

Since $\rho_0 = 1$, it follows from (3.50) that $\rho_1 - \phi_1 \rho_0 = 0$. Thus, $\rho_1 = \phi_1$, and thus $C_1 = 1$. Notice that $\rho_0 = 1 = \phi_1^0$, so

$$\rho_k = \phi_1^k, \quad k \geq 0. \tag{3.52}$$

Since $\rho_k = \rho_{-k}$, the autocorrelation function for an AR(1) process satisfies $\rho_k = \phi_1^{|k|}$, which is the result given earlier in (3.32).

NOTE: In general, for the mth-order difference equation with no repeated roots, the values $C_1, \ldots, C_m$ are chosen in such a way that the starting values, $g_j, g_{j-1}, \ldots, g_{j+1-m}$, also satisfy (3.46) although they do not satisfy (3.44).

NOTE: In Equation 3.77, we will give an expression for (3.49) that provides better insight into the behavior of the autocorrelation function of an AR(p) model.

3.2.6 Spectral Density of an AR(p) Model

Since $\psi(z) = \phi^{-1}(z)$ for an AR(p) process, it follows from (2.8) that the spectrum is given by

$$
\begin{aligned}
P_X(f) &= \sigma_a^2 |\psi(e^{-2\pi i f})|^2 \\
&= \frac{\sigma_a^2}{|\phi(e^{-2\pi i f})|^2} \\
&= \frac{\sigma_a^2}{\left|1 - \phi_1 e^{-2\pi i f} - \phi_2 e^{-4\pi i f} - \cdots - \phi_p e^{-2p\pi i f}\right|^2}.
\end{aligned} \tag{3.53}
$$

Consequently, the spectral density is given by

$$
S_X(f) = \frac{\sigma_a^2}{\sigma_X^2 \left|1 - \phi_1 e^{-2\pi i f} - \phi_2 e^{-4\pi i f} - \cdots - \phi_p e^{-2p\pi i f}\right|^2}, \tag{3.54}
$$

where σ_X^2 is given in (3.43).

Using `GW-WINKS`: The true autocorrelations, process variance, and spectral density for an AR(p) process can be obtained using the `Generate Realization` option in the `Time Series Analysis` menu as discussed in Section 3.1 for the MA(q) case.

3.2.7 AR(2) Model

Understanding the features of an AR(1) and AR(2) model is critical to gaining insight into the more general AR(p) model. We have already discussed the AR(1) model, so, in this section, we consider in some detail the AR(2) model

$$
X_t - \phi_1 X_{t-1} - \phi_2 X_{t-2} = a_t.
$$

3.2.7.1 Autocorrelations of an AR(2) Model

For $k > 0$, the autocorrelation function of an AR(2) model satisfies the difference equation

$$
\rho_k - \phi_1 \rho_{k-1} - \phi_2 \rho_{k-2} = 0.
$$

When the roots are distinct, the autocorrelation function is given by

$$
\rho_k = C_1 r_1^{-k} + C_2 r_2^{-k}, \quad k > 0, \tag{3.55}
$$

where C_1 and C_2 are uniquely determined by ρ_1 and ρ_2. These first two autocorrelations can be obtained by solving the Yule–Walker equations

$$\begin{aligned}\rho_1 &= \phi_1 + \phi_2\rho_1 \\ \rho_2 &= \phi_1\rho_1 + \phi_2,\end{aligned}$$

which gives

$$\begin{aligned}\rho_1 &= \frac{\phi_1}{1-\phi_2} \\ \rho_2 &= \frac{\phi_1^2 + \phi_2 - \phi_2^2}{1-\phi_2}.\end{aligned} \tag{3.56}$$

The roots r_1 and r_2 in (3.55) are (possibly complex) solutions to the characteristic equation

$$1 - \phi_1 z - \phi_2 z^2 = 0.$$

Therefore,

$$r_j = \frac{-\phi_1 \pm \sqrt{\phi_1^2 + 4\phi_2}}{2\phi_2}, \quad j = 1, 2.$$

Note that the characteristic equation can be written in factored form as

$$\begin{aligned}1 - \phi_1 z - \phi_2 z^2 &= \left(1 - r_1^{-1}z\right)\left(1 - r_2^{-1}z\right) \\ &= 1 - \left(r_1^{-1} + r_2^{-1}\right)z + r_1^{-1}r_2^{-1}z^2,\end{aligned}$$

and thus

$$\phi_1 = r_1^{-1} + r_2^{-1} \tag{3.57}$$

$$\phi_2 = -r_1^{-1}r_2^{-1}. \tag{3.58}$$

When X_t is stationary, Theorem 3.2 implies that $\left|r_1^{-1}\right| < 1$ and $\left|r_2^{-1}\right| < 1$, so it follows from (3.58) that $|\phi_2| < 1$.

Additionally, from (3.43) and (3.56), it follows that

$$\sigma_X^2 = \frac{(1-\phi_2)\sigma_a^2}{(1+\phi_2)\left[(1-\phi_2)^2 - \phi_1^2\right]}. \tag{3.59}$$

Case 1: Roots are distinct
As previously mentioned, in this case, ρ_k is given by (3.55) with C_1 and C_2 uniquely determined by (3.56). So it follows that

$$\rho_1 = C_1 r_1^{-1} + C_2 r_2^{-1} = \frac{\phi_1}{1-\phi_2}$$

$$\rho_2 = C_1 r_1^{-2} + C_2 r_2^{-2} = \frac{\phi_1^2 + \phi_2 - \phi_2^2}{1-\phi_2}.$$

Denoting $\alpha_1 = r_1^{-1}$ and $\alpha_2 = r_2^{-1}$ and solving these equations gives

$$C_1 = \frac{\alpha_1(1-\alpha_2^2)}{(\alpha_1-\alpha_2)(1+\alpha_1\alpha_2)}$$

and

$$C_2 = \frac{-\alpha_2(1-\alpha_1^2)}{(\alpha_1-\alpha_2)(1+\alpha_1\alpha_2)},$$

which yields

$$\rho_k = \frac{(1-\alpha_2^2)\alpha_1^{k+1} - (1-\alpha_1^2)\alpha_2^{k+1}}{(\alpha_1-\alpha_2)(1+\alpha_1\alpha_2)}, \quad k \geq 0. \tag{3.60}$$

We consider two cases in which the roots are distinct.

(a) *Real roots, that is,* $\phi_1^2 + 4\phi_2 > 0$.
In this case, the characteristic polynomial can be written in factored form as

$$\phi(z) = (1 - r_1^{-1}z)(1 - r_2^{-1}z),$$

where r_1 and r_2 are real, and ρ_k in (3.60) can be seen to be a mixture of damped exponentials or damped oscillating exponentials that decay to zero as k increases, since $|\alpha_1| < 1$ and $|\alpha_2| < 1$.

(b) *Complex roots, that is,* $\phi_1^2 + 4\phi_2 < 0$.
In this case, the second-order characteristic polynomial is irreducible (i.e., r_1 and r_2 are not real). The roots r_1 and r_2 are complex conjugates and thus so are α_1 and α_2. We use the notation $\alpha_1 = a + bi$ and $\alpha_2 = a - bi$. Now, α_1 and α_2 are given by

$$a \pm bi = \frac{\phi_1}{2} \pm \frac{\sqrt{-(\phi_1^2 + 4\phi_2)}}{2} i.$$

It will be convenient to express α_1 and α_2 in polar form as $\alpha_1 = R(\cos 2\pi f_0 + i \sin 2\pi f_0) = Re^{2\pi f_0 i}$ and $\alpha_2 = Re^{-2\pi f_0 i}$, where $\tan(2\pi f_0) = b/a$ and $R = \sqrt{a^2 + b^2}$. In our case,

$$R = |\alpha_1| = \sqrt{-\phi_2}, \tag{3.61}$$

and $\cos 2\pi f_0 = \frac{a}{\sqrt{a^2+b^2}} = \frac{\phi_1}{2\sqrt{-\phi_2}}$. Substituting into (3.60), we get

$$\rho_k = \left(\sqrt{-\phi_2}\right)^k \left\{ \frac{\sin(k+1)2\pi f_0 + \phi_2 \sin(k-1)2\pi f_0}{(1-\phi_2)\sin 2\pi f_0} \right\}.$$

Letting ψ be such that $\tan \psi = \frac{1-\phi_2}{1+\phi_2} \tan 2\pi f_0$, we can rewrite ρ_k as

$$\rho_k = \left(\sqrt{-\phi_2}\right)^k \left\{ \frac{\sin(2\pi k f_0 + \psi)}{\sin \psi} \right\}, \tag{3.62}$$

where $\cos 2\pi f_0 = \phi_1/\left(2\sqrt{-\phi_2}\right)$. That is, the second factor of (3.62) is periodic with frequency

$$f_0 = \frac{1}{2\pi} \cos^{-1}\left(\frac{\phi_1}{2\sqrt{-\phi_2}} \right). \tag{3.63}$$

So, ρ_k is a damped sinusoidal function, and we call f_0 the *natural frequency* or *system frequency*.

Case 2: Roots are equal

In order for the roots of the characteristic equation to be equal they must be real, and the characteristic polynomial has the factored form $\phi(z) = \left(1 - r_1^{-1} z\right)^2$. From (3.47), ρ_k is given by

$$\rho_k = C_{10} r_1^{-k} + C_{11} k r_1^{-k},$$

and graphically ρ_k has an appearance similar to that for an AR(1).

3.2.7.2 Spectral Density of an AR(2)

The spectral density of an AR(2) model is given by (3.54) to be

$$\begin{aligned} S_X(f) &= \frac{\sigma_a^2}{\sigma_X^2 |1 - \phi_1 e^{-2\pi i f} - \phi_2 e^{-4\pi i f}|^2} \\ &= \frac{\sigma_a^2}{\sigma_X^2 \{1 + \phi_1^2 + \phi_2^2 - 2\phi_1(1-\phi_2)\cos 2\pi f - 2\phi_2 \cos 4\pi f\}}. \end{aligned} \tag{3.64}$$

In the case of complex roots sufficiently close to the unit circle, the peak in $S_X(f)$ occurs near, but not exactly at the system frequency f_0 (see Problem 3.20). Also, if the characteristic equation has at least one real root sufficiently close to the unit circle, then $S(f)$ will have a "peak" at $f=0$ if the root is positive, while $S(f)$ will have a peak at $f=0.5$ if the root is negative.

3.2.7.3 Stationary/Causal Region of an AR(2)

In order for the AR(2) process to be causal and stationary, it must be the case that $|r_1|>1$ and $|r_2|>1$. Clearly, this imposes the constraint that $\phi(z) \neq 0$ for $z=1$, that is, it follows that $1-\phi_1-\phi_2 \neq 0$. However, this condition is not sufficient to guarantee causality and stationarity. Schur's lemma (see Appendix 3.A) gives necessary and sufficient conditions for the roots of a polynomial equation to all lie outside the unit circle. The matrix C described in that lemma in the case of a second-order model can be shown to be

$$C = \begin{bmatrix} 1-\phi_2^2 & -\phi_1(1+\phi_2) \\ -\phi_1(1+\phi_2) & 1-\phi_2^2 \end{bmatrix} \tag{3.65}$$

(see Problem 3.22(a)). By definition, C is positive definite if and only if $1-\phi_2^2>0$ and $(1+\phi_2)^2\{(1-\phi_2)^2-\phi_1^2\}>0$. Thus, it can be shown (Problem 3.22(b)) that the AR(2) process is stationary if and only if ϕ_1 and ϕ_2 satisfy

$$\begin{aligned} |\phi_2| &< 1 \\ \phi_1+\phi_2 &< 1 \\ \phi_2-\phi_1 &< 1. \end{aligned}$$

3.2.7.4 ψ-Weights of an AR(2) Model

Factoring the AR(2) model gives

$$X_t-\phi_1X_{t-1}-\phi_2X_{t-2}=(1-\alpha_1B)(1-\alpha_2B)X_t=a_t, \tag{3.66}$$

where the α_i may be real or complex numbers with $|\alpha_i|<1, i=1,2$. Then, from (3.26) and (3.27)

$$X_t=\sum_{k=0}^{\infty}\psi_k(2)a_{t-k},$$

where

$$\psi_k(2) = \frac{\alpha_2^{k+1} - \alpha_1^{k+1}}{\alpha_2 - \alpha_1}.$$

Now suppose that r_1 and r_2 are complex conjugate roots of the characteristic equation associated with the AR(2) model in (3.66). Then, $\alpha_1 = r_1^{-1} = (a + bi)^{-1}$ and $\alpha_2 = r_2^{-1} = (a - bi)^{-1}$, so it follows that

$$\alpha_1 = \frac{a - bi}{a^2 + b^2}$$

$$\alpha_2 = \frac{a + bi}{a^2 + b^2},$$

where $b \neq 0$. Now letting $R = \sqrt{a^2 + b^2}$, then $\alpha_1 = R^{-1}e^{-i\theta}$ and $\alpha_2 = R^{-1}e^{i\theta}$, where $\theta = \arctan(b/a)$. Then,

$$\begin{aligned}\alpha_2 - \alpha_1 &= 2R^{-1}\left[\frac{e^{i\theta} - e^{-i\theta}}{2}\right] \\ &= 2iR^{-1}\sin\theta,\end{aligned}$$

and

$$\begin{aligned}\psi_k(2) &= R^{-k-1}\left[\frac{e^{i\theta(k+1)} - e^{-i\theta(k+1)}}{2iR^{-1}\sin\theta}\right] \\ &= R^{-k}\frac{\sin\theta(k+1)}{\sin\theta}.\end{aligned}$$

Therefore,

$$X_t = \sum_{k=0}^{\infty} R^{-k}\frac{\sin\theta(k+1)}{\sin\theta}a_{t-k}. \tag{3.67}$$

Example 3.5 AR(2) Models

(a) *Characteristic equation has two positive real roots*
In Figures 3.4a and 3.5a, the autocorrelations and a realization from the process

$$(1 - 0.95B)X_t = a_t \tag{3.68}$$

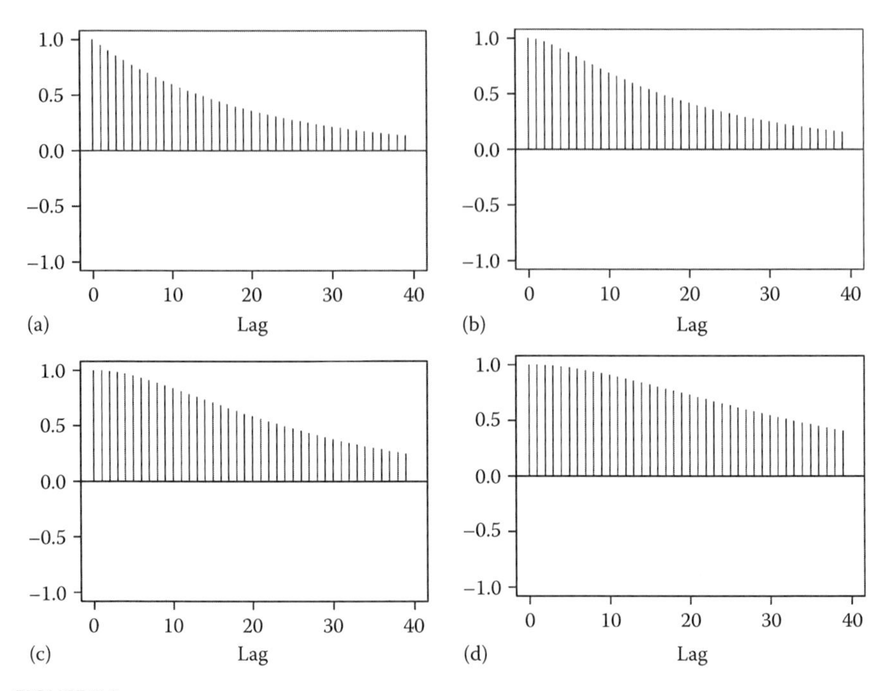

FIGURE 3.8
Autocorrelations of the models $(1 - 0.95B)(1 - \lambda B)X_t = a_t$, (a–d) $\lambda = 0, 0.7, 0.9, 0.95$.

were given. In considering the AR(2) process, it is interesting to examine the effect of introducing an additional first-order factor to the model in (3.68). Figure 3.8 shows autocorrelations for the models

$$(1 - 0.95B)(1 - \lambda B)X_t = a_t, \tag{3.69}$$

for $\lambda =$ 0, 0.7, 0.9, and 0.95, while Figure 3.9 shows realizations of length $n = 100$ with $\sigma_a^2 = 1$ from each model. When $\lambda = 0$ the model in (3.69) is an AR(1) model, and Figures 3.8a and 3.9a were previously shown in Figures 3.4a and 3.5a, respectively, although Figure 3.9a is plotted on a different scale than Figure 3.5a. The realizations in Figure 3.9 were each generated with $\sigma_a^2 = 1$ and using the same seed for the random number generator, so that differences among the realizations are due strictly to the differing autocorrelation functions and process variances. It should be noticed that the existence of the second real root strengthens the behavior seen in Figures 3.4a and 3.5a for the AR(1) model. For example, as λ increases, the autocorrelation functions damp more slowly, and therefore the realizations have more pronounced runs on the vertical axis and a resulting higher variance. The process standard deviations are $\sigma_X = 3.2, 10.0, 26.3$, and 45.3, respectively, and if we consider X_t as

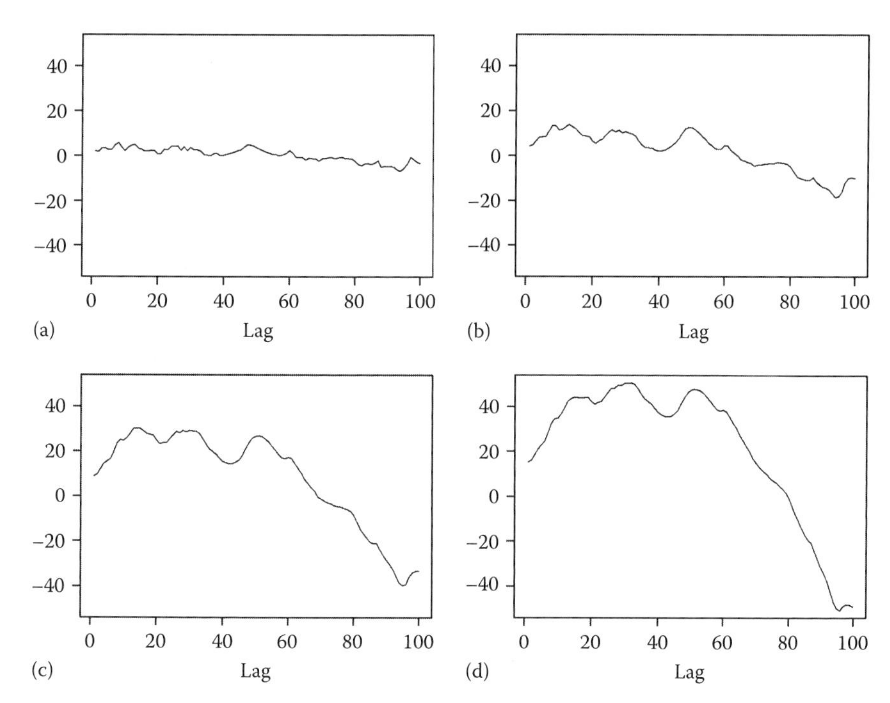

FIGURE 3.9
Realizations from the models $(1-0.95B)(1-\lambda B)X_t = a_t$, (a–d) $\lambda = 0, 0.7, 0.9, 0.95$.

"signal" and a_t as "noise," then the signal-to-noise ratio, σ_X^2/σ_a^2 increases as λ increases. Also note that when $\lambda = 0.95$, the model has repeated roots. The examples given here show behavior of the realizations and autocorrelations as $\lambda \to 0.95$. Clearly, the behavior of the autocorrelations and realizations for the case of equal roots is similar to the behavior when the roots are unequal yet close together. In this book, we typically consider cases in which the roots are not repeated (except for certain seasonal models).

The spectral densities for these processes (not shown) all have a peak at $f = 0$ (that becomes more sharp as λ increases) indicating the lack of periodic behavior. The oscillatory nature of the realizations is essentially due to correlation and is in no way systematic. An AR(2) process cannot have any true "seasonal" or periodic nature unless the characteristic equations has complex roots or at least one negative root.

(b) *Characteristic equation has two negative real roots*
In Figures 3.4b through 3.6b, we demonstrated the periodic nature of a realization from $(1+0.95B)X_t = a_t$. For the AR(2) model

$$(1+0.95B)(1+\lambda B)X_t = a_t, \quad \lambda > 0,$$

the situation is essentially the same as in the case of (3.69) in the sense that the basic oscillating behavior of the process is retained as in the case $(1 + 0.95B)X_t = a_t$, but the variance of the process and the "signal-to-noise ratio" increase as $\lambda \to 1$. In all these cases, the spectral densities have a peak at $f = 0.5$.

(c) *Characteristic equation has a positive and negative real root*
In Figure 3.10, a realization of length $n = 100$, the autocorrelations, and the spectral density are shown for the AR(2) process $(1 - 0.95B)(1 + 0.7B)X_t = a_t$. In this case, the roots of the characteristic equation ($1/0.9 = 1.11$ and $-1/0.7 = -1.43$) are of opposite signs, the realization shows a mixture of wandering and oscillating behavior, the autocorrelations have a damped exponential appearance with a slight oscillation visible at the early lags, and the spectral density has peaks at $f = 0$ and $f = 0.5$. The positive real root, 1.11, is closer to the unit circle than the negative one, and, consequently, the peak in the spectral density at $f = 0$ is higher than at $f = 0.5$. Again, in order for an AR(2) process to have any concentrated periodic or seasonal nature with a period longer than two units, it is necessary for the characteristic equation to have complex roots.

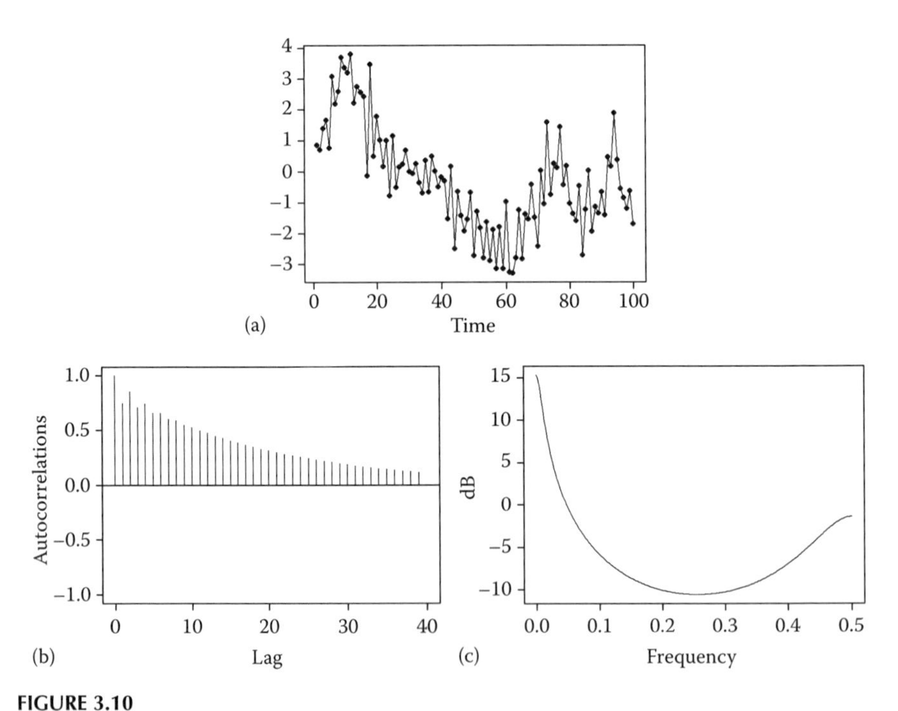

FIGURE 3.10
(a) Realization, (b) autocorrelations, and (c) spectral density for the model $(1 - 0.95B)(1 + 0.7B)X_t = a_t$.

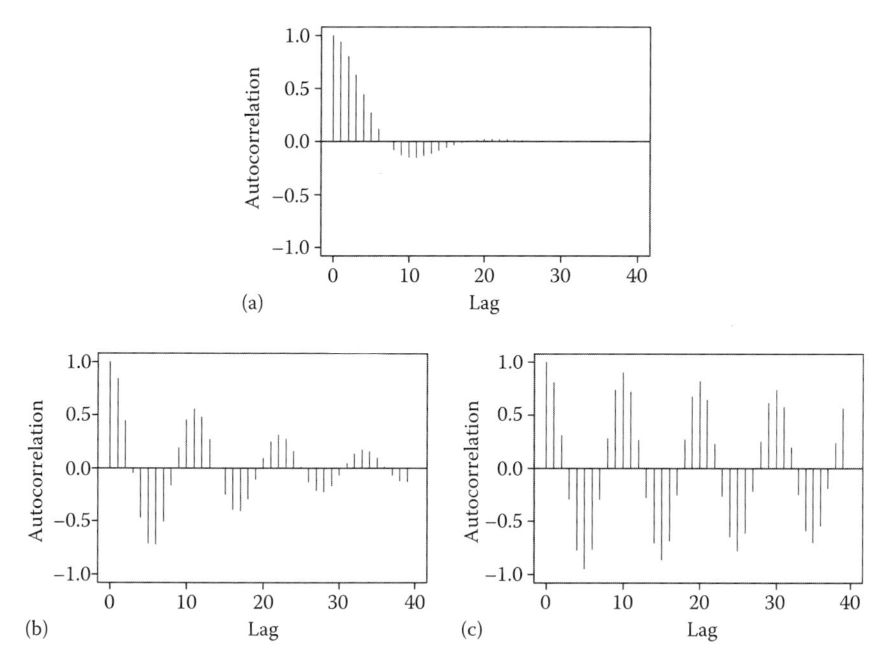

FIGURE 3.11
Autocorrelations for the AR(2) models $X_t - 1.6X_{t-1} - \phi_2 X_{t-2} = a_t$, (a–c) $\phi_2 = -0.7, -0.9, -0.98$.

(d) *Characteristic equation has complex roots*
Figure 3.11 shows autocorrelations for the AR(2) processes

$$\left(1 - 1.6B - \phi_2 B^2\right)X_t = a_t \tag{3.70}$$

for $\phi_2 = -0.7, -0.9$, and -0.98, for which the associated characteristic equation is $1 - 1.6z - \phi_2 z^2 = 0$. This equation has complex roots for all $\phi_2 < -0.64$. By (3.63), the system frequency associated with the operator $1 - 1.6B - \phi_2 B^2$ is

$$f_0 = \frac{1}{2\pi}\cos^{-1}\left(\frac{1.6}{2\sqrt{-\phi_2}}\right). \tag{3.71}$$

Therefore, if $\phi_2 = -1$, then the associated frequency is $f_0 = 0.1024$. Notice that changing ϕ_2 in (3.71) while keeping $\phi_1 = 1.6$ fixed changes the associated system frequency. For the AR(2) models in (3.70), we obtain $f_0 = 0.05,\ 0.09$, and 0.10 for $\phi_2 = -0.7, -0.9$, and -0.98, respectively. The strong periodic nature of the autocorrelations is clear when ϕ_2 is close to -1, and, in this example, it diminishes as $\phi_2 \to -0.7$. Realizations of length $n = 200$ from (3.70) are shown in Figure 3.12, where the effects of the differing autocorrelation

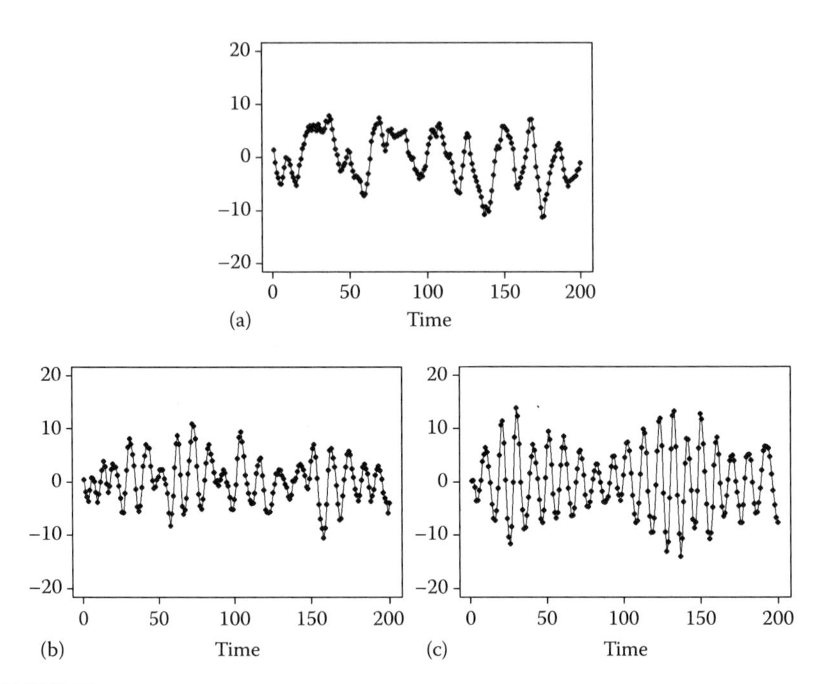

FIGURE 3.12
Realizations from the AR(2) models $X_t - 1.6X_{t-1} - \phi_2 X_{t-2} = a_t$, (a–c) $\phi_2 = -0.7, -0.9, -0.98$.

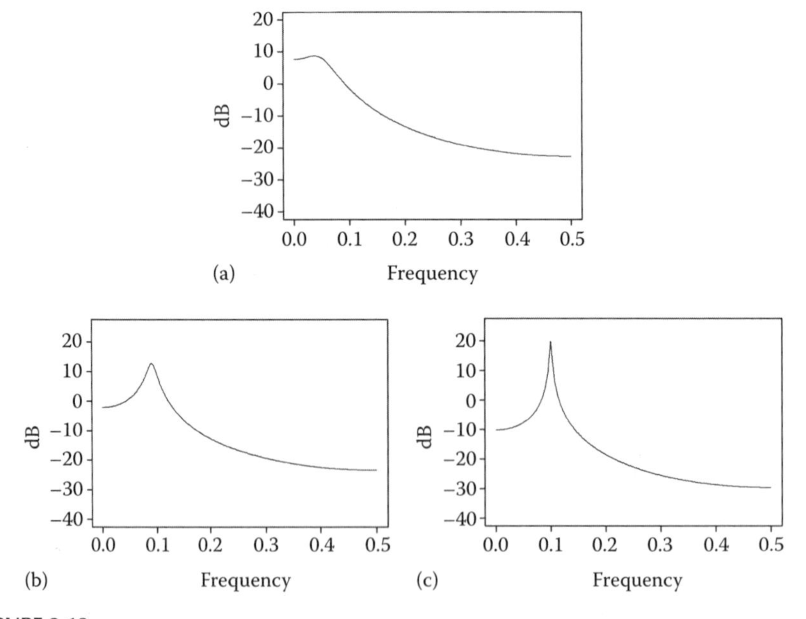

FIGURE 3.13
Spectral densities from the AR(2) models $X_t - 1.6X_{t-1} - \phi_2 X_{t-2} = a_t$, (a–c) $\phi_2 = -0.7$, -0.9, -0.98.

structures can be seen. Specifically, as $\phi_2 \to -1$ the periodicity becomes much more apparent in the realizations, the period becomes shorter, and the process variance increases. In Figure 3.13, the corresponding spectral densities are displayed where it is seen that as $\phi_2 \to -1$, the peak of the spectral density becomes sharper.

As discussed previously, the characteristic equation associated with the AR(2) process $(1 - \phi_1 B - \phi_2 B^2)X_t = a_t$ has complex roots if and only if $\phi_1^2 + 4\phi_2 < 0$. Therefore, it is necessary that $\phi_2 < 0$ in order for the process to have a characteristic equation with complex roots. Moreover, for negative ϕ_2, since $\phi_2 = -(a^2 + b^2)^{-1}$, the process is nonstationary with roots on the unit circle if and only if $\phi_2 = -1$. The behavior shown in this example as roots approach the unit circle is typical.

One final comment about AR(2) models associated with complex conjugate roots of the characteristic equation is that if the system frequency is sufficiently removed from 0 or 0.5, the spectral density will have a peak at $f_s = \frac{1}{2\pi}\cos^{-1}\left(\frac{\phi_1(\phi_2 - 1)}{4\phi_2}\right)$ (see Problem 3.20). It is interesting to note that $f_s \neq f_0$ but as $\phi_2 \to -1$, then $f_s \to f_0$. That is, the peak in the spectral density occurs at a frequency close to but not equal to the system frequency f_0.

3.2.8 Summary of AR(1) and AR(2) Behavior

Before proceeding to further examination of general AR(p) models, we summarize the information about AR(1) and AR(2) models:

(a) *Facts about the stationary AR(1) model* $X_t - \phi_1 X_{t-1} = a_t$.
In this case, the characteristic equation is given by $1 - \phi_1 z = 0$, so that the root, r_1, of the characteristic equation is given by $r_1 = \phi_1^{-1}$. As a consequence, an AR(1) process is stationary if and only if $|\phi_1| < 1$. It is instructive to note that since $\left|r_1^{-1}\right| = |\phi_1|$, the proximity of the root to the unit circle can be determined by simple examination of ϕ_1.

(i) The autocorrelation function of a stationary AR(1) process is given by $\rho_k = \phi_1^k$ for $k \geq 0$. That is, in (3.49), $C_1 = 1$ and $r_1 = \phi_1^{-1}$. The autocorrelation function is a damped exponential if $\phi_1 > 0$ and an oscillating damped exponential if $\phi_1 < 0$.

(ii) As a result of the autocorrelations in (i), realizations from an AR(1) model with $\phi_1 > 0$ tend to be aperiodic with a general "wandering" behavior. When $\phi_1 < 0$ the realizations will tend to oscillate back and forth across the mean.

(iii) The spectral density $S_X(f)$ has a peak at $f = 0$ if $\phi_1 > 0$ and at $f = 0.5$ if $\phi_1 < 0$

(b) *Facts about a stationary AR(2) model* $X_t - \phi_1 X_{t-1} - \phi_2 X_{t-2} = a_t$.
In this case, the characteristic equation is $1 - \phi_1 z - \phi_2 z^2 = 0$, and the features of the AR(2) process depend on the nature of the roots of this quadratic equation.

Case 1: The roots of the characteristic equation are complex.
From (3.61), it follows that $|r_k^{-1}| = \sqrt{-\phi_2}$ where r_k, $k = 1, 2$ are complex conjugate roots of the characteristic equation. This provides a useful measure of how close complex roots are to the unit circle.

(i) In the complex roots case, the general solution in (3.49) satisfied by ρ_k has two terms, $C_2 = C_1^*$, and $r_2 = r_1^*$, where z^* denotes the complex conjugate of z. The autocorrelation function is a damped sinusoidal of the form

$$\rho_k = d\left(\sqrt{-\phi_2}\right)^k \sin(2\pi f_0 k + \psi), \qquad (3.72)$$

where d and ψ are real constants and system frequency $f_0 = \frac{1}{2\pi}\cos^{-1}\left(\frac{\phi_1}{2\sqrt{-\phi_2}}\right)$.

(ii) As a result of (i), if the roots of the characteristic equation are sufficiently close to the unit circle, then realizations from an AR(2) model in this case are pseudo-cyclic with frequency f_0 given in (i), that is, with period $1/f_0$.

(iii) If the system frequency is sufficiently removed from 0 or 0.5, the spectral density will have a peak near f_0 given in (i).

Case 2: The roots of the characteristic equation are real.
In this case the roots, r_1 and r_2, of the characteristic equation are real, and the characteristic equation factors as $(1 - r_1^{-1}z)(1 - r_2^{-1}z) = 0$. Plots of ρ_k will have the appearance of a damped exponential or a mixture of damped exponentials and oscillating exponentials. Realizations from these models will have a mixture of the behaviors induced by their autocorrelations. Depending on the signs of r_1 and r_2, the spectral density will have peaks at $f = 0$, at $f = 0.5$, or at both $f = 0$ and $f = 0.5$.

3.2.9 AR(p) Model

The general behavior of the autocorrelation function (and to a large degree the time series itself) of an AR(p) model can be understood by examining the roots of the characteristic equation. The information summarized earlier for the AR(1) and AR(2) models provides valuable information concerning the characteristics of higher-order AR(p) models. We begin our discussion of AR(p) models by considering two examples.

Example 3.6 An AR(3) Model with a Positive Real Root and Complex Conjugate Roots

Consider the AR(3) model

$$X_t - 2.55X_{t-1} + 2.42X_{t-2} - 0.855X_{t-3} = a_t. \tag{3.73}$$

For this model, the characteristic polynomial $1 - 2.55z + 2.42z^2 - 0.855z^3$ can be factored as $(1 - 0.95z)(1 - 1.6z + 0.9z^2)$, that is, as the product of a linear factor and an irreducible quadratic factor. From this factorization, it follows that the roots of the characteristic equation are $r_1 = 1/0.95 = 1.05$ along with the pair of complex conjugate roots $0.89 \pm 0.57i$. For each of the three roots of the characteristics equation, we have $|r^{-1}| = \sqrt{0.9}$. Based on the factorization of the characteristic equation, it is often useful to write the model in (3.73) in the factored form

$$(1 - 0.95B)(1 - 1.6B + 0.9B^2)X_t = a_t. \tag{3.74}$$

From our discussion of AR(1) and AR(2) models, we know that an AR(1) model based on the first-order factor in (3.74), that is, the model $(1 - 0.95B)Y_t = a_t$, has: (a) realizations that display a wandering or piecewise linear behavior as seen in Figure 3.5a, (b) autocorrelations that are damped exponentials as in Figure 3.4a, and (c) a peak in the spectral density at $f = 0$ as seen in Figure 3.6a. Since the roots of $1 - 1.6z + 0.9z^2 = 0$ are complex, it follows from (3.63) that an AR(2) model based on the second-order factor, $1 - 1.6B + 0.9B^2$, will have: (a) realization that are pseudo-periodic with period of about 11 (1/0.09), (b) autocorrelation with a damped sinusoidal behavior, and (c) a spectral density with a peak near $f_0 = 0.09$. These behaviors are seen in Figures 3.11b, 3.12b, and 3.13b, respectively.

In Figure 3.14, we show a realization of length $n = 200$, where $\sigma_a^2 = 1$, true autocorrelations, and spectral density for the AR(3) model given in (3.73). The AR(1) and AR(2) behaviors shown in the factored form in (3.74) are evident in the realization where we see a wandering behavior (due to the $1-0.95B$ factor) along with a periodic behavior with a period of approximately 11 units induced by the irreducible quadratic factor $1 - 1.6B + 0.9B^2$. These two components of the model can be seen in the autocorrelations where ρ_k has a damped exponential behavior "superimposed" with a damped sinusoid (with a period of about 11 lags) and in the spectral density, which has peaks at $f = 0$ and at about $f = 0.09$.

Example 3.7 An AR(3) Model with a Negative Real Root and Complex Conjugate Roots

In Figure 3.15, we show a realization, where $\sigma_a^2 = 1$, true autocorrelations, and the true spectral density for the AR(3) model

$$X_t - 0.65X_{t-1} - 0.62X_{t-2} + 0.855X_{t-3} = a_t, \tag{3.75}$$

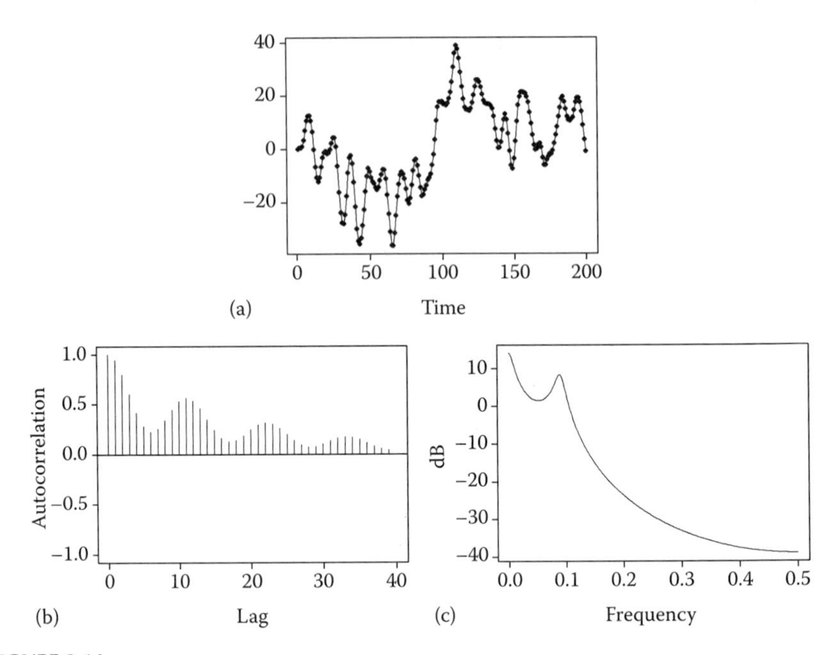

FIGURE 3.14
(a) Realization, (b) autocorrelations, and (c) spectral densities for the AR(3) model $(1 - 0.95B)\,(1 - 1.6B + 0.9B^2)X_t = a_t$.

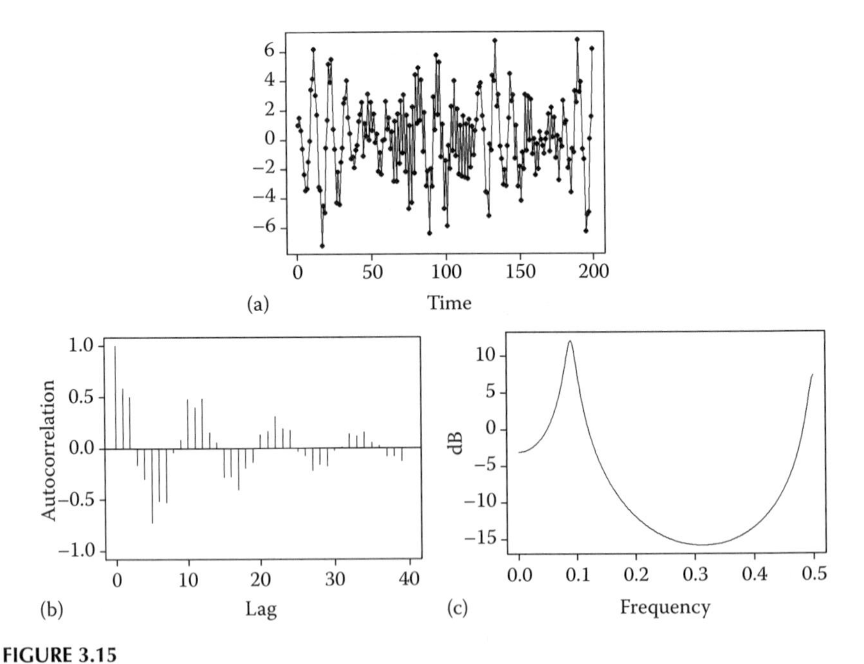

FIGURE 3.15
(a) Realization, (b) autocorrelations, and (c) spectral densities for the AR(3) model $(1 + 0.95B)\,(1 - 1.6B + 0.9B^2)X_t = a_t$.

which can be written in factored form as $(1 + 0.95B)(1 - 1.6B + 0.9B^2)X_t = a_t$. The factored form of the model in (3.75) has the same second-order factor that was in model (3.73) with the difference being that the characteristic equation associated with (3.75) has a negative real root $r_1 = 1/(-0.95) = -1.05$. Again, from the discussion of AR(1) and AR(2) models, we know that an AR(1) model based on the first-order factor $1 + 0.95B$ has: (a) realizations that display a rapidly oscillating behavior as seen in Figure 3.5b, (b) autocorrelations that are oscillating damped exponentials as in Figure 3.4b, and (c) a peak in the spectral density at $f = 0.5$ as seen in Figure 3.6b. As mentioned in the discussion of (3.73), the AR(2) model based on the second-order factor $1 - 1.6B + 0.9B^2$ will have: (a) realizations that are pseudo-periodic in nature with period about 11, (b) autocorrelations that are a damped sinusoidal with a period of about 11 lags, and (c) a spectral density with a peak near $f_0 = 0.09$. Examination of Figure 3.15 shows that the AR(3) realization based on (3.75) has pseudo-periodic behavior with a period of about 11 along with a high-frequency oscillation behavior associated with the negative real root. The autocorrelations primarily show the damped sinusoidal behavior with mild evidence of the oscillating behavior based on the negative real root. However, the spectral density clearly shows peaks at about $f = 0.09$ and $f = 0.5$. The behavior related to the two frequencies in the data is much more visible in the spectral density than in the autocorrelations. This illustrates the point made in Example 1.11 that the spectral density is a better tool than the autocorrelations for detecting and measuring cyclic behavior in data.

3.2.10 AR(1) and AR(2) Building Blocks of an AR(p) Model

In Examples 3.6 and 3.7, we were able to determine the characteristics of the AR(3) models by examining the first and second-order factors of the model. These examples suggest that the key to understanding AR(p) models is to recognize that the AR(1) and AR(2) models serve as "building blocks." The pth-order characteristic equation $\phi(z) = 1 - \phi_1 z - \cdots - \phi_p z^p = 0$ will have real roots, complex roots (which occur as complex conjugates), or a combination of real and complex roots. As a consequence, $\phi(z)$ can be factored as

$$\phi(z) = \prod_{j=1}^{m} \left(1 - \alpha_{1j}z - \alpha_{2j}z^2\right) \tag{3.76}$$

where there are two types of factors.

(i) Factors associated with real roots (linear factors)
Factors of (3.76) associated with real roots are linear factors, that is, $\alpha_{2j} = 0$. The absolute value of the reciprocal of the associated root is

$|r_j^{-1}| = |\alpha_{1j}|$, and, thus, it is necessary that $|\alpha_{1j}| < 1$ in order for the associated root to be outside the unit circle. "System frequencies," f_0, for these real roots are assigned as

$$\begin{aligned} f_0 &= 0 \quad \text{if } \alpha_{1j} > 0 \\ &= 0.5 \quad \text{if } \alpha_{1j} < 0. \end{aligned}$$

(ii) Factors associated with complex roots (quadratic factors)
In this case the factor $1 - \alpha_{1j}z - \alpha_{2j}z^2$ is an irreducible quadratic factor, that is, $a_{1j}^2 + 4a_{2j} < 0$, and the two roots associated with this factor are complex conjugate pairs. For measuring the proximity of a root to the unit circle, it is useful to note that $|r_j^{-1}| = \sqrt{-\alpha_{2j}}$. Recall that $|r_j^{-1}| < 1$ for a stationary process, and the closer $|r_j^{-1}|$ is to 1 the closer these roots are to the nonstationary region. The system frequency associated with the quadratic factor $1 - a_{1j}z - a_{2j}z^2$ is

$$f_{0j} = \frac{1}{2\pi}\cos^{-1}\left(\frac{a_{1j}}{2\sqrt{-a_{2j}}}\right).$$

Thus, an AR(p) model is fundamentally a combination of first and second-order components. Specifically, ρ_k for an AR(p) process behaves as a mixture of damped exponentials and/or damped sinusoids. Each $j = 1, \ldots, m$ in (3.76) corresponds to either a real root or a complex conjugate pair of roots. Without loss of generality, we let $j = 1, \ldots, m_r$ correspond to the real roots and $j = m_r + 1, \ldots, m$ to the pairs of complex conjugate roots. A useful reformulation of the general expression in (3.49) for the autocorrelation of an AR(p) (with no repeated roots of the characteristic equation) can be rewritten as

$$\begin{aligned} \rho_k &= \sum_{j=1}^{m_r} C_j\left(r_j^{-1}\right)^k + \sum_{j=m_r+1}^{m}\left(C_{j1}\left(r_j^{-1}\right)^k + C_{j2}\left(r_j^{*-1}\right)^k\right) \\ &= \sum_{j=1}^{m_r} C_j\alpha_{1j}^k + \sum_{j=m_r+1}^{m} D_j\left(\sqrt{-\alpha_{2j}}\right)^k \sin(2\pi f_{0j}k + \psi_j), \end{aligned} \tag{3.77}$$

where

$$f_{0j} = \frac{1}{2\pi}\cos^{-1}\left(\frac{\alpha_{1j}}{2\sqrt{-\alpha_{2j}}}\right), \tag{3.78}$$

and where $C_j, j = 1, \ldots, m_r$ and D_j and $\psi_j, j = m_r + 1, \ldots, m$ are real constants. Equation 3.77 shows that ρ_k for an AR(p) process behaves like a mixture of damped exponentials and/or damped sinusoids. Thus, the factors of $\phi(z)$ determine the correlation structure of an AR(p) as well as the behavior of its realizations. The understanding of these factors is often critical to understanding the underlying physical phenomenon.

Example 3.8 Autocorrelation Function for Examples 3.6 and 3.7

Based on (3.77), the autocorrelation function for the AR(3) model of Example 3.6 is

$$\rho_k = 0.66(0.95)^k + .3518(0.95)^k \sin(2\pi f_0 k + 1.31), \tag{3.79}$$

where $f_0 = 0.09$. From (3.79), it can be seen that the expression for the autocorrelation function is a mixture of a damped exponential and a damped sinusoidal components which is consistent with our observations concerning Figure 3.14b.

The autocorrelation function for the AR(3) model in Example 3.7, which has a negative real root and the same second-order factor as the model in Example 3.6, is given by

$$\rho_k = 0.139(-0.95)^k + 0.862(0.95)^k \sin(2\pi f_0 k + 1.5), \tag{3.80}$$

where again $f_0 = 0.09$. Note that the coefficient (0.139) of the oscillating damped exponential term in (3.80) is relatively small causing the damped oscillating exponential behavior to not be very visible in a plot of the autocorrelations as was noted in the discussion of Figure 3.15b.

3.2.11 Factor Tables

It is useful to summarize information regarding the first and second-order factors of an AR(p) model, in a *factor table*. For each first-order or irreducible second-order factor of the model the factor table displays the

(i) Roots of the characteristic equation
(ii) Proximity of the roots to the unit circle (by tabling the absolute reciprocal of the roots)
(iii) System frequencies f_{0j}

Factors in the factor table are listed in decreasing order of the proximity of the associated roots to the unit circle since the closer to the unit circle, the stronger the autocorrelation effect. See Woodward et al. (2009).

Example 3.9 Factor Tables for Examples 3.6 and 3.7

The information discussed regarding the AR(3) model in Example 3.6 (Equation 3.73) is summarized in the factor table in Table 3.1. Similarly, Table 3.2 is a factor table describing the factors of model (3.75) in Example 3.7.

Examination of these tables shows:

(a) The processes are stationary AR(3) models, since the roots are outside the unit circle (i.e., the absolute reciprocals of the roots are all less than 1)

(b) In each model, the three roots are all equidistant from the unit circle $\left(\text{i.e., } |r_j^{-1}| = 0.95\right)$

(c) The system frequencies associated with the two factors.

This type of presentation will be used throughout this book. We note that for higher-order models, the factorization must be done numerically.

Using `GW-WINKS`: For a given AR(p) model, you can obtain a factor table by selecting the `Factoring Routines` option in the `Time Series Analysis` menu and selecting `Proceed No Data` (if asked that question). As you enter the coefficients for the model, be careful with signs. For example, the factor table for the AR(3) model in (3.73) would be obtained by specifying the parameter values $\phi_1 = 2.55$, $\phi_2 = -2.42$, and $\phi_3 = 0.855$.

Example 3.10 An AR(4) Model

Consider the AR(4) process given by

$$X_t - 1.15X_{t-1} + 0.19X_{t-2} + 0.64X_{t-3} - 0.61X_{t-4} = a_t. \tag{3.81}$$

TABLE 3.1

Factor Table for the AR(3) Model in Example 3.6

AR Factors	Roots (r_j)	$\lvert r_j^{-1}\rvert$	f_{0j}
$1 - 1.6B + 0.9B^2$	$0.89 \pm 0.57i$	0.95	0.09
$1 - 0.95B$	1.053	0.95	0

TABLE 3.2

Factor Table for the AR(3) Model in Example 3.7

AR Factors	Roots (r_j)	$\lvert r_j^{-1}\rvert$	f_{0j}
$1 - 1.6B + 0.9B^2$	$0.89 \pm 0.57i$	0.95	0.09
$1 + 0.95B$	-1.053	0.95	0.5

TABLE 3.3

Factor Table for the AR(4) Model in Example 3.10

AR Factors	Roots (r_j)	$\lvert r_j^{-1} \rvert$	f_{0j}
$1 - 0.95B$	1.05	0.95	0
$1 - B + 0.8B^2$	$0.62 \pm 0.93i$	0.90	0.16
$1 + 0.8B$	−1.25	0.80	0.5

Although inspecting the coefficients in (3.81) tells us nothing about the nature of the model, much can be learned about the model characteristics by examining the factors. The characteristic polynomial can be factored as

$$\begin{aligned}\phi(z) &= 1 - 1.15z + 0.19z^2 + 0.64z^3 - 0.61z^4 \\ &= (1 - 0.95z)(1 + 0.8z)(1 - z + 0.8z^2).\end{aligned}$$

Information concerning the factors of $\phi(z)$ is given in the factor table in Table 3.3.

Since $\lvert r_j^{-1} \rvert < 1$ for all roots, the model is causal and stationary. In the factor table, it can be seen that there are two real roots, one positive and one negative, and a complex conjugate pair of roots. The positive real root is associated with damped exponential autocorrelations, wandering behavior in the realizations, and a peak in the spectral density at $f = 0$. The negative real root produces oscillating damped exponential autocorrelations, oscillating behavior in realizations, and a peak in the spectral density at $f = 0.5$. The complex conjugate roots are related to damped sinusoidal behavior in the autocorrelations with a system frequency (using (3.63)) of $f = 0.16$, pseudo-periodic behavior in the data with periods of about 6 (i.e., $1/0.16$), and a peak in the spectral density at about $f = 0.16$.

In Figure 3.16, we show a realization of length $n = 200$ where $\sigma_a^2 = 1$, autocorrelations, and the spectral density for the AR(4) model in (3.81). The autocorrelations have a damped sinusoidal appearance (with period about 6) along a damped exponential path with very little evidence of the 0.5 frequency behavior associated with the factor $(1 + 0.8B)$. The realization shows a pseudo-cyclic behavior of period about six produced by the factor $(1 - B + 0.8B^2)$ along a wandering path produced by the $(1 - 0.95B)$ factor. The realization shows very slight evidence of the high-frequency behavior associated with the factor $1 + 0.8B$. The spectral density shows peaks at about $f = 0$, 0.16, and 0.5 as would be expected based on the factor table.

It is very important to realize that the behavior associated with roots closest to the unit circle tends to be dominant. In this example, the root 1.05 associated with the factor $(1 - 0.95B)$ is the closest to the unit circle while the root -1.25 associated with $(1 + 0.8B)$ is the furthest removed from the unit circle. We have already noted that the behavior associated with the factor $1 - 0.95B$ is clearly evident in the realization and the autocorrelations while

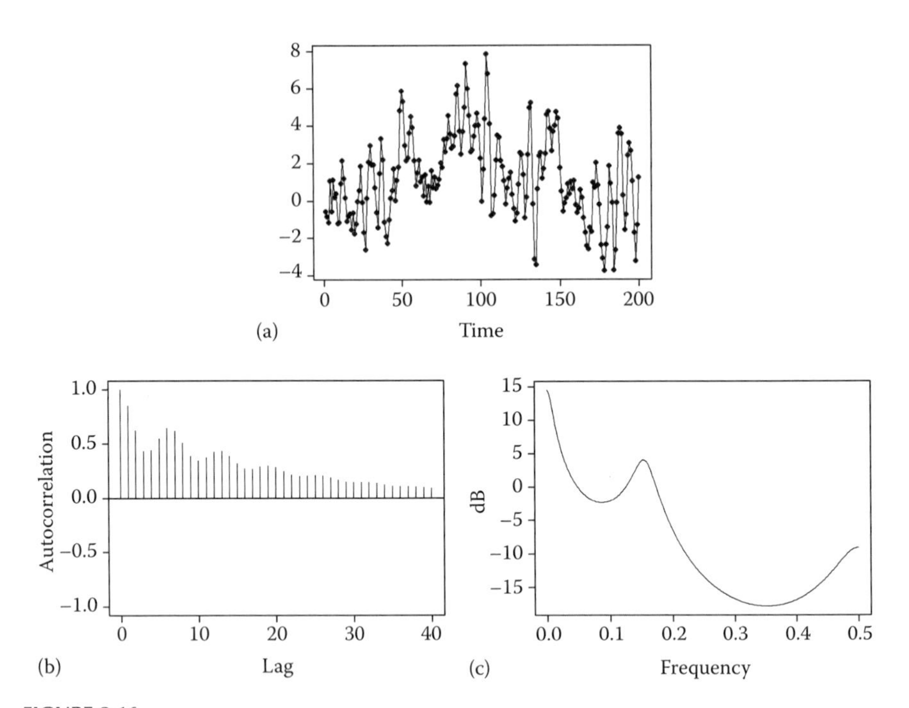

FIGURE 3.16
(a) Realization, (b) autocorrelations, and (c) spectral densities for the AR(4) model $(1-0.95B)(1+0.8B)(1-B+0.8B^2)X_t = a_t$.

that associated with $1 + 0.8B$ is not apparent. In the spectral density, we see peaks close to all three system frequencies, but the peak at $f = 0$ is much higher than is the peak at $f = 0.5$ and $f = 0.16$. We will explore the domination of roots close to the unit circle further in Example 3.12 and in Chapter 5.

Example 3.11 An AR(4) Model with All Roots Complex

We next consider the AR(4) model

$$X_t - 0.3X_{t-1} - 0.9X_{t-2} - 0.1X_{t-3} + 0.8X_{t-4} = a_t,$$

whose factor table is given in Table 3.4. In the factor table, it can be seen that the characteristic equation has two pairs of complex conjugate roots, one

TABLE 3.4

Factor Table for the AR(4) Model in Example 3.11

AR Factors	Roots (r_j)	$\lvert r_j^{-1} \rvert$	f_{0j}
$1-1.8B+0.95B^2$	$0.95 \pm 0.39i$	0.97	0.06
$1+1.5B+0.85B^2$	$-0.89 \pm 0.63i$	0.92	0.4

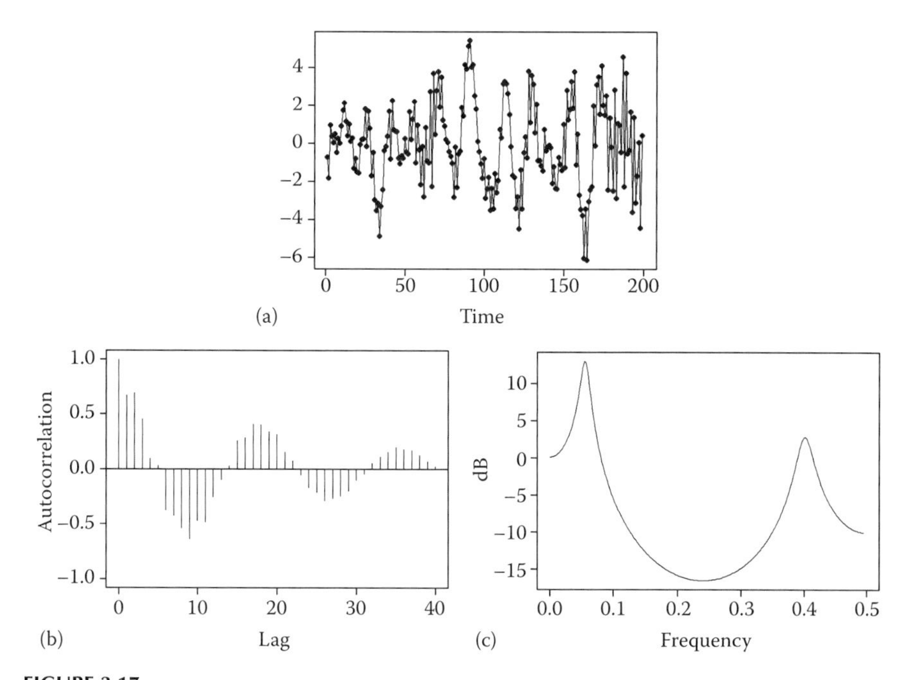

FIGURE 3.17
(a) Realization, (b) autocorrelations, and (c) spectral densities for the AR(4) model $(1 - 1.8B + 0.95B^2)(1 + 1.5B + 0.85B^2)X_t = a_t$.

associated with a system frequency of $f = 0.06$ and the other with a frequency of $f = 0.4$. It can also be seen that the roots associated with $f = 0.06$ are closer to the unit circle. Thus, the realizations should have pseudo-periodic behavior with a more dominant period of about 17 (1/0.06) and a higher-frequency component with period of about 2.5 (1/0.4). We also expect two peaks in the spectral density, at about $f = 0.06$ and $f = 0.4$.

In Figure 3.17, we show a realization of length $n = 200$, where $\sigma_a^2 = 1$, the autocorrelations, and the spectral density for the AR(4) model in this example. The realization has the anticipated pseudo-periodic behavior with a period of about 17 and a high-frequency component. There are peaks in the spectral density at about $f = 0.06$ and $f = 0.4$ as expected. The autocorrelation is a damped sinusoidal with a period of about 17 with a slight indication of the higher-frequency behavior. Again, we see that the spectral density and the factor table give a much clearer picture of the periodic behavior.

Example 3.12 Dominance of Roots near the Unit Circle

In this example, we return to the AR(3) model in Example 3.6. We first consider the AR(3) models

$$(1 - 0.95B)(1 - 1.6B - \lambda_2 B^2)X_t = a_t$$

where $\lambda_2 = -0.7$, -0.9 (as in Example 3.6), and -0.98. We do not show the factor tables for these three models, but it should be noted that $|r_j^{-1}| = 0.95$ for the real root while for $\lambda_2 = -0.7, -0.9$, and -0.98 the complex roots have $|r_j^{-1}| = 0.84, 0.95$, and 0.99, respectively. Thus, the complex roots and real root are equally close to the unit circle when $\lambda_2 = -0.9$; however, when $\lambda_2 = -0.7$, the real root is closer to the unit circle than are the complex roots, and when $\lambda_2 = -0.98$, the complex roots are closer. So, based on the discussion preceding Example 3.11, we expect the real root to dominate the cyclic behavior when $\lambda_2 = -0.7$, and we expect the cyclic behavior associated with the complex roots to overshadow the real root induced wandering behavior when $\lambda_2 = -0.98$. In Figures 3.18 through 3.20, we show autocorrelations, realizations (based on $\sigma_a^2 = 1$), and spectral densities for these three models. Figures 3.18b, 3.19b, and 3.20b are based on $\lambda_2 = -0.9$ in which case the behavior imposed on the realization, autocorrelations, and spectral density by both the real and complex roots is clearly visible. However, in Figures 3.18a, 3.19a, and 3.20a, it can be seen that when $\lambda_2 = -0.7$, the autocorrelations look like a damped exponential with very little indication of the cyclic behavior. The realizations have a wandering behavior with only a very mild pseudo-periodic behavior, and the peak at $f = 0$ dominates the spectral

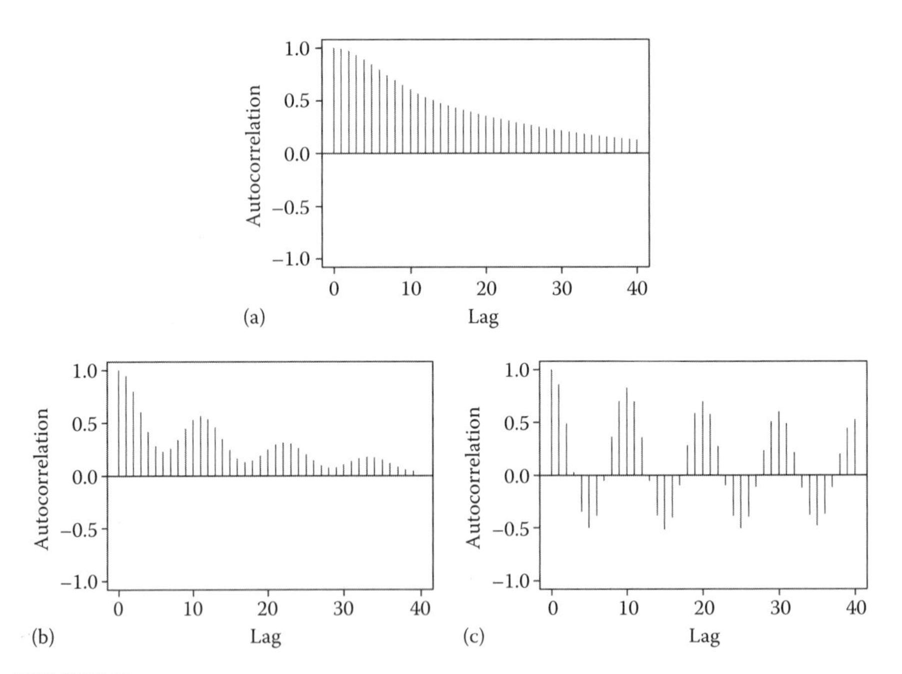

FIGURE 3.18
Autocorrelations for the AR(3) models $(1 - 0.95B)(1 - 1.6B - \lambda_2 B^2)X_t = a_t$, (a–c) $\lambda_2 = -0.7$, -0.9, -0.98.

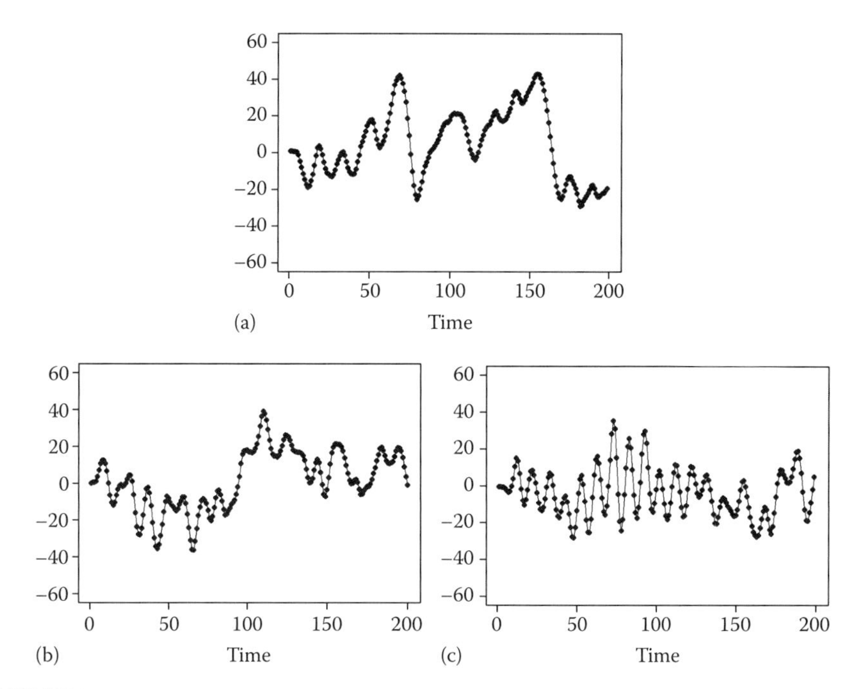

FIGURE 3.19
Realizations from the AR(3) models $(1-0.95B)(1-1.6B-\lambda_2B^2)X_t=a_t$, (a–c) $\lambda_2=-0.7,-0.9,-0.98$.

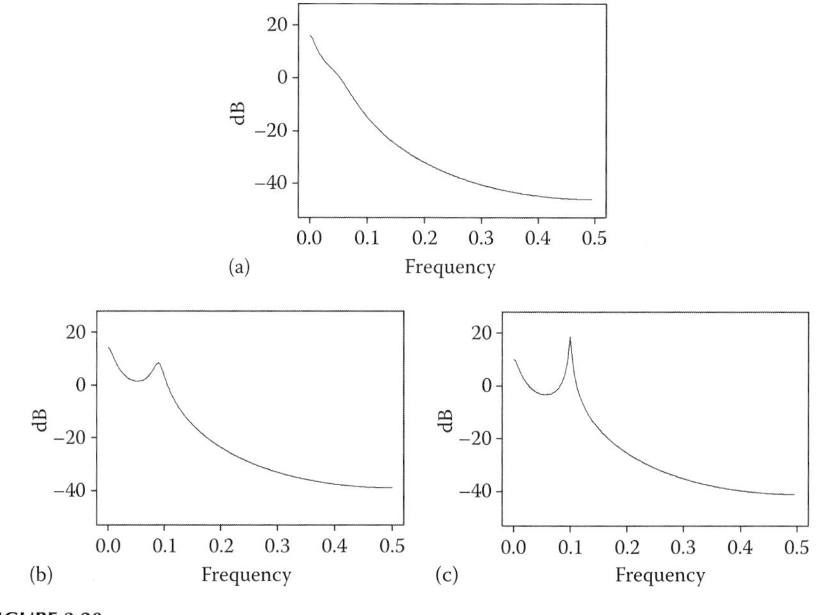

FIGURE 3.20
Spectral densities for the AR(3) models $(1-0.95B)(1-1.6B-\lambda_2B^2)X_t=a_t$, (a–c) $\lambda_2=-0.7$, -0.9, -0.98.

density with no peak showing associated with the complex roots. Correspondingly, in Figures 3.18c, 3.19c, and 3.20c, we observe that when $\lambda_2 = -0.98$, the autocorrelations look primarily like a damped sinusoidal with very little detectable damped exponential appearance, the realization has a distinct pseudo-periodic behavior with only a mild tendency to wander, while the peak at about $f = 0.1$ is the most dominant peak in the spectral density although the peak at $f = 0$ is visible.

It is clear from Example 3.12 that much of the behavior of an AR process can be traced to its dominant factors, that is, those closest to the unit circle. The factor table is an invaluable tool for inspecting the factors of an AR model, for understanding its composition, and ultimately for interpreting the physical process generating the data. As we discuss the analysis of actual time series data in later chapters, the validity of these remarks will become apparent.

3.2.12 Invertibility/Infinite-Order Autoregressive Processes

Under certain conditions, a stationary AR(p) process can be written in the form of a general linear process, that is, as an infinite-order MA. A question of interest then arises concerning when an MA(q) process can be expressed as an infinite-order AR process. In operator notation, the question is, "When can $X_t = \theta(B)a_t$ be written as $\theta^{-1}(B)X_t = a_t$?" If we expand $\theta^{-1}(z)$ in powers of z as

$$\theta^{-1}(z) = \sum_{j=0}^{\infty} \pi_j z^j, \tag{3.82}$$

and proceed formally to define $\theta^{-1}(B)$ by replacing z by B in (3.82), then application of the operator term by term yields

$$\sum_{j=0}^{\infty} \pi_j X_{t-j} = a_t, \tag{3.83}$$

which is a process of infinite AR form.

Definition 3.4 If an MA(q) process X_t can be expressed as in (3.83) with $\sum_{j=0}^{\infty} |\pi_j| < \infty$, then X_t is said to be *invertible*.

The condition $\sum_{j=0}^{\infty} |\pi_j| < \infty$ assures that $\sum_{j=0}^{\infty} \pi_j X_{t-j}$ is mean square convergent (see Theorem 3.1). The following theorem provides necessary and sufficient conditions for an MA(q) process to be invertible.

Theorem 3.3

Let X_t be the MA(q) process $X_t = \theta(B)a_t$. Then X_t is invertible if and only if the roots of $\theta(z) = 0$ all lie outside the unit circle.

Proof:

See Appendix 3.B.

In order to determine whether or not a given MA(q) process is invertible, we recommend examining a factor table in which the factors of $\theta(B)$ are given.

Using `GW-WINKS`: Factor tables for MA components can be obtained as in the AR case by using the `Factoring Routines` option in the `Time Series Analysis` menu of `GW-WINKS`.

3.2.13 Two Reasons for Imposing Invertibility

When using an MA(q) model to analyze time series data, we will restrict our attention to invertible models. There are two major reasons for this restriction.

(i) *Invertibility assures that the present events are associated with the past in a sensible manner.*
We rewrite the MA(1) model in (3.9) with $\mu = 0$ in the form $a_t = X_t + \theta_1 a_{t-1}$. Using this form of the model, it follows that $a_{t-1} = X_{t-1} + \theta_1 a_{t-2}$, so that

$$\begin{aligned} a_t &= X_t + \theta_1(X_{t-1} + \theta_1 a_{t-2}) \\ &= X_t + \theta_1 X_{t-1} + \theta_1^2 a_{t-2} \\ &= X_t + \theta_1 X_{t-1} + \theta_1^2 X_{t-2} + \theta_1^3 a_{t-3}. \end{aligned}$$

Continuing, we obtain $a_t = X_t + \theta_1 X_{t-1} + \theta_1^2 X_{t-2} + \cdots + \theta_1^k X_{t-k} + \theta_1^{k+1} a_{t-k-1}$, and thus

$$X_t = -\theta_1 X_{t-1} - \theta_1^2 X_{t-2} - \cdots - \theta_1^k X_{t-k} + a_t - \theta_1^{k+1} a_{t-k-1}. \tag{3.84}$$

If $|\theta_1| > 1$ the process is not invertible and (3.84) shows that in this case, the present, X_t, is increasingly dependent on the distant past. For most problems, this is not a physically acceptable model.

(ii) *Imposing invertibility removes model multiplicity.*
We again consider the MA(1) model in (3.9). From (3.11), the autocorrelation function, ρ_k, is zero for $k > 1$ and for $k=1$ is given by $\rho_1 = \dfrac{-\theta_1}{1+\theta_1^2}$. Now consider the MA(1) process given by

$$X_t = \left(1 - \frac{1}{\theta_1}B\right)a_t. \tag{3.85}$$

For the process in (3.85), it follows that $\rho_k = 0, k > 1$, and ρ_1 is given by

$$\begin{aligned}\rho_1 &= \frac{-\frac{1}{\theta_1}}{1+\frac{1}{\theta_1^2}} \\ &= \frac{-\theta_1}{1+\theta_1^2}.\end{aligned}$$

Thus, for example, if $\theta_1 = 0.5$, then the two MA(1) processes $X_t = (1-0.5B)a_t$ and $X_t = (1-2B)a_t$ have the same autocorrelations. Notice, however, that only $X_t = (1-0.5B)a_t$ is invertible. In general, if $\rho_k, k = 1, 2, \ldots$ are the autocorrelations of an MA(q) process, there are several different MA(q) processes having these autocorrelations, but only one of these processes is invertible (see Problem 3.8). Thus, by restricting attention to invertible MA(q) processes, a unique model is associated with a given set of MA(q) autocorrelations.

3.3 Autoregressive–Moving Average Processes

In this section, we combine the definitions of MA(q) and AR(p) processes to define the ARMA(p,q) model, that is, the autoregressive-moving average model of orders p and q.

Definition 3.5 Suppose $\{X_t;\ t = 0, \pm 1, \pm 2, \ldots\}$ is a causal, stationary, and invertible process satisfying the difference equation

$$X_t - \mu - \phi_1(X_{t-1} - \mu) - \cdots - \phi_p(X_{t-p} - \mu) = a_t - \theta_1 a_{t-1} - \cdots - \theta_q a_{t-q}, \tag{3.86}$$

which in operator notation is $\phi(B)(X_t - \mu) = \theta(B)a_t$, where $\phi(B)$ and $\theta(B)$ have no common factors, $\theta_q \neq 0$ and $\phi_p \neq 0$. Then, X_t will be called an ARMA process of orders p and q.

As before, without loss of generality, we will let $\mu = 0$. It is clear that the ARMA(p,q) models contain the MA(q) and AR(p) processes as special cases with $p = 0$ and $q = 0$, respectively. When $p \neq 0$ and $q \neq 0$, the resulting model is sometimes referred to as a *mixed model.*

An ARMA(p,q) process can formally be written in the general linear process form, that is, as the infinite-order MA

$$\begin{aligned} X_t &= \phi^{-1}(B)\theta(B)a_t \\ &= \psi(B)a_t, \end{aligned} \tag{3.87}$$

and as an infinite-order AR

$$\theta^{-1}(B)\phi(B)X_t = a_t. \tag{3.88}$$

Theorem 3.4 will give conditions under which the forms in (3.87) and (3.88) are valid.

The reason for the requirement that there are no common factors in the definition of an ARMA(p,q) process can be illustrated using the model

$$(1 - 0.8B)X_t = (1 - 0.8B)a_t. \tag{3.89}$$

The model in (3.89) can be expressed as

$$\begin{aligned} X_t &= (1 - 0.8B)^{-1}(1 - 0.8B)a_t \\ &= a_t, \end{aligned}$$

so that X_t in (3.89) is white noise. Whenever $\phi(B)$ and $\theta(B)$ share common factors, the model $\phi(B)X_t = \theta(B)a_t$ can be reduced in a similar manner to a lower-order model.

It will be demonstrated in the following material that the AR and MA components, that is, $\phi(B)$ and $\theta(B)$, of an ARMA(p,q) model introduce characteristics into the model of the same type as those in the AR(p) or MA(q) models separately. Some of these characteristics may be apparent in the ARMA(p,q) model. On the other hand, as should be clear from the preceding white noise example, some of the individual characteristics may not be apparent due to near cancellation, etc. We recommend examination of the factor tables of the AR and MA characteristic polynomials in order to obtain an understanding of a given ARMA(p,q) model. Inspection of these factor tables illuminates the component contributions to the model along with any near cancellations.

Example 3.13 ARMA(1,1) with Nearly Canceling Factors

Consider the ARMA(1,1) model

$$(1 - 0.95B)X_t = (1 - 0.8B)a_t. \tag{3.90}$$

Figure 3.21 shows a realization, the autocorrelations, and the spectral density from the model in (3.90). Recall that in Example 3.2, the AR(1) model

$$(1 - 0.95B)X_t = a_t \tag{3.91}$$

was considered. In Figures 3.4a, 3.5a, and 3.6a, we observed that this AR(1) model has damped exponential autocorrelations ($\rho_k = 0.95^k$), realizations with a wandering behavior that reflect the high autocorrelation among neighboring values, and a spectral density with a peak at $f = 0$. In the current ARMA(1,1) example, the autocorrelations have a damped exponential behavior given by $\rho_k = 0.3(0.95^k)$, so that even at small lags, the correlation is not substantial. In the realization in Figure 3.21a, there is some indication of a wandering behavior but with much less attraction between adjacent or near adjacent values. The spectral density in Figure 3.21c is

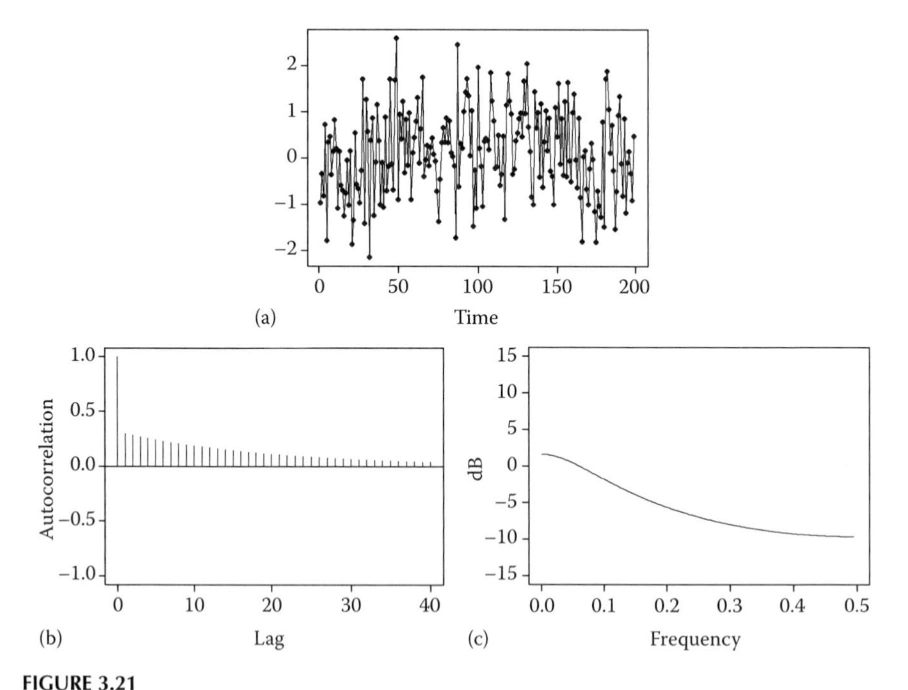

FIGURE 3.21
(a) Realization, (b) autocorrelations, and (c) spectral densities for the ARMA(1,1) model $(1 - 0.95B)X_t = (1 - 0.8B)a_t$.

plotted on the same scale as Figure 3.6a to emphasize that although the spectral density for this ARMA(1,1) model has a maximum at $f = 0$, there is much less power concentrated around the frequency $f = 0$. In fact, the realization, autocorrelations, and spectral density in Figure 3.21 are approaching the corresponding behaviors of white noise, demonstrating the fact that the MA term in (3.90) has nearly canceled the effect of the AR operator. Clearly, as $\theta_1 \to 0.95$, the ARMA(1,1) model $(1 - 0.95B)X_t = (1 - \theta_1)a_t$ approaches white noise.

3.3.1 Stationarity and Invertibility Conditions for an ARMA(*p*,*q*) Model

Theorem 3.4 that follows is analogous to Theorems 3.2 and 3.3 and is given here without proof. Theorem 3.4(i) indicates that, as might be expected, the stationarity of a process satisfying (3.86) depends on the AR operator $\phi(B)$. Additionally, Theorem 3.4(ii) gives conditions on the MA operator $\theta(B)$ of a process satisfying (3.86) that are necessary and sufficient for X_t to have a "meaningful" infinite AR representation, that is, to be invertible.

Theorem 3.4

Let X_t be a causal process specified by (3.86). Then, X_t is an ARMA(p,q) process if and only if the following two conditions are met:

(i) The roots $r_j, j = 1, 2, \ldots, p$ of the characteristic equation $\phi(z) = 0$ all lie outside the unit circle (stationarity).

(ii) The roots of $\theta(z) = 0$ lie outside the unit circle (invertibility).

3.3.2 Spectral Density of an ARMA(*p*,*q*) Model

The power spectrum of the stationary ARMA(p,q) process is obtained from (3.87), since

$$\begin{aligned}\psi(z) &= \phi^{-1}(z)\theta(z)\\ &= \left(1 - \phi_1 z - \cdots - \phi_p z^p\right)^{-1}\left(1 - \theta_1 z - \theta_2 z^2 - \cdots - \theta_q z^q\right),\end{aligned}$$

and hence

$$P_X(f) = \sigma_a^2 \frac{\left|1 - \theta_1 e^{-2\pi i f} - \theta_2 e^{-4\pi i f} - \cdots - \theta_q e^{-2q\pi i f}\right|^2}{\left|1 - \phi_1 e^{-2\pi i f} - \phi_2 e^{-4\pi i f} - \cdots - \phi_p e^{-2p\pi i f}\right|^2}.$$

The spectral density is given by

$$S_X(f) = \frac{\sigma_a^2}{\sigma_X^2} \frac{\left|1 - \theta_1 e^{-2\pi i f} - \theta_2 e^{-4\pi i f} - \cdots - \theta_q e^{-2q\pi i f}\right|^2}{\left|1 - \phi_1 e^{-2\pi i f} - \phi_2 e^{-4\pi i f} - \cdots - \phi_p e^{-2p\pi i f}\right|^2}. \tag{3.92}$$

System frequencies of the AR part of the ARMA(p,q) model will tend to introduce peaks in the spectral density similar to those for the corresponding AR(p) model. However, system frequencies of the MA part tend to produce troughs or dips in the spectral density at the corresponding system frequencies. The peaks in a spectral density for an ARMA(p,q) (or AR(p)) model are caused by small values in the denominator of (3.92) which occur near the system frequencies of the AR part of the model. Correspondingly, small values in the numerator of (3.92) near the system frequencies of the MA part will produce the tendency for dips or troughs to appear in the spectral density.

Using `GW-WINKS`: The true autocorrelations, process variance and spectral density for a given ARMA(p,q) model can be obtained using the `Generate Realizations` option of the `Time Series Analysis` menu in the `GW-WINKS` program. This option was discussed in more detail in Section 3.1.

3.3.3 Factor Tables and ARMA(*p*,*q*) Models

According to Theorem 3.4 the stationarity and invertibility of an ARMA(p,q) model can be tested by examining the factors of $\phi(z)$ and of $\theta(z)$ respectively. As in the AR(p) case, factor tables are a useful tool for understanding these factors, and this can be accomplished using the `Factoring Routines` options on the `Time Series Analysis` menu in `GW WINKS`. The factor tables for the AR components of an ARMA(p,q) model are interpreted as in the AR(p) case. That is, system frequencies indicate natural frequencies in the model, and spectral densities will tend to have peaks near these system frequencies. The key to understanding factor tables for an ARMA(p,q) process is to realize that the factor tables for the MA components are interpreted in an opposite manner. Specifically, the MA system frequencies indicate frequencies whose impact has been weakened in the total model. Spectral densities will tend to have dips rather than peaks in the vicinity of the MA system frequencies. (See the discussion following (3.92).) As an example note that the spectral density in Figure 3.3a for the MA(1) model $X_t = (1 - 0.99B)a_t$ has a dip or trough at $f = 0$ whereas the spectral density for $(1 - 0.99B)X_t = a_t$ would have a sharp peak at $f = 0$. Another example of this effect is the fact that in (3.90) the AR factor $1 - 0.95B$ and the MA factor $1 - 0.8B$ are each associated with system frequency $f = 0$. In Example 3.13 we illustrated the fact that the MA factor $1 - 0.8B$ weakened the behavior of the AR factor $1 - 0.95B$ in the ARMA(1,1) model in (3.90). We use the term "system frequencies" to indicate the frequencies displayed in the factor table whether the factor is an AR or MA factor. Examples 3.14 and 3.15 further illustrate the use of factor tables to aid the understanding of ARMA(p,q) models.

Example 3.14 An ARMA(2,1) Model

Consider the ARMA(2,1) model given by

$$(1 - 1.6B + 0.9B^2)X_t = (1 - 0.8B)a_t. \tag{3.93}$$

Recall that autocorrelations, a realization (based on $\sigma_a^2 = 1$) and the spectral density for the AR(2) model $(1 - 1.6B + 0.9B^2)X_t = a_t$ were given in Figures 3.11b, 3.12b, and 3.13b. In Table 3.5, we show the factor table for the model in (3.93), which is an ARMA(2,1) model, since all roots associated with AR and MA factors are outside the unit circle. As discussed, the effect of the MA factor $1 - 0.8B$ is to reduce frequency content at $f = 0$.

In Figure 3.22, we show a realization, autocorrelations, and spectral density for the ARMA(2,1) model in (3.93) where $\sigma_a^2 = 1$. In comparing Figure 3.22 with Figures 3.11b through 3.13b, we note similar behavior in the realization and autocorrelations. Both spectral densities have a peak at about $f = 0.1$, but it should be noted that the spectral densities differ at and around $f = 0$. Specifically, the spectral density in Figure 3.22c for the ARMA (2,1) model is much lower at $f = 0$ than is the corresponding spectral density for the AR(2) model $(1 - 1.6B + 0.9B^2)X_t = a_t$. Thus, the effect of the MA factor $1 - 0.8B$ is to reduce the spectral content at $f = 0$ and cause the frequency content to be more focused about $f = 0.1$. Thus, the effect of the MA factor $1 - 0.8B$ in Examples 3.13 and 3.14 is to remove or reduce frequency behavior at $f = 0$

Example 3.15 An ARMA(4,3) Model

Consider the ARMA(4,3) model

$$X_t - 0.3X_{t-1} - 0.9X_{t-2} - 0.1X_{t-3} + 0.8X_{t-4} = a_t + 0.9a_{t-1} + 0.8a_{t-2} + 0.72a_{t-3} \tag{3.94}$$

TABLE 3.5

Factor Table for the ARMA(2,1) Model in (3.93)

	Roots (r_j)	$\lvert r_j^{-1} \rvert$	f_{0j}
AR factor			
$1 - 1.6B + 0.9B^2$	$0.89 \pm 0.57i$	0.95	0.09
MA factor			
$1 - 0.8B$	1.25	0.8	0

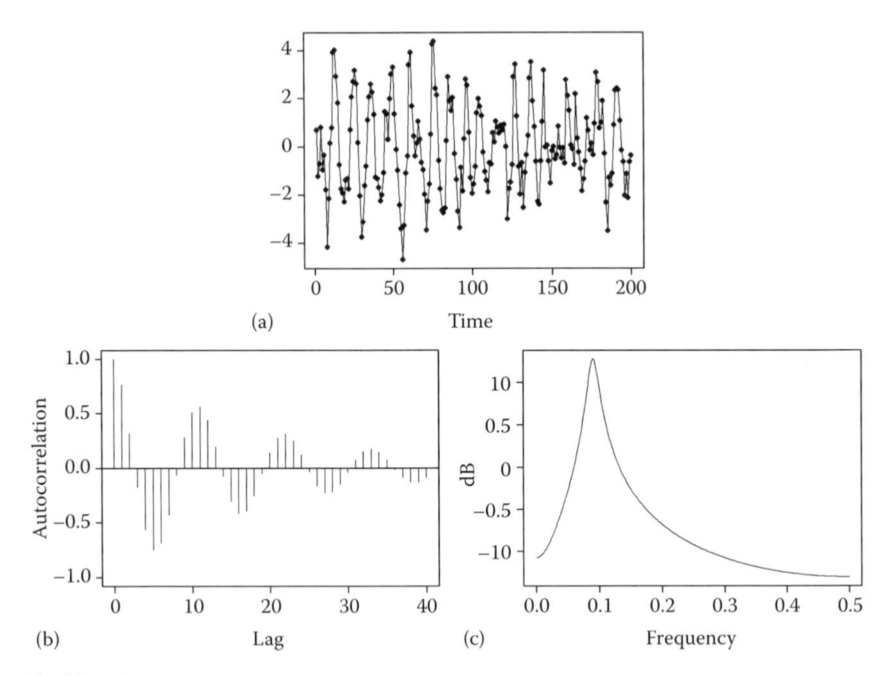

FIGURE 3.22
(a) Realization, (b) autocorrelations, and (c) spectral densities for the ARMA(2,1) model $(1 - 1.6B + 0.9B^2)X_t = (1 - 0.8B)a_t$.

which can be factored as

$$(1 - 1.8B + 0.95B^2)(1 + 1.5B + 0.85B^2)X_t = (1 + 0.9B)(1 + 0.8B^2)a_t. \quad (3.95)$$

The factor table is shown in Table 3.6, and it should be noted that the AR factor is the one previously considered in the AR(4) model in Example 3.11. Figure 3.23 displays a realization of length $n = 200$ from the ARMA(4,3)

TABLE 3.6

Factor Table for the ARMA(4,3) Model in (3.94) and (3.95)

	Roots (r_j)	$\lvert r_j^{-1}\rvert$	f_{0j}
AR factors			
$1 - 1.8B + 0.95B^2$	$0.95 \pm 0.39i$	0.97	0.06
$1 + 1.5B + 0.85B^2$	$-0.89 \pm 0.63i$	0.92	0.40
MA factors			
$1 + 0.9B$	-1.11	0.9	0.50
$1 + 0.8B^2$	$\pm 1.12i$	0.89	0.25

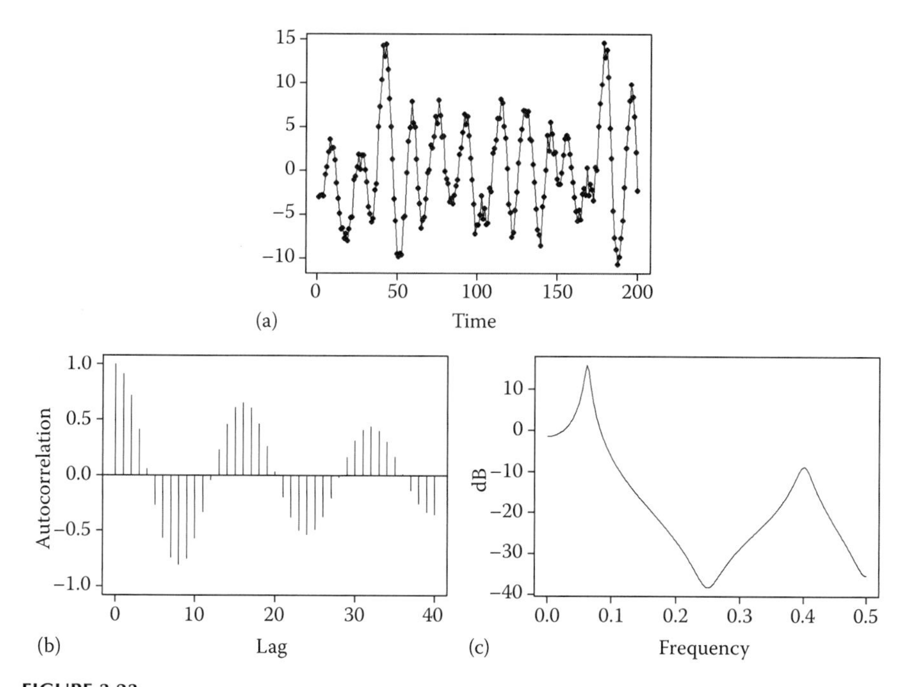

FIGURE 3.23
(a) Realization, (b) autocorrelations, and (c) spectral densities for the ARMA(4,3) model $(1 - 1.8B + 0.95B^2)\ (1 + 1.5B + 0.85B^2)X_t = (1 + 0.9B)(1 + 0.8B^2)a_t$.

model in (3.94) and (3.95) with $\sigma_a^2 = 1$ along with the autocorrelations and spectral density. Comparing the spectral density in Figure 3.23c with that in Figure 3.17c, it can be seen that the MA factor $1 + 0.8B^2$ produced a dip at $f = 0.25$. The MA factor $1 + 0.9B$ created a dip at $f = 0.5$, which had the effect of lowering the peak in the spectral density at $f = 0.4$. The effect of the MA factors $1 + 0.9B$ and $1 + 0.8B^2$ can be seen in the realization, and autocorrelations where the high-frequency behavior visible in Figure 3.17 is much less apparent in Figure 3.23.

3.3.4 Autocorrelations of an ARMA(p,q) Model

Multiplying each term in (3.86) (without loss of generality using the zero mean form) by X_{t-k} ($k \geq 0$) and taking expectations yields

$$\begin{aligned}\gamma_k - \phi_1\gamma_{k-1} - \cdots - \phi_p\gamma_{k-p} &= E[X_{t-k}a_t] - \theta_1 E[X_{t-k}a_{t-1}] - \cdots - \theta_q E[X_{t-k}a_{t-q}] \\ &= \gamma_{Xa}(k) - \theta_1\gamma_{Xa}(k-1) - \cdots - \theta_q\gamma_{Xa}(k-q),\end{aligned} \tag{3.96}$$

where

$$\gamma_{Xa}(m) = E[X_{t-m}a_t]$$
$$= E\left(\sum_{j=0}^{\infty} \psi_j a_{t-m-j} a_t\right). \quad (3.97)$$

From (3.97), it follows that

$$\gamma_{Xa}(m) = 0, \quad m > 0$$
$$= \psi_{|m|}\sigma_a^2, \quad m \leq 0. \quad (3.98)$$

Thus, for $k \geq 0$

$$\gamma_k = \phi_1\gamma_{k-1} + \cdots + \phi_p\gamma_{k-p} - \sigma_a^2(\theta_k\psi_0 + \cdots + \theta_q\psi_{q-k}), \quad (3.99)$$

where $\theta_0 = -1$, and the parenthetical expression is taken to be zero if $k > q$. Specifically, γ_q is given by

$$\gamma_q = \phi_1\gamma_{q-1} + \cdots + \phi_p\gamma_{q-p} - \sigma_a^2\theta_q, \quad (3.100)$$

and γ_k does not satisfy a homogeneous equation, since $\sigma_a^2\theta_q \neq 0$. However, for $k > q$,

$$\gamma_k - \phi_1\gamma_{k-1} - \cdots - \phi_p\gamma_{k-p} = 0 \quad (3.101)$$

and

$$\rho_k - \phi_1\rho_{k-1} - \cdots - \phi_p\rho_{k-p} = 0, \quad (3.102)$$

so that γ_k and ρ_k satisfy a pth-order linear homogeneous difference equation with constant coefficients for $k > q$. Since we have assumed that $\phi(z)$ and $\theta(z)$ share no common factors, p is the lowest-order homogeneous linear difference equation with constant coefficients satisfied by ρ_k. The solution to (3.102) can be made to satisfy p linear boundary conditions obtained from (3.99) for $k = q, q-1, \ldots, q-p+1$. Thus, for $k > q-p$, it follows that

$$\rho_k = \sum_{i=1}^{v}\sum_{j=0}^{m_i-1} C_{ij}k^j r_i^{-k}, \quad (3.103)$$

or in the case of no repeated roots of the characteristic equation,

$$\rho_k = C_1 r_1^{-k} + \cdots + C_p r_p^{-k}, \quad (3.104)$$

where the C_{ij} and C_j are constants uniquely determined from (3.99) with $k = q, q-1, \ldots, q-p+1$.

The autocovariance function can also be found as the solution of a system of nonhomogeneous linear difference equations. Letting $k = 0, 1, \ldots, p$ in (3.99) gives the $p+1$ linear equations in the $p+1$ unknowns $\gamma_0, \gamma_1, \ldots, \gamma_p$

$$\begin{aligned}
\gamma_0 &= \phi_1\gamma_1 + \phi_2\gamma_2 + \cdots + \phi_p\gamma_p - \sigma_a^2(-1 + \theta_1\psi_1 + \cdots + \theta_q\psi_q) \\
\gamma_1 &= \phi_1\gamma_0 + \phi_2\gamma_1 + \cdots + \phi_p\gamma_{p-1} - \sigma_a^2(\theta_1 + \theta_2\psi_1 + \cdots + \theta_q\psi_{q-1}) \\
&\vdots \\
\gamma_v &= \phi_1\gamma_{v-1} + \phi_2\gamma_{v-2} + \cdots + \phi_v\gamma_{v-p} - \sigma_a^2(\theta_v + \theta_{v+1}\psi_1 + \cdots + \theta_q\psi_{q-v})
\end{aligned} \tag{3.105}$$

where $v = \max(p,q)$ and $\theta_j = 0$ whenever $j > q$. The ψ-weights can be obtained from the ϕ_j and θ_j using a recursive algorithm to be given shortly. For $k > v$, γ_k can be solved by simply using the recursion in (3.101).

Note that (3.103) shows that $\rho_k \to 0$ (and equivalently $\gamma_k \to 0$) as $k \to \infty$ for an ARMA(p,q) process. From Theorem 1.1, it thus follows that, in the case of an ARMA(p,q) process, $\mathbf{\Gamma}_k$ and $\boldsymbol{\rho}_k$, defined in (1.2) and (1.3), respectively, are positive definite for each integer $k > 0$.

In the following example, we give closed form expressions for $\gamma_0, \ldots, \gamma_q$ of an ARMA(1,1), ARMA(2,1), and ARMA(1,2).

Example 3.16 Autocovariances for Specific ARMA(p,q) Models

(a) *ARMA(1,1) model*
Consider the ARMA(1,1) model given by

$$X_t - \phi_1 X_{t-1} = a_t - \theta_1 a_{t-1}. \tag{3.106}$$

From (3.99), we obtain

$$\begin{aligned}
\gamma_0 &= \phi_1\gamma_{-1} - \sigma_a^2(\theta_0\psi_0 + \theta_1\psi_1) \\
\gamma_1 &= \phi_1\gamma_0 - \sigma_a^2(\theta_1\psi_0) \\
\gamma_k &= \phi_1\gamma_{k-1}, \quad k > 1,
\end{aligned} \tag{3.107}$$

where $\theta_0 = -1$ and $\psi_0 = 1$. Thus, if $k > 1$, then ρ_k satisfies the linear homogeneous difference equation

$$\rho_k - \phi_1\rho_{k-1} = 0, \tag{3.108}$$

and from (3.104) it follows that

$$\rho_k = C_1(1/\phi_1)^{-k}$$
$$= C_1\phi_1^k. \tag{3.109}$$

To find C_1 note that from (3.108) $\rho_2 = \phi_1\rho_1$, and from (3.109) $\rho_2 = C_1\phi_1^2$. So $\phi_1\rho_1 = C_1\phi_1^2$, that is, $C_1 = \rho_1\phi_1^{-1}$, and (3.109) becomes

$$\rho_k = \rho_1\left(\phi_1^{k-1}\right) \tag{3.110}$$

for $k \geq 2$. Now, in order to use (3.107) for $k = 0$ and 1 it is necessary to find ψ_1. For an ARMA(1,1), $\psi_1 = \phi_1 - \theta_1$ (see (3.116)). So (3.107) becomes

$$\begin{aligned} \gamma_0 &= \phi_1\gamma_1 - \sigma_a^2(-1 + \theta_1(\phi_1 - \theta_1)) \\ \gamma_1 &= \phi_1\gamma_0 - \sigma_a^2\theta_1, \end{aligned} \tag{3.111}$$

and $\gamma_k = \phi_1\gamma_{k-1}, k > 1$. Solving (3.111) for γ_0 and γ_1 yields

$$\gamma_0 = \frac{(1 - 2\theta_1\phi_1 + \theta_1^2)}{1 - \phi_1^2}\sigma_a^2$$

and

$$\gamma_1 = \frac{(\phi_1 - \theta_1)(1 - \phi_1\theta_1)}{1 - \phi_1^2}\sigma_a^2,$$

so that

$$\rho_1 = \frac{(\phi_1 - \theta_1)(1 - \phi_1\theta_1)}{1 - 2\phi_1\theta_1 + \theta_1^2}. \tag{3.112}$$

The ARMA(1,1) model of Example 3.13 revisited
Consider again the ARMA(1,1) model $(1 - 0.95B)X_t = (1 - 0.8B)a_t$ in (3.90). From (3.112), we obtain $\rho_1 = 0.3$, and from (3.110), it follows that $\rho_k = 0.3(0.95^{k-1})$ for $k \geq 2$. Note that while ρ_1 does not satisfy the linear homogeneous difference equation in (3.108), it does satisfy (3.110) since $\rho_1 = 0.3(0.95^0) = 0.3$.

(b) *ARMA(2,1) model*
Consider the ARMA(2,1) model given by

$$(1 - \phi_1 B - \phi_2 B^2)X_t = (1 - \theta_1 B)a_t.$$

For this model,

$$\gamma_0 = \frac{2\phi_1\theta_1 - (1-\phi_2)(1+\theta_1^2)}{(1+\phi_2)(\phi_1^2 - (1-\phi_2)^2)}\sigma_a^2, \tag{3.113}$$

and

$$\gamma_1 = \frac{(\phi_1 - \theta_1)(1-\phi_1\theta_1) + \phi_2^2\theta_1}{(1+\phi_2)((1-\phi_2)^2 - \phi_1^2)}\sigma_a^2. \tag{3.114}$$

For $k > 1, \gamma_k$ can be obtained using the recursion $\gamma_k = \phi_1\gamma_{k-1} + \phi_2\gamma_{k-2}$.

(c) *ARMA(1,2) model*
For the ARMA(1,2) model

$$(1-\phi_1 B)X_t = (1 - \theta_1 B - \theta_2 B^2)a_t,$$

it can be shown that

$$\gamma_0 = \frac{1 - 2\phi_1\theta_1 - 2\phi_1\theta_2(\phi_1 - \theta_1) + \theta_1^2 + \theta_2^2}{1-\phi_1^2}\sigma_a^2$$

$$\gamma_1 = \frac{(1 - \theta_2 - \phi_1^2\theta_2 - \phi_1 - \theta_1)(\phi_1 - \theta_1) + \phi_1\theta_2^2}{1-\phi_1^2}\sigma_a^2$$

and

$$\gamma_2 = \frac{\phi_1\left[(1 - \theta_2 - \phi_1^2\theta_2 - \phi_1\theta_1)(\phi_1 - \theta_1) + \phi_1\theta_2^2\right] - \theta_2 + \phi_1^2\theta_2}{1-\phi_1^2}\sigma_a^2.$$

Again, the autocovariances for $k > 2$ can be found using the recursive formula $\gamma_k = \phi_1\gamma_{k-1}$.

3.3.5 ψ-Weights of an ARMA(p,q)

We have shown that an ARMA(p,q) process can be expressed in the linear filter form, $X_t = \sum_{k=0}^{\infty}\psi_k a_{t-k}$. In this section, we discuss calculation of the ψ-weights using three methods: (a) equating coefficients, (b) division, and (c) finding a general expression using Theorem 3.1.

(a) *Finding ψ-weights by equating coefficients*
Since a stationary ARMA(p,q) process can be written in the form $X_t = \psi(B)a_t$, where $\psi(B) = \phi^{-1}(B)\theta(B)$, we have the relationship $\phi(B)\psi(B) = \theta(B)$, or in the expanded form

$$\left(1-\phi_1 B-\cdots-\phi_p B^p\right)\left(1+\psi_1 B+\psi_2 B^2+\cdots\right)=\left(1-\theta_1 B-\cdots-\theta_q B^q\right). \tag{3.115}$$

Equating the coefficients of B^j in (3.115) displays the relationship among the ϕ's, θ's, and ψ's. In particular, equating coefficients of B^j yields

$$\begin{aligned}
&\psi_1 - \phi_1 = -\theta_1\\
&\psi_2 - \phi_1\psi_1 - \phi_2 = -\theta_2\\
&\vdots\\
&\psi_j - \phi_1\psi_{j-1} - \phi_2\psi_{j-2} - \cdots - \phi_j = -\theta_j,
\end{aligned} \tag{3.116}$$

where $\psi_0 = 1, \theta_k = 0$ for $k > q$, and $\phi_k = 0$ for $k > p$. The equations in (3.116) can be rearranged to provide a recursive scheme for calculating the ψ-weights. Namely,

$$\begin{aligned}
&\psi_1 = \phi_1 - \theta_1\\
&\psi_2 = \phi_1\psi_1 + \phi_2 - \theta_2\\
&\vdots\\
&\psi_j = \phi_1\psi_{j-1} + \phi_2\psi_{j-2} + \cdots + \phi_j - \theta_j.
\end{aligned} \tag{3.117}$$

Notice from (3.117) that for $j > \max(p-1, q)$, then ψ_j satisfies the pth-order homogeneous difference equation

$$\psi_j = \phi_1\psi_{j-1} + \phi_2\psi_{j-2} + \cdots + \phi_p\psi_{j-p} \tag{3.118}$$

or

$$\phi(B)\psi_j = 0.$$

(b) *Finding ψ-weights using division*
A very easy way to calculate a few ψ-weights is to use division based on the fact that properties of the operators, for example, $\phi(B)$ and $\theta(B)$, can be conveniently found using the algebraic equivalents $\phi(z)$ and $\theta(z)$. For example, we can express $\phi(B)X_t = \theta(B)a_t$ as

$X_t = \phi^{-1}(B)\theta(B)a_t$, and the coefficients of the operator $\phi^{-1}(B)\theta(B)$ can be found using division based on the algebraic equivalent $\phi^{-1}(z)\theta(z) = \theta(z)/\phi(z)$. As an example, consider the ARMA(2,1) model $(1 - 1.6B + 0.9B^2)X_t = (1 - 0.8B)a_t$. The ψ-weights are given by

$$\frac{1 - 0.8z}{1 - 1.6z + 0.9z^2} = \sum_{k=0}^{\infty} \psi_k z^k,$$

where $\psi_0 = 1$. Calculations show that

$$\frac{1 - 0.8z}{1 - 1.6z + 0.9z^2} = 1 + 0.8z + 0.38z^2 - 0.112z^3 \cdots.$$

So, $\psi_1 = 0.8, \psi_2 = 0.38, \psi_3 = -0.112$, etc.

(c) *A general expression for the ψ-weights*
Let $\psi_k(p,q)$ denote the ψ-weights of the ARMA(p,q) process $\phi(B)X_t = \theta(B)a_t$, and let $\psi_k(p)$ be the ψ-weights of the AR(p) process with the same AR operator $\phi(B)$. Then the ψ-weights $\psi_k(p,q)$ are given by

$$\begin{aligned}\psi_k(p,q) &= \psi_k(p) - \theta_1\psi_{k-1}(p) - \cdots - \theta_q\psi_{k-q}(p) \\ &= \theta(B)\psi_k(p) \qquad (3.119)\end{aligned}$$

where the $\psi_k(p)$ are given in Theorem 3.1. The derivation of this result will be left as an exercise (see Problem 3.31).

3.3.6 Approximating ARMA(p,q) Processes Using High-Order AR(p) Models

We have discussed the fact that the stationary and invertible ARMA(p,q) model $\phi(B)X_t = \theta(B)a_t$ can be written as the infinite-order AR process $\theta^{-1}(B)\phi(B)X_t = a_t$. Thus, an ARMA($p$,$q$) model can often be well approximated using a high-order AR(p) model. For this reason, some time series analysts use the approach of always fitting AR models to data since parameter estimation and model identification tend to be more difficult when MA parameters are included in a model. On the other hand, some analysts prefer the ARMA(p,q) model since it often contains fewer parameters (is more *parsimonious*) than does an AR model fit to the same data set. In this text, we will cover parameter estimation, forecasting, etc., in the generality provided by the ARMA(p,q) model. However, in practice, we will often restrict ourselves to AR(p) models.

An example of this trade-off between a parsimonious ARMA(p,q) model and a high-order AR(p) model is illustrated in Example 4.1, where theoretical justification is used to show that an ARMA(2,2) model is a good approximating model for harmonic signal-plus-noise data. However, in that example, we also show that an AR(7) model also works well as an approximating model.

3.4 Visualizing Autoregressive Components

Another useful tool for understanding the features of an AR(p) realization is the decomposition of the realization into additive components. Specifically, letting x_t denote a realization from an AR(p) model for which there are no repeated roots of the characteristic equation, then x_t can be written as $x_t = x_t^{(1)} + x_t^{(2)} + \cdots + x_t^{(m)}$ where m, as given in (3.76), is the number of real roots plus the number of complex conjugate pairs of roots of the characteristic equation. The jth additive component $x_t^{(j)}$

- Corresponds to the jth factor in the factor table
- Is a realization from an AR(1) model for first-order factors
- Is a realization from an ARMA(2,1) model for irreducible second-order factors

For example, consider the AR(3) model

$$(1 - \alpha_1 B)(1 - \beta_1 B - \beta_2 B^2)X_t = a_t, \tag{3.120}$$

with one real root and one complex conjugate pair of roots (i.e., $m = 2$). The model in (3.120) can be written in inverted form as

$$X_t = \frac{1}{(1 - \alpha_1 B)(1 - \beta_1 B - \beta_2 B^2)} a_t. \tag{3.121}$$

Recalling that operators can be inverted using properties of the corresponding algebraic quantities, we use partial fractions to write

$$\frac{1}{(1 - \alpha_1 z)(1 - \beta_1 z - \beta_2 z^2)} = \frac{A}{1 - \alpha_1 z} + \frac{C_1 + C_2 z}{1 - \beta_1 z - \beta_2 z^2}, \tag{3.122}$$

from which we obtain

$$\begin{aligned} X_t &= \frac{1}{(1 - \alpha_1 B)(1 - \beta_1 B - \beta_2 B^2)} a_t \\ &= \left(\frac{A}{1 - \alpha_1 B} + \frac{C_1 + C_2 B}{1 - \beta_1 B - \beta_2 B^2} \right) a_t \\ &= \left(\frac{A}{1 - \alpha_1 B} \right) a_t + \left(\frac{C_1 + C_2 B}{1 - \beta_1 B - \beta_2 B^2} \right) a_t, \end{aligned} \tag{3.123}$$

for real constants A, C_1, and C_2. Letting $X_t^{(1)} = \frac{A}{1-\alpha_1 B} a_t$ and $X_t^{(2)} = \frac{C_1 + C_2 B}{1 - \beta_1 B - \beta_2 B^2} a_t$, it follows that $(1-\alpha_1 B)X_t^{(1)} = Aa_t$ and $(1-\beta_1 B - \beta_2 B^2)X_t^{(2)} = (C_1 + C_2 B)a_t = \left(1 + \frac{C_2}{C_1} B\right)(C_1 a_t)$. That is, $x_t^{(1)}$ is a realization from an AR(1) process where the AR factor is the first factor in the factored form of (3.120), and $x_t^{(2)}$ is a realization from an ARMA(2,1) process where the AR factor is the second-order factor in the factored form of (3.120). In Example 10.14, we discuss the use of state-space formulation to actually construct these components. The important point to be made here is that each component corresponds to a row in the factor table.

Example 3.17 Additive Components of the AR(3) Model in (3.73)

As an example, consider the AR(3) model $X_t - 2.55X_{t-1} + 2.42X_{t-2} - 0.855X_{t-3} = a_t$ previously given in (3.73) which, in factored form is $(1-0.95B)(1-1.6B+0.9B^2)X_t = a_t$. It was pointed out earlier that realizations from this model have a wandering behavior produced by the positive real root ($f = 0$) and cyclic behavior with frequency $f = 0.09$ (i.e., periods of about 11) because of the complex conjugate roots. The factor table for this model is given in Table 3.1. Figure 3.24 shows the realization given previously in Figure 3.14a along with the additive ARMA(2,1) and AR(1) components obtained using the state-space approach to be discussed in Section 10.6.7. We note the following:

- The ARMA(2,1) component in Figure 3.24b contains the pseudo-cyclic portion of the realization associated with the complex conjugate root listed in the first row of the factor table in Table 3.1.
- The wandering behavior in the realization associated with the positive real root listed on the second row of the factor table in Table 3.1 is captured by the AR(1) component shown in Figure 3.24c.

These types of decompositions are an instructive aid for visualizing the underlying structure of AR(p) models as well as understanding the associated physical phenomena.

Using `GW-WINKS`: Decompositions of AR realizations can be obtained using the `Factoring Routines` option in the `Time Series Analysis` menu. For a data set showing in the current GW data editor, you can request Burg estimates of order p that you specify and then request that a factor table be calculated. After the factor table is displayed, select the option `Create Additive Components`. The `GW-WINKS` program will produce columns in the data matrix associated with components 1 through m where m is the number of lines in the factor table. These components will have variable names `COMP1, COMP2`, etc.

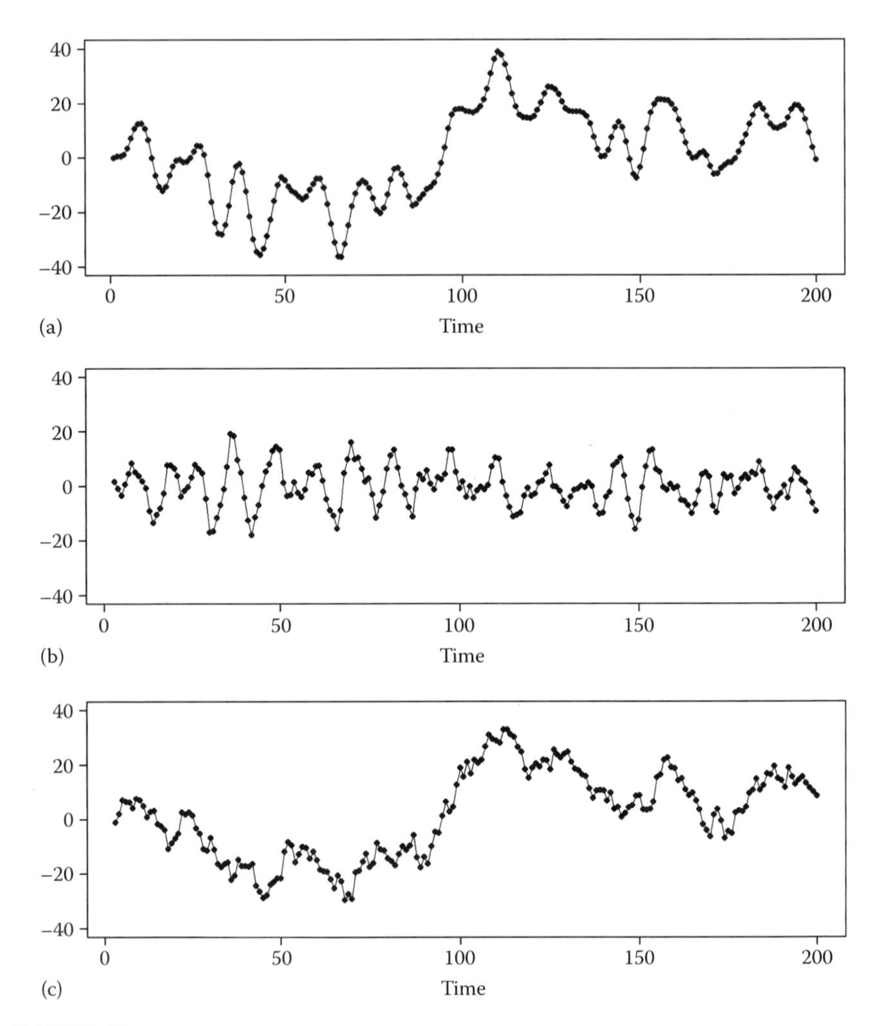

FIGURE 3.24
Realization and additive components for the AR(3) model in (3.76) whose factor table is given in Table 3.1. (a) Realization. (b) Component 1: $f = 0.09$. (c) Component 2: $f = 0$.

3.5 Seasonal ARMA$(p,q) \times (P_s,Q_s)_s$ Models

Several modifications of the ARMA(p,q) model have been found to be useful. In Chapter 5, we will discuss ARMA-type models that are nonstationary due to roots of the characteristic equation on the unit circle. One modification of the ARMA(p,q) is designed to take into account seasonal behavior in data. For example, quarterly data (e.g., breaking the year into four 3 month groups) will often have seasonal characteristics implying that there is a natural correlation structure between, X_t and X_{t-4}, that is, between the current quarter and the same quarter last year. In this case, it makes sense to include factors in the model such as $\Phi(B^4) = 1 - \Phi_1 B^4 - \Phi_2 B^8 - \cdots - \Phi_P B^{4P}$, that take into account this four-quarter per year relationship. For example, we might have $\Phi(B^4) = 1 - 0.9B^4$ or $\Phi(B^4) = 1 - B^4 + 0.8B^8$, that is, the models $X_t = 0.9X_{t-4} + a_t$ and $X_t = X_{t-4} - 0.8X_{t-8} + a_t$, which specify that this quarter's performance is a linear combination of the same quarter in previous years plus noise. In general, we use s to denote the "season" (i.e., for monthly data, $s = 12$ and for quarterly data $s = 4$). A (purely) stationary seasonal model is given by

$$\Phi(B^s)X_t = \Theta(B^s)a_t, \tag{3.124}$$

where all the roots of $1 - \Phi_1 z^s - \Phi_2 z^{2s} - \cdots - \Phi_{P_s} z^{P_s s} = 0$ and $1 - \Theta_1 z^s - \Theta_2 z^{2s} - \cdots - \Theta_{Q_s} z^{Q_s s} = 0$ lie outside the unit circle.

Example 3.18 A Purely Seasonal Stationary Model

In this example, we consider the model

$$(1 - 0.8B^4)X_t = a_t \tag{3.125}$$

in more detail. Note that (3.125) is an AR(4) model with the coefficients $\phi_1 = \phi_2 = \phi_3 = 0$, and $\phi_4 = 0.8$ However, it is also a seasonal model with $s = 4$ and $P_s = 1$, that is, as a seasonal model it is first order. In order to examine the nature of the model in (3.125), we generate a realization of length $n = 20$, which we interpret to be quarterly data for a 5 year period. The realization along with the true autocorrelations and AR-spectral density is shown in Figure 3.25. In Figure 3.25a, the vertical lines break the realization into years, and it can be seen that there is a strong correlation between a particular quarter and that same quarter the next year. For example, quarters 1,3, and 4 are typically relatively high while in each year quarter 2 shows a dip. The autocorrelations vividly illustrate the sense in which the model describes seasonal (in this case quarterly) behavior. In Figure 3.25b, it is seen that the only nonzero autocorrelations are at lags that are multiples of four. For example, there is no correlation between adjacent months. The plot of autocorrelations also illustrates the first-order seasonal nature of the model since the autocorrelations at lags that are multiples of four damp exponentially. The spectral density in Figure 3.25c shows peaks at frequencies 0, 0.25, and 0.5. This behavior is also illustrated in the factor table in Table 3.7 where it can also be seen that (3.125) is a stationary model since all of the roots of the characteristic equation are outside the unit circle.

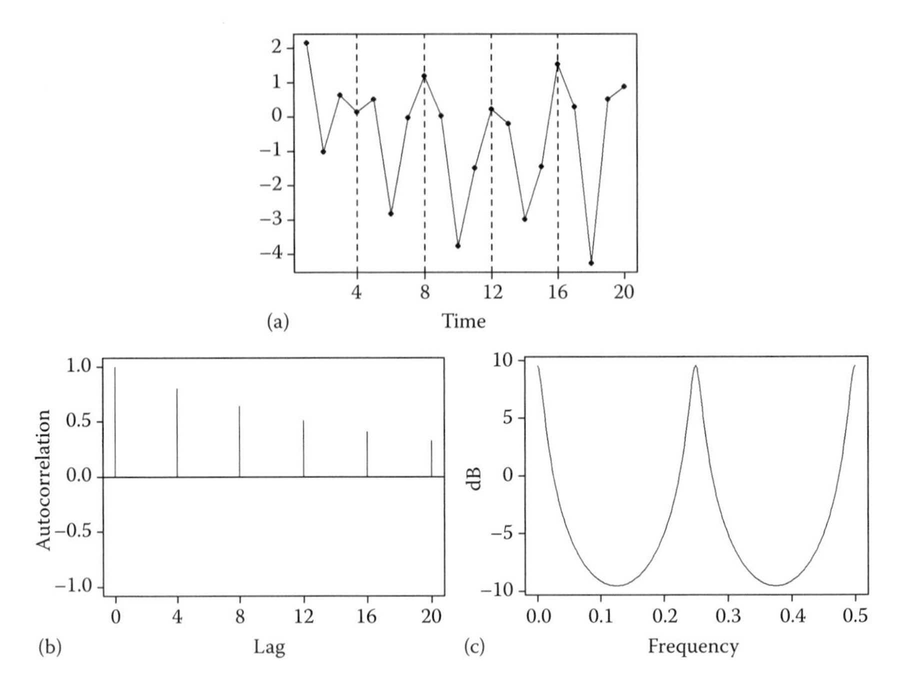

FIGURE 3.25
(a) Realization, (b) autocorrelations, and (c) spectral densities for the seasonal model $(1 - 0.8B^4)X_t = a_t$.

TABLE 3.7
Factor Table for Seasonal Model $(1 - 0.8B^4)X_t = a_t$

AR Factor	Roots (r_j)	$\lvert r_j^{-1}\rvert$	f_{0j}
$1 - 0.95B$	1.06	0.95	0.00
$1 + 0.95B$	−1.06	0.95	0.50
$1 + 0.89B^2$	$\pm 1.06i$	0.95	0.25

Using `GW-WINKS`: Data from the seasonal model in (3.125) and the factor table in Table 3.7 can be obtained using `Generate Realization` option on the `Time Series Analysis` menu and specifying an AR(4) with coefficients 0, 0, 0, and 0.8 respectively. The `Factoring Routines` option could also be used.

One feature of the model (3.125) that makes it less than ideal for practical situations is that there is very likely autocorrelation between adjacent or nearly adjacent quarters in addition to the quarterly behavior modeled by (3.125). Consequently, the purely seasonal model in (3.124) requires modification to allow for this type of behavior. The *multiplicative seasonal ARMA model*

$$\Phi(B^s)\phi(B)X_t = \Theta(B^s)\theta(B)a_t, \tag{3.126}$$

where the roots of $\phi(z) = 1 - \phi_1 z - \cdots - \phi_p z^p = 0$ and $\theta(z) = 1 - \theta_1 z - \cdots - \theta_q z^q = 0$ all lie outside the unit circle, is widely used for this purpose. The model in (3.126) is denoted by ARMA$(p,q) \times (P_s, Q_s)_s$.

Example 3.19 A Multiplicative Seasonal Stationary ARMA(1,0) × (1,0)$_s$ Model

As an example of a multiplicative seasonal ARMA model, we consider the model

$$(1 - 0.7B)(1 - 0.8B^4)X_t = a_t, \tag{3.127}$$

which is a modification of the seasonal model in (3.125). A realization of length $n = 20$ from this model along with the true autocorrelations and spectral density is shown in Figure 3.26. Figure 3.26a shows that the factor $1 - 0.7B$ induces within-year correlation, that is, all quarters in year one have lower values than the corresponding quarters in year two. This within-year correlation is also apparent in the autocorrelations in Figure 3.26b where there is a general damping behavior of autocorrelations with relatively higher correlations at lags that are multiples of four. The factor table for (3.127) (not shown) is the same as that shown in Table 3.7 except for an additional factor, $1 - 0.7B$, associated with a system frequency at $f = 0$. As a

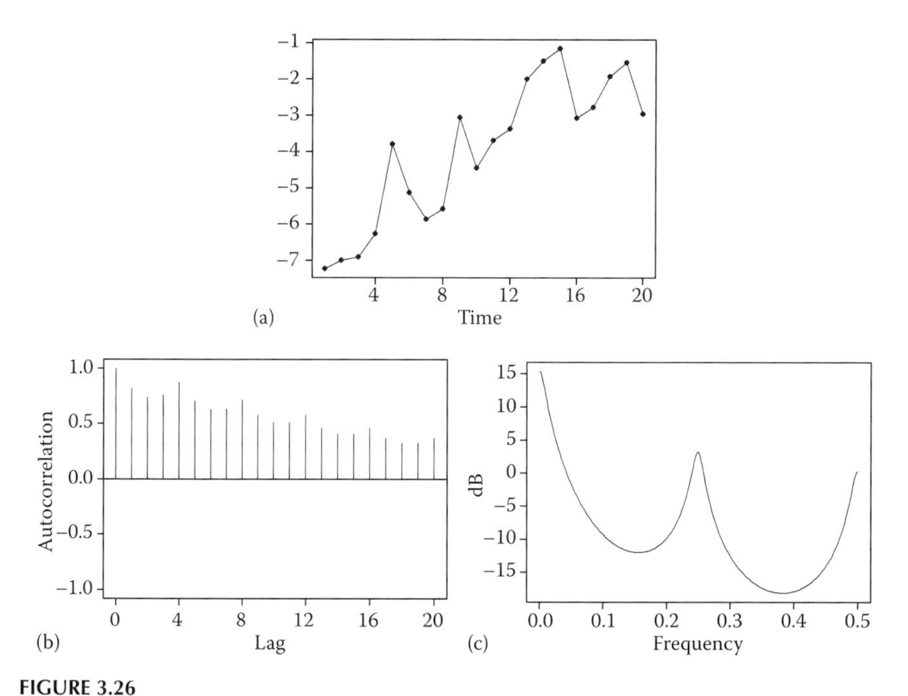

FIGURE 3.26
(a) Realization, (b) autocorrelations, and (c) spectral densities for the multiplicative seasonal model $(1 - 0.7B)(1 - 0.8B^4)X_t = a_t$.

result, the spectral density in Figure 3.26c has a similar behavior to that of Figure 3.25c except that the peak at $f = 0$ is higher.

We return to the multiplicative seasonal models in Chapter 5 when we discuss nonstationary models. In fact, the most common use of the seasonal model is the case in which the seasonal components of the model have roots on the unit circle.

3.6 Generating Realizations from ARMA(p,q) Processes

The study of time series analysis using ARMA(p,q) models often involves the simulation of finite realizations, $\{x_t, t = 1, \ldots, n\}$, from known ARMA models to provide information concerning the characteristics of typical realizations, autocorrelation functions, and spectral densities. These realizations are also useful in Monte Carlo studies, bootstrap resampling, etc. For these reasons, we include a brief discussion concerning the generation of ARMA(p,q) time series realizations.

3.6.1 MA(q) Model

Generating realizations from the MA(q) process in (3.1) is relatively simple. For known $\theta_k, k = 1, \ldots, q, \mu$ and σ_a^2, realization values are obtained from

$$\begin{aligned} x_1 &= \mu + a_1 - \theta_1 a_0 - \cdots - \theta_q a_{1-q} \\ x_2 &= \mu + a_2 - \theta_1 a_1 - \cdots - \theta_q a_{2-q} \\ &\vdots \end{aligned} \tag{3.128}$$

where the a_t values are generated using a random number generator for generating uncorrelated deviates with mean zero and variance σ_a^2. From (3.128), it is clear that the generation is initiated by obtaining the q uncorrelated starting values $a_0, a_{-1}, \ldots, a_{-(q-1)}$.

3.6.2 AR(2) Model

As an example of the complications that arise in computing appropriate starting values for stimulating realizations from ARMA(p,q) models with $p > 0$, we consider the AR(2) model with $\mu = 0$ and σ_a^2. For the AR(2) model, the observations are generated from

$$
\begin{aligned}
x_1 &= \mu(1-\phi_1-\phi_2)+\phi_1 x_0+\phi_2 x_{-1}+a_1 \\
x_2 &= \mu(1-\phi_1-\phi_2)+\phi_1 x_1+\phi_2 x_0+a_2 \\
&\vdots
\end{aligned}
\tag{3.129}
$$

From (3.129), we see that a white noise value a_1 and values x_0 and x_{-1} are required in order to initiate the simulation. Recall that the random variables X_0 and X_{-1} are uncorrelated with a_1 but are correlated with each other. Assuming normally distributed noise, x_0 and x_{-1} are values of a bivariate normal random variable (X_0, X_{-1}) that has mean μ and variance–covariance matrix,

$$
\Sigma = \frac{\sigma_a^2}{1-\phi_1\rho_1-\phi_2\rho_2}\begin{bmatrix} 1 & \dfrac{\phi_1}{1-\phi_2} \\ \dfrac{\phi_1}{1-\phi_2} & 1 \end{bmatrix},
$$

(See (3.43) and (3.56)).

3.6.3 General Procedure

It is apparent from the preceding AR(2) example that finding appropriate starting values can become difficult for higher-order models. For this reason, we typically use approximate techniques that are designed to be relatively insensitive to starting values. Since $\rho_k \to 0$ as $k \to \infty$ for a stationary ARMA(p,q) process, it follows that X_t is nearly uncorrelated with the starting values for sufficiently large t. That is, for $\varepsilon > 0$, there is a K such that $|\rho_k| < \varepsilon$ for $k > K$. One conservative technique is to choose K such $|r_{\max}^{-1}|^k < \varepsilon$ where $|r_{\max}^{-1}|$ is the maximum absolute reciprocal of the roots, r_j, of the characteristic equation $\phi(z) = 0$. In Table 3.8, we show the "warm-up"

TABLE 3.8

Warm-Up Length, K, for Various Values of ε and $|r_{\max}^{-1}|$

ε	$\lvert r_{\max}^{-1}\rvert$	K
0.01	0.90	44
0.05	0.90	29
0.01	0.95	90
0.05	0.95	59
0.01	0.99	459
0.05	0.99	299
0.01	0.999	4603
0.05	0.999	2995

length, K, sometimes referred to as the *induction period,* required for various values of ε and $|r_{\max}^{-1}|$.

A procedure for generating a realization of length n from an ARMA(p,q) model is to use arbitrarily selected starting values, generate a realization of length $n+K$ based on these starting values, and retain only the last n values as the realization $x_t, t = 1, \ldots, n$. According to the preceding discussion, these last n values are "nearly uncorrelated" with the starting values. This procedure is used in the `GW-WINKS` program where the starting values a_t and x_t are set equal to their unconditional expectations of zero and μ, respectively.

Using `GW-WINKS`: Realizations from ARMA models can be obtained using the `Generate Realization` option in the `Time Series Analysis` menu. The warm-up length is set to 2500, and realizations generated by `GW-WINKS` are based on normally distributed a_t.

3.7 Transformations

It will at times be useful to apply a transformation to time series data. In this section, we cover the basic issues regarding memoryless and AR transformations.

3.7.1 Memoryless Transformations

Sometimes, it will be helpful to transform each observation in order to produce data that more nearly satisfy underlying assumptions. In Figure 1.25a, we showed the logarithm of the number of airline passengers for the classical airline data. Figure 3.27a shows the "raw airline data" without applying the logarithmic transformation, and in Figure 3.27b, we show the log airline data as previously shown in Figure 1.27a. The "raw" airline

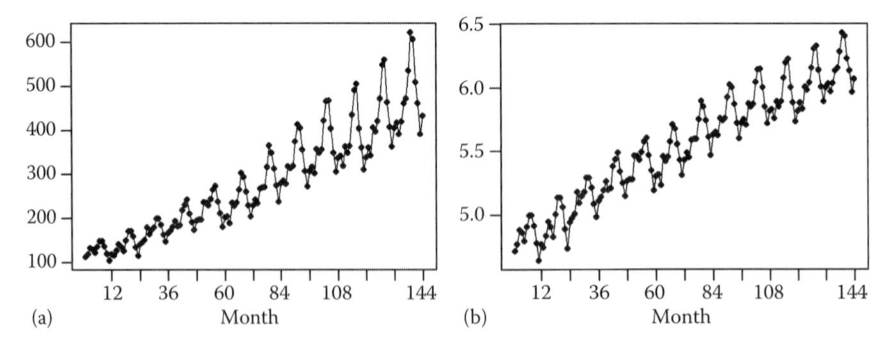

FIGURE 3.27
(a) Airline data before and (b) after making a logarithmic transformation.

data in Figure 3.27a do not appear to be a realization from a stationary process, since the variance seems to be increasing with time. In such cases, a variance stabilizing transformation such as the logarithmic transformation could be applied to produce the series $Y_t = \log(X_t)$ such as the one in Figure 3.27b that more nearly satisfies the equal variance condition of stationarity. Techniques for parameter estimation, model identification, and forecasting the airline data, that will be given in chapters to follow, are more appropriately applied to Y_t. The logarithmic transformation is an example of a *memoryless transformation*. Box and Cox (1964) introduced a general class of memoryless transformations that have been used on time series data. For a given λ, such that $0 < \lambda \leq 1$, the transformation is specified by

$$Y_t = \frac{X_t^\lambda - 1}{\lambda}. \tag{3.130}$$

The logarithmic transformation applied to the airline data is a limiting case of this class as $\lambda \to 0$, which along with the square root (a scaled version of (3.130) when $\lambda = 1/2$) are among the more popular transformations of this type. It should be noted that in order to apply the logarithmic, square root, or any other of the Box-Cox transformations, it is necessary that $x_t > 0$, $t = 1, \ldots, n$.

Using `GW-WINKS`: You can apply transformations using the `Edit` menu. For example, a logarithmic transformation can be obtained by choosing the `New Variable, Common Transformations` option, and then selecting by `Natural Log` or by `Log(Base 10)`. A new column in the spreadsheet will be created containing the transformed data.

3.7.2 Autoregressive Transformations

A second type of transformation, referred to here as an *autoregressive transformation*, is also useful. Consider the AR(1) model

$$(1 - 0.95B)X_t = a_t,$$

and let Y_t be defined by

$$\begin{aligned} Y_t &= X_t - 0.95X_{t-1} \\ &= (1 - 0.95B)X_t. \end{aligned} \tag{3.131}$$

It follows that Y_t is white noise since

$$\begin{aligned} Y_t &= (1 - 0.95B)X_t \\ &= a_t. \end{aligned}$$

Thus, the AR model itself can be viewed as a transformation to white noise or as a "whitening filter." In fact, this is one of the primary applications of the AR model. Note that application of transformation (3.131) to a realization of length n produces a "new" realization of length $n-1$ given by

$$Y_1 = X_2 - 0.95X_1$$
$$Y_2 = X_3 - 0.95X_2$$
$$\vdots$$
$$Y_{n-1} = X_n - 0.95X_{n-1}.$$

In Figure 3.28a, a realization of length $n = 100$ is displayed from the AR(1) process, X_t, given in (3.131). This realization was previously shown in Figure 3.5a. In Figure 3.28b, we show the corresponding realization of length $n = 99$ for Y_t. The realization for X_t displays the behavior discussed in Section 3.2 for an AR(1) process whose characteristic equation has a positive root. The realization for Y_t in Figure 3.28b has the appearance of white noise.

More generally for the ARMA(p,q) model

$$\phi(B)X_t = \theta(B)a_t, \tag{3.132}$$

we consider the problem of determining the model satisfied by Y_t, where

$$Y_t = \alpha(B)X_t,$$

and $\alpha(B)$ is of order k, that is, $\alpha(B) = 1 - \alpha_1 B - \cdots - \alpha_k B^k$. Thus, $X_t = \alpha^{-1}(B)Y_t$, and (3.132) can be rewritten as

$$\phi(B)\alpha^{-1}(B)Y_t = \theta(B)a_t.$$

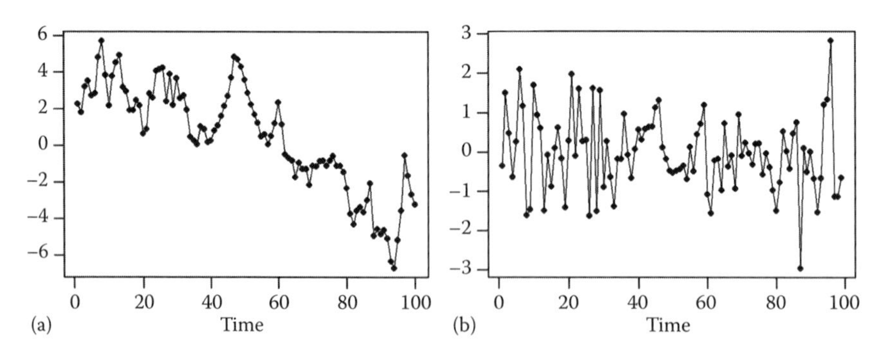

FIGURE 3.28
Effect of transforming realization from an AR(1) model. (a) $X_t - 0.95X_{t-1} = a_t$. (b) $Y_t = X_t - 0.95X_{t-1}$.

There are two cases of primary interest.

Case (1) $\alpha(B)$ is a factor of $\phi(B)$
In this case, $k \leq p$ and $\phi(B)$ can be factored as $\phi_1(B)\alpha(B)$, so that formally

$$\begin{aligned}\phi(B)\alpha^{-1}(B) &= \phi_1(B)\alpha(B)\alpha^{-1}(B) \\ &= \phi_1(B),\end{aligned}$$

that is $\phi_1(B)Y_t = \theta(B)a_t$, and Y_t satisfies an ARMA $(p - k, q)$ model.

Case (2) $\alpha(B)$ is not a factor of $\phi(B)$ and has no common subfactors
In this case

$$\phi(B)\alpha^{-1}(B)Y_t = \theta(B)a_t,$$

or

$$\phi(B)Y_t = \alpha(B)\theta(B)a_t,$$

that is, Y_t is ARMA $(p, q + k)$.

Example 3.20 An Autoregressive Transformation in Which $\alpha(B)$ Is a Factor of $\phi(B)$

Consider the AR(4) model previously discussed in Example 3.10. In factored form, this model is given by $(1 - 0.95B)(1 - B + 0.8B^2)$ $(1 + 0.8B)X_t = a_t$. In Figure 3.29a and b, we show a realization from this AR(4) model and the theoretical spectral density. These plots were previously shown in Figure 3.16a and c, respectively. In the factor table given in Table 3.3, it can be seen that these factors, in the order given, are associated with system frequencies $f_0 = 0$, 0.16, and 0.5, respectively, which is consistent with the spectral density in Figure 3.29b. We now transform to obtain $Y_t = \alpha(B)X_t = (1 - B + 0.8B^2)X_t$. In this case, it is seen that Y_t satisfies the model $(1 - 0.95B)(1 + 0.8B)Y_t = a_t$, that is, the transformed model is an AR (2). In Figure 3.29c, we show the transformed data $y_t = (1 - B + 0.8B^2)x_t$ and, in Figure 3.29d, we show the Parzen spectral density estimate of the transformed data. In these figures, it can be seen that the transformation acts as a *band-stop* or *notch filter* discussed in Section 2.4. That is, the transformation removes the effect of the factor $1 - B + 0.8B^2$, which from a practical standpoint removes the frequency behavior in the neighborhood of $f_0 = 0.16$.

Using `GW-WINKS`: You can perform an autoregressive transformation using the `New Variable, Common Transformations` on the `Edit` menu, and then choosing `by Autoregressive transformation`. The parameters of the transformation are separated by commas. For example, to apply the transformation $Y_t = (1 - B + 0.8B^2)X_t$ in Example 3.20, you would enter "1,–0.8" (without the quotes) in the coefficient entry box.

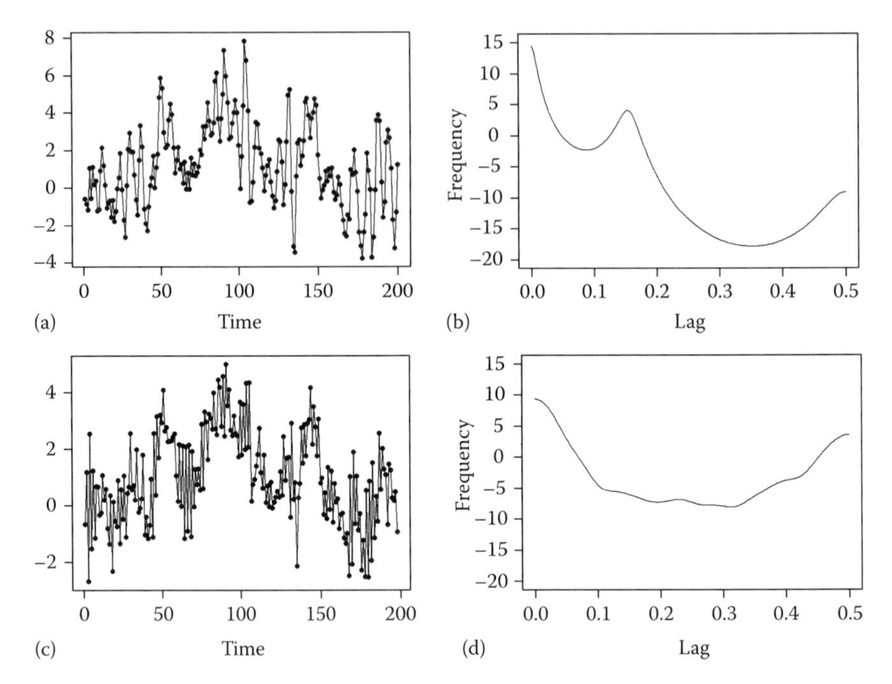

FIGURE 3.29
(a) Realization from the AR(4) model $(1 - .95B)(1 - B + 0.8B^2)(1 - 0.8B)X_t = a_t$, (b) spectral density of the AR(4) model, (c) data in (a) transformed by $Y_t = (1 - B + 0.8B^2)X_t$, and (d) Parzen spectral density estimator of the data in (c).

Example 3.21 Autoregressive Transformations

Let X_t be given by

$$(1 - 2.5B + 2.34B^2 - 0.8B^3)X_t = (1 + 0.8B)a_t$$

which in factored form is

$$(1 - 0.9B)(1 - 1.6B + 0.9B^2)X_t = (1 + 0.8B)a_t.$$

If $Y_t = (1 - 0.9B)X_t$, then

$$(1 - 1.6B + 0.9B^2)Y_t = (1 + 0.8B)a_t$$

and Y_t satisfies an ARMA(2,1) model illustrating Case (1). However, if X_t is given by

$$(1 - 1.3B + 0.8B^2)X_t = (1 + 0.8B)a_t$$

and $Y_t = (1 - 0.9B)X_t$, then since $1 - 0.9B$ is not a factor of $(1 - 1.3B + 0.8B^2)$, it follows that

$$\begin{aligned}(1 - 1.3B + 0.8B^2)Y_t &= (1 + 0.8B)(1 - 0.9B)a_t \\ &= (1 - 0.1B - 0.72B)a_t,\end{aligned}$$

and consequently, Y_t is an ARMA(2,2) process exemplifying Case (2).

3.A Appendix: Proofs of Theorems

Proof of Theorem 3.1

AR(1) Case: We obtained the result for the AR(1) in (3.23) and the discussion that preceded it.

AR(2) Case: We begin the proof by showing the complete development for an AR(2) model and then discuss the extension to the general AR(p) case. We consider the AR(2) model

$$\phi(B)X_t = (1 - \alpha_1 B)(1 - \alpha_2 B)X_t = a_t,$$

where α_1 and α_2 are real or complex, $\alpha_1 \neq \alpha_2$, and $|\alpha_i| < 1, i = 1, 2$. Using ∇ to denote the operator $1 - B$, we have

$$\alpha^t \nabla \alpha^{-t} X_t \equiv (1 - \alpha B)X_t.$$

So,

$$\alpha_1^t \nabla \left(\frac{\alpha_2}{\alpha_1}\right)^t \nabla \alpha_2^{-t} X_t = (1 - \alpha_1 B)(1 - \alpha_2 B)X_t,$$

and it follows that

$$\nabla \left(\frac{\alpha_2}{\alpha_1}\right)^t \nabla \alpha_2^{-t} X_t = \alpha_1^{-t} a_t.$$

Then, summing both sides gives

$$\sum_{k=-\infty}^{t} \nabla \left(\frac{\alpha_2}{\alpha_1}\right)^k \nabla \alpha_2^{-k} X_k = \sum_{k=-\infty}^{t} \alpha_1^{-k} a_k,$$

so

$$\left(\frac{\alpha_2}{\alpha_1}\right)^t \nabla\alpha_2^{-t}X_t - \lim_{k\to-\infty}\left(\frac{\alpha_2}{\alpha_1}\right)^k \nabla\alpha_2^{-k}X_k = \sum_{k=-\infty}^{t}\alpha_1^{-k}a_k,$$

and

$$\lim_{k\to-\infty}\left(\frac{\alpha_2}{\alpha_1}\right)^k \nabla\alpha_2^{-k}X_k = 0$$

with probability 1. Thus,

$$\nabla\alpha_2^{-t}X_t = \left(\frac{\alpha_1}{\alpha_2}\right)^t \sum_{k=-\infty}^{t}\alpha_1^{-k}a_k,$$

and similarly

$$\begin{aligned} X_t &= \alpha_2^t \sum_{j=-\infty}^{t}\left(\frac{\alpha_1}{\alpha_2}\right)^j \sum_{k=-\infty}^{j}\alpha_1^{-k}a_k \\ &= \alpha_2^t \sum_{k=-\infty}^{t}\left[\sum_{j=k}^{t}\left(\frac{\alpha_1}{\alpha_2}\right)^j\right]\alpha_1^{-k}a_k. \end{aligned}$$

Since

$$\sum_{j=k}^{t}\left(\frac{\alpha_1}{\alpha_2}\right)^j = \frac{\left(\frac{\alpha_1}{\alpha_2}\right)^k - \left(\frac{\alpha_1}{\alpha_2}\right)^{t+1}}{1-\frac{\alpha_1}{\alpha_2}},$$

it follows that

$$X_t = \sum_{k=-\infty}^{t}\frac{\alpha_2^{t-k+1}-\alpha_1^{t-k+1}}{\alpha_2-\alpha_1}a_k.$$

Then, letting $\ell = t-k$, we obtain

$$\begin{aligned} X_t &= \sum_{\ell=0}^{\infty}\frac{\alpha_2^{\ell+1}-\alpha_1^{\ell+1}}{\alpha_2-\alpha_1}a_{t-\ell}. \\ &= \sum_{\ell=0}^{\infty}\psi_\ell a_{t-\ell}. \end{aligned}$$

Note that this produces a closed form formula for the ψ-weights of an AR(2) given by

$$\psi_\ell = \frac{\alpha_2^{\ell+1} - \alpha_1^{\ell+1}}{\alpha_2 - \alpha_1}.$$

AR(p) case: Suppose X_t satisfies the AR(p) model $\phi(B)X_t = a_t$. Note that we can always write

$$\phi(B) = \prod_{i=1}^{p} (1 - \alpha_i B),$$

where the α_i's (the reciprocals of the roots of the characteristic equation $\phi(r) = 0$) are real or complex. Further assume that $\alpha_i \neq \alpha_j$ when $i \neq j$. Then, the AR(p) model can be written as

$$\phi(B)X_t = (1 - \alpha_1 B)(1 - \alpha_2 B) \cdots (1 - \alpha_p B)X_t = a_t,$$

which becomes

$$\alpha_1^t \nabla \alpha_1^{-t} \alpha_2^t \nabla \alpha_2^{-t} \cdots \alpha_p^t \nabla \alpha_p^{-t} X_t = a_t. \tag{3.A.1}$$

Further assuming $|\alpha_i| < 1, i = 1, \ldots, p$, then using the procedure described in the AR(1) and AR(2) cases yields the result.

Proof of Theorem 3.2

(Necessity) Suppose the roots of the characteristic equation all lie outside the unit circle. Then, we need to show that $\sum_{\ell=0}^{\infty} |\psi_\ell(p)| < \infty$. Now if $p = 1$, then $\psi_\ell(p) = \alpha_1^\ell$, and since $|\alpha_1| < 1$, then $\sum_{\ell=0}^{\infty} |\psi_\ell(p)| = \sum_{\ell=0}^{\infty} |\alpha_1|^\ell < \infty$. For $p > 1$, we rewrite (3.27) (b) as

$$\sum_{k=1}^{p-1} c_k \left(\alpha_p^{\ell+1} - \alpha_{p-k}^{\ell+1}\right),$$

and note that the coefficients,

$$c_k = \frac{\alpha_{p-k}^{p-2}}{(\alpha_p - \alpha_{p-k}) \prod_{\substack{h=1 \\ h \neq p-k}}^{p-1} (\alpha_{p-k} - \alpha_h)},$$

do not depend on ℓ and are finite-valued for each k, since there are no repeated roots. So,

$$\begin{aligned}\sum_{\ell=0}^{\infty}|\psi_\ell(p)| &= \sum_{\ell=0}^{\infty}\left|\sum_{k=1}^{p-1}c_k\left(\alpha_p^{\ell+1}-\alpha_{p-k}^{\ell+1}\right)\right| \\ &\le \sum_{\ell=0}^{\infty}\sum_{k=1}^{p-1}|c_k|\left(\left|\alpha_p^{\ell+1}\right|+\left|\alpha_{p-k}^{\ell+1}\right|\right) \\ &= \sum_{k=1}^{p-1}\sum_{\ell=0}^{\infty}|c_k|\left(\left|\alpha_p^{\ell+1}\right|+\left|\alpha_{p-k}^{\ell+1}\right|\right) \\ &= \sum_{k=1}^{p-1}|c_k|\sum_{\ell=0}^{\infty}\left|\alpha_p^{\ell+1}\right|+\sum_{k=1}^{p-1}|c_k|\sum_{\ell=0}^{\infty}\left|\alpha_{p-k}^{\ell+1}\right|.\end{aligned}$$

Since $\sum_{\ell=0}^{\infty}\left|\alpha_p^{\ell+1}\right| < \infty$ and $\sum_{\ell=0}^{\infty}\left|\alpha_{p-k}^{\ell+1}\right| < \infty$ if the roots of the characteristic equation are all outside the unit circle, then $\sum_{l=0}^{\infty}\psi_l(p)$ converges absolutely. It follows that the right side of (3.26) converges in mean square, and X_t is a causal stationary solution of (3.25).

(Sufficiency) See Theorem 3.1.1 Brockwell and Davis (1991).

Schur's lemma: A necessary and sufficient condition that the roots of the algebraic equation

$$\phi(z) = 1-\phi_1 z-\phi_2 z^2-\cdots-\phi_p z^2 = 0$$

all lie outside the unit circle is that the quadratic form

$$\sum_{j=0}^{p-1}\left(\left\{\sum_{k=0}^{p-j-1}\phi_k Y_{j+k}\right\}^2-\left\{\sum_{k=0}^{p-j-1}\phi_{p-k}Y_{j+k}\right\}^2\right), \tag{3.A.2}$$

where $\phi_0 = -1$ is positive definite. By expanding (3.A.2), one can see that the symmetric matrix, C, associated with (3.A.2) is given by

$$C = (c_{jk}),$$

where

$$c_{jk} = \sum_{h=1}^{\min(j,k)}(\phi_{j-h}\phi_{k-h}-\phi_{p+h-j}\phi_{p+h-k}).$$

Proof:

See Henrici (1974).

Proof of Theorem 3.3

(Sufficiency) The fact that all roots of $\theta(z) = 0$ lie outside the unit circle implies that $\theta(z) \neq 0$ for $|z| \leq 1$, and, hence, there exists an $\varepsilon > 0$ such that $\theta^{-1}(z)$ has a convergent power series expansion for $|z| < 1 + \varepsilon$. So, from (3.82),

$$|\theta^{-1}(z)| = \left| \sum_{j=0}^{\infty} \pi_j z^j \right| < \infty \tag{3.A.3}$$

for $|z| < 1 + \varepsilon$. Now, from (3.A.3) it follows that $\pi_j(1 + \varepsilon/2)^j \to 0$ as $j \to 0$. Thus, there is an $M > 0$, such that $\left|\pi_j(1 + \varepsilon/2)^j\right| < M$ for all $j = 0, 1, \ldots$. Thus, since

$$|\pi_j| < |M(1 + \varepsilon/2)^{-j}|$$

and

$$\sum_{j=0}^{\infty} |M(1 + \varepsilon/2)^{-j}| < \infty,$$

it follows that $\sum_{j=0}^{\infty} |\pi_j| < \infty$, and thus that X_t is invertible. It can be shown rigorously that under the conditions of the theorem, $\sum_{j=0}^{\infty} \pi_j X_{t-j} = a_t$, that is, $\theta^{-1}(B)$ as defined in (3.82) is the inverse operator of $\theta(B)$ (see Proposition 3.1.2 Brockwell and Davis, 1991).

(Necessity) Suppose X_t is invertible, that is, $\sum_{j=0}^{\infty} \pi_j X_{t-j} = a_t$ where $\sum_{j=0}^{\infty} |\pi_j| < \infty$. By essentially reversing the steps used in the proof of sufficiency, it can be shown that $\theta^{-1}(z) < \infty$ for $|z| \leq 1$, that is, $\theta(z) \neq 0$ for $|z| \leq 1$. Thus, all of the roots of $\theta(z) = 0$ are outside the unit circle.

Exercises

Applied Problems

If you use `GW-WINKS` to work the following problems, you will find the `Generate Realization`, `Factoring Routines`, `Sample Autocorrelations`, and `Spectrum` options on the `Time series Analysis` menu to be useful.

3.1 Generate and plot a realization of length $n = 200$ from the MA(3) process

$$X_t - 25 = a_t - 0.95a_{t-1} + 0.9a_{t-2} - 0.855a_{t-3}$$

where $\sigma_a^2 = 1$. Plot the following:
(a) The realization generated
(b) The theoretical autocorrelations
(c) The theoretical spectral density
Discuss the information available in the plots. Also find:
(d) μ and σ_X^2
(e) $\hat{\mu}$ and $\hat{\sigma}_X^2$ for the realization in (a)

3.2 Repeat Problem 3.1 for realizations from the AR(p) models:
(i) $X_t - 0.95X_{t-1} = a_t$ where $\sigma_a^2 = 1$.
(ii) $(X_t - 10) - 1.5(X_{t-1} - 10) + 0.9(X_{t-2} - 10) = a_t$ where $\sigma_a^2 = 1$.

3.3 Determine whether the following AR(p) models are stationary and explain your answers.
(a) $X_t - 1.5X_{t-1} + X_{t-2} - 0.25X_{t-3} = a_t$
(b) $X_t - 2X_{t-1} + 1.76X_{t-2} - 1.6X_{t-3} + 0.77X_{t-4} = a_t$
(c) $X_t = a_t - 2a_{t-1} + 1.76a_{t-2} - 1.6a_{t-3} + 0.77a_{t-4}$
(d) $X_t - 1.9X_{t-1} + 2.3X_{t-2} - 2X_{t-3} + 1.2X_{t-4} - 0.4X_{t-5} = a_t$

3.4 Using the same random number seed in each case (i.e., in `GW-WINKS` by using the same negative seed), generate realizations of length 200 from the AR(1) processes $(1 - \phi_1 B)X_t = a_t$ for $\phi_1 = \pm 0.9, \pm 0.5$ and $\sigma_a^2 = 1$.
(a) Plot the true autocorrelations.
(b) Plot the true spectral density.
(c) Find σ_X^2.
(d) Plot the realizations.
(e) Plot the sample autocorrelations.
(f) For $\phi_1 = 0.9$, repeat steps (a)–(c) with $\sigma_a^2 = 10$. What differences and similarities do you observe?

3.5 Using (3.46), find the autocorrelation function for the AR(3) process

$$X_t - 2.3X_{t-1} + 2.1X_{t-2} - 0.7X_{t-3} = a_t.$$

Use this expression to calculate the autocorrelations for lags 1, 2, 3, 5, and 10. Check your answers using the `Generate Realizations` option in the `GW-WINKS` program. Also, find γ_0 if $\sigma_a^2 = 2$. *Hint*: You can use `GW-WINKS Factoring Routines` option to find the roots and factors of the third-order polynomial.

3.6 Consider the model

$$X_t - X_{t-1} + 0.26X_{t-2} + 0.64X_{t-3} - 0.576X_{t-4} = a_t - 2.4a_{t-1} + 2.18a_{t-2} - 0.72a_{t-3}$$

(a) This specifies an ARMA(p,q) model. What are p and q? *Hint*: Factor the polynomials.

(b) Using the result in (a), write the equation of the corresponding ARMA(p,q) model.

(c) Show that the model (as expressed in (b)) is stationary and invertible.

3.7 Consider the following models:

(a) $X_t - 0.1X_{t-1} + 0.5X_{t-2} + 0.08X_{t-3} - 0.24X_{t-4} = a_t$

(b) $X_t - 1.3X_{t-1} + 0.4X_{t-2} = a_t - 1.9a_{t-1}$

(c) $X_t - 1.9X_{t-1} = a_t - 1.3a_{t-1} + 0.4a_{t-2}$

(d) $X_t - 2.95X_{t-1} + 3.87X_{t-2} - 2.82X_{t-3} + 0.92X_{t-4} = a_t - 0.9a_{t-1}$

(e) $(1 - B - 0.49B^2 + 0.9B^3 - 0.369B^4)X_t = (1 + B + B^2 + 0.75B^3)a_t$

For each process:

(i) Determine whether the process is stationary.

(ii) Determine whether the process is invertible.

(iii) Plot the true autocorrelations and spectral density for the processes that are both stationary and invertible. Discuss how the features of the model as shown in the factor table are visually displayed in the autocorrelations and spectrum.

3.8 Show that each of the following MA(2) processes is not invertible. In each case, find an invertible MA(2) process that has the same autocorrelations as X_t. Use `GW-WINKS` to verify that the autocorrelations are the same.

(a) $X_t = a_t - 1.1a_{t-1} + 1.8a_{t-2}$

(b) $X_t = a_t - 2a_{t-1} + 1.5a_{t-2}$

3.9 Consider the model

$$(1 - 1.61B + 1.91B^2 - 1.61B^3 + 0.91B^4)(X_t + 75) = a_t.$$

Generate a realization of length $n = 200$ from this model where $\sigma_a^2 = 10$. Transform the data to obtain a realization on $Y_t = X_t + X_{t-2}$.

(a) Plot the following:

(i) Realization and transformed realization

(ii) Sample autocorrelations from original and transformed data

(iii) Spectral density estimates (Tukey window with same truncation point) from original and transformed data

(b) Discuss the features of the original model and the effect of the transformation as displayed on the three types of plots.

(c) Interpret the effect of the transformation by examining the model itself.

3.10 Let X_t be an ARMA(3,2) process specified by

$$X_t + 0.1X_{t-1} - 0.4X_{t-2} - 0.45X_{t-3} = 25 + a_t - 0.4a_{t-1} - 0.6a_{t-2}$$

Let $Y_t = X_t - 0.3X_{t-1} - 0.54X_{t-2}$.

(a) What ARMA model does Y_t satisfy?

(b) Is Y_t stationary? Invertible?

3.11 Consider the ARMA(4,3) model

$$(1 - B - 0.49B^2 + 0.9B^3 - 0.369B^4)X_t = (1 + B + B^2 + 0.75B^3)a_t$$

(a) Verify that this is a stationary and invertible process.
(b) Find $\psi_i, i = 1, 2, \ldots, 10$.

3.12 Data file SUN.XLS in the GW-WINKS library of data files contains the yearly sunspot averages for the 176 year period 1749–1924.

(a) Using GW-WINKS print plots of the sunspot data, sample autocorrelations, and a spectral density estimator for this data set.

The following two models (the mean is ignored here and is assumed to be zero, i.e., $X_t = Z_t - \bar{Z}$ where Z_t is the actual sunspot data) have been proposed for the mean adjusted sunspot data:

(i) $X_t - 1.42X_{t-1} + 0.73X_{t-2} = a_t$
(ii) $X_t - 1.23X_{t-1} + 0.47X_{t-2} + 0.14X_{t-3} - 0.16X_{t-4} + 0.14X_{t-5} - 0.07X_{t-6} + 0.13X_{t-7} - 0.21X_{t-8} = a_t$

Using GW-WINKS, for each of these models:

(b) Generate a realization of length $n = 176$.
(c) Plot the realization and the true autocorrelations and true spectral density for model.
(d) Discuss the following:
   (1) What feature of each of the models produces the dominant 10–11 year period?
   (2) What is the major difference between the two models concerning this periodic component?
   (3) Based on these plots, which model appears to be the better model for the sunspot data? (Clearly neither model accounts for the fact that there is more variability in the peaks than the troughs. Ignoring this problem that both models have, which seems better and why?)

3.13 Generate realizations of length $n = 100$ from the following stationary AR(p) models

(i) $(1 - 2.2B + 2.1B^2 - 0.8B^3)(X_t - 25) = a_t$
(ii) $(1 - 0.42B - 0.34B^2 + 0.91B^3)(X_t + 10) = a_t$
(iii) $(1 - 2.09B + 2.56B^2 - 1.96B^3 + 0.88B^4)(X_t - 100) = a_t$

For each of these realizations

(a) Plot the realization along with the additive AR components (see Section 3.4).
(b) Discuss your results relating each additive component to one of the lines in the factor table.

NOTE: To best visualize the contribution of each component, use the same vertical scale on each of the plots associated with a given realization.

3.14 Use the `Factoring Routines` module in `GW-WINKS` to fit an AR (8) model to the sunspot data in SUN.XLS using Burg estimates. Then create the additive AR components and plot the data along with the first four components. Discuss the contributions of each component.

Theoretical Problems

3.15 Show that the function $f(u) = |u|/(1 + u^2)$ has a maximum of 0.5, and thus for an MA(1) process, $\max|\rho_1| = 0.5$.

3.16 (a) For an MA(2), show that $|\rho_1| \le \sqrt{2}/2$ and $|\rho_2| \le 0.5$.
(b) For an MA(q) process, show that $|\rho_q| \le 0.5$.

3.17 Let X_t be a stationary zero-mean time series and define $Y_t = X_t - 0.4X_{t-1}$ and $W_t = X_t - 2.5X_{t-1}$. Express the autocovariance function of Y_t and W_t in terms of the autocovariance function of X_t and show that Y_t and W_t have the same autocovariance functions.

3.18 Using the formula in (3.63) (not `GW-WINKS`):
(a) Find the system frequency f_0 associated with the AR(2) process $X_t - 1.3X_{t-1} + 0.85X_{t-2} = a_t$
(b) Find ϕ_1, so that the AR(2) model $X_t - \phi_1 X_{t-1} + 0.9X_{t-2} = a_t$ has system frequency $f_0 = 0.4$.

3.19 Consider the AR(2) model $X_t - \phi_1 X_{t-1} - \phi_2 X_{t-2} = a_t$ for which the roots, r_1, r_2 of the characteristic equation are complex conjugates whose reciprocals are $a \pm bi$. Show that $\phi_2 = -1/|r_j|^2$ and $\phi_1 = 2a$.

3.20 (a) Let X_t satisfy the AR(2) model $X_t - \phi_1 X_{t-1} - \phi_2 X_{t-2} = a_t$ for which the roots of the characteristic equation are complex conjugates. Show that

(i) If $|\phi_1| < \dfrac{-4\phi_2}{1 - \phi_2}$, then the peak in the AR(2) spectral density occurs at

$$f_s = \frac{1}{2\pi}\cos^{-1}\left(\frac{\phi_1(\phi_2 - 1)}{4\phi_2}\right).$$

(ii) If $\phi_1 > \dfrac{-4\phi_2}{1 - \phi_2}$, then the peak in the spectral density occurs at frequency 0.

(iii) If $\phi_1 < \dfrac{4\phi_2}{1 - \phi_2}$, then the peak in the spectral density occurs at frequency 0.5.

(b) Suppose $|\phi_1| < \dfrac{-4\phi_2}{1 - \phi_2}$ for a given ϕ_1 and ϕ_2. Then if ϕ_1 stays fixed as $|\phi_2| \to 1$, show $f_s \to f_0$ approaches f_0, the system frequency.

(c) For the AR(2) model $X_t - 1.42X_{t-1} + 0.73X_{t-2} = a_t$ fit to the sunspot data, find f_s and f_0.

3.21 Consider the AR(2) model given in factored form by $(1 - 0.8B)(1 + 0.5B)(X_t - 50) = a_t$, where $\sigma_a^2 = 5$.
(a) Use the Yule–Walker equations to find ρ_1 and ρ_2.
(b) Find C_1 and C_2 in the general expression for ρ_k.

$$\rho_k = C_1(r_1)^{-k} + C_2(r_2)^{-k}.$$

(c) Use the expression in (b) to find ρ_8.

3.22 Schur's lemma (see Appendix 3.A) gives necessary and sufficient conditions for the roots of a polynomial equation to all lie outside the unit circle.
(a) Verify that the matrix C described in that lemma in the case of a second-order model is given by

$$C = \begin{bmatrix} 1 - \phi_2^2 & -\phi_1(1 + \phi_2) \\ -\phi_1(1 + \phi_2) & 1 - \phi_2^2 \end{bmatrix}.$$

(b) By definition, C is positive definite if and only if $1 - \phi_2^2 > 0$ and $(1 + \phi_2)^2\{(1 - \phi_2)^2 - \phi_1^2\} > 0$. Using this, show that the AR(2) process is stationary if and only if ϕ_1 and ϕ_2 satisfy

$$|\phi_2| < 1$$
$$\phi_1 + \phi_2 < 1$$
$$\phi_2 - \phi_1 < 1.$$

3.23 (a) Equation 3.60 gives a formula for the autocorrelation function for an AR(2) with distinct roots. Show that in the case of complex conjugate roots, this can be written

$$\rho_k = \left(\sqrt{-\phi_2}\right)^k \left\{\frac{\sin 2\pi(k+1)f_0 + \phi_2 \sin 2\pi(k-1)f_0}{(1 - \phi_2)\sin 2\pi f_0}\right\},$$

where f_0 is defined in (3.63).
(b) Show that the autocorrelation function in (a) can be simplified to the form in (3.62).

3.24 Consider an AR(2) model with complex roots, $r_j = a \pm bi$. Let $\alpha_1 = r_1^{-1} = (a + ib)^{-1}$, $\alpha_2 = r_2^{-1} = (a - ib)^{-1}$, and $R = \sqrt{a^2 + b^2}$ where $b \neq 0$. Thus, $\alpha_1 = R^{-1}e^{-i\xi}$ and $\alpha_2 = R^{-1}e^{i\xi}$, where $\xi = \arctan b/a$
(a) Using (3.27), show that $\psi_\ell(2) = \dfrac{\alpha_2^{\ell+1} - \alpha_1^{\ell+1}}{\alpha_2 - \alpha_1}$.
(b) From (a) show that

$$\psi_\ell(2) = R^{-\ell-1}\left[\frac{e^{i\xi(\ell+1)} - e^{-i\xi(\ell+1)}}{2iR^{-1}\sin\xi}\right]$$
$$= R^{-\ell}\frac{\sin \xi(\ell+1)}{\sin \xi}.$$

3.25 Use (3.27) to show that the ψ-weights for an AR(3) are given by the following:

$$\psi_\ell(3) = \frac{\alpha_2\left(\alpha_3^{\ell+1} - \alpha_2^{\ell+1}\right)}{(\alpha_3 - \alpha_2)(\alpha_2 - \alpha_1)} + \frac{\alpha_1\left(\alpha_3^{\ell+1} - \alpha_1^{\ell+1}\right)}{(\alpha_3 - \alpha_1)(\alpha_1 - \alpha_2)}.$$

3.26 Consider the ARMA(1,2) process $(1 - 0.8B)X_t = (1 - 0.5B + 0.9B^2)a_t$ with $\sigma_a^2 = 1$. Find γ_0, γ_1, and ρ_5 from first principles.

3.27 Show that a stationary ARMA(p,q) process is ergodic for μ.

3.28 Let a_t and w_t be two uncorrelated white noise processes, each with mean zero and variance 1. Define $Y_t - 50 = a_t + (1 - 0.7B)w_t$.
(a) Show that Y_t is covariance stationary.
(b) What ARMA(p,q) process has the same autocorrelations as Y_t? Specify, p, q, and the other model parameters.

3.29 The "inverse autocovariance" at lag k of a stationary and invertible time series, X_t, is given by $\gamma_k^{(I)} = \int_{-0.5}^{0.5} e^{2\pi i f k} \frac{1}{P_X(f)}\, df$, where $P_X(f)$ is the spectrum of X_t. The inverse autocorrelation at lag k is given by $\rho_k^{(I)} = \gamma_k^{(I)}/\gamma_0^{(I)}$. Find $\gamma_0^{(I)}$ and $\rho_2^{(I)}$ for the MA(1) process $X_t = a_t - 0.9a_{t-1}$, where $\sigma_a^2 = 4$.

3.30 Let X_t satisfy the ARMA(2,2) model $(1 - 1.2B + 0.95B^2)X_t = (1 + 1.2B + 0.95B^2)a_t$.
(a) Find ψ_1, ψ_2, and ψ_3 for this model.
(b) Write down a mathematical expression for the spectral density of X_t.
(c) Specify the model satisfied by $Y_t = X_t + X_{t-2}$.

3.31 Let X_t be an ARMA(p,q) process with

$$\phi(B)X_t = \theta(B)a_t$$

where a_t is zero mean white noise and

$$\phi(B) = 1 - \phi_1 B - \cdots - \phi_p B^p = \prod_{j=1}^{p}(1 - \alpha_j B)$$

$$\theta(B) = 1 - \theta_1 B - \cdots - \theta_q B^q = \prod_{j=1}^{q}(1 - \beta_j B).$$

Further suppose $\alpha_j \neq \alpha_\ell$ if $j \neq \ell$. Show that the ψ-weights for X_t, denoted by $\psi_k(p,q)$, are given by

$$\begin{aligned}\psi_k(p,q) &= \psi_k(p) - \theta_1\psi_{k-1}(p) - \cdots - \theta_q\psi_{k-q}(p)\\ &= \theta(B)\psi_k(p),\end{aligned}$$

where the $\psi_k(p)$ are the ψ-weights for the AR(p) model $\phi(B)X_t = a_t$, which are given by (3.27).

3.32 Let $X_t - 0.9X_{t-1} = a_t + 0.5a_{t-1}$. Use Problem 3.31 to show that the ψ-weights, $\psi_k(1,1)$, for this process are given by $\psi_0(1,1) = 1$ and $\psi_k(1,1) = (1.4)(0.9)^{k-1}$, for $k \geq 1$.

3.33 ψ-weights for an ARMA(2,1) with repeated roots:

(a) Consider the ARMA(2,1) model $(1 - \alpha B)^2 X_t = (1 + B)a_t$. In this case, the ψ-weight formula in (3.27) does not hold, since $\alpha_1 = \alpha_2 = \alpha$. By taking the limit in (3.27) as $\alpha_2 \to \alpha_1$, show that in this case, $\psi_k(2,1) = (k+1)\alpha_1^k + k\alpha_1^{k-1}$.

(b) Use the results in (a) to find the ψ-weights for the model $(1 - B + 0.25B^2)X_t = (1 + B)a_t$.

3.34 Suppose X_t is an invertible ARMA(p,q) process satisfying $\phi(B)X_t = \theta(B)a_t$, and let the inverted process be written as

$$\sum_{j=0}^{\infty} \pi_j X_{t-j} = a_t.$$

Show that the π_j are determined by the equations

$$\pi_j - \sum_{k=1}^{\min(q,j)} \theta_k \pi_{j-k} = -\phi_j, \ldots, \quad j = 0, 1, \ldots$$

where we define $\phi_0 = -1, \theta_k = 0$ for $k > q$ and $\phi_j = 0$ for $j > p$.

4

Other Stationary Time Series Models

The ARMA models discussed in Chapter 3 are used extensively for modeling stationary time series data and finding corresponding forecasts and spectral estimates. In this chapter, we discuss two other types of models that are also used to model stationary data. In Section 4.1, we discuss the harmonic models and harmonic signal-plus-noise models that are used extensively in the engineering field. Section 4.2 discusses the Autoregressive Conditional Heteroscedasticity (ARCH) models and the generalized ARCH (GARCH) models that have been developed in the field of econometrics.

4.1 Stationary Harmonic Models

In many problems, the physical system generating the data will impose a fixed period in the data. Mean temperatures in Pennsylvania, for example, have a clear annual cycle imposed on them by the rotation of the Earth. Other examples include radio and television signals, certain types of seismic waves, and cardiograms. For such processes, a harmonic signal-plus-noise model may be appropriate. However, as we shall see, an ARMA model may also be of value as an approximating model in these cases.

Figure 4.1a shows a realization from the AR(2) model

$$(1 - 1.55B + 0.8B^2)X_t = a_t, \tag{4.1}$$

where a_t is $N(0,1)$ white noise and Figure 4.1b shows a realization from the "signal-plus-noise" model

$$Y_t = 4.4\cos\left(\frac{2\pi t}{12} + U\right) + N_t, \tag{4.2}$$

for $t = 1, 2, \ldots, 200$, where N_t is $N(0,1)$ white noise, U is a uniform $[0, 2\pi]$ random variable, and where N_t and U are independent. Inspection of (4.1) and (4.2) along with the associated realizations shows some similarities and some differences. The range of the data seems to be about the same, the variances of X_t and Y_t are the same, and the ratio of the process variance to

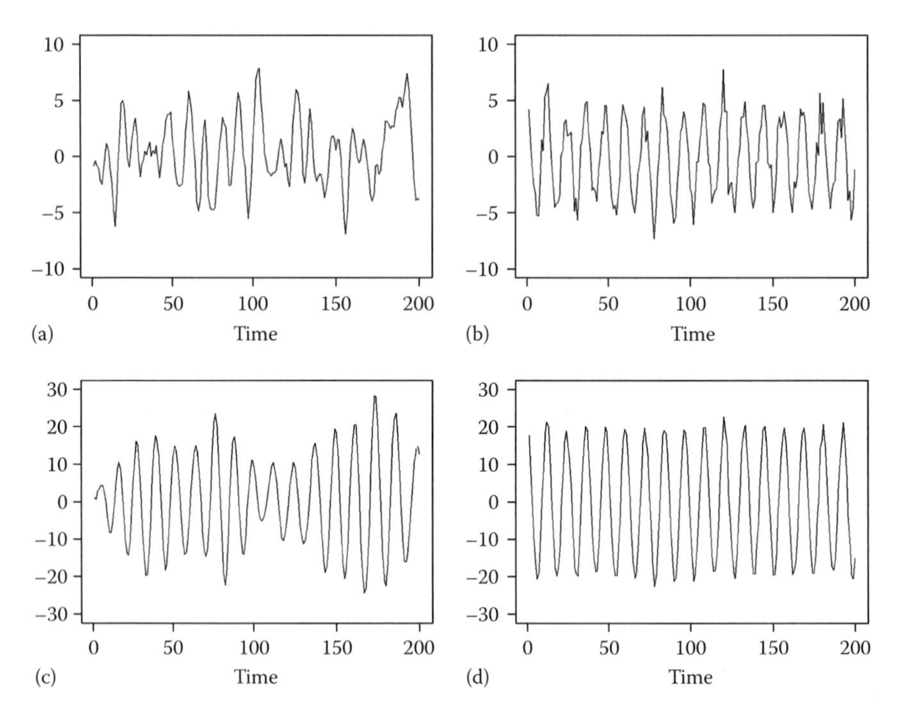

FIGURE 4.1
Realizations from AR(2) and signal-plus-noise models. (a) $X_t - 1.55X_{t-1} + 0.8X_{t-2} = a_t$. (b) $Y_t = 4.4\cos(2\pi t/12 + U) + N_t$. (c) $X_t - 1.72X_{t-1} + 0.99X_{t-2} = a_t$. (d) $Y_t = 20\cos(2\pi t/12 + U) + N_t$.

noise variance is 10.7. That is, $\sigma_X^2/\sigma_a^2 = \sigma_Y^2/\sigma_N^2 = 10.7$. Also, both processes have a natural frequency of $f = 1/12$. To see this notice that the system frequency associated with the AR(2) model of (4.1) given by (3.63) with $\phi_1 = 1.55$ and $\phi_2 = -0.8$ is $f_0 \approx 1/12$. In order to further compare these two types of models, Figure 4.1c and d shows realizations from

$$(1 - 1.72B + 0.99B^2)X_t = a_t \tag{4.3}$$

and

$$Y_t = 20\cos\left(\frac{2\pi t}{12} + U\right) + N_t. \tag{4.4}$$

These are analogous to (4.1) and (4.2), but, in these realizations, the noise plays a much smaller role. This can be seen by noting that for the models in (4.3) and (4.4), $\sigma_X^2/\sigma_a^2 = \sigma_Y^2/\sigma_N^2 = 201$, so realizations from these two models should more clearly show the underlying "signal." Figure 4.1c shows that the AR(2) realization is more variable in amplitude, and, thus, the period appears not to be as stable as that in the signal-plus-noise realization in Figure 4.1d.

In Chapter 6, we will see that the best forecast functions (in the sense of least squares) for $X_{t+\ell}$, given data to time t based on the AR(2) processes in (4.1) and (4.3), are given by

$$\hat{X}_{t+\ell} = CR^{-1}\cos\left(\frac{2\pi\ell}{12} + D\right),$$

where C and D are random variables determined by the values of X_t and X_{t-1} and R is a damping factor.

One way to interpret the distinction between data modeled by the AR(2) models in (4.1) and (4.3) and the harmonic component models in (4.2) and (4.4) is that the AR(2) models allow random amplitude and phase shifts (or at least respond quickly to them as a model). Consequently, best forecasts eventually damp to the mean. However, the harmonic model is characterized by a fixed phase and amplitude for any given realization. AR(p) models with complex roots of the characteristic equation produce realizations with random amplitude and phase shift, and so the AR(p) model may serve as an alternative to the harmonic models for those cases in which the researcher wants to be more responsive to apparent shifts in the amplitude and phase of the data. This will be further illustrated in Section 6.10.

4.1.1 Pure Harmonic Models

We begin by considering purely harmonic discrete parameter models

$$Y_t = \sum_{j=1}^{m} A_j \cos(2\pi f_j t + U_j), \tag{4.5}$$

in which there is no additive noise component, where the A_j and f_j are constants, and the U_j are independent random variables each distributed as uniform $[0, 2\pi]$. We first consider the case $m = 1$ that was considered in the continuous parameter case in Examples 1.3 and 1.7. In Example 1.3, it was shown that Y_t is stationary with $E[Y_t] = 0$, variance $\sigma_Y^2 = A_1^2/2$, and autocovariance function

$$\gamma_k = \frac{A_1^2}{2}\cos 2\pi f_1 k, \tag{4.6}$$

from which it immediately follows that

$$\rho_k = \cos 2\pi f_1 k. \tag{4.7}$$

Unlike the autocorrelation function for an ARMA(p,q) process, ρ_k in (4.7) does not damp, and, in fact, $|\rho_k| = 1$ if $2\pi f_1 k$ is an integer. In the discrete parameter case, the spectrum defined in (1.25) is given by

$$P_Y(f) = \sum_{k=-\infty}^{\infty} e^{-2\pi i f k} \frac{A_1^2}{2} \cos 2\pi f_1 k, \tag{4.8}$$

which does not converge. In order to calculate the spectrum in this case, it is necessary to use the spectral distribution function (see Priestley, 1993). We will not cover this topic here, except to note that using this approach, the spectrum can be shown to be

$$P_Y(f) = \frac{A_1^2}{2}\left(\frac{\Delta(f - f_1) + \Delta(f + f_1)}{2}\right),$$

and hence

$$S_Y(f) = \frac{\Delta(f - f_1) + \Delta(f + f_1)}{2}, \tag{4.9}$$

where $\Delta(x)$ is the Kronecker delta function given by

$$\begin{aligned} \Delta(x) &= 1, \quad x = 0 \\ &= 0, \quad x \neq 0. \end{aligned}$$

As usual, we only plot the spectral plots for $f \in [0, 0.5]$, and Figure 4.2 shows the spectral density (not plotted in dB) for the process in (4.5) with $m = 1$. It is clear that realizations from (4.5) with $m = 1$ will be perfectly sinusoidal, which is consistent with γ_k and $P_Y(f)$, and although the series has random phase (due to U), there is no phase shifting in the sense of an AR process or otherwise within a realization.

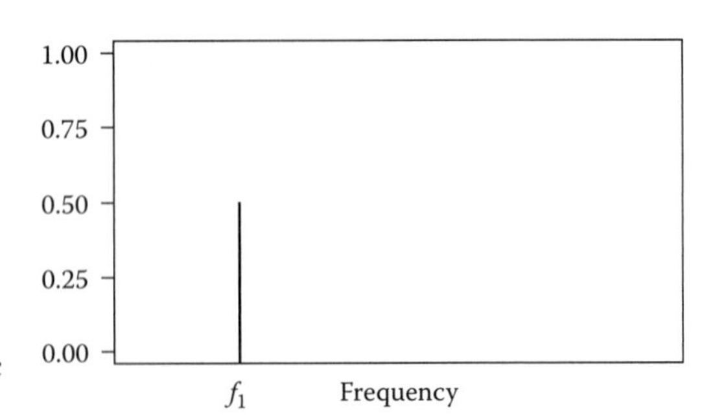

FIGURE 4.2
True spectral density for the harmonic model in (4.5) with $m = 1$.

The autocorrelations and spectral density for the full model in (4.5) are given by

$$\rho_k = \frac{\sum_{j=1}^{m} A_j^2 \cos 2\pi f_j k}{\sum_{j=1}^{m} A_j^2}$$

and

$$S_Y(f) = \frac{1/2 \sum_{j=1}^{m} A_j^2 \left(\Delta(f - f_j) + \Delta(f + f_j)\right)}{\sum_{j=1}^{m} A_j^2},$$

respectively. Realizations from models such as (4.5) will be sums of sinusoids and will display multiple periodicities. Similarly, a plot of the true spectral density (for positive f) will be zero except for m spikes at $f_j, j = 1, \ldots, m$. Spectra such as these are said to be discrete or line spectra as contrasted with the smooth or continuous spectra associated with ARMA(p,q) models.

4.1.2 Harmonic Signal-plus-Noise Models

We now consider the signal-plus-noise model

$$\begin{aligned} Y_t &= \sum_{j=1}^{m} A_j \cos(2\pi f_j t + U_j) + N_t \\ &= H_t + N_t, \end{aligned} \qquad (4.10)$$

for $t = 0, \pm 1, \ldots$, where H_t is referred to as the harmonic component or signal and N_t is a stationary noise process. As in (4.5), the A_j and f_j are constants, and the U_j s are uncorrelated random variables each distributed as uniform $[0, 2\pi]$. Additionally, H_t and N_t are uncorrelated processes. It follows at once that Y_t in (4.10) is a stationary process, since it is the sum of uncorrelated stationary processes.

For our purpose here, we consider the case $m = 1$, for which it can be shown (Problem 1.12) that when N_t is discrete white noise, γ_k is given by

$$\begin{aligned} \gamma_k &= \frac{A_1^2}{2} \cos 2\pi f_1 k, \quad k \neq 0 \\ &= \frac{A_1^2}{2} + \sigma_N^2, \quad k = 0. \end{aligned} \qquad (4.11)$$

Thus, ρ_k is given by

$$\rho_k = \frac{\frac{A_1^2}{2}\cos 2\pi f_1 k}{\frac{A_1^2}{2} + \sigma_N^2}, \quad k \neq 0$$
$$= 1, \quad k = 0.$$

Since H_t and N_t are uncorrelated, it follows that the spectrum is given by

$$P_Y(f) = \frac{A_1^2}{4}\left(\Delta(f - f_1) + \Delta(f + f_1)\right) + \sigma_N^2,$$

and the spectral density is

$$S_Y(f) = \frac{\frac{A_1^2}{4}\left(\Delta(f - f_1) + \Delta(f + f_1)\right) + \sigma_N^2}{\frac{A_1^2}{2} + \sigma_N^2}.$$

A realization from the model

$$Y_t = 4.4\cos\left(\frac{2\pi t}{12} + U\right) + N_t, \tag{4.12}$$

where N_t is discrete white noise with unit variance, was previously shown in Figure 4.1b. The autocorrelations and spectrum (not in dB) for this model are plotted in Figure 4.3 where it can also be seen that the autocorrelations

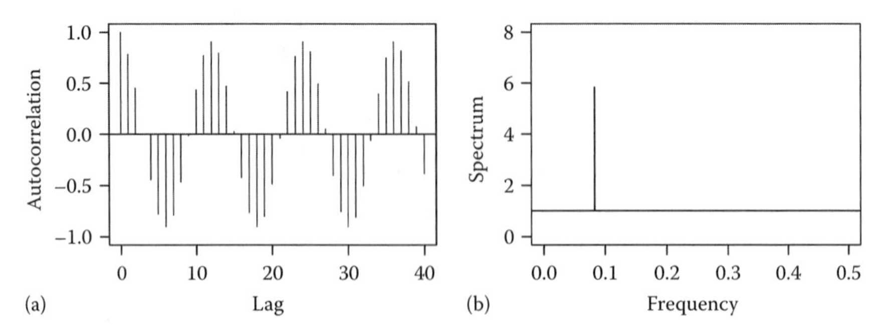

FIGURE 4.3
(a) Theoretical autocorrelations and (b) spectrum for the model $Y_t = 4.4\cos\left(\frac{2\pi t}{12} + U\right) + N_t$.

do not damp out, but in this case, $\max_{k\neq 0}|\rho_k| = 0.9$. The spectrum shows a finite spike of $A_1^2/4 + \sigma_N^2 = 5.84$ at $f = 1/12$ along with a flat spectrum of $P_Y(f) = 1$ for $f \neq 1/12$ associated with the white noise component. Spectra such as the one in Figure 4.3b are called mixed spectra, since they are mixtures of the discrete and continuous spectra encountered up to this point.

Using `GW-WINKS`: A realization from the model

$$X_t = A_1 \cos(2\pi f_1 t) + A_2 \cos(2\pi f_2 t) + N_t \tag{4.13}$$

where N_t is Gaussian white noise can be obtained using the `Generate Series` option on the `Time Series Analysis` menu. Select the `Signal + Noise` button and then you will be asked to enter c_1, f_1, c_2, f_2, and white noise variance σ_N^2 along with the series length and seed.

4.1.3 ARMA Approximation to the Harmonic Signal-plus-Noise Model

We have seen that realizations from the harmonic signal-plus-noise model and ARMA(p,q) models can be similar in nature. However, as we discussed, there are distinct differences between the two models (4.1) and (4.2). Nevertheless, in many cases, the ARMA(p,q) or high-order autoregressive can serve as a very good approximation to the signal-plus-noise model. The primary differences between an ARMA model and a harmonic signal-plus-noise are the random phase and amplitude nature of ARMA realizations. However, since phase information is ''lost'' in the power spectrum (and the amplitude information, random or nonrandom, is ''captured'' by the variance), it is not surprising that the ARMA model can serve as a good model for estimating the spectrum of harmonic signals.

Example 4.1 Approximating a Harmonic Signal-Plus-Noise Model with an ARMA(2,2) Model

A single harmonic signal-plus-noise model such as (4.12) can often be well approximated by an ARMA(2,2) model as the following development shows. Let Y_t be given by (4.10) with $m = 1$ with $N_t = a_t$, and consider $\phi(B)Y_t$, where

$$\phi(B) = 1 - 2\cos(2\pi f_1)B + B^2.$$

It follows that

$$\begin{aligned}\phi(B)Y_t &= \phi(B)(H_t + a_t)\\ &= \phi(B)H_t + \phi(B)a_t,\end{aligned} \tag{4.14}$$

where H_t is given in (4.10). But $\phi(B)H_t = 0$ (Problem 4.2) from which it follows that $\phi(B)Y_t = \phi(B)a_t$. In practice, $\phi(B)$ must be estimated, and, in place of (4.14), we have

$$\hat{\phi}(B)Y_t = \varepsilon_t + \hat{\phi}(B)a_t, \tag{4.15}$$

where

$$\varepsilon_t = \hat{\phi}(B)A_1 \cos(2\pi f_1 t + U_1).$$

If ε_t is sufficiently small, which it often is, then it produces only a small perturbation to $\hat{\phi}(B)a_t$ and (4.14) is well approximated by

$$\hat{\phi}(B)Y_t = \hat{\theta}(B)Z_t$$

where Z_t is white noise, $\hat{\phi}(B) \approx \hat{\theta}(B)$ (but $\hat{\phi}(B) \neq \hat{\theta}(B)$), and the roots of $\hat{\phi}(z) = 0$ are very near the unit circle.

Figure 4.4 shows realizations of length $n = 200$ from the ARMA(2,2) process

$$(1 - 1.72B + 0.99B^2)X_t = (1 - 1.37B + 0.72B^2)a_t, \tag{4.16}$$

and of the harmonic signal-plus-noise model

$$Y_t = 3.57\cos\left(\frac{2\pi t}{12} + U\right) + Z_t, \tag{4.17}$$

where a_t and Z_t are both $N(0,1)$ white noise and $\sigma_X^2 = \sigma_Y^2 = 7.4$. The two realizations are similar, but the amplitudes associated with the ARMA(2,2) realization are more variable than those for the signal-plus-noise model. In Table 4.1, we see the factor table for the ARMA(2,2) model in (4.16) where it can be seen that the autoregressive and moving average factors have very

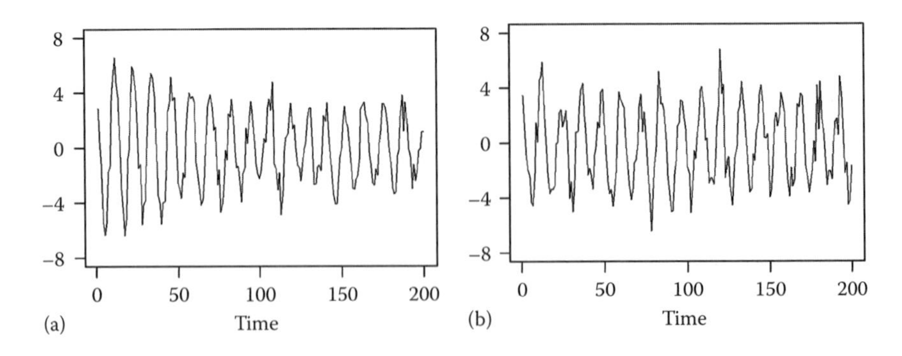

FIGURE 4.4
(a) Realizations from the ARMA(2,2) in (4.16) and (b) the harmonic signal-plus-noise model in (4.17).

TABLE 4.1

Factor Table for the ARMA(2,2) Model in (4.16)

	Roots (r_j)	$\lvert r_j^{-1}\rvert$	f_{0j}
AR factors			
$1 - 1.72B + 0.99B^2$	$0.87 \pm 0.51i$	0.995	0.08
MA factors			
$1 - 1.37B + 0.72B^2$	$0.95 \pm 0.70i$	0.85	0.10

similar system frequencies (0.08 and 0.10). The spectral density for the signal-plus-noise model in (4.17) is a single spike at $f = .083$ along with a flat spectrum at other frequencies associated with the white nose. In Figure 4.5a, we show the spectral density for the ARMA(2,2) model in (4.16), and it can be seen that there is a sharp peak at about $f = 0.08$. In Figure 4.5b, we show the spectral density associated with the AR(2) model $(1 - 1.72B + 0.99B^2)X_t = a_t$. This spectral density also has a peak at about $f = 0.08$, but the difference between Figure 4.5a and b is that the peak in

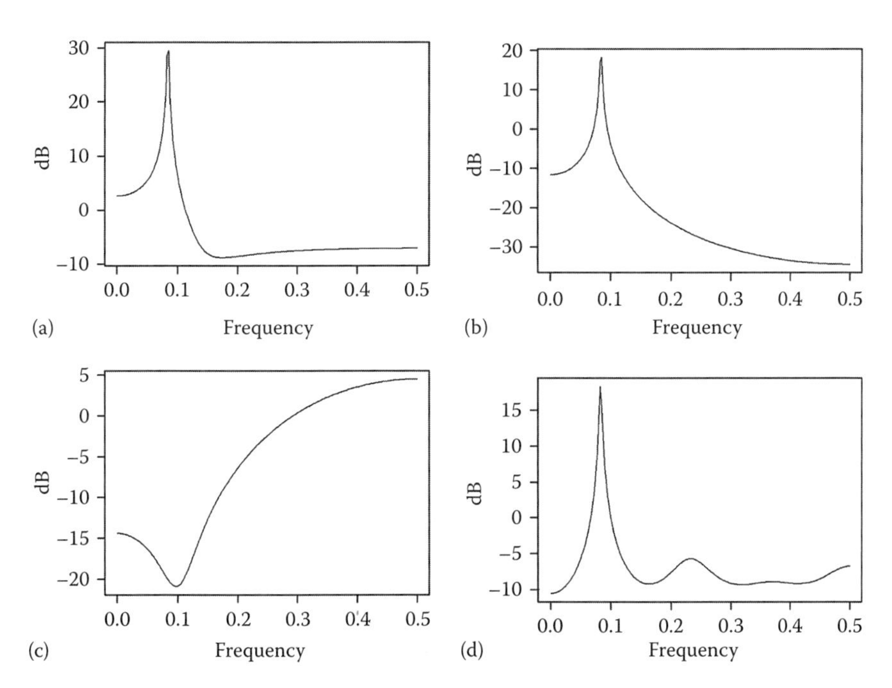

FIGURE 4.5
True spectral plots for (a) ARMA(2,2) model, (b) AR(2) model, (c) MA(2) model, and (d) AR(7) model.

Figure 4.5a is sharper with less frequency content near the peak frequency. In Figure 4.5c, we show the spectral density for the MA(2) model $X_t = (1 - 1.37B + 0.72B^2)a_t$. This spectral density has a dip at about $f = 0.1$. The effect of the moving average component of the ARMA(2,2) model in (4.16) is to reduce the frequency content in a neighborhood just above $f = 0.08$ causing the sharp, nearly spike-like peak in the ARMA(2,2) spectral density in Figure 4.5a. In Figure 4.5d, we fit a high-order AR(p) model with $p = 7$ to model the data in Figure 4.4a. It can be seen that the peak is quite sharp, similar to Figure 4.5a, but there are some secondary peaks associated with the other model components. This is an example of the extent to which a high-order autoregressive can be used to approximate the behavior of an ARMA(p,q) model.

These ideas can be extendable to higher order models, and the general harmonic signal-plus-noise model of (4.10) where N_t is white noise can be approximated by an ARMA($2m$,$2m$) process.

4.2 ARCH and GARCH Processes

In our previous study of ARMA(p,q) processes, we have simply required the noise to be white noise. That is, for example, in the ARMA(p,q) model

$$(1 - \phi_1 B - \cdots - \phi_p B^p)X_t = (1 - \theta_1 B - \cdots - \theta_q B)a_t, \qquad (4.18)$$

the noise, a_t, has only been required to be white, that is, uncorrelated, identically distributed random variables with zero mean (see Definition 1.6). However, it is not uncommon to have data for which the noise appears more variable or volatile over certain sections of the time axis than over others. For example, consider the data in Figure 4.6a that show the DOW Jones Daily Rate of Returns from October 1, 1928 to December 31, 2010. Clearly, the data look much more variable in some time regions than others. For example, the 1930s was a period of continuing volatility. Isolated extremes are also evident, most noticeably on the 1929 stock market crash on October 29, 1929 and Black Monday, October 19, 1987. Similar behavior is seen in Figure 4.7b, which shows seismic data from an earthquake known as the Massachusetts Mountain Earthquake. In this case, there is an extended period of higher variability around $t = 100$.

In both realizations in Figure 4.6, it appears that even though the noise a_t may be white, the conditional variance, $\sigma^2_{t|t-1}$, of a_t given $a_{t-1}, a_{t-2}, \ldots$, that is, $E\left[a_t^2 | a_{t-k}, k = 1,2,\ldots\right]$, depends on t. Thus, although the a_t are uncorrelated, the a_t^2 may be correlated. This is possible, since the a_t are only uncorrelated and not necessarily independent. Time series for which the

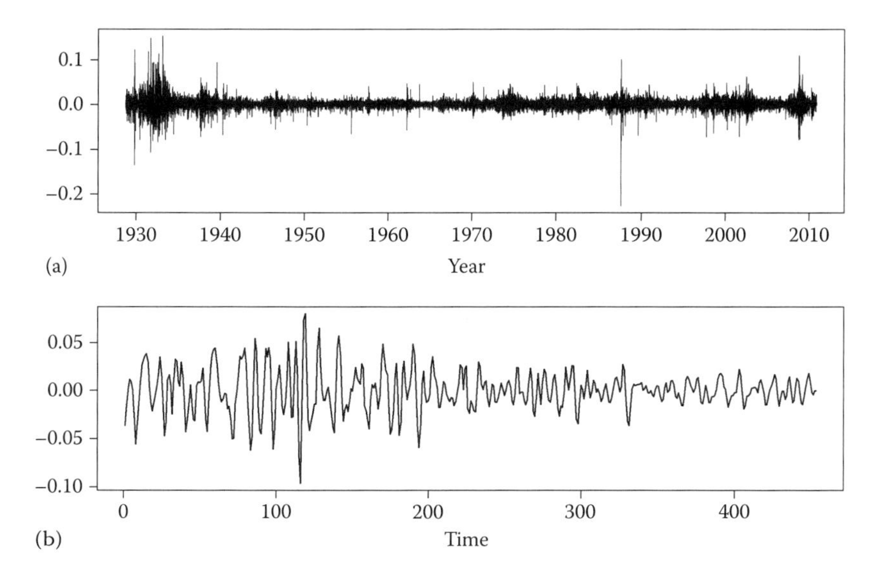

FIGURE 4.6
(a) DOW Jones daily rate of return from October 1, 1928 through December 31, 2010 and (b) Seismic data from the Lg wave measured on the Massachusetts Mountain Earthquake.

conditional variance of the noise, $\sigma^2_{t|t-1}$, depends on t are said to be *volatile*, and the conditional variance is often referred to as the *conditional volatility*.

NOTES: If the white noise $a_t \sim N(0, \sigma^2_a)$, then both the a_t's and the a^2_t's are uncorrelated (and independent). Also note that it is often stressed in statistics courses that correlation is only a measure of linear correlation. This setting provides a good example of the impact of data being uncorrelated but not independent.

4.2.1 ARCH Processes

To address the conditional volatility behavior discussed earlier, Engle (1982) introduced the ARCH model. To clarify our remarks and demonstrate the existence of such processes. We first consider the simplest ARCH process, the ARCH(1) model.

4.2.1.1 The ARCH(1) Model

Let $\sigma^2_{t|t-1}$ be the conditional variance of a_t, and suppose

$$\begin{aligned} a_t &= \sigma_{t|t-1}\varepsilon_t \\ \sigma^2_{t|t-1} &= \alpha_0 + \alpha_1 a^2_{t-1}, \end{aligned} \tag{4.19}$$

where $0 \le \alpha_1 < 1$, and $\{\varepsilon_t\}$ is a sequence of independent and identically distributed (i.i.d.), zero mean, unit variance random variables that are independent of the $a_{t-k}, k=1,2,\ldots$. Then a_t is said to satisfy an ARCH(1) process. Note that the unit variance of ε_t is required, so that

$$\begin{aligned} E\left[(\sigma_{t|t-1}\varepsilon_t)^2|a_{t-k}, k=1,2,\ldots\right] &= \sigma^2_{t|t-1}E\left[\varepsilon_t^2|a_{t-k}, \quad k=1,2,\ldots\right] \\ &= \sigma^2_{t|t-1}E\left[\varepsilon_t^2\right] = \sigma^2_{t|t-1}. \end{aligned} \tag{4.20}$$

It is also easily shown (see Shumway and Stoffer, 2006) that the a_t are uncorrelated with a conditional variance that depends on t. Note from (4.19) that

$$a_t^2 - \left(\alpha_0 + \alpha_1 a_{t-1}^2\right) = \sigma^2_{t|t-1}\left(\varepsilon_t^2 - 1\right), \tag{4.21}$$

and thus that

$$a_t^2 = \alpha_0 + \alpha_1 a_{t-1}^2 + w_t, \tag{4.22}$$

where $w_t = \sigma_t^2\left(\varepsilon_t^2 - 1\right)$ and $E[w_t] = 0$. Thus, a_t^2 satisfies a non-Gaussian AR(1) process with

$$\begin{aligned} \sigma_a^2 &= E\left[a_t^2\right] \\ &= \alpha_0 + \alpha_1 E\left[a_{t-1}^2\right] + E[w_t] \\ &= \alpha_0 + \alpha_1\sigma_a^2. \end{aligned}$$

Therefore,

$$\sigma_a^2 = \frac{\alpha_0}{1-\alpha_1}.$$

Example 4.2 ARCH(1) Model

Figure 4.7 shows four realizations from ARCH(1) models, where ε_t is normal white noise. The ARCH coefficients for each model are $\alpha_1 = 0, 0.4, 0.7$, and 0.9, respectively, and the α_0 values are 1, 0.6, 0.3, and 0.1, respectively, so that the processes each have the same unconditional variance. Note that when $\alpha_1 = 0$, the a_t's are simply Gaussian white noise. It can be seen that as α_1 gets closer to 1, the realizations in Figure 4.7 show regions of isolated more extreme variability. The histograms in Figure 4.8 for the data in Figure 4.7a and d show that when $\alpha_1 = 0.9$ (Figure 4.8b) the data are more heavy tailed than the Gaussian white noise case (Figure 4.8a). Figure 4.9 shows the conditional

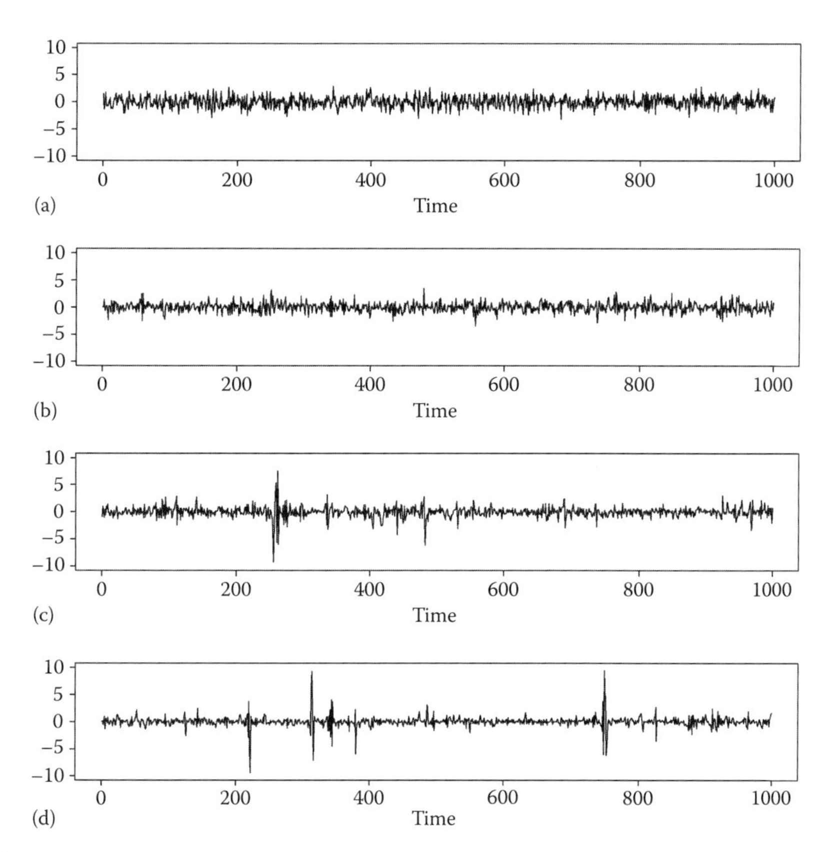

FIGURE 4.7
Realizations from ARCH(1) models, where $\alpha_1 = 0$ (a), 0.4 (b), 0.7 (c), and 0.9 (d).

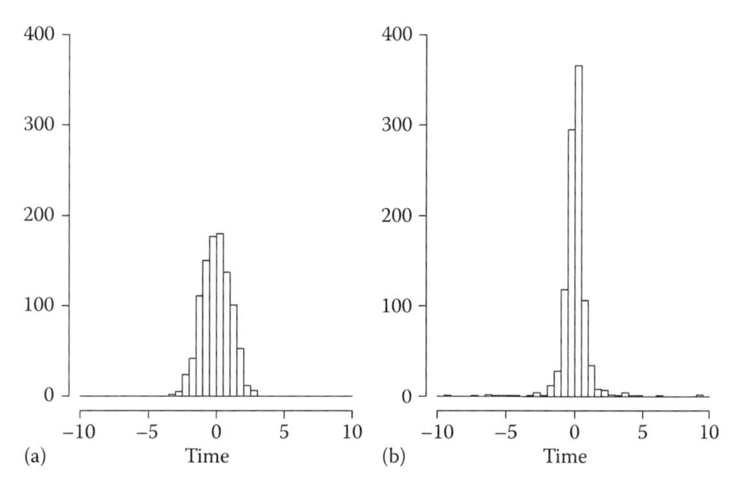

FIGURE 4.8
Histograms for the realizations in Figure 4.7 with (a) $\alpha_1 = 0$ and (b) $\alpha_1 = 0.9$.

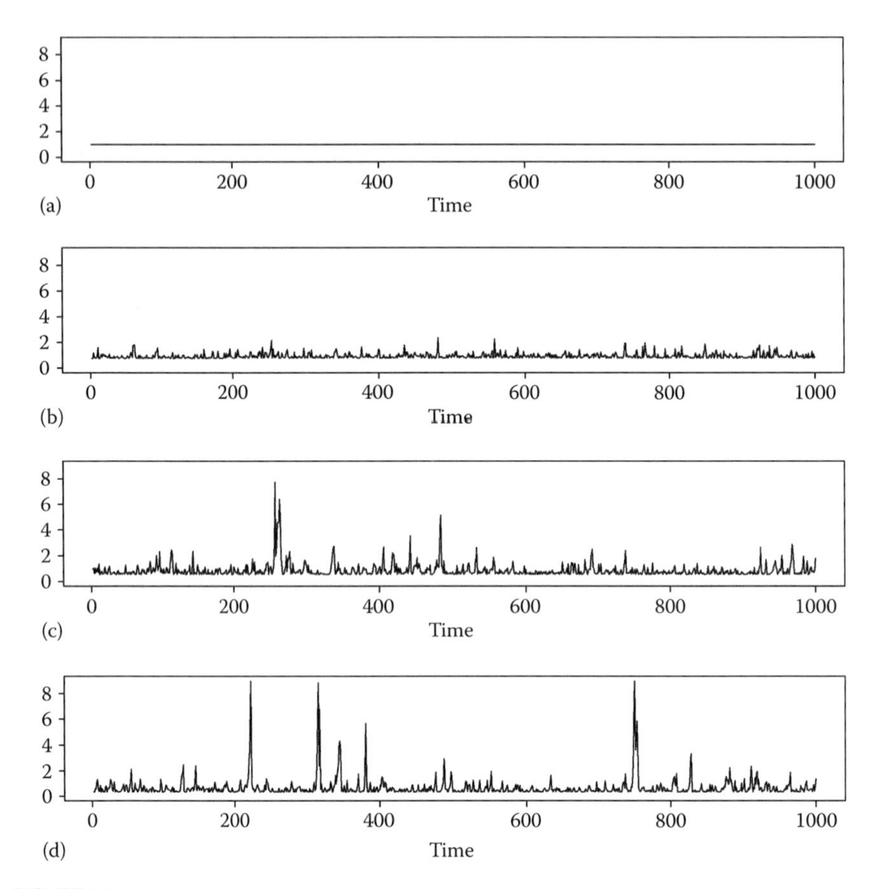

FIGURE 4.9
Conditional standard deviations for the ARCH(1) realizations in Figure 4.7. (a) $\alpha_1 = 0$. (b) $\alpha_1 = 0.4$. (c) $\alpha_1 = 0.7$. (d) $\alpha_1 = 0.9$.

standard deviations, calculated using (4.19) for the realizations in Figure 4.7. These plots emphasize the increasing tendency for extreme changes in conditional variability as $\alpha_1 \to 1$.

4.2.1.2 The ARCH(q_0) Model

The ARCH(1) model can be extended as follows. Let

$$a_t = \sigma_{t|t-1}\varepsilon_t$$
$$\sigma^2_{t|t-1} = \alpha_0 + \alpha_1 a^2_{t-1} + \cdots + \alpha_{q_0} a^2_{t-q_0}, \tag{4.23}$$

where, as in the ARCH(1) case, the ε_t's are i.i.d., zero mean, unit variance random variables. The process, $\{a_t\}$, in (4.23) is referred to as an ARCH(q_0) process. Recall from (4.18) that by specifying the white noise process, a_t, to be an ARCH process, we are able to effectively model the volatility suggested by the data in Figure 4.6a and b. For the data in Figure 4.6a, $p = 0$ and $q = 0$ in (4.18), and the volatility of the white noise itself is the problem of interest. Economic applications, such as the returns data, originally motivated the introduction of the ARCH model. On the other hand, it is generally the case for seismic data such as in Figure 4.6b that p and q are not all zero, in which case, the ARCH noise driver allows us to capture the effect of volatility in the driving noise process on the ARMA(p,q) process, X_t.

The ARCH(q_0) model in (4.23) can be expressed in a form analogous to (4.21). That is, if a_t satisfies the ARCH(q_0) model in (4.23), then it can be written as

$$a_t^2 = \alpha_0 + \alpha_1 a_{t-1}^2 + \alpha_2 a_{t-2}^2 + \cdots + \alpha_{q_0} a_{t-q_0}^2 + w_t, \tag{4.24}$$

that is, a_t^2 satisfies an AR(q_0) process with non-Gaussian white noise. From (4.24), it follows that for an ARCH(q_0) process, the unconditional variance can be written as follows:

$$\sigma_a^2 = \frac{\alpha_0}{1 - \alpha_1 - \cdots - \alpha_{q_0}}. \tag{4.25}$$

4.2.2 The GARCH(p_0,q_0) Process

The ARCH(q_0) model in (4.23) has been generalized by Bollerslev (1986) and Taylor (1986) to the GARCH(p_0,q_0) model to include lagged terms of the conditional variance in the model. Specifically, for the GARCH(p_0,q_0) model, (4.23) is generalized to

$$\begin{aligned}\sigma_{t|t-1}^2 &= \alpha_0 + \beta_1 \sigma_{t-1|t-2}^2 + \cdots + \beta_{p_0} \sigma_{t-p_0|t-p_0-1}^2 \\ &\quad + \alpha_1 a_{t-1}^2 + \alpha_2 a_{t-2}^2 + \cdots + \alpha_{q_0} a_{t-q_0}^2,\end{aligned} \tag{4.26}$$

where $a_t = \sigma_{t|t-1}\varepsilon_t$, and where the ε_t's are i.i.d., zero mean, unit variance random variables. Bollerslev (1986) introduced constraints on the parameters that ensure stationarity, and Nelson and Cao (1992) presented looser constraints. Corresponding to (4.24), an alternative expression is given by

$$\begin{aligned}a_t^2 &= \alpha_0 + (\alpha_1 + \beta_1)a_{t-1}^2 + \cdots + (\alpha_{\max(p,q)} + \beta_{\max(p,q)})a_{t-\max(p,q)}^2 \\ &\quad + w_t - \beta_1 w_{t-1} - \cdots - \beta_p w_{t-p}.\end{aligned} \tag{4.27}$$

From (4.27), the unconditional variance is given by

$$\sigma_a^2 = \frac{\alpha_0}{1 - (\alpha_1 + \beta_1) - \cdots - (\alpha_{\max(p,q)} + \beta_{\max(p,q)})}. \quad (4.28)$$

For details, see Box et al. (2008), Cryer and Chan (2008), and Shumway and Stoffer (2006).

Example 4.3 ARCH(4), ARCH(8), and GARCH(1,1) Models

Bera and Higgins (1993) showed that the higher the order of the ARCH operator, the more persistent the shocks tend to be. This is illustrated in Figure 4.10a and b, where we show realizations from the ARCH(4) model

$$\sigma_{t|t-1}^2 = 0.1 + 0.36a_{t-1}^2 + 0.27a_{t-2}^2 + 0.18a_{t-3}^2 + 0.09a_{t-4}^2, \quad (4.29)$$

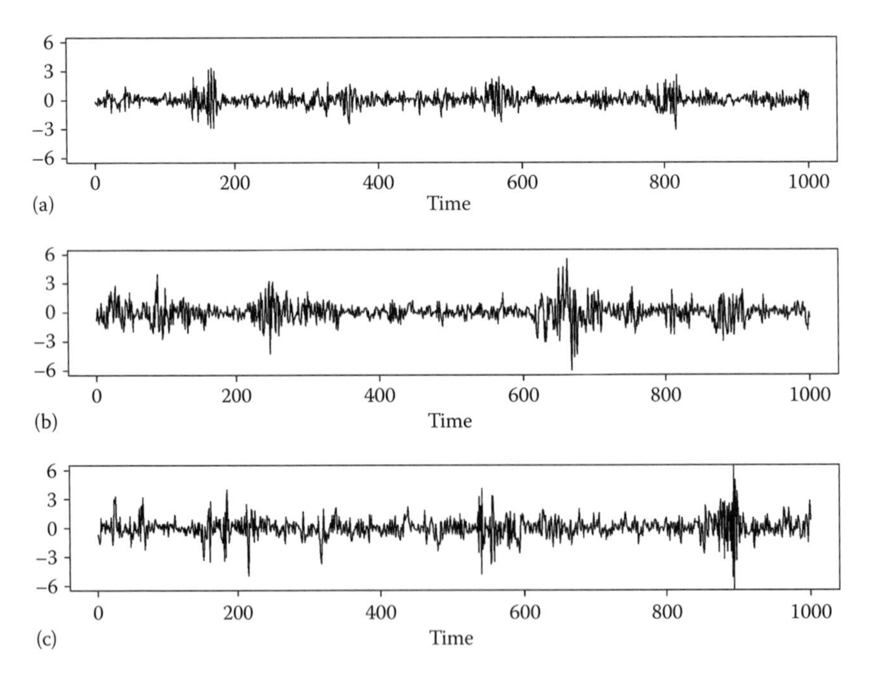

FIGURE 4.10
Realizations from (a) the ARCH(4) model in (4.29), (b) the ARCH(8) model in (4.30), and (c) the GARCH(1,1) model in (4.31).

and the ARCH(8) model

$$\sigma^2_{t|t-1} = 0.1 + 0.2a^2_{t-1} + 0.175a^2_{t-2} + 0.15a^2_{t-3} + 0.125a^2_{t-4} + 0.1a^2_{t-5} + 0.075a^2_{t-6} + 0.05a^2_{t-7} + 0.025a^2_{t-8}, \quad (4.30)$$

respectively. Finally, in Figure 4.10c, we show a realization from the GARCH (1,1) model

$$\sigma^2_{t|t-1} = 0.1 + 0.45a_{t-1} + 0.45\sigma^2_{t-1|t-2}. \quad (4.31)$$

The coefficients in these models are selected, so that the unconditional variance is unity in each case. In each of the realizations in Figure 4.10, there is a tendency for the shocks to last longer than in the ARCH(1) models in Figure 4.7.

For discussion regarding the estimation of parameters and model identification of ARCH and GARCH models, see Box et al. (2008) and Cryer and Chan (2008).

4.2.3 AR Processes with ARCH or GARCH Noise

We briefly consider the effect of ARCH and GARCH noise on autoregressive processes. For the discussion here, we consider the AR(2) process

$$X_t - 1.95X_{t-1} + 0.98X_{t-2} = a_t. \quad (4.32)$$

Figure 4.11a shows a Gaussian white noise realization and Figure 4.11b shows a realization from (4.32) where a_t is the noise sequence in Figure 4.11a. While there is no volatility in the noise sequence, the realization in Figure 4.11b shows oscillation in amplitudes, which is consistent with realizations from AR(2) models with complex roots near the unit circle. Figure 4.11c shows a realization from an ARCH(1) model with $\alpha_0 = 0.1$ and $\alpha_1 = 0.9$, and Figure 4.11d shows a realization from the AR(2) model in (4.32), where a_t is the ARCH(1) noise realization n Figure 4.11c. The ARCH (1) realization shows fairly isolated spikes at about $t = 700$ and $t = 1150$, which initiate an abrupt increase in amplitude of the corresponding output process in Figure 4.11d, followed by a relatively slow decay. This behavior is similar to that seen in Figure 4.6 in the DOW returns data in the 1930s and the seismic data around $t = 100$. In Figure 4.11e, we show a realization from the GARCH(1,1) model in (4.31), and, in Figure 4.11f, we show a realization from the AR(2) model in (4.32) using the GARCH(1,1) realization as the driving noise sequence. The shocks in the GARCH(1,1) sequence tend to be longer lasting but smaller in amplitude than those of the ARCH(1) sequence. Correspondingly, the AR(2) realization in Figure 4.11f shows more subtle response to regions of volatility in the noise.

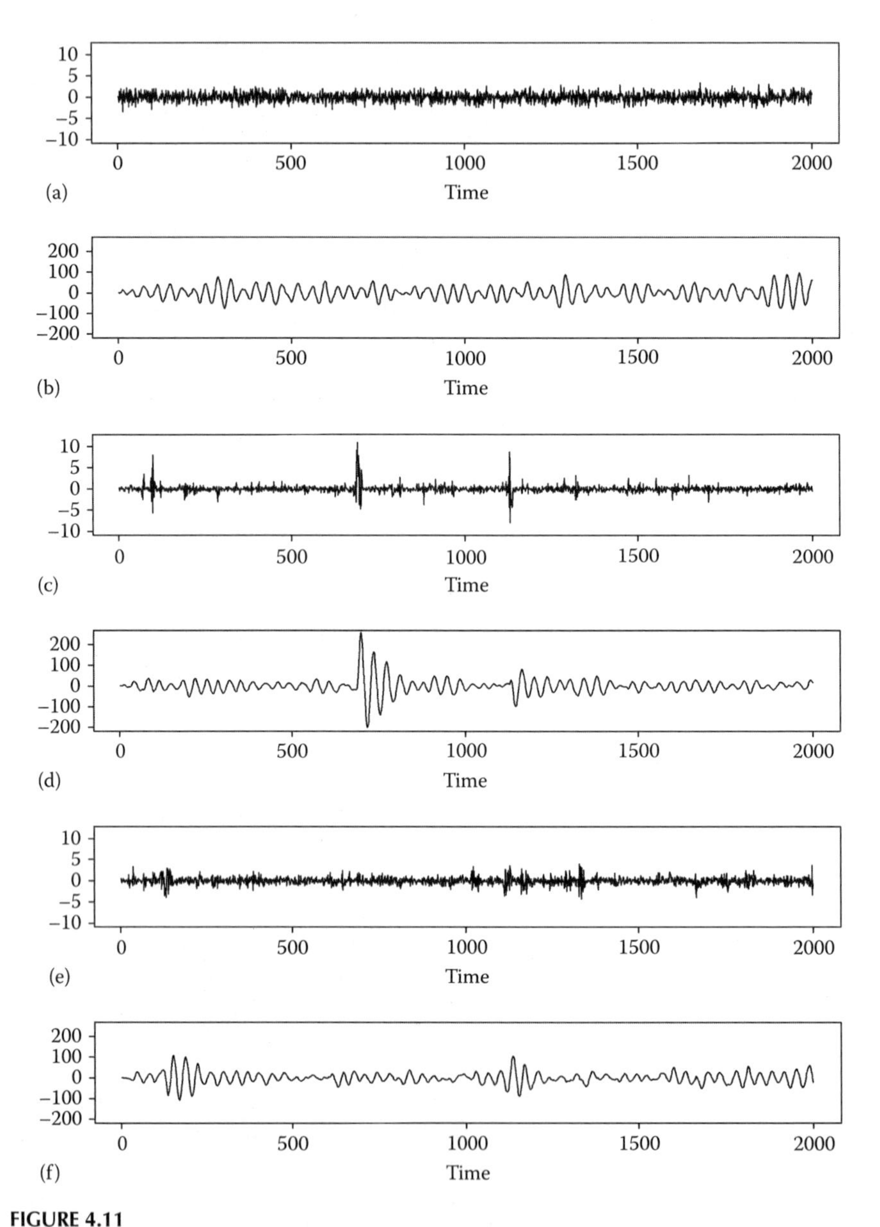

FIGURE 4.11
(a) Realization of Gaussian white noise; (b) realization of AR(2) process with Gaussian white noise in (a); (c) realization of ARCH(1) noise process; (d) realization of AR(2) process with ARCH(1) errors in (c); (e) realization of GARCH(1,1) noise process; (f) realization of AR(2) process with GARCH(1,1) noise errors in (e).

Exercises

4.1 Consider the two models

(i) $Y_t = 4.4\cos(2\pi(1/12)t + \psi) + N_t$, where $\sigma_N^2 = 1$ and $\psi = 0$

(ii) $(1 - 1.643B + 1.6B^2 - 1.15B^3 + 0.63B^4)X_t = a_t$, where $\sigma_a^2 = 1$.

(a) Factor the AR(4) model in (ii) to show that it has a system frequency close to $f = 1/12$.

(b) Generate and plot realizations of length $n = 200$ from each of the two models. Comment on similarities and differences.

(c) Plot spectral estimates of the two realizations using the Parzen window with the same truncation point M for each. Comment on similarities and differences.

(d) Repeat (b) and (c) but this time with $\sigma_N^2 = \sigma_a^2 = 5$. It will be helpful to use the same seed when generating your realizations, so that the only differences will be from the change in noise variance.

(e) How has changing the noise variances affected the signal-to-noise ratios for each model and why? As a part of your explanation, use Var(signal)/Var(Noise) as a measure of the signal-to-noise ratio and find the signal-to-noise ratios for the two models (i.e., find σ_Y^2/σ_N^2 for (i) and σ_X^2/σ_a^2 for (ii)).

(f) Find an AR(4) model for which the signal-to-noise ratio is the same as that for model (i) when $\sigma_N^2 = 5$. *Hint:* Move the roots associated with $f = 1/12$ either further from or closer to the unit circle.

4.2 Let H_t be defined as in Problem 1.12. Show that if $\phi(B)$ is defined by

$$\phi(B) = 1 - 2\cos(2\pi f_1)B + B^2,$$

then $\phi(B)H_t = 0$.

4.3 Write a program for generating data from the ARCH(1) model and generate realizations of length 1000 similar to those shown in Figure 4.7 for the cases: $\alpha_1 = 0, 0.6, 0.8$, and 0.98. In each case, let ε_t be zero mean, unit variance white noise. Also, choose α_0, so that in each case, the unconditional process variance is 1. Discuss the effect of different α_1 values. Explain why α_1 is constrained to be less than one for an ARCH (1) model.

4.4 Write a program for generating realizations of length 1000 from ARCH(2) processes with (α_1, α_2) values: (0.4, 0.4), (0.1, 0.8), and (0.8, 0.1). In each case, let ε_t be zero mean, unit variance white noise. Also, choose α_0, so that in each case, the unconditional process variance is 1. Discuss the effect of changes in α_1 and α_2.

4.5 Write a program for generating realizations of length 1000 from GARCH(1,1) processes with (α_1, β_1) values: (0.4, 0.4), (0.1, 0.8), and (0.8, 0.1). In each case, let ε_t be zero mean, unit variance white noise. Also, choose α_0, so that in each case, the unconditional process variance is 1. Discuss the effect of changes in α_1 and α_2.

4.6 Write a program for generating realizations of length 1000 from AR(2) models with ARCH(1) errors. In each case, let ε_t be zero mean, unit variance white noise and scale the ARCH(1) noise to have unit variance. Generate from each of the following scenarios:
(a) $\phi_1 = 1.96, \phi_2 = -0.98, \alpha_1 = 0$
(b) $\phi_1 = 1.96, \phi_2 = -0.98, \alpha_1 = 0.5$
(c) $\phi_1 = 1.96, \phi_2 = -0.98, \alpha_1 = 0.95$
(d) $\phi_1 = 0.95, \phi_2 = 0, \alpha_1 = 0$
(e) $\phi_1 = 0.95, \phi_2 = 0, \alpha_1 = 0.5$
(f) $\phi_1 = 0.95, \phi_2 = 0, \alpha_1 = 0.98$
Discuss and explain the behavior observed.

5

Nonstationary Time Series Models

In Chapters 3 and 4, we considered stationary processes. However, many or even most processes of interest tend to have some type of nonstationary or near-nonstationary behavior. Referring to the definition of stationarity in Definition 1.5, a process can be nonstationary by having (1) a nonconstant mean, (2) a nonconstant or infinite variance, or (3) an autocorrelation function that depends on time. The processes we discuss in this chapter will have one or more of these properties. In Chapters 12 and 13, we discuss techniques for analyzing time series having property (3). In this chapter, we will discuss the following:

- Deterministic signal-plus-noise models
- ARIMA, ARUMA, and nonstationary multiplicative seasonal models
- Random walk models with and without drift
- Time-varying frequency (TVF) models (or time-varying autocorrelation [TVA] models)

5.1 Deterministic Signal-plus-Noise Models

In this section, we consider models of the form

$$X_t = s_t + Z_t, \tag{5.1}$$

where s_t is a deterministic signal and Z_t is a zero-mean stationary noise component that may or may not be uncorrelated.

In this book, we will model Z_t using an AR(p) model although other models (e.g., ARMA(p,q)) could be used. We consider two types of signals: (a) "trend" signals, such as $s_t = a + bt$ or $s_t = a + b_1 t + b_2 t^2, \ldots$, and (b) harmonic signals, such as $s_t = C\cos(2\pi ft + U)$. We emphasize that signal-plus-noise models are quite restrictive, and they should only be considered if there are physical reasons to believe that a deterministic signal is present.

5.1.1 Trend-Component Models

In this section, we consider the model

$$X_t = a + bt + Z_t, \tag{5.2}$$

where a and b are real constants and Z_t is a zero-mean AR(p) process.

Clearly, X_t is nonstationary, since the mean of X_t is $\mu_t = a + bt$, which depends on t if $b \neq 0$. It can be easily shown that $\text{Var}(X_t) = \sigma_Z^2$, which is constant and finite, and that the autocovariance $\gamma(t_1, t_2)$ only depends on $t_2 - t_1$ because Z_t is stationary. A model, such as the one in (5.2), is often fit to data by first estimating a and b and then analyzing the residuals $\hat{Z}_t = X_t - \hat{a} - \hat{b}t$ as a stationary process. Of course, polynomial trends or other types of trends could be considered, but these will not be discussed here.

If the goal of the analysis of a particular time series is to determine whether there is a deterministic linear trend, a common practice is to fit a model such as (5.2) and test the null hypothesis $H_0 : b = 0$. As an example, for the global temperature data (see Figure 1.24a), we might ask the question whether the observed increase in temperatures over the past 160 years should be predicted to continue. If Z_t is uncorrelated white noise, then the testing of $H_0 : b = 0$ is routinely accomplished using the results of simple linear regression. However, the uncorrelated errors are a key assumption of this type of regression analysis and, as a consequence, should not be used in the presence of correlated errors. We will discuss the topic of trend testing in time series data in Section 9.3. This is a difficult problem, and as Example 5.1 shows, certain stationary time series models without a trend component can produce realizations with trending behavior.

Example 5.1 Global Temperature Data

To emphasize the issues involved in testing for trend, we consider modeling the global temperature data shown previously in Figure 1.24a. An AR(4) model fit to the global temperature series is

$$X_t - 0.66X_{t-1} + 0.02X_{t-2} - 0.10X_{t-3} - 0.24X_{t-4} = a_t, \tag{5.3}$$

where, for convenience, we use $\mu = 0$ and $\sigma_a^2 = 1$ in the following discussion rather than the corresponding estimates from the temperature data. The factor table for the model in (5.3) is given in Table 5.1. It is important to note that

TABLE 5.1

Factor Table for the AR(4) Model in (5.3)

AR Factors	Roots (r_j)	$\left\|r_j^{-1}\right\|$	f_{0j}
$1 - 0.99B$	1.01	0.99	0
$1 - 0.2B + 0.455B^2$	$0.22 \pm 1.47i$	0.67	0.23
$1 + 0.53B$	−1.88	0.53	0.5

this model does not contain a deterministic component and thus accounts for the trending behavior via the correlation structure in the model. From the factor table, it is clear that the model is stationary since all the roots are greater than one in absolute value (i.e., all the $|r_j^{-1}|$ values are less than one). Table 5.1 shows that the dominant feature of the model is a first-order factor associated with a positive real root quite close to the unit circle. As discussed previously, such a factor is related to aimless wandering behavior in realizations. The other factors contribute less pronounced higher frequency behavior, so we expect the wandering behavior associated with the positive real root to dominate realizations. In Figure 5.1, we show four realizations of length $n = 150$ from the AR(4) model in (5.3). The wandering/random trending behavior is visible in all realizations as is the presence of more modest high frequency behavior. Figure 5.1a and c shows realizations that trend up and down, respectively, in a manner similar to that of the temperature series. Figure 5.1b and d shows random "trends" that change direction, once in Figure 5.1b, and several times in Figure 5.1d. Figure 5.1a and c shows that the model in (5.3) is capable of generating realizations that have the appearance of the temperature data, but Figure 5.1b and d shows that these "trends" may turn around abruptly.

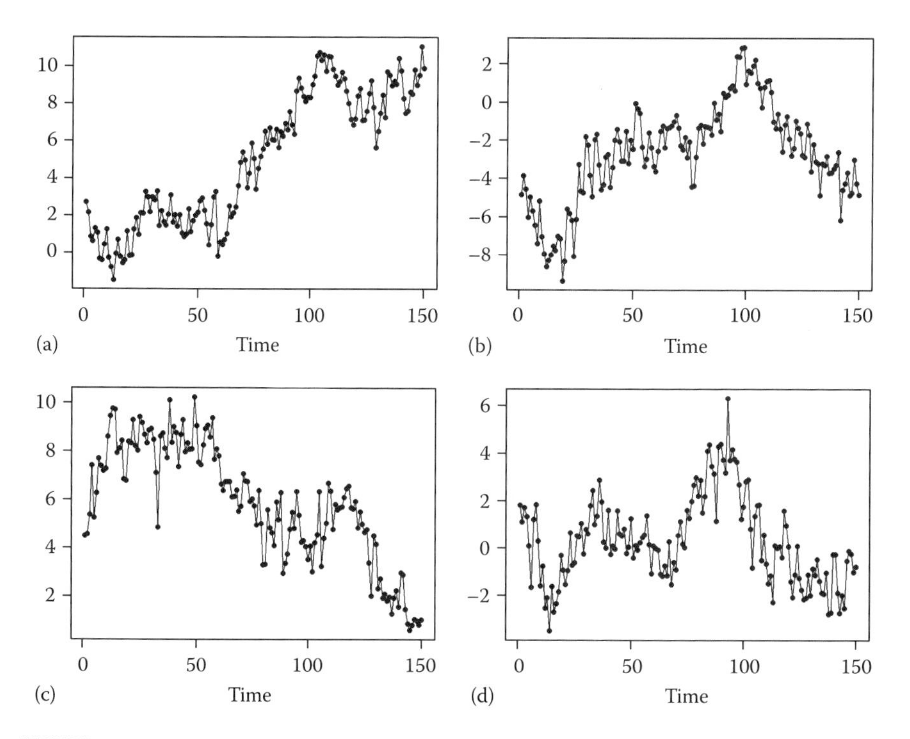

FIGURE 5.1
Four realizations from the AR(4) model (5.3) fit to the global temperature data. (a) Realization 1, (b) Realization 2, (c) Realization 3, and (d) Realization 4.

The point of this example is to illustrate the fact that trending behavior can be produced by the autocorrelation structure alone, and as a consequence, the problem of testing for the existence of a real or deterministic trend of the form in (5.2) is quite difficult. See, for example, Woodward and Gray (1993, 1995) and Woodward et al. (1997). We return to this problem in Chapter 9. Another important point is that forecasts (to be studied in Chapter 6) from the stationary model in (5.3) do not predict the increasing trend to continue because subsequent observations are just as likely to begin going down as they are to continue going up as Figure 5.1b and d illustrate. However, forecasts using the model in (5.2) with nonzero b will forecast an existing trend to continue. We will refer to trends produced by models of type (5.2) as *deterministic trends* while those produced by stationary models, such as (5.3), will be called *random trends*.

5.1.2 Harmonic Component Models

In this section, we consider the case in which s_t is a function of the form $s_t = C\cos(2\pi ft + U)$. In Chapters 1 and 4, we have discussed models with harmonic components and have studied them as a type of stationary process. The stationarity requires that the phase, U, is a random variable. However, in physical situations in which one realization is available, the phase is constant throughout the only realization observed. For this reason, it is often the case that in signal-plus-noise models with harmonic components, the harmonic term, such as $s_t = C\cos(2\pi ft + U)$ is considered to be a deterministic signal as in the case of the trend-plus-noise examples.

Consider the case of the signal-plus-noise model with one harmonic component

$$X_t = C\cos(2\pi ft + U) + Z_t, \tag{5.4}$$

which is usually written as

$$X_t = A\cos(2\pi ft) + B\sin(2\pi ft) + Z_t, \tag{5.5}$$

where $A = C\cos U$ and $B = -C\sin U$ using trigonometric identities. In practice, if X_t satisfies (5.5), then it is analyzed as if it is a nonstationary process, in which case the parameters A, B, and f are often estimated as a first step, and then the process

$$\hat{Z}_t = X_t - \hat{A}\cos(2\pi\hat{f}t) - \hat{B}\sin(2\pi\hat{f}t) \tag{5.6}$$

is modeled as a stationary process (see Bhansali 1979). While the noise term in Chapter 4 is often taken to be white noise, this is not a requirement, and $\hat{Z}_t$ is typically modeled as an AR(p) or ARMA(p,q) process to allow for autocorrelation.

5.2 ARIMA(*p*,*d*,*q*) and ARUMA(*p*,*d*,*q*) Processes

The ARIMA(p,d,q) and ARUMA(p,d,q) models defined in Definition 5.1 are useful nonstationary extensions of the ARMA(p,q) model.

Definition 5.1

(a) The *autoregressive integrated moving average process* of orders p, d, and q (ARIMA(p,d,q)) is a process, X_t, whose differences $(1 - B)^d X_t$ satisfy a ARMA(p,q) model (that is, a stationary model) where d is a non-negative integer. We use the notation

$$\phi(B)(1 - B)^d X_t = \theta(B)a_t, \tag{5.7}$$

where all the roots of $\phi(z) = 0$ and $\theta(z) = 0$ are outside the unit circle, and $\phi(z)$ and $\theta(z)$ have no common factors.

(b) The *autoregressive unit root moving average process* of orders p, d, and q (ARUMA(p,d,q)) is an extension of the ARIMA(p,d,q) model to include processes, X_t, for which the process $\lambda(B)X_t$ satisfies an ARMA(p,q) model where $\lambda(B) = 1 - \lambda_1 B - \cdots - \lambda_d B^d$ is an operator whose characteristic equation has all its roots (real and/or complex) on the unit circle. We use the notation

$$\phi(B)\lambda(B)(X_t - \mu) = \theta(B)a_t \tag{5.8}$$

to denote such a process.

It is worth noting that $\phi(B)\lambda(B)\mu = \phi(1)\lambda(1)\mu$, and therefore, we can write (5.8) in the form

$$\phi(B)\lambda(B)X_t = \phi(1)\lambda(1)\mu + \theta(B)a_t. \tag{5.9}$$

The term $\phi(1)\lambda(1)\mu$ is called the *moving average constant*, and if $\lambda(B) = (1 - B)^d$, then $\lambda(1) = 0$ and so $\phi(1)\lambda(1)\mu = 0$. This is not true in general for an ARUMA(p,d,q) model, that is, if $1 - B$ is not a factor of $\lambda(B)$, then $\phi(1)\lambda(1) \neq 0$.

Example 5.2 An ARUMA(2,2,2) Process

Consider the process

$$X_t + 0.3X_{t-1} + 0.6X_{t-2} + 0.3X_{t-3} - 0.4X_{t-4} = 9 + a_t + 0.5a_{t-1} + 0.9a_{t-2},$$

which in factored form is given by

$$(1 - 0.5B)(1 + 0.8B)(1 + B^2)(X_t - 5) = (1 + 0.5B + 0.9B^2)a_t. \tag{5.10}$$

From the factored form in (5.10), it can be seen that X_t satisfies an ARUMA (2,2,2) with $\mu = 9/(\phi(1)\lambda(1)) = 5$ since $Y_t = (1 + B^2)(X_t - 5)$ satisfies the equation

$$(1 - 0.5B)(1 + 0.8B)Y_t = a_t + 0.5a_{t-1} + 0.9a_{t-2},$$

and Y_t is stationary with $E[Y_t] = 0$.

5.2.1 Extended Autocorrelations of an ARUMA(p,d,q) Process

Although it is not possible to directly define the autocovariances of an ARUMA(p,d,q) process with $d > 0$, it is possible to define an "extended" autocorrelation of such a process in a meaningful way. To this end, we provide the following definition.

Definition 5.2 Let $r_1, r_2, \ldots, r_{p+d}$ be the roots of $\phi(r) = 0$, and let $\rho_k = \rho(k, r_1, r_2, \ldots, r_{p+d}, \theta)$ denote the autocorrelations of a stationary ARMA (p+d,q) process. Now suppose that the first d of the r_j approach the unit circle uniformly, that is, $|r_j| = |r_m|$ for $j, m \leq d$. We denote these d roots by $r_1, r_2, \ldots, r_d$. We define the *extended autocorrelation* of an ARUMA(p,d,q) process as ρ_k^*, where ρ_k^* is the limit of ρ_k as the roots r_j, $j = 1, \ldots, d$ approach the unit circle uniformly. That is

$$\rho_k^* = \lim_{|r_1| = |r_2| = \cdots = |r_d| \to 1} \rho(k, r_1, r_2, \ldots, r_{p+d}, \theta). \tag{5.11}$$

Example 5.3 AR(3) Models with Roots Close to +1

In Example 3.12, we illustrated the fact that the behavior of a (stationary) ARMA(p,q) process tends to be dominated by the roots of $\phi(r) = 0$ that are near the unit circle. Figure 5.2 shows the autocorrelations from the AR(3) models

$$(1 - \alpha_1 B)(1 - 1.2B + 0.8B^2)X_t = a_t, \tag{5.12}$$

where $\alpha_1 = 0.8$, 0.95, and 0.995 respectively. The factor table is shown in Table 5.2 where it can be seen that there is a positive real root $\left(\alpha_1^{-1}\right)$ and a pair of complex conjugate roots associated with the system frequency $f = 0.13$. Figure 5.2 shows true autocorrelations for these models while Figure 5.3 shows realizations of length $n = 200$ from the three models. Again, as in Example 3.12, it should be noticed that as $\alpha_1 \to 1$ (i.e., as the positive real root of the characteristic equation approaches the unit circle) the behavior

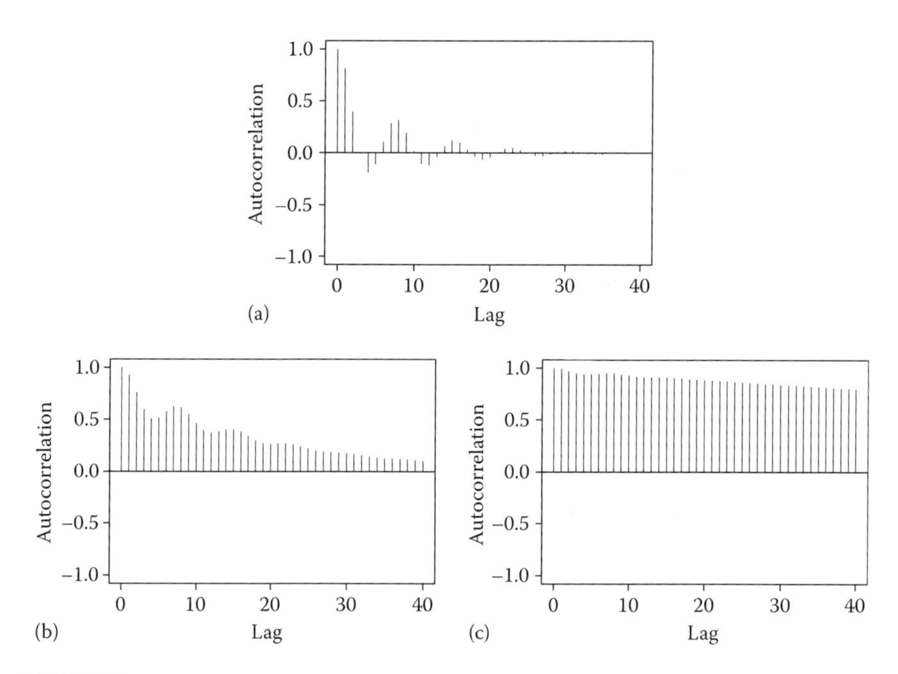

FIGURE 5.2
True autocorrelations for the AR(3) model in (5.12) with (a) $\alpha_1 = 0.8$, (b) $\alpha_1 = 0.95$, and (c) $\alpha_1 = 0.995$.

TABLE 5.2
Factor Table for the AR(3) Model in (5.12)

AR Factors	Roots (r_j)	$\lvert r_j^{-1} \rvert$	f_{0j}
$1 - \alpha_1 B$	α_1^{-1}	α_1	0
$1 - 1.2B + .8B^2$	$0.75 \pm 0.83i$	0.89	0.13

associated with the positive real root becomes more dominant in the autocorrelations and in the realizations. Specifically regarding the autocorrelations in Figure 5.2, we note that both the damped exponential (first order) behavior associated with the positive real root and the damped sinusoidal (second order) behavior associated with the pair of complex conjugate roots are visible when $\alpha_1 = 0.8$. However, as $\alpha_1 \to 1$, the damped exponential (first order) behavior tends to dominate the autocorrelations, to the point that in Figure 5.2c, the autocorrelations simply have the appearance of a damped exponential. This suggests that in this case, the extended autocorrelations, ρ_k^*, will be totally dominated by the first order behavior, that is, they will satisfy a first-order difference equation.

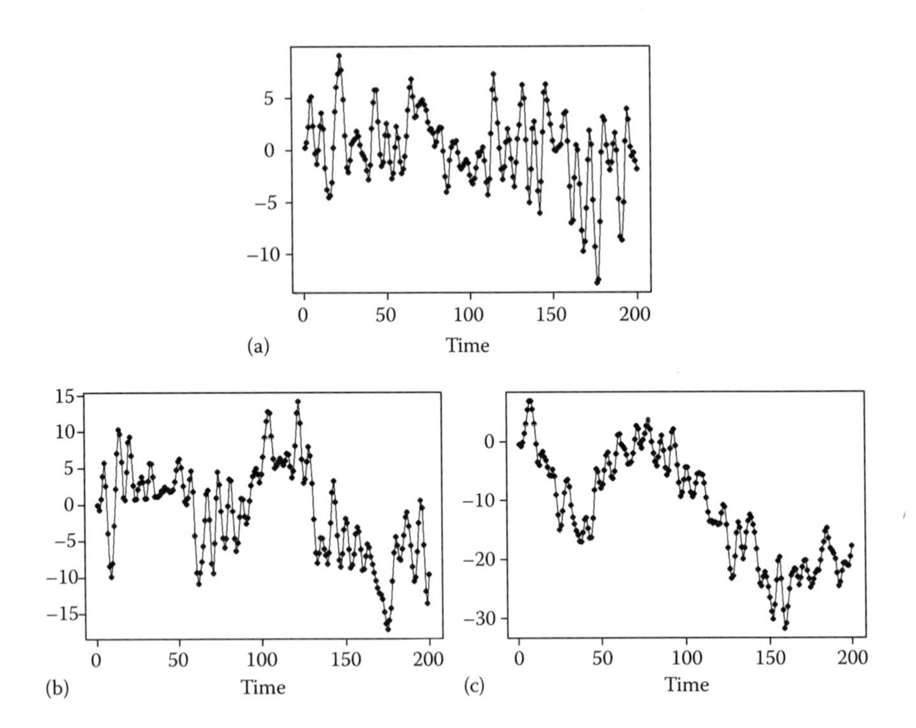

FIGURE 5.3
Realizations from the AR(3) model in (5.12) with (a) $\alpha_1 = 0.8$, (b) $\alpha_1 = 0.95$, and (c) $\alpha_1 = 0.995$.

Example 5.4 Near Nonstationary Model with Complex Roots

Consider the AR(3) model

$$(1 - 0.8B)(1 - 1.34B + 0.995B^2)X_t = a_t, \tag{5.13}$$

whose factor table is given in Table 5.3. Model (5.13) is similar to (5.12) in that the real root corresponds to $\alpha_1 = 0.8$, and the coefficients of the second-order factor are designed so that the associated system frequency is $f = 0.13$. However, the difference is that the pair of complex roots is much closer to the unit circle in (5.13) than they are in (5.12). Specifically, for these roots $|r_j^{-1}| = 0.9975$ while for (5.12), $|r_j^{-1}| = 0.89$. Figure 5.4 shows the true autocorrelations and a realization of length $n = 200$ for the model in (5.13). In this case, the autocorrelations and realization are almost entirely dominated by the second-order

TABLE 5.3
Factor Table for the AR(3) Model in (5.13)

AR Factors	Roots (r_j)	$\lvert r_j^{-1}\rvert$	f_{0j}
$1 - 1.34B + 0.995B^2$	$0.67 \pm 0.74i$	0.9975	0.13
$1 - 0.8B$	1.25	0.8	0

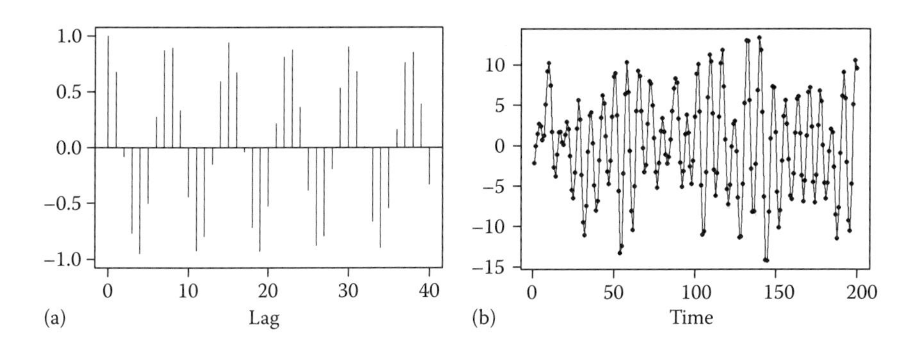

FIGURE 5.4
(a) Autocorrelation and (b) realization from the model in (5.13).

behavior associated with the pair of complex roots close to the unit circle. This behavior was previously observed in Example 3.12, but the point to be made here is the implication that the limiting autocorrelations ρ_k^* (i.e., for the case in which the complex roots are on the unit circle) will exhibit second-order behavior that "ignores" the first-order behavior induced by the positive real root, 0.8^{-1}.

As might be deduced from this example, there is a fundamental difference between the behavior of ρ_k^* for a nonstationary process and of ρ_k for a stationary process. That is, in a stationary process, ρ_k is influenced to some degree by all of the factors in the model, both autoregressive and moving average. However, in a nonstationary process, ρ_k^* is affected only by the roots on the unit circle, and in fact, only by those of highest multiplicity. In this regard, we have the following theorem that was first suggested by Gray et al. (1978) and proven in general by Findley (1978) and Quinn (1980).

Theorem 5.1 (Findley–Quinn Theorem)

Suppose that the following are true:

(a) X_t is an ARMA(p+d,q) with d roots "close" to the unit circle
(b) Of the d roots in (a), j have highest multiplicity ℓ

Then, ρ_k^* (the limiting value of ρ_k as the d roots approach the unit circle uniformly) exists, and for each $k = 0, \pm 1, \pm 2, \ldots$, ρ_k^* satisfies the linear homogeneous difference equation of order j.

$$(1 - \zeta_1 B - \cdots - \zeta_j B^j)\rho_k^* = 0, \tag{5.14}$$

where j is specified in (b). The roots of $\zeta(z) = 0$ are all on the unit circle (i.e., have modulus 1), and $\zeta(B)$ is the operator formed from the product of the nonstationary factors of highest multiplicity in $\phi(B)$.

Example 5.5 Extended Autocorrelations for Selected ARUMA Models

Below we demonstrate the results in Theorem 5.1 by listing several ARUMA processes along with the difference equation satisfied by ρ_k^*. In each case $\phi(B)$ is a stationary operator and $\theta(B)$ is an invertible operator.

1. $\phi(B)(1 - B)X_t = \theta(B)a_t$
   In this case, $d = 1$, $j = 1$, and $\ell = 1$.
   ρ_k^* satisfies $(1 - B)\rho_k^* = 0 \Rightarrow \rho_k^* = 1, k = 0, \pm1, \pm2, \ldots$
2. $\phi(B)(1 - B^4)X_t = \theta(B)a_t$.
   $(1 - B^4) = (1 - B)(1 + B)(1 + B^2)$ so that $d = 4$, $j = 4$, and $\ell = 1$.
   ρ_k^* satisfies $(1 - B^4)\rho_k^* = 0 \Rightarrow \rho_k^* = 1, k = 4m, m = 0, \pm1, \pm2, \ldots$
   $= 0$, elsewhere
3. $\phi(B)(1 - B)(1 - B^2)X_t = \theta(B)a_t$.
   $(1 - B)(1 - B^2) = (1 - B)(1 - B)(1 + B)$ so that $d = 3$, $j = 1$, and $\ell = 2$.
   ρ_k^* satisfies $(1 - B)\rho_k^* = 0 \Rightarrow \rho_k^* = 1, k = 0, \pm1, \pm2, \ldots$
4. $\phi(B)(1 - B)(1 - \sqrt{3}B + B^2)X_t = \theta(B)a_t$.
   In this case $d = 3$, $j = 3$, and $\ell = 1$.
   ρ_k^* satisfies $(1 - B)(1 - \sqrt{3}B + B^2)\rho_k^* = 0 \Rightarrow \rho_k^* = 0.65 + 0.35\cos\frac{\pi k}{6}$
5. $\phi(B)(1 - B)(1 - \sqrt{3}B + B^2)^2X_t = \theta(B)a_t$.
   In this case $d = 5$, $j = 2$, and $\ell = 2$.
   ρ_k^* satisfies $(1 - \sqrt{3}B + B^2)\rho_k^* = 0 \Rightarrow \rho_k^* = \cos\frac{\pi k}{6}$

Figure 5.5 shows the sample autocorrelations associated with the realizations in Figures 5.3c and 5.4b. The behavior of the sample autocorrelations is very similar to that observed for the true autocorrelations in Figures 5.2c and 5.4a, which suggests that the sample autocorrelations behave asymptotically like ρ_k^*. Useful criteria under which this asymptotic behavior holds have not been established, although in all examples that we consider, $\hat{\rho}_k$ behaves like ρ_k^* when the roots are near the unit circle. It should be noted that the sample autocorrelations in Figure 5.5 damp more quickly than do their true counterparts. This can be attributed to the formula used to calculate $\hat{\gamma}_k$ in (1.19) that tends to shrink sample autocorrelations toward zero. Therefore, the fact that there is some damping in the sample autocorrelations should not necessarily lead to a stationary model.

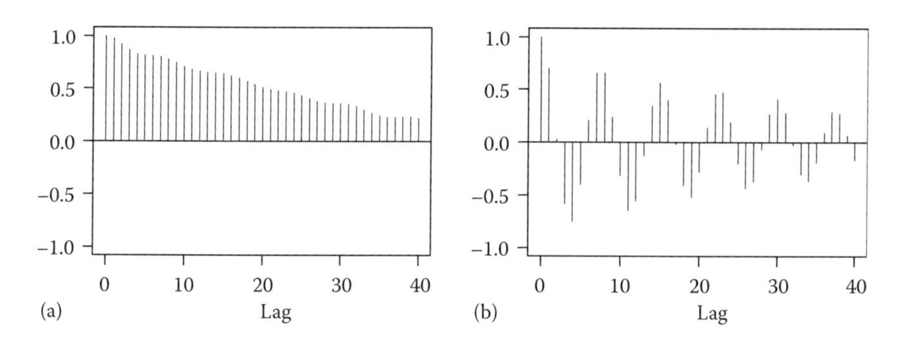

FIGURE 5.5
Sample autocorrelation for the realizations in (a) Figures 5.3c and (b) 5.4b.

5.2.2 Cyclical Models

In Example 5.5, it is seen that for an ARUMA(p,d,q) process, the limiting autocorrelations, ρ_k^*, do not decay to zero. In some cases they stay fixed at 1, and in others they are periodic. In particular, if all of the d roots of $\lambda(r) = 0$ in an ARUMA(p,d,q) process are equal to 1, the process is noncyclical. On the other hand, if at least one of the d roots of highest multiplicity is -1 or not real, then we will refer to the process as a *cyclical* ARUMA(p,d,q) process. The simplest cyclical ARUMA process is $(1 + B)X_t = a_t$. In this case, $\rho_k^* = (-1)^k$ so that the autocorrelation is periodic with system frequency $f_0 = 0.5$.

Example 5.6 An ARUMA(1,2,0) Model

Consider the ARUMA(1,2,0) process

$$(1 - 0.8B)(1 - 1.34B + B^2)X_t = a_t. \tag{5.15}$$

Note that the model in (5.15) is the nonstationary counterpart of the near-nonstationary model in (5.13).

The autocorrelations and realization in Figure 5.4 can be considered to essentially be from the ARUMA(2,1,0) model in (5.15). As noted previously, the sample autocorrelations from a realization from an ARUMA(p,d,q) model will damp (even though there is no damping in the true autocorrelations) because of the formula for calculating $\hat{\gamma}_k$. Figure 5.4 shows that the true autocorrelations and the realization behave much as they would for a harmonic component model with period equal to 7.7 (1/0.13).

Using `GW-WINKS`: It should be noted that if you use the `Generate Realization` option on the `Time Series Analysis` menu in `GW-WINKS` to generate a realization or true autocorrelations from (5.15), you must move the roots slightly outside the unit circle, as in (5.13), to avoid numerical problems. For example, you should replace $1 - 1.34B + B^2$ in (5.15) with an operator, such as $1 - 1.34B + 0.999B^2$ when you enter the coefficients in `GW-WINKS`.

5.3 Multiplicative Seasonal ARUMA$(p,d,q)\times(P_s,D_s,Q_s)_s$ Process

As we have extended the ARMA(p,q) model to allow roots on the unit circle, we also extend the multiplicative seasonal model to include roots on the unit circle. A zero-mean process, X_t, is said to follow a multiplicative seasonal ARUMA$(p,d,q) \times (P_s,D_s,Q_s)_s$ model if it can be written in the form

$$\phi(B)\lambda(B)\Phi(B^s)\lambda_s(B^s)X_t = \theta(B)\Theta(B^s)a_t, \tag{5.16}$$

where the roots of $\phi(z) = 0$, $\Phi(z^s) = 0$, $\theta(z) = 0$, and $\Theta(z^s) = 0$ all lie outside the unit circle, while the roots of $\lambda(z) = 0$ and $\lambda_s(z^s) = 0$ lie on the unit circle.

Using Theorem 5.1, it follows that for the multiplicative seasonal model $\phi(B)\Phi(B^s)(1 - B^s)X_t = \theta(B)\Theta(B^s)a_t$, for which the only roots on the unit circle are the roots of $1 - z^s = 0$, then the limiting autocorrelations are given by

$$\begin{aligned} \rho_k^* &= 1, \quad k = 0, \pm s, \pm 2s, \ldots \\ &= 0, \quad \text{elsewhere.} \end{aligned}$$

In this example $\lambda_s(B^s) = 1 - B^s$ and $D_s = 1$.

A simple example of a nonstationary multiplicative seasonal model is given by

$$\phi(B)(1 - B^4)(X_t - \mu) = \theta(B)a_t, \tag{5.17}$$

for which the limiting autocorrelations are given by

$$\begin{aligned} \rho_k^* &= 1, \quad k = 0, \pm 4, \pm 8, \ldots \\ &= 0, \quad \text{elsewhere.} \end{aligned}$$

In model (5.17) $d = 0$ and $D_4 = 1$.

5.3.1 Factor Tables for Seasonal Models of the Form (5.17) with $s = 4$ and $s = 12$

Although the seasonal order, s, can take on any integer value greater than one in applications, it is often the case that $s = 4$ (for quarterly data) and $s = 12$ (for monthly data). In the case of $s = 4$, we have $(1 - B^4) = (1 - B)(1 + B)(1 + B^2)$ so that the roots of the characteristic equation $1 - z^4 = 0$ are ± 1 and $\pm i$. Specifically, the factor table associated with the model $(1 - B^4)X_t = a_t$ is given in Table 5.4. In Table 5.5, we show the more complicated factor table for $(1 - B^{12})X_t = a_t$. The roots of $1 - z^{12} = 0$ are the 12 roots of unity that are shown in Table 5.5 along with the associated factors

TABLE 5.4

Factor Table for $(1 - B^4)X_t = a_t$

AR Factors	Roots (r_j)	$\lvert r_j^{-1} \rvert$	f_{0j}
$1 - B$	1	1	0
$1 + B$	−1	1	0.5
$1 + B^2$	$\pm i\cos(\pi/2)$	1	0.25

TABLE 5.5

Factor Table for $(1 - B^{12})X_t = a_t$

AR Factors	Roots (r_j)	$\lvert r_j^{-1} \rvert$	f_{0j}
$1 - B$	1	1	0
$1 - \sqrt{3}B + B^2$	$\cos(\pi/6) \pm i\sin(\pi/6)$	1	0.083
$1 - B + B^2$	$\cos(\pi/3) \pm i\sin(\pi/3)$	1	0.167
$1 + B^2$	$\pm i\sin(\pi/2)$	1	0.25
$1 + B + B^2$	$\cos(2\pi/3) \pm i\sin(2\pi/3)$	1	0.333
$1 + \sqrt{3}B + B^2$	$\cos(5\pi/6) \pm i\sin(5\pi/6)$	1	0.417
$1 + B$	−1	1	0.5

and system frequencies. When analyzing data for which a seasonal model with $s = 4$ or $s = 12$ is a candidate, it will be useful to refer to Tables 5.4 and 5.5 to see how the factors of the estimated model agree with those expected for the seasonal models represented in these two tables. This will be further discussed in Section 8.5.3.

Example 5.7 The Airline Model

The monthly data for the number of international airline passengers (log scale) shown in Figure 1.25a have been widely analyzed as a seasonal data set with a trending behavior. A variety of models have been proposed for this data set, and as a group, these are referred to as "airline models." The model we will use here for this purpose is an ARUMA(p,1,q) × $(P_{12}, 1, Q_{12})_{12}$ model of the form

$$\phi(B)\Phi(B^{12})(1 - B)(1 - B^{12})(X_t - \mu) = \theta(B)\Theta(B^{12})a_t. \tag{5.18}$$

Note that this model has 13 roots on the unit circle: a nonstationary factor $1 - B$ and a 12 month nonstationary seasonal factor $1 - B^{12}$. A factor table associated with the characteristic equation $(1 - z)(1 - z^{12}) = 0$ would be the same as the one in Table 5.5, except that it would include two factors of $1 - B$.

Thus, in Theorem 5.1 we have $d=13$, $j=1$, $\ell=2$. Notice that the limiting autocorrelations, ρ_k^*, of such a process are given by

$$\rho_k^* \equiv 1, \tag{5.19}$$

since the root of 1 is repeated twice. These ideas will be discussed in more detail in Section 8.3.5.3 as we analyze seasonal data.

5.4 Random Walk Models

Random walk models have been used for describing the behavior of the stock market, the path traveled by a molecule as it moves through a liquid, and many other applications. In this section, we will define *random walk* and *random walk with drift* models and briefly discuss their properties.

5.4.1 Random Walk

Let $X_t, t = 1, 2, \ldots$ be defined by

$$X_t = X_{t-1} + a_t, \tag{5.20}$$

where $X_0 = 0$ and a_t, $t = 1, 2, \ldots$ is a sequence of independent and identically distributed random variables with mean zero and variance σ_a^2. The name, "random walk," comes from the fact that at time $t-1$, the movement to time t is totally determined by the random quantity a_t. It is sometimes convenient to write the model in (5.20) as

$$X_t = \sum_{k=1}^{t} a_k, \tag{5.21}$$

for $t = 1, 2, \ldots$. From (5.21), it can be seen that $E(X_t) = 0$, and the variance of X_t is given by

$$\mathrm{Var}(X_t) = t\sigma_a^2. \tag{5.22}$$

Thus, the random walk model in (5.20) is nonstationary since $\mathrm{Var}(X_t)$ is not constant. It can also be shown that if $1 \le t_1 \le t_2$, then

$$\mathrm{Cov}(X_{t_1}, X_{t_2}) = t_1\sigma_a^2 \tag{5.23}$$

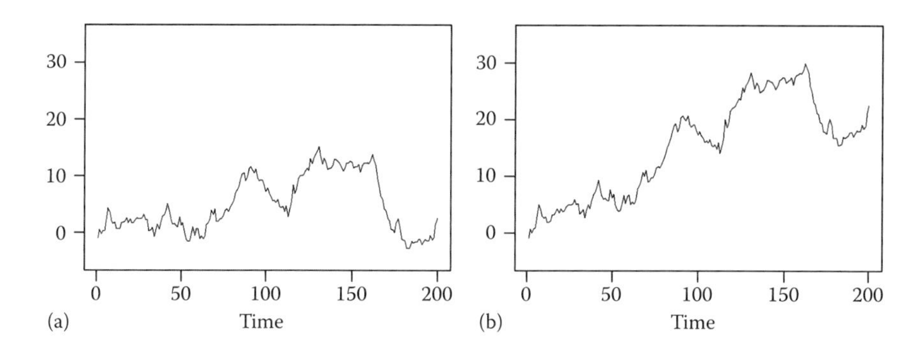

FIGURE 5.6
Realizations from (a) random walk and (b) random walk with drift models.

and

$$\rho_X(t_1, t_2) = \sqrt{\frac{t_1}{t_2}}, \tag{5.24}$$

(see Problem 5.9). Consequently, the autocorrelation between X_{t_1} and X_{t_2} depends on time, and the random walk process does not satisfy the stationarity conditions for either variance or autocorrelations. The impact of (5.22) through (5.24) on realizations is that the variability increases with time while simultaneously the autocorrelation between X_t and X_{t+k}, for fixed k, increases as t increases. Figure 5.6a shows a realization of length $n = 200$ from a random walk model with $\sigma_a^2 = 1$. The random progression in time is apparent along with the tendency for the variability to increase as t increases.

5.4.2 Random Walk with Drift

An extension of the model in (5.20) is the *random walk with drift* model given by

$$X_t = \alpha + X_{t-1} + a_t, \tag{5.25}$$

where α is a nonzero constant called the *drift*, and again $X_0 = 0$. Based on model (5.25), at time $t - 1$, the movement to time t is α plus the random a_t. It follows that $E(X_t) = \alpha t$, and, thus, the process mean increases or decreases in time depending upon whether α is positive or negative, respectively. Figure 5.6b displays a realization of length $n = 200$ for a random walk with drift $\alpha = 0.1$ and $\sigma_a^2 = 1$ using the same random sequence $a_1, a_2, \ldots, a_{200}$ that was used in Figure 5.5a. The upward drift is evident.

5.5 G-Stationary Models for Data with Time-Varying Frequencies

Figure 1.26a shows a bat echolocation signal that has strong cyclic behavior but for which it was previously noted that the cycle length increases with time. This is an example of data with time-varying frequencies (TVF), a property shared by a variety of other natural signals, such as chirps, Doppler signals, and seismic signals. Figure 5.7a shows seismic surface wave data from a mining explosion (collected at seismic station MNTA in Wyoming) that consist of 700 measurements with a sampling rate of 0.1 s. Examination of the data shows that the periodic behavior is compacting in time, that is, the dominant frequency in the data is increasing with time. Figure 5.7b shows a Doppler signal that dramatically demonstrates frequency behavior that is decreasing with time. These signals are not well modeled using ARMA(p,q) models that are designed to model stationary behavior. Historically, TVF data have been analyzed using windowing techniques that break the data into windows sufficiently small that "stationarity" within the window is a reasonable approximation. However, if the frequencies are changing sufficiently fast (as in the Doppler signal), then such a procedure is problematic at best.

In Chapter 12, we discuss the short term Fourier transform and wavelet techniques for dealing with TVF data. If the frequencies in the data are changing in time in a smooth manner as is the case in the three TVF realizations shown in Figures 1.26a and 5.7, then the data can be modeled using G-stationary models. These are models for which the data are non-stationary, but an appropriate transformation of the time axis produces stationary data that can be modeled using stationarity-based techniques. G-stationary models are the topic of Chapter 13, and further discussion will be deferred until then.

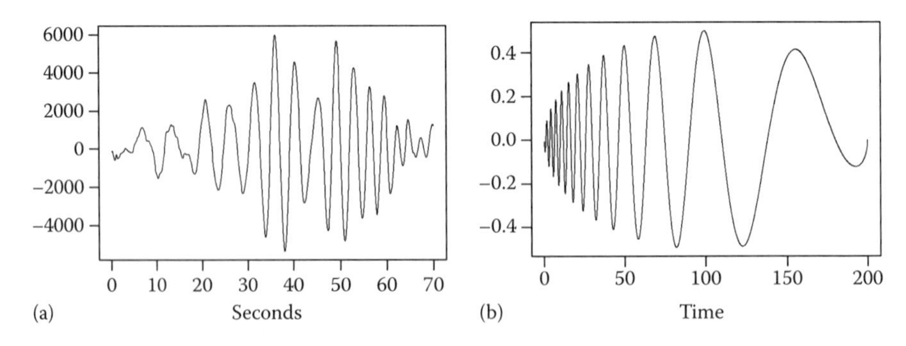

FIGURE 5.7
Two examples of TVF data: (a) MNTA seismic data and (b) Doppler data.

Exercises

Applied Problems

If you use `GW-WINKS` to work out the following problems, you will find the `Generate Realization`, `Factoring Routines`, and `Sample Autocorrelations` options on the `Time Series Analysis` menu to be useful.

5.1 This problem uses the global temperature data shown in Figure 1.24a and given in HADLEY.XLS.

(a) Plot the data.
(b) Using least squares methods, fit a regression line $X_t = a + bt + Z_t$. What are $\hat{a}$ and $\hat{b}$ for the fitted model?
(c) Find and plot Z_t, the residuals from the least squares fit.
(d) Does Z_t have the appearance of stationary data? Discuss.
(e) Find and plot the sample autocorrelations for X_t and Z_t. Compare and contrast these two sets of sample autocorrelations.

5.2 Generate and plot four realizations from the AR(4) model in (5.3). Do any of these realizations have a "linear-trending" appearance?

5.3 This problem involves generating realizations from the ARUMA(3,2,1) model

$$(1 - B)(1 + B)(1 - \alpha_1 B)(1 + B + 0.5B^2)X_t = a_t,$$

where $|\alpha_1| < 1$. Recall that when using `GW-WINKS`, the factors $1-B$ and $1+B$ should be entered using nearly nonstationary factors, such as $1 - 0.9999B$ and $1 + 0.9999B$, respectively.

(a) Generate realizations of length $n = 500$ for the cases $\alpha_1 = 0.01, 0.5, 0.9, 0.999$.
(b) For each α_1, plot the realization, true autocorrelations, and sample autocorrelations.
(c) Do the plots of the true autocorrelations behave in the way suggested by the Findley–Quinn Theorem?
(d) Do the sample autocorrelations behave in a manner analogous to that displayed by the true autocorrelations?
(e) How is the behavior in the autocorrelations manifested in the realizations?

5.4 For each of these four ARUMA(p,d,q) models, find p, d, and q.

(i) $(1 - 3B + 4.5B^2 - 5B^3 + 4B^4 - 2B^5 + 0.5B^6)X_t = (1 - 1.7B + 0.8B^2)a_t$
(ii) $(1 - 0.5B + 0.3B^2 - 0.95B^3 + 0.3B^4 - 0.35B^5 + 0.2B^6)X_t = a_t$
(iii) $(1-1.5B+1.3B^2+0.35B^3-B^4+1.35B^5-0.7B^6+0.4B^7)X_t=(1+0.9B^2)a_t$
(iv) $(1 - 0.5B - 0.5B^2 - B^4 + 0.5B^5 + 0.5B^6)X_t = (1 - 0.81B^2)a_t$

5.5 Find and print the factor table for $(1 - B^6)X_t = a_t$. What are the 6 roots of unity?

Theoretical Problems

5.6 Verify the following from Example 5.5:

(a) $(1 - B)\rho_k^* = 0 \;\Rightarrow\; \rho_k^* = 1,\, k = 0,\, \pm 1,\, \pm 2, \ldots$

(b) $(1 - B^4)\rho_k^* = 0 \;\Rightarrow\; \rho_k^* = 1, \quad k = 4m, m = \pm 1,\, \pm 2, \ldots$
$\qquad\qquad\qquad\qquad\qquad = 0, \quad \text{otherwise}$

(c) $(1 - \sqrt{3}B + B^2)\rho_k^* = 0 \;\Rightarrow\; \rho_k^* = \cos\frac{\pi k}{6}$

(d) $(1 - B)(1 - \sqrt{3}B + B^2)\rho_k^* = 0 \;\Rightarrow\; \rho_k^* = 0.65 + 0.35\cos\frac{\pi k}{6}$

5.7 Consider the following ARUMA(*p*,*d*,*q*) model:

$$(1 - B^2)(1 - B^3)(1 + B)(1 - 1.6B + 0.6B^2)(1 + 0.8B^2)(X_t - 10) = (1 + 0.2B + 0.9B^2)a_t.$$

(a) Identify *p*, *d*, and *q*.
(b) Identify *j* and ℓ in Theorem 5.1.
(c) What difference equation does ρ_k^* satisfy? What is ρ_6^*?

5.8 Consider the second-order operator $1 - \lambda_1 B - \lambda_2 B^2$ whose characteristic equation $1 - \lambda_1 z - \lambda_2 z^2 = 0$ has a pair of complex roots on the unit circle. Show that in this case $\lambda_2 = -1$.

5.9 Verify Equations 5.23 and 5.24 in the notes.

5.10 Find the mean and variance for a random walk with drift.

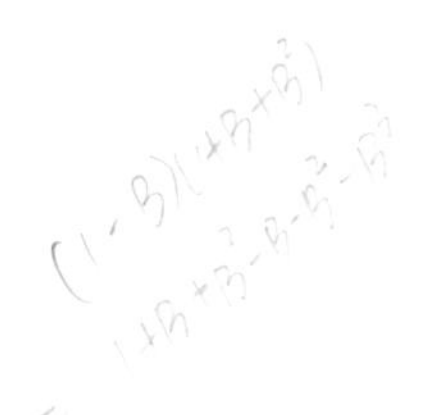

6

Forecasting

In this chapter, we discuss techniques for forecasting the future behavior of a time series, given present and past values. ARMA models are popular tools for obtaining such forecasts. Forecasts found using ARMA models differ from forecasts based on curve fitting in a fundamental way. Given a realization, such as the sunspot data in Figure 1.23a, one might forecast into the future based on a mathematical curve fit to the data, that is, based on the underlying assumption that future behavior follows some deterministic path with only random fluctuations. However, for data, such as the sunspot data, it is unlikely that the use of such a deterministic function for forecasting is a reasonable approach. On the other hand, the ARMA forecasting techniques to be discussed in this chapter are based on the assumption that the transition to the future is guided by its correlation to the past. When there is no physical model for connecting the past with the future, the correlation-based approach seems reasonable.

The forecasting or prediction problem
The basic problem that we address is that of finding an "optimal" predictor for $X_{t_0+\ell}$ given observations $X_1, \ldots, X_{t_0}$ from a known ARMA model. We use the notation:

- $\tilde{X}_{t_0}(\ell)$ is a predictor of $X_{t_0+\ell}$ given data up to X_{t_0}.
- t_0 is referred to as the *forecast origin* (in most cases t_0 is the "present" time).
- ℓ is the lead time, that is, the number of time units (*steps ahead*) into the future to be forecast.

The "best" or "optimal" predictor will be in the mean square sense. That is, we will search for a predictor, $\tilde{X}_{t_0}(\ell)$, of $X_{t_0+\ell}$ that minimizes $E\left[\left(X_{t_0+\ell} - \tilde{X}_{t_0}(\ell)\right)^2\right]$.

6.1 Mean Square Prediction Background

This section briefly reviews some of the key results related to the problem of finding optimal predictors in the mean square sense. Theorem 6.1 shows the fundamental role that conditional expectation plays in the search for a best mean square predictor. From probability theory, it is known that if X and Y are random variables, the conditional expectation of Y given X, denoted $E(Y|X)$, is a function of X (Chung 2001). Theorem 6.1, below, states that $E(Y|X)$ is closest to Y in the mean square sense among all square integrable random variables that are functions of X.

Theorem 6.1

Suppose X and Y are random variables with $E(X^2) < \infty$ and $E(Y^2) < \infty$. If $h(X)$ is a random variable such that $E[h(X)^2] < \infty$, then

$$E\left[(Y - E(Y|X))^2\right] \leq E\left[(Y - h(X))^2\right].$$

Proof:

See Brockwell and Davis (1991) and Billingsley (1995).

The result in Theorem 6.1 extends to the case in which $X_1, X_2, \ldots, X_{t_0}$ are random variables such that $E\left(X_j^2\right) < \infty$, $j = 1, \ldots, t_0$. In this case, the conditional expectation $E[X_{t_0+\ell}|X_1, \ldots, X_{t_0}]$ is the best predictor of $X_{t_0+\ell}$ in the mean square sense among all square integrable functions of the random variables $X_1, X_2, \ldots, X_{t_0}$. Thus, since our goal is to forecast $X_{t_0+\ell}$ based on $X_1, \ldots, X_{t_0}$, it follows that $E[X_{t_0+\ell}|X_1, \ldots, X_{t_0}]$ is the optimal predictor. Although $E[X_{t_0+\ell}|X_1, \ldots, X_{t_0}]$ is the desired optimal predictor, that is, the gold standard, it is not necessarily linear and may be very difficult to obtain in practice. As a consequence, we will choose to restrict our attention to "linear predictors" and search for "optimal" predictors of the form

$$Z_{t_0} = a_1 X_{t_0} + a_2 X_{t_0-1} + \cdots + a_{t_0} X_1. \tag{6.1}$$

Theorem 6.2 (sometimes called the *Projection Theorem*) says that the best linear predictor of the form (6.1) is the projection of $X_{t_0+\ell}$ onto the hyperplane spanned by $X_1, \ldots, X_{t_0}$.

Theorem 6.2 (Projection Theorem)

If X_t is a stationary time series, then $Z_{t_0}^*$ is the best linear predictor of $X_{t_0+\ell}$ (has minimum mean square error among linear predictors of the form (6.1)) if and only if

$$E\left[(X_{t_0+\ell} - Z_{t_0}^*)X_j\right] = 0, \quad j = 1, 2, \ldots, t_0.$$

Proof:

See Appendix 6.A.

We let $\tilde{X}_{t_0}(\ell)$ denote the linear predictor, $Z^*_{t_0}$, in Theorem 6.2. Theorem 6.3 shows how to compute the coefficients $a^*_1, \ldots, a^*_{t_0}$ of $\tilde{X}_{t_0}(\ell)$, where

$$\begin{aligned}\tilde{X}_{t_0}(\ell) &= a^*_1 X_{t_0} + a^*_2 X_{t_0-1} + \cdots + a^*_{t_0} X_1 \\ &= \sum_{k=1}^{t_0} a^*_k X_{t_0-k+1}.\end{aligned} \tag{6.2}$$

Theorem 6.3

If X_t is a stationary process with autocovariance function γ_k, then $a^*_1, \ldots, a^*_{t_0}$ in (6.2) are the solutions to the system of equations

$$\mathbf{g}_\ell = \mathbf{\Gamma a}^*, \tag{6.3}$$

where $g_\ell = (\gamma_\ell, \gamma_{\ell+1}, \ldots, \gamma_{\ell+t_0-1})'$, $\mathbf{a}^* = (a^*_1, a^*_2, \ldots, a^*_{t_0})'$, and

$$\mathbf{\Gamma} = \begin{pmatrix} \gamma_0 & \gamma_1 & \cdots & \gamma_{t_0-1} \\ \gamma_1 & \gamma_0 & \vdots & \gamma_{t_0-2} \\ \vdots & & & \\ \gamma_{t_0-1} & \gamma_{t_0-2} & \cdots & \gamma_0 \end{pmatrix}.$$

Proof:

See Problem 6.8.

The typical application of these results is the case $t_0 = n$ (i.e., the realization length). In this case, (6.3) can be an extremely large system of equations, and these forecasts can be difficult to obtain consequently, in the next section, we will discuss the so-called Box–Jenkins forecasts. The Box–Jenkins forecasts are based on an ARMA model and are linear predictors based on the (unattainable) assumption that the random variables X_t, $t = t_0, t_{0-1}, \ldots$ (i.e., the infinite past) are available. Nevertheless, forecasts based on the assumption of the availability of the infinite past are widely used in practice because they are easy to compute and are intuitively attractive, and for a moderately large t_0, the results are close to those obtained by solving (6.3). In this chapter, we focus on the Box–Jenkins style forecasts. These forecasts can be computed using the GW-WINKS software.

6.2 Box–Jenkins Forecasting for ARMA(p,q) Models

We return to the problem of finding the best predictor of $X_{t_0+\ell}$, but now we assume that we are given the infinite past $X_{t_0}, X_{t_0-1}, \ldots$, and that the data are from a *known* ARMA(*p*,*q*) model. That is, we assume that the model orders and parameters are known. In a practical situation, the model orders and parameters will be estimated, and, thus, the forecast errors will differ somewhat from those we present here. However, in most cases, this difference is small. As discussed in Section 6.1, "best" will be taken to be in the mean square sense, and the best such predictor is the conditional expectation $E[X_{t_0+\ell}|X_{t_0}, X_{t_0-1}, \ldots]$. As mentioned earlier, the conditional expectation is often difficult to obtain, but it turns out that it can be readily computed in the current setting.

Since X_t is an ARMA(*p*,*q*) process, then because of its stationarity and invertibility we can write

$$X_t - \mu = \sum_{j=0}^{\infty} \psi_j a_{t-j} \tag{6.4}$$

and

$$a_t = \sum_{j=0}^{\infty} \pi_j (X_{t-j} - \mu), \tag{6.5}$$

respectively. Based on (6.4) and (6.5), it follows that "knowing" $X_k, X_{k-1}, \ldots$ is equivalent to knowing $a_k, a_{k-1}, \ldots$. Thus,

$$E[X_t|X_k, X_{k-1}, \ldots] = E[X_t|a_k, a_{k-1}, \ldots]. \tag{6.6}$$

For notational convenience, we denote the conditional expectation in (6.6) as $E_k[X_t]$. Similarly,

$$E[a_t|a_k, a_{k-1}, \ldots] = E[a_t|X_k, X_{k-1}, \ldots], \tag{6.7}$$

which will be denoted $E_k[a_t]$. Since $E[a_t] = 0$ and a_{t_1} and a_{t_2} are uncorrelated for $t_1 \neq t_2$, it follows that

$$\begin{aligned} E_k[a_t] &= 0, \quad t > k \\ &= a_t, \quad t \leq k. \end{aligned}$$

Thus, whenever $t > k$,

$$E_k[X_t] = E_k\left[\mu + \sum_{j=0}^{\infty} \psi_j a_{t-j}\right]$$
$$= \mu + \sum_{j=t-k}^{\infty} \psi_j a_{t-j}.$$

Our interest is in $E_{t_0}[X_{t_0+\ell}]$ when $\ell > 0$, which is given by

$$E_{t_0}[X_{t_0+\ell}] = E_{t_0}\left[\mu + \sum_{j=0}^{\infty} \psi_j a_{t_0+\ell-j}\right]$$
$$= \mu + \sum_{j=\ell}^{\infty} \psi_j a_{t_0+\ell-j}, \tag{6.8}$$

and the desired forecast $\hat{X}_{t_0}(\ell) \equiv E_{t_0}[X_{t_0+\ell}]$ is given in (6.9) below.

General Linear Process Form of the Best Forecast Equation

$$\hat{X}_{t_0}(\ell) = \mu + \sum_{j=\ell}^{\infty} \psi_j a_{t_0+\ell-j}. \tag{6.9}$$

Note that the forecast equation in (6.9) is not attainable in practice, but is given here because of its usefulness in deriving the properties of the best forecasts. Several properties of $\hat{X}_{t_0}(\ell)$ are given in the next section.

6.3 Properties of the Best Forecast $\hat{X}_{t_0}(\ell)$

In this section, we give several properties of the "best" predictor, $\hat{X}_{t_0}(\ell)$, when X_t is a stationary, invertible ARMA(p,q) process. We provide derivations of property (b) in the following, and leave the others as exercises.

Properties of $\hat{X}_{t_0}(\ell)$

(a) In the current setting, the best predictor, given by the conditional expectation, is a *linear* function of $X_{t_0}, X_{t_0-1}, \ldots$

(b) $\hat{X}_{t_0}(\ell) = X_{t_0+\ell}$ when $\ell \le 0$.

This useful (but not surprising) result follows from the fact that

$$
\begin{aligned}
\hat{X}_{t_0}(\ell) &= E_{t_0}[X_{t_0+\ell}] \\
&= E_{t_0}\left[\mu + \sum_{j=0}^{\infty} \psi_j a_{t_0+\ell-j}\right] \\
&= \mu + \sum_{j=0}^{\infty} \psi_j a_{t_0+\ell-j},
\end{aligned} \tag{6.10}
$$

since whenever $\ell \leq 0$, it follows that $t_0 + \ell - j \leq t_0$ for all $j \geq 0$. Property (b) follows by noting that $\hat{X}_{t_0}(\ell)$, given in (6.10), is simply the general linear process expression for $X_{t_0+\ell}$.

(c) The ℓ-step ahead forecast errors, $e_{t_0}(\ell) = X_{t_0+\ell} - \hat{X}_{t_0}(\ell)$, are given by

$$
\begin{aligned}
e_{t_0}(\ell) &= a_{t_0+\ell} + \psi_1 a_{t_0+\ell-1} + \cdots + \psi_{\ell-1} a_{t_0+1} \\
&= \sum_{j=0}^{\ell-1} \psi_j a_{t_0+\ell-j}.
\end{aligned} \tag{6.11}
$$

(d) The ℓ-step ahead forecast errors have expectation zero, that is,

$$E[e_{t_0}(\ell)] = 0, \quad \ell \geq 1. \tag{6.12}$$

Because of (6.12), we say that the forecasts are unbiased.

(e) The 1-step ahead forecast errors are the a_t's (sometimes called the residuals), that is,

$$e_{t_0}(1) = a_{t_0+1}. \tag{6.13}$$

(f) The variance of the forecast errors is given by

$$\mathrm{Var}[e_{t_0}(\ell)] = \sigma_a^2 \sum_{j=0}^{\ell-1} \psi_j^2. \tag{6.14}$$

(g) The covariance between forecast errors satisfies

$$\mathrm{Cov}(e_{t_0}(\ell), e_{t_0-j}(\ell)) = 0, \quad j \geq \ell.$$

Whenever $j = 0, 1, \ldots, \ell - 1$, the forecast errors may be correlated.

(h) $e_{t_0}(\ell_1)$ and $e_{t_0}(\ell_2)$ are correlated, that is, forecast errors from the same origin are correlated.

(i) The "best MSE forecast" for $\sum_{k=1}^{m} \alpha_k X_{t_0+k}$ from origin t_0 is given by $\sum_{k=1}^{m} \alpha_k \hat{X}_{t_0}(k)$.

As an example, if X_t represents monthly sales, then (i) states that from the forecast origin t_0, the "best" forecast for next quarter's sales, that is, for $X_{t_0+1} + X_{t_0+2} + X_{t_0+3}$ is $\hat{X}_{t_0}(1) + \hat{X}_{t_0}(2) + \hat{X}_{t_0}(3)$.

6.4 π-Weight Form of the Forecast Function

A stationary and invertible process can be written in the π-weight form

$$\sum_{j=0}^{\infty} \pi_j (X_{t-j} - \mu) = a_t, \tag{6.15}$$

which was previously given in (3.83). Therefore, since $\pi_0 = 1$, (6.15) can be written as

$$X_t = \mu - \sum_{j=1}^{\infty} \pi_j (X_{t-j} - \mu) + a_t,$$

which leads to the forecasting formula

$$\hat{X}_{t_0}(\ell) = \mu - \sum_{j=1}^{\infty} \pi_j (\hat{X}_{t_0}(\ell - j) - \mu), \tag{6.16}$$

where, as before, $\hat{X}_{t_0}(\ell) = X_{t_0+\ell}$, for $\ell \le 0$. The form in (6.16) is similar to the ψ-weight form in (6.9) in that it involves an infinite summation and is based on the unrealistic assumption that $X_{t_0}, X_{t_0-1}, \ldots, X_1, X_0, X_{-1}, \ldots$ are known. Thus, the following truncated version of the forecast function is used in practice.

Truncated Version of the Forecast Equation

$$\hat{X}_{t_0}^{(T)}(\ell) = \overline{X} - \sum_{j=1}^{t_0-1+\ell} \pi_j \left(\hat{X}_{t_0}^{(T)}(\ell - j) - \overline{X} \right). \tag{6.17}$$

For example, if $t_0 = n$, then using (6.17), it follows that

$$\hat{X}_n^{(T)}(1) = \overline{X} - \pi_1(X(n) - \overline{X}) - \pi_2(X(n-1) - \overline{X}) - \cdots - \pi_n(X(1) - \overline{X})$$
$$\hat{X}_n^{(T)}(2) = \overline{X} - \pi_1\left(\hat{X}_n^{(T)}(1) - \overline{X}\right) - \pi_2(X(n) - \overline{X}) - \cdots - \pi_n(X(2) - \overline{X}),$$
$$\vdots$$

NOTES: The π-weight version of the forecast function is appropriately used when the π-weights damp sufficiently quickly so that the π_j's are essentially zero for $j > t_0 - 1 + \ell$, which should generally hold for stationary and invertible ARMA processes when t_0 is sufficiently large. This form of the forecast function is useful for forecasting based on long memory models (see Section 11.6).

6.5 Forecasting Based on the Difference Equation

If X_t is a stationary, invertible ARMA(p,q) process, then (6.9) is an expression for the best predictor of $X_{t_0+\ell}$ given $X_{t_0}, X_{t_0-1}, \ldots$, in terms of the general linear process or "infinite moving average" representation of the process. The forecast, $\hat{X}_{t_0}(\ell)$, can also be written in terms of the difference equation satisfied by the data, a form that is more useful in practice. Suppose, X_t satisfies $\phi(B)\, X_t = \theta(B)a_t$, and hence that $\phi(B)X_{t_0+\ell} = \theta(B)a_{t_0+\ell}$. Thus,

$$X_{t_0+\ell} - \phi_1 X_{t_0+\ell-1} - \cdots - \phi_p X_{t_0+\ell-p} = a_{t_0+\ell} - \theta_1 a_{t_0+\ell-1} - \cdots - \theta_q a_{t_0+\ell-q}.$$

Taking conditional expectations, we obtain

$$E_{t_0}\left[X_{t_0+\ell} - \phi_1 X_{t_0+\ell-1} - \cdots - \phi_p X_{t_0+\ell-p}\right] = E_{t_0}\left[a_{t_0+\ell} - \theta_1 a_{t_0+\ell-1} - \cdots - \theta_q a_{t_0+\ell-q}\right]. \tag{6.18}$$

This can be written as

$$\begin{aligned}\hat{X}_{t_0}(\ell) &= \phi_1 \hat{X}_{t_0}(\ell-1) + \phi_2 \hat{X}_{t_0}(\ell-2) + \cdots + \phi_p \hat{X}_{t_0}(\ell-p)\\ &\quad + E_{t_0}[a_{t_0+\ell}] - \theta_1 E_{t_0}[a_{t_0+\ell-1}] - \cdots - \theta_q E_{t_0}[a_{t_0+\ell-q}],\end{aligned} \tag{6.19}$$

where

$$\hat{X}_{t_0}(k) = X_{t_0+k}, \quad k \le 0$$

and

$$E_{t_0}[a_{t_0+k}] = 0, \qquad k > 0$$
$$= a_{t_0+k}, \quad k \leq 0. \tag{6.20}$$

If $E[X_t] = \mu \neq 0$, then (6.19) holds for $X_t - \mu$, and we have the following for $\ell \geq 1$.

Difference Equation Form of the Best Forecast Equation

$$\hat{X}_{t_0}(\ell) = \sum_{j=1}^{p} \phi_j \hat{X}_{t_0}(\ell - j) - \sum_{j=1}^{q} \theta_j E_{t_0}[a_{t_0+\ell-j}] + \left(1 - \sum_{j=1}^{p} \phi_j\right)\mu. \tag{6.21}$$

Before proceeding with the implementation of this result, we first note that

$$X_{t_0+\ell} = \sum_{j=1}^{p} \phi_j X_{t_0+\ell-j} + a_{t_0+\ell} - \sum_{j=1}^{q} \theta_j a_{t_0+\ell-j} + \left(1 - \sum_{j=1}^{p} \phi_j\right)\mu. \tag{6.22}$$

Thus, if data are available up to time t_0, we obtain $\hat{X}_{t_0}(\ell)$ in (6.21) by the heuristically sensible procedure of replacing each term in (6.22) by either

(a) Its known value as of time t_0
(b) Its forecast, or expected value, given data to time t_0

Although (6.21) has an appeal over (6.9) from a practical standpoint, there are still difficulties with its implementation. Specifically, in a practical situation:

(a) $X_1, X_2, \ldots, X_{t_0}$ have been observed.
(b) Model parameters have been estimated.
(c) The a_t are unknown.

Consequently, (6.21) cannot be immediately applied, and in practice, we use (6.21) with the following modifications:

(i) The estimated model is assumed to be the true model (and μ is estimated by $\overline{X}$).
(ii) The a_t are estimated from the data.

The comment in (i) relates to the fact mentioned at the beginning of this chapter, that our forecasting formulas are based on a known model.

The estimation of a_t proceeds by noting that for an ARMA(p,q) model,

$$X_t = \sum_{j=1}^{p} \phi_j X_{t-j} + a_t - \sum_{j=1}^{q} \theta_j a_{t-j} + \left(1 - \sum_{j=1}^{p} \phi_j\right)\mu,$$

which can be rewritten as

$$a_t = X_t - \sum_{j=1}^{p} \phi_j X_{t-j} + \sum_{j=1}^{q} \theta_j a_{t-j} - \left(1 - \sum_{j=1}^{p} \phi_j\right)\mu.$$

So, for $t \leq t_0$, $E_{t_0}[a_t]$ $(=a_t)$ is estimated by

$$\hat{a}_t = \begin{cases} 0, & t \leq p \\ X_t - \sum_{j=1}^{p} \phi_j X_{t-j} + \sum_{j=1}^{q} \theta_j \hat{a}_{t-j} - \left(1 - \sum_{j=1}^{p} \phi_j\right)\overline{X}, & t = p+1, \ldots, t_0. \end{cases} \tag{6.23}$$

Moreover, from (6.20), since $E_{t_0}[a_t] = 0$ for $t > t_0$, we define

$$\hat{a}_t = 0, \quad t > t_0.$$

This leads to the following.

Basic Difference Equation Form for Calculating Forecasts from an ARMA(p,q) Model

$$\hat{X}_{t_0}(\ell) = \sum_{j=1}^{p} \phi_j \hat{X}_{t_0}(\ell - j) - \sum_{j=\ell}^{q} \theta_j \hat{a}_{t_0+\ell-j} + \left(1 - \sum_{j=1}^{p} \phi_j\right)\overline{X}. \tag{6.24}$$

Notice that the $\hat{a}_t$'s enter into the calculation of the forecasts only when there are moving average terms in the model. The use of (6.24) for calculating forecasts is illustrated in the following two examples. It should be noted that an improved method of calculating the $\hat{a}_t$'s, based on backcasting, is discussed in Section 7.4.

Example 6.1 Forecasts from an AR(1) Model

Consider the AR(1) model

$$(1 - 0.8B)(X_t - \mu) = a_t. \tag{6.25}$$

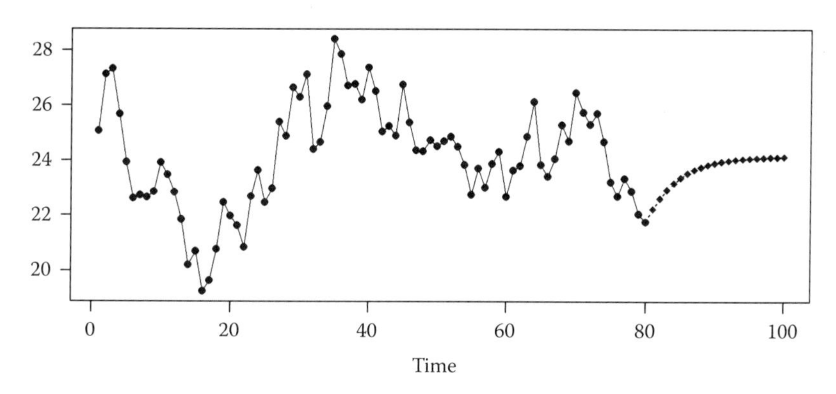

FIGURE 6.1
Forecasts using the AR(1) model.

We observe $X_1, X_2, \ldots, X_{t_0}$, and our goal is to estimate or forecast $X_{t_0+\ell}$. From (6.24), we obtain

$$\hat{X}_{t_0}(\ell) = 0.8\hat{X}_{t_0}(\ell - 1) + 0.2\overline{X}.$$

Thus, the forecasts are obtained by the following simple recursion:

$$\begin{aligned}
\hat{X}_{t_0}(1) &= 0.8\hat{X}_{t_0} + 0.2\overline{X} \\
&= 0.8X_{t_0} + 0.2\overline{X} \\
\hat{X}_{t_0}(2) &= 0.8\hat{X}_{t_0}(1) + 0.2\overline{X} \\
&\vdots \\
\hat{X}_{t_0}(\ell) &= 0.8\hat{X}_{t_0}(\ell - 1) + 0.2\overline{X}.
\end{aligned}$$

Figure 6.1 shows the forecasts $\hat{X}_{80}(\ell)$ for $\ell = 1, \ldots, 20$ based on a realization of length 80 from model (6.25) with $\mu = 25$. For this realization, $\overline{X} = 24.17$. The general form of the forecasts for large ℓ is discussed in Section 6.6.

Example 6.2 Forecasts from an ARMA(2,1) Model

Consider the ARMA(2,1) model given by

$$(1 - 1.2B + 0.6B^2)(X_t - 50) = (1 - 0.5B)a_t. \tag{6.26}$$

TABLE 6.1

Realization of Length 25 and Associated Estimates of a_t from $(1 - 1.2B + 0.6B^2)(X_t - 50) = (1 - 0.5B)a_t$

t	X_t	$\hat{a}_t$	T	X_t	$\hat{a}_t$
1	49.49	0.00	14	50.00	−0.42
2	51.14	0.00	15	49.70	−0.61
3	49.97	−1.72	16	51.73	1.77
4	49.74	−0.42	17	51.88	0.49
5	50.35	0.42	18	50.88	−0.11
6	49.57	−0.81	19	48.78	−1.22
7	50.39	0.70	20	49.14	0.51
8	51.04	0.65	21	48.10	−1.36
9	49.93	−0.78	22	49.52	0.59
10	49.78	0.08	23	50.06	−0.23
11	48.37	−1.38	24	50.66	0.17
12	48.85	−0.03	25	52.05	1.36
13	49.85	0.22			

Table 6.1 displays a realization X_t, $t = 1, 2, \ldots, 25$ generated from this model with $\sigma_a^2 = 1$. According to (6.24), forecasts from $t_0 = 25$ are given by

$$\begin{aligned}\hat{X}_{25}(\ell) &= \phi_1\hat{X}_{25}(\ell-1) + \phi_2\hat{X}_{25}(\ell-2) - \sum_{j=\ell}^{q}\theta_j\hat{a}_{25+\ell-j} + (1-\phi_1-\phi_2)\overline{X},\\ &= 1.2\hat{X}_{25}(\ell-1) - 0.6\hat{X}_{25}(\ell-2) - 0.5\hat{a}_{25} + 0.4\overline{X}, \quad \text{for } \ell = 1\\ &= 1.2\hat{X}_{25}(\ell-1) - 0.6\hat{X}_{25}(\ell-2) + 0.4\overline{X}, \quad \text{for } \ell > 1.\end{aligned} \tag{6.27}$$

Thus, it is seen that only $\hat{a}_{25}$ is needed in order to obtain the forecasts. The $\hat{a}_t$'s are obtained recursively using (6.23), and in order to calculate $\hat{a}_{25}$, we must first compute $\hat{a}_t$, $t = 1, \ldots, 24$. These estimates are also shown in Table 6.1. Notice that $\hat{a}_1$ and $\hat{a}_2$ are both zero since, by (6.23) $\hat{a}_t = 0$ for $t \le p$. Using the fact that $\overline{X} = 50.04$ for the realization in Table 6.1, it follows that

$$\begin{aligned}\hat{a}_3 &= X_3 - 1.2X_2 + 0.6X_1 + 0.5\hat{a}_2 - 50.04(0.4)\\ &= -1.72,\end{aligned}$$

and the calculation of the remaining $\hat{a}_t$'s proceeds in a similar manner. The effect of setting $\hat{a}_1$ and $\hat{a}_2$ equal to zero is minimal, especially for cases in which t_0 is much larger than p. This can be seen by examining the expression for $\hat{X}_{t_0}(\ell)$ given in (6.9) in terms of the ψ-weights and recalling that $\psi_j \to 0$ as $j \to \infty$ for a stationary model. We will return to the topic of calculating residuals in Section 7.4.

Finally, using (6.27), forecasts are calculated as

$$\begin{aligned}
\hat{X}_{25}(1) &= 1.2\hat{X}_{25}(0) - 0.6\hat{X}_{25}(-1) - 0.5\hat{a}_{25} + 0.4\overline{X} \\
&= 1.2X_{25} - 0.6X_{24} - 0.5\hat{a}_{25} + 0.4\overline{X} \\
&= 1.2(52.05) - 0.6(50.66) - 0.5(1.36) + 0.4(50.04) \\
&= 51.40, \\
\hat{X}_{25}(2) &= 1.2\hat{X}_{25}(1) - 0.6\hat{X}_{25}(0) + 0.4\overline{X} \\
&= 1.2(51.40) - 0.6(52.05) + 0.4(50.04) \\
&= 50.46, \\
&\vdots
\end{aligned}$$

The forecasts $\hat{X}_{25}(\ell)$, $\ell = 1, \ldots, 4$ are given in Table 6.2 and $\hat{X}_{25}(\ell)$, $\ell = 1, \ldots, 20$ are shown in Figure 6.2.

TABLE 6.2

Forecasts from $t_0 = 25$ for the Realization in Table 6.1

ℓ	$\hat{X}_{25}(\ell)$
1	51.40
2	50.46
3	49.73
4	49.42

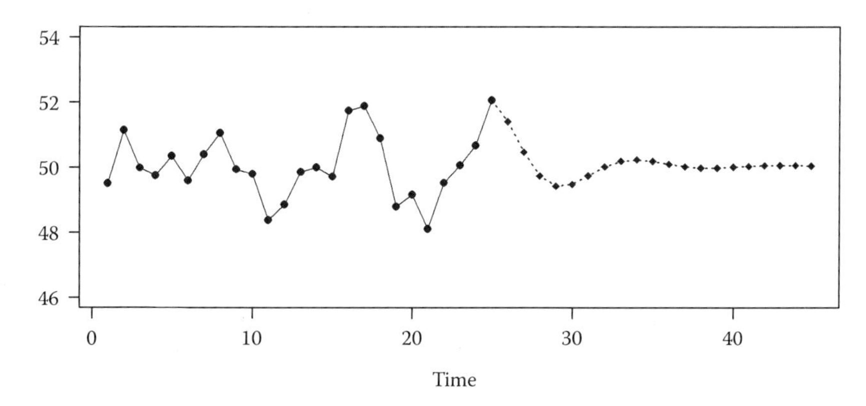

FIGURE 6.2
Forecasts using the ARMA(2,1) model $(1 - 1.2B + 0.6B^2)(X_t - 50) = (1 - 0.5B)a_t$.

Using `GW-WINKS`: Forecasts can be found by selecting the `Forecasting` option in the `Time Series Analysis` menu. You are asked to select a data set for which forecasts are needed. When you see the Forecast entry screen, the default model with which to obtain forecasts is the model most recently obtained using the `Estimate Parameters` option in the `Time Series Analysis` menu. However, you can enter the model orders and parameters if you do not want to use the defaults. For forecasts based on stationary models, the selections for `Difference`, `Seasonal Difference`, and `Nonstationary operators` should be 0 or blank. You have the option of forecasting N steps beyond the end of the series or the last N steps, and in each case you can specify N. The written output reports forecasts, lower, and upper 95% prediction limits. When forecasting the last N steps, the program also reports the actual value of the series. The GW-WINKS data sheet contains the forecasts, lower, and upper limits. Forecasts can be graphed by selecting `Graph` at the top of the Output Viewer screen immediately after calculating the forecasts. Graphs can also be obtained by selecting `Line/Time Series Plots` on the `Graphs/Charts` menu, choosing the data, forecasts, and forecasts limits (as desired), and then selecting `Overlay Plots`. Note that GW-WINKS uses the backcasting procedure discussed in Section 7.4 for computing the residuals.

6.6 Eventual Forecast Function

The basic formula for calculating forecasts, given in (6.24), can be written as

$$\begin{aligned}\hat{X}_{t_0}(\ell) - \overline{X} &= \sum_{j=1}^{p} \phi_j \left\{ \hat{X}_{t_0}(\ell - j) - \overline{X} \right\} - \sum_{j=\ell}^{q} \theta_j \hat{a}_{t_0+\ell-j} \\ &= \sum_{j=1}^{p} \phi_j \left\{ \hat{X}_{t_0}(\ell - j) - \overline{X} \right\}, \quad \text{for } \ell > q. \end{aligned} \tag{6.28}$$

Thus, from (6.28), it is seen that $\hat{X}_{t_0}(\ell) - \overline{X}$ satisfies a linear homogeneous difference equation of order p with constant coefficients for $\ell > q$. In fact in (6.28), we see that $\hat{X}_{t_0}(\ell) - \overline{X}$ satisfies the same difference equation satisfied by ρ_k, that is, $\phi(B)g_\ell = 0$. Thus, for a given stationary ARMA(p,q) model, when $\ell > q$ the forecast function will have the same fundamental behavior as the autocorrelation function, that is, damped exponential, damped sinusoid, or a mixture of damped exponentials and/or sinusoids. When $\ell > q$, we call $\hat{X}_{t_0}(\ell)$, the *eventual forecast function*. Two properties of the eventual forecast function are of particular interest.

First, from (3.45) and L'Hospital's rule, it can be seen that the solution, g_k, of a linear homogeneous difference equation with constant coefficients has the property that $g_k \to 0$ as $k \to \infty$ when the roots of the characteristic equation are all outside the unit circle. It follows that for a (stationary) ARMA(p,q) model, $\hat{X}_{t_0}(\ell) - \overline{X} \to 0$ as $\ell \to \infty$. Thus, in this case, $\hat{X}_{t_0}(\ell) \to \overline{X}$ as ℓ increases. For stationary models, this is reasonable since $\rho_\ell \to 0$ zero as ℓ increases, that is, $X_{t_0+\ell}$ is nearly uncorrelated with X_{t_0}, $X_{t_0-1}, \ldots$, for sufficiently large ℓ. Thus, it makes intuitive sense that the best prediction of $X_{t_0+\ell}$ from origin t_0 should approach the mean of the process which, based on the given realization, is estimated by $\overline{X}$.

Secondly, it is important to note that the eventual form of the forecast function depends on starting values. Specifically, when $\ell > q$, then $\hat{X}_{t_0}(\ell) - \overline{X}$ satisfies the difference equation in (6.28), the solution of which depends on the p starting values $\hat{X}_{t_0}(q-p+1), \ldots, \hat{X}_{t_0}(q)$ where $\hat{X}_{t_0}(k) = X_{t_0+k}$ for $k \le 0$. The characteristic equation $1 - 1.2z + 0.6z^2 = 0$ associated with the ARMA (2,1) model of Example 6.2 has complex roots. The forecasts shown in Figure 6.2 have the damped sinusoidal behavior typical of autocorrelations from such a model. The values $\hat{X}_{25}(0) = X_{25} = 52.05$ and $\hat{X}_{25}(1) = 51.40$ serve as starting values for the forecasts, which can be observed to damp toward $\overline{X} = 50.04$ as ℓ increases.

Example 6.1 Revisited

For the AR(1) model of Example 6.1, it follows from (6.28) that $g_\ell = \hat{X}_{80}(\ell) - \overline{X}$ satisfies the first-order difference equation

$$g_\ell - 0.8g_{\ell-1} = 0, \tag{6.29}$$

for $\ell > 0$. From the discussion of linear difference equations in Section 3.2.5 it follows that

$$g_\ell = C_1(0.8)^\ell,$$

that is,

$$\hat{X}_{80}(\ell) - \overline{X} = C_1(0.8)^\ell, \tag{6.30}$$

for $\ell > 0$. The constant C_1 is determined by the starting value $\hat{X}_{80}(0) = X_{80} = 21.77$. That is,

$$\begin{aligned} C_1 &= g_0 \\ &= X_{80} - \overline{X} \\ &= 21.77 - 24.17 \\ &= -2.4. \end{aligned}$$

6.7 Probability Limits for Forecasts

In the preceding sections, techniques were described for obtaining optimal forecasts of $X_{t_0+\ell}$ given information up to time t_0. In addition to finding the forecast $\hat{X}_{t_0}(\ell)$, it is also important to assess the uncertainty associated with the predictions.

Recall from (6.11) that the forecast error associated with the prediction of $X_{t_0+\ell}$, given data to time t_0, is given by

$$\begin{aligned} e_{t_0}(\ell) &= X_{t_0+l} - \hat{X}_{t_0}(\ell) \\ &= \sum_{k=0}^{\ell-1} \psi_k a_{t_0+\ell-k}, \end{aligned}$$

where $\psi_0 = 1$. If the white noise is normal with zero mean and variance σ_a^2, then $e_{t_0}(\ell)$ is also normal, and from (6.12) and (6.14), it follows that $e_{t_0}(\ell)$ has zero mean and variance:

$$\mathrm{Var}[e_{t_0}(\ell)] = \sigma_a^2 \sum_{k=0}^{\ell-1} \psi_k^2. \tag{6.31}$$

Thus

$$\frac{e_{t_0}(\ell)}{\sigma_a \left\{ \sum_{k=0}^{\ell-1} \psi_k^2 \right\}^{1/2}} \sim N(0,1),$$

from which it follows that

$$\mathrm{Prob}\left\{ -z_{1-a/2} \le \frac{X_{t_0+\ell} - \hat{X}_{t_0}(\ell)}{\sigma_a \left\{ \sum_{k=0}^{\ell-1} \psi_k^2 \right\}^{1/2}} \le z_{1-a/2} \right\} = 1 - \alpha, \tag{6.32}$$

where z_β denotes the $\beta \times 100$ percentile of the standard normal distribution. Using (6.32), a $(1-\alpha) \times 100\%$ prediction interval for $X_{t_0+\ell}$ is given by

$$\hat{X}_{t_0}(\ell) \pm z_{1-\alpha/2} \sigma_a \left\{ \sum_{k=0}^{\ell-1} \psi_k^2 \right\}^{1/2}. \tag{6.33}$$

That is, given data to time t_0, the probability that $X_{t_0+\ell}$ will fall within the limits given in (6.33) is $1-\alpha$. It is important to notice that the limits in (6.33) are for individual forecasts. Thus, for example, if a 95% prediction interval is obtained for X_{t_0+5} using (6.33), then there is a 95% chance that the value of X_{t_0+5} will fall within the specified limits. *It is not true*, for example, that there is a 95% chance that the future values $X_{t_0+\ell}$, $\ell = 1, \ldots, 5$ will simultaneously fall within their respective limits. In a practical situation, the white noise variance must be estimated from the data, and the ψ-weights are computed from the parameter estimates.

The prediction interval in (6.33) has the intuitively reasonable property that the half widths,

$$z_{1-\alpha/2}\sigma_a \left\{ \sum_{k=0}^{\ell-1} \psi_k^2 \right\}^{1/2}, \tag{6.34}$$

are nondecreasing as ℓ increases. For a stationary process, as $\ell \to \infty$,

$$\begin{aligned} \sigma_a \left(\sum_{k=0}^{\ell-1} \psi_k^2 \right)^{1/2} &\to \sigma_a \left(\sum_{k=0}^{\infty} \psi_k^2 \right)^{1/2} \\ &= \sigma_X(<\infty). \end{aligned}$$

Example 6.2 Revisited

Consider again the ARMA(2,1) model of Example 6.2. From (3.116), it follows that the ψ-weights are given by

$$\psi_1 = \phi_1 - \theta_1 = 1.2 - 0.5 = 0.7$$

and for $k > 1$,

$$\begin{aligned} \psi_k &= \phi_1\psi_{k-1} + \phi_2\psi_{k-2} \\ &= 1.2\psi_{k-1} - 0.6\psi_{k-2}. \end{aligned}$$

Thus, since $\psi_0 = 1$, it follows that

$$\begin{aligned} \psi_2 &= 1.2\psi_1 - 0.6\psi_0 \\ &= 1.2(0.7) - 0.6(1) \\ &= 0.24, \end{aligned}$$

and the calculation of ψ_k for $k > 2$ follows similarly. For the ARMA(2,1) model, the 95% prediction interval is given by

$$\hat{X}_{t_0}(\ell) \pm 1.96\sigma_a \left\{ \sum_{k=0}^{\ell-1} \psi_k^2 \right\}^{1/2},$$

where in our problem, $\sigma_a^2 = 1$. As an example, the half width of the interval for $\ell = 3$ is given by

$$1.96\sigma_a \left\{ \sum_{k=0}^{2} \psi_k^2 \right\}^{1/2} = 1.96(1)(1 + 0.7^2 + 0.24^2)^{1/2}$$
$$= 2.44.$$

In practical situations, σ_a^2 (and the model coefficients) will not be known and must be estimated from the data. Thus, an approximate 95% prediction interval is given by

$$\hat{X}_{t_0}(\ell) \pm 1.96\hat{\sigma}_a \left\{ \sum_{k=0}^{\ell-1} \psi_k^2 \right\}^{1/2},$$

where $\hat{\sigma}_a^2$ is

$$\hat{\sigma}_a^2 = \frac{1}{n-p} \sum_{t=p+1}^{n} \hat{a}_t^2$$
$$= \frac{1}{23} \sum_{t=3}^{25} \hat{a}_t^2$$
$$= 0.7497,$$

and $n = t_0 = 25$. The ψ_k's are the ψ-weights from the estimated model. The half width of the approximate 95% prediction interval for $\ell = 3$ is given by

$$1.96\hat{\sigma}_a \left\{ \sum_{k=0}^{2} \psi_k^2 \right\}^{1/2} = 1.96\sqrt{.7497}(1 + 0.7^2 + 0.24^2)^{1/2}$$
$$= 2.11.$$

We will present approximate prediction intervals obtained by replacing σ_a in (6.33) by $\hat{\sigma}_a$ (and ψ-weights based on the estimated model), and without further comment will refer to these as prediction intervals. In Table 6.3, we show the forecasts, $\hat{X}_{25}(\ell)$, and half widths of the 95% prediction intervals for $\ell = 1, \ldots, 4$.

In Figures 6.3 and 6.4, we redisplay the realizations and forecasts shown in Figures 6.1 and 6.2 along with 95% prediction limits. It can be seen that the half widths of the prediction intervals increase as ℓ increases, but tend to level off as ℓ becomes large, converging to $1.96\sigma_X$, if in fact the true σ_a and model parameters were used. In our case, using $\hat{\sigma}_a$, the half widths converge to 3.38 and 2.27 for the AR(1) and ARMA(2,1) models, respectively.

TABLE 6.3

Forecasts and 95% Prediction Interval Half Widths for the ARMA(2,1) Realization in Table 6.1

ℓ	Forecast	Half Width
1	51.39	1.70
2	50.45	2.07
3	49.72	2.11
4	49.40	2.12

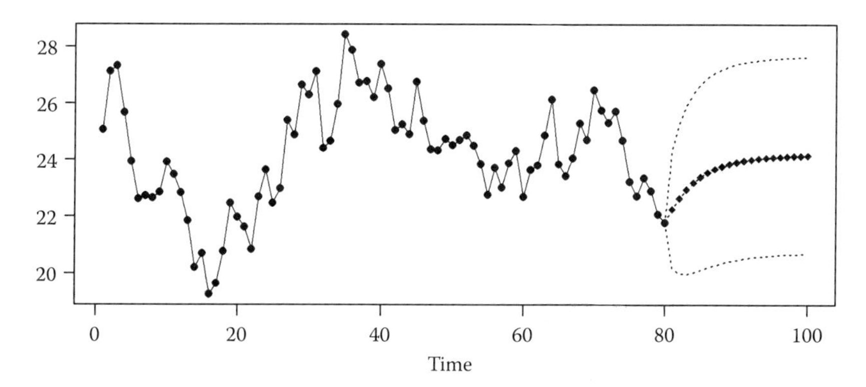

FIGURE 6.3
Forecasts in Figure 6.1 along with 95% prediction limits.

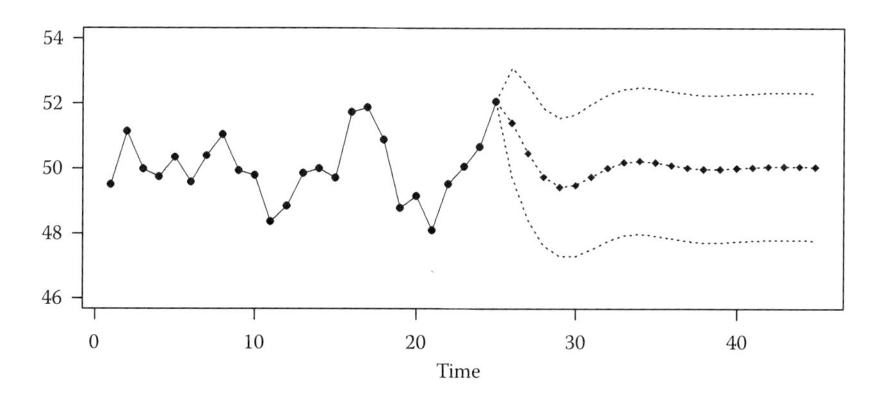

FIGURE 6.4
Forecasts in Figure 6.2 along with 95% prediction limits.

6.8 Forecasts Using ARUMA(p,d,q) Models

Although the forecasting results obtained in Sections 6.1 through 6.7 are for ARMA(p,q) models, which are by definition stationary, the ARUMA(p,d,q) models of Section 5.2 are also an important class of models for which forecasts are of interest. Suppose that X_t follows an ARUMA(p,d,q) model

$$\phi(B)\lambda(B)(X_t - \mu) = \theta(B)a_t, \tag{6.35}$$

where $\phi(z)$ is a pth-order polynomial for which all the roots of $\phi(z)=0$ lie outside the unit circle. In addition, $\lambda(z)$ is a dth-order polynomial ($d>0$) with all roots of $\lambda(z)=0$ on the unit circle. Let $\phi^{(u)}(B)=\phi(B)\lambda(B)$, so that $\phi^{(u)}(B)$ is a $(p+d)$th-order operator with coefficients, say $\phi_1^{(u)}, \phi_2^{(u)}, \ldots, \phi_{p+d}^{(u)}$. Consider (6.35) as a limit of stationary models

$$\phi(B)\lambda_m(B)(X_t^{(m)} - \mu) = \theta(B)a_t,$$

where the d roots of $\lambda_m(z)=0$ are outside the unit circle but approach the roots of $\lambda(z)=0$ (on the unit circle) as $m \to \infty$. From (6.21), we obtain the "limiting" forecast equation

$$\hat{X}_{t_0}(\ell) = \sum_{j=1}^{p+d} \phi_j^{(u)} \hat{X}_{t_0}(\ell - j) - \sum_{j=\ell}^{q} \theta_j E_{t_0}[a_{t_0+\ell-j}] + \left(1 - \sum_{j=1}^{p+d} \phi_j^{(u)}\right)\mu.$$

As in the stationary case, in practice, we use the forecast formula

$$\hat{X}_{t_0}(\ell) = \sum_{j=1}^{p+d} \phi_j^{(u)} \hat{X}_{t_0}(\ell - j) - \sum_{j=\ell}^{q} \theta_j \hat{a}_{t_0+\ell-j} + \left(1 - \sum_{j=1}^{p+d} \phi_j^{(u)}\right)\overline{X}, \tag{6.36}$$

where the $\hat{a}_t$ s are given by

$$\begin{aligned}\hat{a}_t &= 0, \quad t \le p+d \\ &= X_t - \sum_{j=1}^{p+d} \phi_j^{(u)} X_{t-j} + \sum_{j=1}^{q} \theta_j \hat{a}_{t-j} - \left(1 - \sum_{j=1}^{p+d} \phi_j^{(u)}\right)\overline{X},\end{aligned} \tag{6.37}$$

$t = p+d+1, \ldots, t_0$. It also follows that $g_\ell = \hat{X}_{t_0}(\ell) - \overline{X}$ satisfies the $(p+d)$th-order difference equation $\phi^{(u)}(B)g_\ell = 0$. However, since X_t given in (6.35) follows a nonstationary model, the general linear process form

$$X_t - \mu = \sum_{j=0}^{\infty} \psi_j a_{t-j},$$

obtained by algebraically performing the inversions involved in

$$\phi^{-1}(z)\lambda^{-1}(z)\theta(z) = \psi(z), \tag{6.38}$$

may not converge in a meaningful sense. Thus, the expression in (6.11) for forecast error, $e_{t_0}(\ell)$, is not immediately valid. Again, we consider X_t in (6.35) to be a limit of stationary processes, $\phi(B)\lambda_m(B)(X_t^{(m)} - \mu) = \theta(B)a_t$, where the roots of $\lambda_m(z) = 0$ are outside the unit circle for each m, yet approach the roots of $\lambda(z) = 0$ as $m \to \infty$. Then, for each m, $X_t^{(m)}$ is a stationary process, producing forecasts from t_0 whose forecast error is given by (6.11). As $m \to \infty$, the forecast errors associated with $X_t^{(m)}$ approach

$$e_{t_0}(\ell) = \sum_{k=0}^{\ell-1} \psi_k a_{t_0+\ell-k}, \tag{6.39}$$

where the ψ_k's in (6.39) are obtained using the algebraic inversions indicated in (6.38). We use (6.39) as the ℓ-step forecast error associated with forecasts, $\hat{X}_{t_0}(\ell)$, found using (6.36) for an ARUMA model. It follows, however, that the associated error variance,

$$\text{Var}[e_{t_0}(\ell)] = \sigma_a^2 \sum_{k=0}^{\ell-1} \psi_k^2, \tag{6.40}$$

has the reasonable property that $\text{Var}[e_{t_0}(\ell)] \to \infty$ as $\ell \to \infty$. We conclude this section by considering forecasts from a few specific ARUMA(p,d,q) models.

Example 6.3 Forecasts from an ARIMA(0,1,0) Model

Consider the ARIMA(0,1,0) model

$$(1 - B)X_t = a_t. \tag{6.41}$$

From (6.36), forecasts, $\hat{X}_{t_0}(\ell)$, are given by

$$\begin{aligned}\hat{X}_{t_0}(\ell) &= \hat{X}_{t_0}(\ell - 1) + (1 - 1)\overline{X} \\ &= \hat{X}_{t_0}(\ell - 1).\end{aligned}$$

Note that since the characteristic equation $\lambda(z) = 1 - z = 0$ has a root of $+1$, then $\overline{X}$ plays no role in the forecasts. The eventual forecast function satisfies the difference equation $g_\ell = g_{\ell-1}$ for $\ell > 0$, where $g_\ell = \hat{X}_{t_0}(\ell) - \overline{X}$. The root of the characteristic equation $1 - z = 0$ is $r_1 = 1$, so that solutions of this difference equation are given by $g_\ell = C_1(1)^\ell = C_1$, where $C_1 = g_0 = X_{t_0} - \overline{X}$. Thus,

$$\begin{aligned}\hat{X}_{t_0}(1) &= \hat{X}_{t_0}(0) = X_{t_0}, \\ \hat{X}_{t_0}(2) &= \hat{X}_{t_0}(1) = X_{t_0},\end{aligned}$$

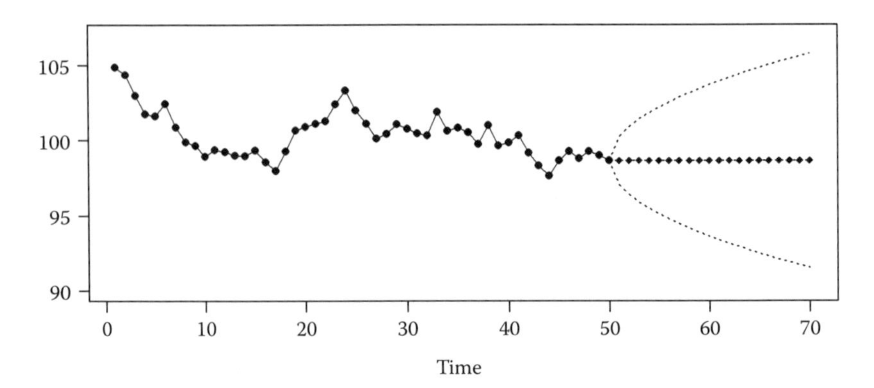

FIGURE 6.5
Forecasts and 95% prediction limits from the ARIMA(0,1,0) model $(1 - B)(X_t - 100) = a_t$.

and in general

$$\hat{X}_{t_0}(\ell) = X_{t_0}, \quad \ell = 1, 2, \ldots.$$

Thus, these forecasts display no tendency to return to $\overline{X}$ as in the stationary case, but due to the random walk nature of data from this model, the best forecast of the future is the last value observed. For large ℓ, however, these forecasts may be quite poor, as evidenced by the fact that $\text{Var}[e_{t_0}(\ell)]$ grows without bound as ℓ increases. Forecasts and 95% prediction limits for a realization of length $n = 50$ from the model in (6.41) with $\sigma_a^2 = 1$, are given in Figure 6.5. For this realization, $\overline{X} = 100.4$ and $\hat{X}_{50}(\ell) = X_{50} = 98.6$ for $\ell \geq 1$.

Example 6.4 Forecasts from an ARIMA(1,1,0) Model

Consider the ARIMA(1,1,0) model

$$(1 - 0.8B)(1 - B)X_t = a_t. \tag{6.42}$$

Upon multiplication of the factors in (6.42), we obtain

$$(1 - 1.8B + 0.8B^2)X_t = a_t,$$

so that $\phi_1^{(u)} = 1.8$ and $\phi_2^{(u)} = -0.8$. From (6.36), forecasts from this model are given by

$$\begin{aligned}\hat{X}_{t_0}(\ell) &= 1.8\hat{X}_{t_0}(\ell - 1) - 0.8\hat{X}_{t_0}(\ell - 2) + (1 - 1.8 + 0.8)\overline{X} \\ &= 1.8\hat{X}_{t_0}(\ell - 1) - 0.8\hat{X}_{t_0}(\ell - 2).\end{aligned}$$

Again, $\overline{X}$ plays no role in the forecasts, and

$$\hat{X}_{t_0}(1) = 1.8X_{t_0} - 0.8X_{t_0-1},$$
$$\hat{X}_{t_0}(2) = 1.8\hat{X}_{t_0}(1) - 0.8X_{t_0},$$

and, in general, for $\ell > 0$,

$$\hat{X}_{t_0}(\ell) = 1.8\hat{X}_{t_0}(\ell - 1) - 0.8\hat{X}_{t_0}(\ell - 2). \tag{6.43}$$

These forecasts and associated 95% prediction limits are shown in Figure 6.6 for a realization of length $n = 50$.

Although (6.43) furnishes a convenient method for computing forecasts, the general behavior of the forecasts can often best be observed by examining the solution of the difference equation satisfied by the forecast function. Forecasts, $g_\ell = \hat{X}_{t_0}(\ell) - \overline{X}$, satisfy the difference equation $g_\ell - 1.8g_{\ell-1} + 0.8g_{\ell-2} = 0$ for $\ell > 0$. Since the roots of $1 - 1.8z + 0.8z^2 = 0$ are $r_1 = 1/0.8$ and $r_2 = 1$, a general solution of the difference equation is

$$\begin{aligned} g_\ell &= C_1(0.8)^\ell + C_2(1)^\ell \\ &= C_1(0.8)^\ell + C_2. \end{aligned} \tag{6.44}$$

For the realization shown in Figure 6.6, $X_{t_0} = X_{50} = 144.67$, $X_{t_0-1} = X_{49} = 145.6$, $\overline{X} = 149.1$, and it can be shown (see Problem 6.2) that $C_1 = 3.75$ and $C_2 = -8.2$. Thus, as ℓ increases, $g_\ell = \hat{X}_{t_0}(\ell) - \overline{X} \rightarrow C_2 = -8.2$. That is, forecasts, $\hat{X}_{t_0}(\ell)$, approach the horizontal line $y = C_2 + \overline{X} = 140.9$. The forecasts damp exponentially to a limit, but this limit is neither $\overline{X}$ nor X_{t_0}.

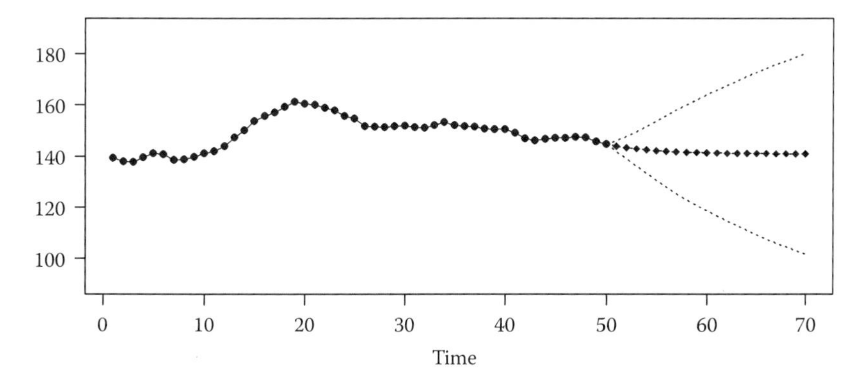

FIGURE 6.6
Forecasts and 95% prediction limits from the ARIMA(1,1,0) model $(1 - 0.8B)(1 - B)X_t = a_t$.

Example 6.5 Forecasts from an ARIMA(0,2,0) Model

Consider the ARIMA(0,2,0) model

$$(1 - B)^2 X_t = a_t. \tag{6.45}$$

The characteristic equation associated with this model has two roots of +1, and forecasts are given by

$$\begin{aligned}\hat{X}_{t_0}(\ell) &= 2\hat{X}_{t_0}(\ell - 1) - \hat{X}_{t_0}(\ell - 2) + (1 - 2 + 1)\bar{X} \\ &= 2\hat{X}_{t_0}(\ell - 1) - \hat{X}_{t_0}(\ell - 2). \end{aligned} \tag{6.46}$$

A realization of length $n = 50$ from the model in (6.45) is given in Figure 6.7, along with forecasts obtained from (6.46). For this realization, $X_{50} = -573.9$ and $X_{49} = -572.7$, and forecasts are

$$\begin{aligned}\hat{X}_{50}(1) &= 2\hat{X}_{50}(0) - \hat{X}_{50}(-1) \\ &= 2X_{50} - X_{49} \\ &= 2(-573.9) + 572.7 \\ &= -575.1, \\ \hat{X}_{50}(2) &= 2\hat{X}_{50}(1) - \hat{X}_{50}(0) \\ &= 2(-575.1) + 573.9 \\ &= -576.3, \\ &\vdots \end{aligned}$$

Realizations from (6.45) differ from those from the ARIMA(0,1,0) model in (6.41), which has a single root of +1, in that upward and downward trends tend to be more persistent. From (3.45), the eventual forecast function based

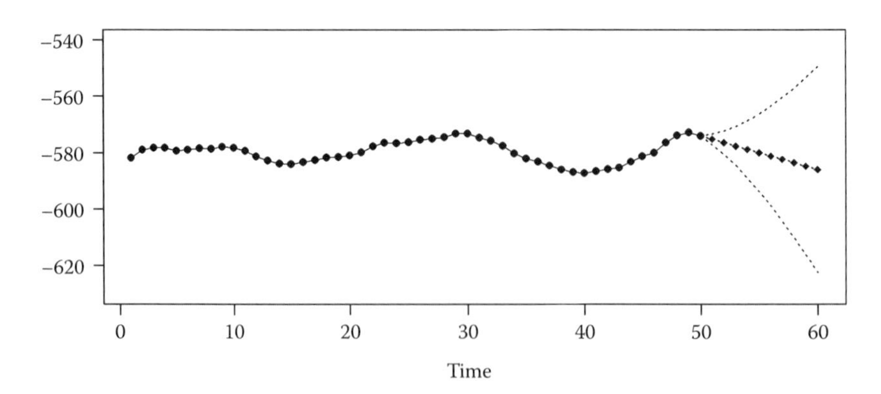

FIGURE 6.7
Realization of length 50 from $(1 - B)^2(X_t - 200) = a_t$, forecasts, and 95% limits.

on the ARIMA(0,2,0) model in (6.45) can be seen to be of the form $g_\ell = \hat{X}_{t_0}(\ell) - \overline{X} = C_1 + \ell C_2$. In our case,

$$\begin{aligned} C_1 + 1C_2 &= \hat{X}_{50}(1) - \overline{X} \\ &= -575.1 + 579.7 \\ &= 4.6 \end{aligned}$$

and

$$\begin{aligned} C_1 + 2C_2 &= \hat{X}_{50}(2) - \overline{X} \\ &= -576.3 + 579.7 \\ &= 3.4. \end{aligned}$$

Solving for C_1 and C_2, we obtain $C_1 = 5.8$ and $C_2 = -1.2$. Thus, the forecast function is given by

$$\begin{aligned} \hat{X}_{50}(\ell) &= \overline{X} + C_1 + \ell C_2 \\ &= -573.9 - 1.2\ell, \end{aligned}$$

which is the straight line through the points $\hat{X}_{50}(0) = X_{50}$ and $\hat{X}_{50}(-1) = X_{49}$. In general, it can be shown (see Problem 6.10) that forecasts $\hat{X}_{t_0}(\ell)$ from (6.46) all lie on a straight line that passes through X_{t_0} and X_{t_0-1}. That is, the forecasts predict that the current trend, upward or downward, will continue. Figure 6.7 shows that even though there had been an upward trend in the data from X_{40} through X_{49}, the fact that $X_{50} < X_{49}$ caused the forecasts to decrease with ℓ, that is, predict a decreasing (linear) trend. Had we obtained forecasts using forecast origin $t_0 = 49$, the forecasts would have predicted an increasing trend. As with the ARIMA(0,1,0) model of (6.39), forecast errors increase without bound as ℓ increases, which, as this example illustrates, is reasonable. Because of the persistent trending behavior in realizations from (6.45), forecasts predict that near future trending behavior is likely to reflect the immediate past, but forecasts of $X_{t_0+\ell}$ for ℓ reasonably large are very unreliable. The forecast limits in Figure 6.7 reflect the lack of certainty that the trending behavior will continue as predicted.

It should be noted that the more general ARIMA(p,2,q) model

$$\phi(B)(1 - B)^2(X_t - \mu) = \theta(B)a_t,$$

will have forecasts for which the eventual forecast function is a straight line. In this case, the stationary autoregressive and moving average components in the model have a bearing on the forecasts for small lead times and the slope of the eventual forecast function.

Using `GW-WINKS`: To obtain forecasts from a nonstationary model containing a factor $(1-B)^d$ where d is a positive integer, enter the value of d in the `Difference` entry box on the `Forecasting` entry screen. For the forecasts shown in Figure 6.6, you should specify $p=1$, $q=0$, and enter 0.8 as the autoregressive parameter. To obtain forecasts shown in Figures 6.5 and 6.7, you should specify $p=1$, $q=0$, and enter 0 as the autoregressive parameter. Note that you should specify the original (undifferenced) series as the data set to be used in the forecasting.

6.9 Forecasts Using Multiplicative Seasonal ARUMA Models

In this section, we briefly extend the forecasting technique for ARUMA models discussed in Section 6.8 to the more general case of the nonstationary multiplicative seasonal model

$$\phi(B)\lambda(B)\Phi(B^s)\lambda_s(B^s)(X_t-\mu)=\theta(B)\Theta(B^s)a_t, \tag{6.47}$$

where the roots of $\phi(z)=0$, $\Phi(z^s)=0$, $\theta(z)=0$, and $\Theta(z^s)=0$ all lie outside the unit circle, while the roots of $\lambda(z)=0$ and $\lambda_s(z^s)=0$ lie on the unit circle. We let

$$\phi^{(u)}(B)=\phi(B)\lambda(B)\Phi(B^s)\lambda_s(B^s),$$

so that $\phi^{(u)}(B)$ is a $(p+d+P_s+D_s)$th-order operator with coefficients, say $\phi_1^{(u)},\phi_2^{(u)},\ldots,\phi_{p+d+P_s+D_s}^{(u)}$. Viewing (6.47) as a limit of stationary models whose characteristic equation roots approach the roots associated with the nonstationary components of (6.47), we obtain a forecast equation analogous to (6.36).

Example 6.6 Forecasts from a Basic Seasonal Model

Consider the simple nonstationary seasonal model

$$(1-B^s)(X_t-\mu)=a_t, \tag{6.48}$$

which was previously discussed in Section 5.3. That is, in this case $p=d=P_s=q=Q_s=0$ and $D_s=1$. Realizations from (6.48) will be characterized by a strong positive correlation between observations separated in time by multiples of s. Forecasts from (6.48) are given by

$$\begin{aligned}\hat{X}_{t_0}(\ell)&=\hat{X}_{t_0}(\ell-s)+(1-1)\overline{X}\\&=\hat{X}_{t_0}(\ell-s),\end{aligned} \tag{6.49}$$

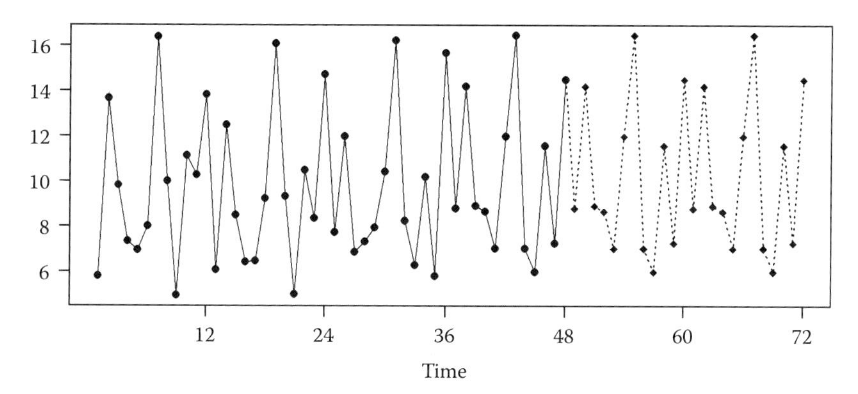

FIGURE 6.8
Realization and forecasts from the model $(1 - B^{12})(X_t - 10) = a_t$.

which take the form

$$\begin{aligned} X_{t_0-s+1} &= \hat{X}_{t_0}(1) = \hat{X}_{t_0}(s+1) = \hat{X}_{t_0}(2s+1) = \cdots \\ X_{t_0-s+2} &= \hat{X}_{t_0}(2) = \hat{X}_{t_0}(s+2) = \hat{X}_{t_0}(2s+2) = \cdots \\ &\vdots \\ X_{t_0} &= \hat{X}_{t_0}(s) = \hat{X}_{t_0}(2s) = \hat{X}_{t_0}(3s) = \cdots. \end{aligned}$$

Thus, for the simple seasonal model in (6.48), the last s data values are repeatedly forecast into the future. Forecasts, $\hat{X}_{48}(\ell)$, $\ell = 1, 2, \ldots, 24$ based on the model

$$(1 - B^{12})(X_t - 10) = a_t, \tag{6.50}$$

for a realization of length $n = 48$ are shown in Figure 6.8, where the nature of the seasonality of the data and the forecasts is clear.

Forecasts from the more general seasonal model

$$\phi(B)\Phi(B^s)(1 - B^s)(X_t - \mu) = \theta(B)\Theta(B^s)a_t,$$

for which the only nonstationary components are seasonal, that is, $d = 0$, will behave similarly, but the forecasts in this case will not be a precise copy of the last s data values observed.

Example 6.7 Forecasting Airline Data

In Figure 1.25a, we showed the Airline Data examined by Box et al. (2008). These data consist of the natural logarithms of 144 monthly airline passenger counts, and a seasonal behavior can be clearly seen. That is, there is a strong relationship between the number of passengers in the current month and the number of passengers during the same month in the previous year, etc. Also, for the years represented, there seems to be a persistent upward trend.

In Section 5.3, we discussed models of the form

$$\phi(B)\Phi(B^s)(1-B)(1-B^{12})(X_t-\mu)=\theta(B)\Theta(B^s)a_t, \tag{6.51}$$

and referred to these as "airline models." Since +1 is a root of $1-z^{12}=0$, the characteristic equation associated with the autoregressive component of (6.51) has two roots of +1. That is, the autoregressive operator has a factor of $(1-B)^2$. Thus, in addition to the seasonal behavior associated with the model in Example 6.6, realizations from (6.51) will tend to have persistent upward or downward trends much as we encountered in Example 6.5 for the model $(1-B)^2(X_t-\mu)=a_t$. Box et al. (2008) model the airline data with

$$(1-B)(1-B^{12})(X_t-\mu)=(1-0.4B)(1-0.6B^{12})a_t, \tag{6.52}$$

which is a special case of (6.51). Other special cases of the model in (6.51) have been considered for the airline data. For example, Gray and Woodward (1981) considered the multiplicative seasonal ARUMA(12,1,0)×(0,1,0)$_{12}$ model

$$\phi_1(B)(1-B)(1-B^{12})(X_t-\mu)=a_t, \tag{6.53}$$

where

$$\begin{aligned}\phi_1(B) &= 1+0.36B+0.05B^2+0.14B^3+0.11B^4-0.04B^5-0.09B^6+0.02B^7\\ &\quad -0.02B^8-0.17B^9-0.03B^{10}+0.10B^{11}+0.38B^{12}.\end{aligned} \tag{6.54}$$

Parzen (1980) proposed the ARUMA(1,0,0)×(1,1,0)$_{12}$ model of the form

$$(1-0.74B)(1+0.38B^{12})(1-B^{12})(X_t-\mu)=a_t. \tag{6.55}$$

Figure 6.9 shows forecasts, based on forecast origin $t_0 = 108$, from the multiplicative seasonal ARUMA(12,1,0)×(0,1,0)$_{12}$ model obtained by Gray and Woodward (1981). The data are shown as circles separated by solid lines and forecasts for the last 36 values are shown as diamonds separated by dotted lines. These forecasts do an excellent job of forecasting both the seasonality and the trending behavior in the data.

Using `GW-WINKS`: To obtain the forecasts shown in Figure 6.8 from the purely 12-order seasonal model, use the `Forecast` option on the `Time Series Analysis` menu and again specify $p=1$, $q=0$, and enter 0 as the autoregressive parameter. The 12th-order difference is specified by entering 12 in the `Seasonal Difference` entry box. To obtain the forecasts shown in Figure 6.9, specify $p=12$, $q=0$, and enter the 12 autoregressive parameters as specified in (6.54). Enter 1 as the `Difference` and 12 as the `Seasonal Difference`. Note that you should specify the original series as the data set to be used in the forecasting.

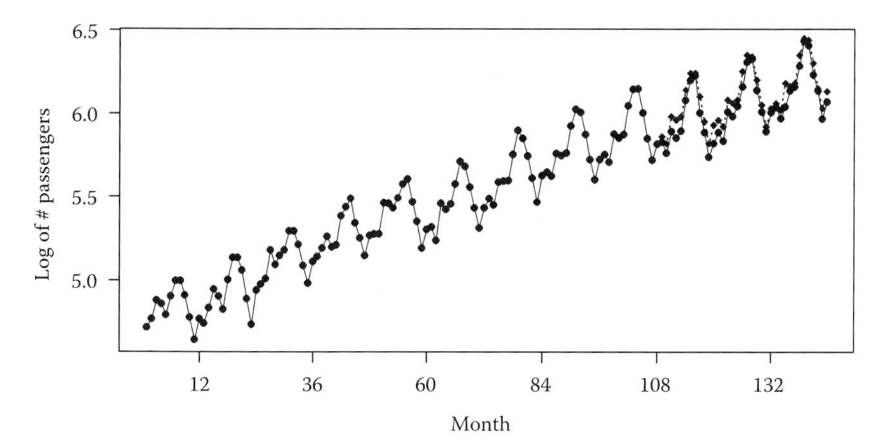

FIGURE 6.9
Forecasts of the log airline data using $t_0 = 108$ using $(1 - B^{12})(1 - B)\phi_1(B)(X_t - \mu) = a_t$.

6.10 Forecasts Based on Signal-plus-Noise Models

We next consider the signal-plus-noise model

$$X_t = s_t + Z_t, \tag{6.56}$$

discussed in Section 5.1 where s_t is a deterministic signal, such as a linear or sinusoidal component, and Z_t is a stationary process that follows an ARMA(p,q) model. Forecasts from these models can be obtained with slight modifications from the techniques considered previously. Consider, for example, the case in which forecasts $\hat{X}_{t_0}(\ell)$ are desired based on a realization collected from the model in (6.56), where $s_t = b_0 + b_1 t$, and b_0 and b_1 are considered to be known. Since $Z_t = X_t - b_0 - b_1 t$, is a stationary process, forecasts $\hat{Z}_{t_0}(\ell)$ can be computed based on the ARMA(p,q) model satisfied by Z_t, and then converted to forecasts for $X_{t_0+\ell}$ using the formula $\hat{X}_{t_0}(\ell) = b_0 + b_1 t + \hat{Z}_{t_0}(\ell)$. Of course, in practice, typically neither, b_0, b_1, nor the parameters of the ARMA(p,q) model satisfied by Z_t are known. As has been the practice throughout this chapter, forecasts and forecast errors are calculated based on the assumption that the model parameters are known, although these parameters will in all likelihood need to be estimated from the data. Parameter estimation will be discussed in Chapter 7.

The current setting in which the line (estimated or true) is assumed to be the true signal, then forecast limits for $\hat{X}_{t_0}(\ell)$ are $b_0 + b_1 t + \hat{Z}_{t_0}(\ell) \pm 1.96\hat{\sigma}_a \left(\sum_{k=0}^{\ell-1} \psi_k^2\right)^{1/2}$. The ψ-weights and $\hat{\sigma}_a^2$ are based on the ARMA(p,q)

model fit to Z_t. It also follows that the same strategy will produce forecasts and forecast errors for any deterministic signal s_t. In this case

$$\hat{X}_{t_0}(\ell) = s_t + \hat{Z}_{t_0}(\ell). \tag{6.57}$$

Note that based on the preceding discussion, the eventual forecast functions from the models

$$\phi(B)(1 - B)^2 X_t = \theta(B)a_t \tag{6.58}$$

and

$$X_t = b_0 + b_1 t + Z_t, \tag{6.59}$$

are straight lines. Considering the case in which b_0 and b_1 are estimated from the data, note that when modeling using (6.59), the entire realization up to time t_0 is used to estimate the slope of the eventual forecast line. However, as noted in Example 6.5, the slope of the eventual forecasts based on (6.58) is determined by the behavior in the neighborhood of t_0. For example, in Example 6.5, we showed that if $p = q = 0$, then the forecasts from (6.58) follow a line determined entirely by X_{t_0} and X_{t_0-1}. Thus, the forecasts from (6.58) are more *adaptable* than are those from (6.59). The forecast errors based on (6.58) increase without bound, which illustrates the need for caution due to the fact that the current upward trend could turn downward. On the other hand, forecast error limits based on (6.59) for large lags tend to $\pm 1.96\sigma_Z$. However, limits based on (6.59) will be incorrect (too narrow) if the assumption that the signal $b_0 + b_1 t$ is appropriate for extrapolating into the future is not valid. Problems associated with the use of a functional form to extrapolate into the future are well known and are discussed in most introductory texts in statistics.

Example 6.8 Cyclical Forecasts from ARUMA and Signal-Plus-Noise Models

As a final example, we consider the distinction between the forecasts from the models

$$\phi(B)(1 - \lambda_1 B + B^2)(X_t - \mu) = \theta(B)a_t \tag{6.60}$$

and

$$X_t = A_1 + A_2 \cos(2\pi f_0 t + A_3) + Z_t, \tag{6.61}$$

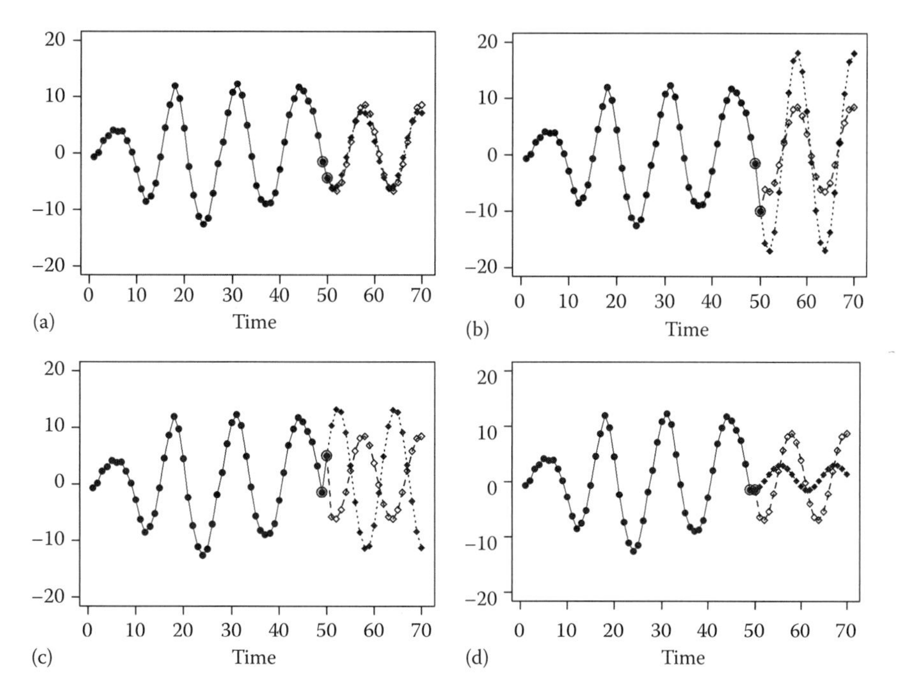

FIGURE 6.10
Effect of starting values x_{49} and x_{50} (circled values in each plots) on forecasts using models (6.60) (solid diamonds separated by dotted lines) and (6.61) (unfilled diamonds separated by dashed lines). Four different values for x_{50} are considered in (a) through (d).

where $1 - \lambda_1 z + z^2 = 0$ has a pair of complex roots on the unit circle with the associated system frequency f_0, and where Z_t is zero-mean white noise. We consider the case in which $f_0 = 1/12$ for both models, in which case realizations from these models will be cyclical with a period of about 12. The first 50 points in the four panels of Figure 6.10 illustrate a typical realization that might be observed from either model. The eventual forecast function from each model has the form

$$\hat{X}_{t_0}(\ell) = B_1 + B_2 \cos(2\pi\ell/12 + B_3). \tag{6.62}$$

Forecasts using (6.62) depend on a mean level, an amplitude (B_2), and a phase shift (B_3). From (6.57), it is seen that eventual forecasts from (6.61) are based on the signal, s_t, that is typically estimated from the data. That is, if (6.61) is the assumed model, then the parameters B_1, B_2, and B_3 are estimated from the data by fitting a sinusoid to the t_0 values in the realization.

For comparison purposes, when forecasting from (6.60) with $p=q=0$ and $\lambda_1 = \sqrt{3}$ for which $f_0 = 1/12$, the constants B_2 and B_3, that is, the amplitude and phase, are determined by the last two observed data values, and are random variables. Figure 6.10a shows a realization of length $n = t_0 = 50$ from this ARUMA(0,2,0) model along with forecasts for the next 20 values using models (6.60) and (6.61). Data values $x_{49} = -1.46$ and $x_{50} = -4.4$ are circled. In this case, the forecasts from each model are very similar and are visually difficult to distinguish from each other in Figure 6.10a. The extent to which the forecasts depend on the last two data values is not obvious in Figure 6.10a, and in order to illustrate this effect, in panels (b) through (d), we show forecasts from the two models resulting from changing x_{50}. In Figure 6.10b, we set $x_{50} = -10$, which results in forecasts with much larger amplitudes using (6.60). The forecasts using (6.61) are very similar to those in Figure 6.10a. That is, the changing of one data value did not have a substantial effect on the fit of (6.61) to the data. In Figure 6.10c, $x_{50} = 5$ ($>x_{49}$), and forecasts based on (6.60) use the information to predict that the sinusoidal behavior is "on the rise," and consequently, $\hat{X}_{50}(1) > x_{50}$. In essence, the phase shift is dramatically changed in Figure 6.10c compared to Figure 6.10a and b. Again, the fact that one data value has been altered does not have a dramatic effect on forecasts based on (6.61). Finally, in Figure 6.10d, we set $x_{49} = x_{50}$, and the forecasts using (6.60) have much smaller amplitude and are based on an "interpretation" that the value at which the signal leveled out (i.e., -1.46) is the current minimum of the sinusoidal curve. Again, the effect of changing X_{50} has a much smaller effect on forecasts from (6.61). Summarizing, it is seen that forecasts from the ARUMA(p,d,q) model are *much more adaptive* than are those from the signal-plus-noise model. Whether this is desirable in a particular physical situation is a decision to be made by the analyst.

As in the case in which s_t is a line, long-term forecasts based on the models (6.60) and (6.61) should be interpreted with caution. Forecasts from each model have a fixed phase shift and amplitude even though these values are determined very differently as is illustrated earlier. Consider, for example, the sunspot data shown in Figure 1.23a. Using either of these two models to produce long-term forecasts of future sunspot activity is problematic for at least two reasons. First, the sunspot oscillations are asymmetric with more variability in the peaks than in the troughs. Sinusoidal forecasts will not be able to account for this behavior and may, for example, result in forecasts of negative sunspot activity, which makes no sense. Tong (1990) uses a nonlinear time series model to model the asymmetry. However, a more subtle feature of the data is that there are small changes in the period length through time. As a consequence, long-term forecasts based on a fixed phase, no matter how obtained, may eventually begin to predict a trough where a peak occurs, etc. For this reason, forecasts that eventually damp to the mean (based on a stationary model fit) may be more reflective of the information available concerning sunspot values in the more distant future.

Using GW-WINKS: The forecasts based on (6.60) with $\lambda_1 = \sqrt{3} = 1.732$ and $p = q = 0$, are obtained by specifying $p = 1$, $q = 0$, entering 0 as the autoregressive parameter, and entering `1.732, −1` (be sure to include the comma) in the `Nonstationary operators` entry box. Again, note that you should specify the original series as the data set to be used in the forecasting.

6.A Appendix

Theorem 6.2 (Projection Theorem)

If X_t is a stationary time series, then $Z_{t_0}^*$ is the best linear predictor of $X_{t_0+\ell}$ based on X_t, $t = 1, 2, \ldots, t_0$ (has minimum mean square error among square integrable linear predictors) if and only if

$$E\left[(X_{t_0+\ell} - Z_{t_0}^*)X_j\right] = 0, \quad j = 1, 2, \ldots, t_0.$$

($\Leftarrow$) Let $Z_{t_0}^*$ be such that $E\left[(X_{t_0+\ell} - Z_{t_0}^*)X_j\right] = 0, j = 1, \ldots, t_0$, and let Z_{t_0} be an arbitrary linear combination of X_t, $t = 1, 2, \ldots, t_0$. Then it follows that

$$\begin{aligned}
E\left[(X_{t_0+\ell} - Z_{t_0})^2\right] &= E\left[(X_{t_0+\ell} - Z_{t_0}^* + Z_{t_0}^* - Z_{t_0})^2\right] \\
&= E\left[(X_{t_0+\ell} - Z_{t_0}^*)^2\right] + E\left[(Z_{t_0}^* - Z_{t_0})^2\right] \\
&\quad + 2E\left[(X_{t_0+\ell} - Z_{t_0}^*)(Z_{t_0}^* - Z_{t_0})\right] \\
&\geq E\left[(X_{t_0+\ell} - Z_{t_0}^*)^2\right],
\end{aligned}$$

since the cross-product term $E[(X_{t_0+\ell} - Z_{t_0}^*) \quad (Z_{t_0}^* - Z_{t_0})] = 0$ and $E\left[(Z_{t_0}^* - Z_{t_0})^2\right] \geq 0$. The fact that the cross-product term is zero can be shown using properties of the conditional expectation (see Billingsley 1995).

($\Rightarrow$) To show that $A \Rightarrow B$, we will prove the contrapositive, that is, $\sim B \Rightarrow \sim A$. Thus, we assume that $E\left[(X_{t_0+\ell} - Z_{t_0}^*)X_j\right] = a \neq 0$ for some j, and will show that $Z_{t_0}^*$ does not have minimum MSE by producing a linear combination, Z_t, of X_t, $t = 1, 2, \ldots, t_0$ with smaller MSE than $Z_{t_0}^*$. That is, we will show that $E\left[(X_{t_0+\ell} - Z_{t_0}^*)^2\right] \geq E\left[(X_{t_0+\ell} - Z_{t_0})^2\right]$. Suppose $E\left[(X_{t_0+\ell} - Z_{t_0}^*)X_j\right] = a \neq 0$ and let

$$Z_{t_0} = Z_{t_0}^* + \frac{aX_j}{E\left[X_j^2\right]}.$$

Then

$$
\begin{aligned}
E\left[(X_{t_0+\ell}-Z_{t_0})^2\right] &= E\left[(X_{t_0}-Z_{t_0}^*+Z_{t_0}^*-Z_{t_0})^2\right] \\
&= E\left[(X_{t_0}-Z_{t_0}^*-(Z_{t_0}-Z_{t_0}^*))^2\right] \\
&= E\left[(X_{t_0+\ell}-Z_{t_0}^*)^2\right]+E\left[(Z_{t_0}-Z_{t_0}^*)^2\right] \\
&\quad -2E[(X_{t_0+\ell}-Z_{t_0}^*)(Z_{t_0}-Z_{t_0}^*)] \\
&= E\left[(X_{t_0+\ell}-Z_{t_0}^*)^2\right]+E\left[\left(\frac{aX_j}{E[X_j^2]}\right)^2\right]-2E\left[(X_{t_0+\ell}-Z_{t_0}^*)\frac{aX_j}{E[X_j^2]}\right]. \\
&= E\left[(X_{t_0+\ell}-Z_{t_0}^*)^2\right]+\frac{a^2E[X_j^2]}{(E[X_j^2])^2}-\frac{2a}{E[X_j^2]}a \\
&= E\left[(X_{t_0+\ell}-Z_{t_0}^*)^2\right]-\frac{a^2}{E[X_j^2]}.
\end{aligned}
$$

Since $a \neq 0$, it follows that

$$
E\left[(X_{t_0+\ell}-Z_{t_0})^2\right] \leq E\left[(X_{t_0+\ell}-Z_{t_0}^*)^2\right],
$$

and the result follows.

Exercises

Applied Problems

If you use `GW-WINKS` to work the following problems, you will find the `Forecasting` and `Generate Realization` options on the `Time Series Analysis` menu to be useful.

6.1 A realization of length $n=10$ from the ARMA(4,2) model

$$
(X_t-\mu)+1.1(X_{t-1}-\mu)+1.78(X_{t-2}-\mu)+0.88(X_{t-3}-\mu)+0.64(X_{t-4}-\mu) \\
= a_t-0.2a_{t-1}+0.9a_{t-2}
$$

where $\mu=40$ is given in the following table.

t	X_t	t	X_t
1	40.0	6	48.7
2	30.5	7	39.2
3	49.8	8	31.7
4	38.3	9	46.1
5	29.3	10	42.4

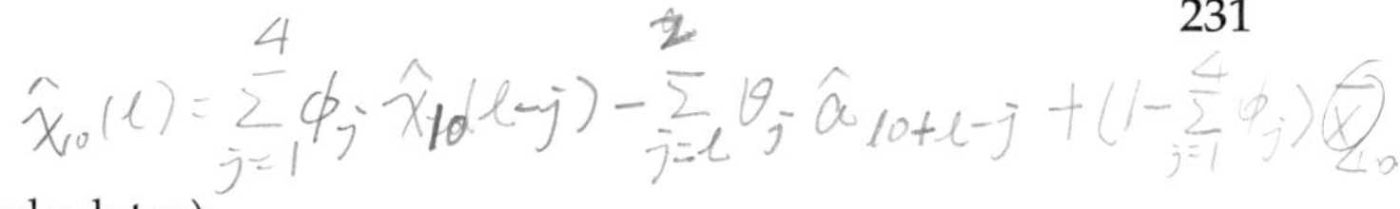

(a) By hand (calculator):
  (i) Obtain forecasts $\hat{X}_{10}(\ell)$, $\ell = 1, 2, 3, 4$ using (6.24).
  (ii) Find 95% prediction intervals for the forecasts in (a).
(b) Repeat (a) parts (i) and (ii) using a statistical forecasting package

6.2 For the realization in Figure 6.6, $X_{t_0} = X_{50} = 144.7$ and $X_{t_0-1} = X_{49} = 145.6$ while $\overline{X} = 149.1$. Show that the general solution of the forecast function in (6.44) has constants $C_1 = 3.75$ and $C_2 = -8.2$. Find $\hat{X}_{t_0}(\ell)$, $\ell = 1, \ldots, 3$ using this general solution. Also, find these forecasts using (6.43).

6.3 Using `GW-WINKS` generate a realization of length $n = 100$ from the ARMA(3,1) model

$$(1 - 0.8B)(1 - B + 0.9B^2)(X_t - 50) = (1 + 0.5B)a_t.$$

Do the following:
(a) Forecast the next 20 steps beyond the end of the series showing 95% probability limits.
(b) Find (mathematically) the eventual forecast function, and describe its behavior.
(c) Forecast the last 10 values of the series and discuss the behavior of the forecasts and how well the forecasts perform.
(d) Redo part (c) this time showing the 95% probability limits. How many of the actual last 10 values stay within these limits?

6.4 Generate a realization of length $n = 100$ from the ARUMA(1,5,0) model

$$(1 - B)(1 - B^4)(1 + 0.5B)(X_t - 50) = a_t.$$

NOTE: If you use the `GW-WINKS` program, you will need to specify the nonstationary factors as being "nearly nonstationary" in order to generate the realizations.

Do the following:
(a) Forecast the next 10 steps beyond the end of the series.
(b) Forecast the last 10 values of the series and describe the behavior of the forecasts and how well the forecasts perform.
(c) Redo part (b) this time showing the 95% probability limits. How many of the actual last 10 values stay within these limits?

6.5 Figure 6.9 shows forecasts of the last 36 values of the log-airline data in `AIRLOG.XLS` using the ARUMA(12,13,0) model $(1 - B)\ (1 - B^{12})\phi_1(B)$ $(X_t - \mu) = a_t$ where $\phi_1(B)$ is the stationary 12th-order operator given in (6.54). Using the same model, find and plot forecasts, such as those in Figure 6.9 for the last 12, 24, 36, and 48 values. Discuss the features of the associated forecasts.

6.6 Forecast 3 years (36 months) beyond the end of the log-airline data (a) using the ARUMA(12,13,0) model used in Problem 6.5, and (b) using the ARUMA(13,12,0) model in (6.55) obtained by Parzen. Compare and contrast the associated forecasts. What feature of the two models accounts for the differing behaviors of the two eventual forecast functions?

6.7 Models A and B that follow are the models given in Problem 3.12 as models that have been proposed for the annual sunspot data from 1749 to 1924 found in `SUN.XLS`. Model C is the nonstationary model analogous to (b) obtained by moving the complex roots associated with the dominant 10–11 year cycle to the unit circle.

(A) $(1 - 1.42B + 0.73B^2)(X_t - 44.78) = a_t$

(B) $(1 - 1.23B + 0.47B^2 + 0.14B^3 - 0.16B^4 + 0.14B^5 - 0.07B^6$
$+ 0.13B^7 - 0.21B^8)(X_t - 44.78) = a_t$

(C) $(1 - 1.646B + B^2)(1 + 0.37B + 0.11B^2 - 0.03B^3 - 0.29B^4 - 0.31B^5$
$- 0.27B^6)(X_t - 44.78) = a_t$

(a) For each of these models, find forecasts for the next 30 years.
 (i) Plot these forecasts and determine, on the basis of the forecasts, which model appears to be the better model for the sunspot data?
 (ii) What features in the two models account for the rather substantial differences in the forecasts?
 (iii) On the basis of this comparison, which model do you prefer? Which seems to be the poorest model?

(b) One method for assessing the quality of a model is to examine how well the model forecasts the last several data values in the data set. Using the three models
 (i) Compute and plot the forecasts for the last 50 data values (i.e., plot the realization along with the forecasts of the last 50 values).
 (ii) Discuss the quality of the forecasts. That is, how close were the forecasts to the actual values, and how well did the forecasts stay in phase with the actual cycles?
 (iii) On the basis of this comparison, which model do you prefer? Which seems to be the poorest model?

Theoretical Problems

6.8 Prove Theorem 6.3. That is, show that if X_t is a stationary process with autocovariance function γ_k, then the quantities $a_1^*, \ldots, a_{t_0}^*$ in the definition of $\tilde{X}_{t_0}(\ell)$ in (6.2) are the solutions to the system

$$\begin{pmatrix} \gamma_\ell \\ \gamma_{\ell+1} \\ \vdots \\ \gamma_{\ell+t_0-1} \end{pmatrix} = \begin{pmatrix} \gamma_0 & \gamma_1 & \cdots & \gamma_{t_0-1} \\ \gamma_1 & \gamma_2 & \cdots & \gamma_{t_0-2} \\ \vdots & & & \vdots \\ \gamma_{t_0-1} & \gamma_{t_0-2} & \cdots & \gamma_0 \end{pmatrix} \begin{pmatrix} a_1^* \\ a_2^* \\ \vdots \\ a_{t_0}^* \end{pmatrix}.$$

6.9 Verify the following properties associated with the forecasts $\hat{X}_{t_0}(\ell)$.

(a) The 1-step ahead forecast errors are the a_t's, that is, $e_{t_0}(1) = a_{t_0+1}$.

(b) The forecast errors have expectation zero, that is, $E[e_{t_0}(\ell)] = 0$, $\ell \geq 1$.

(c) The variance of the forecast errors is given by $\text{Var}[e_{t_0}(\ell)] = \sigma_a^2 \sum_{j=0}^{\ell-1} \psi_j^2$.

(d) The covariance between forecast errors satisfies

$$\text{Cov}[e_{t_0}(\ell), e_{t_0-j}(\ell)] = 0, \quad j \geq \ell.$$

Thus, whenever $j = 0, 1, \ldots, \ell - 1$, the forecast errors may be correlated.

(e) $e_{t_0}(\ell_1)$ and $e_{t_0}(\ell_2)$ are correlated. That is, forecast errors from the same forecast origin are correlated. Specifically, find an expression for $\text{Cov}[e_{t_0}(\ell_1), e_{t_0}(\ell_2)]$ where $\ell_2 > \ell_1$.

(f) The "best MSE forecast" for $\sum_{k=1}^{m} \alpha_k X_{t_0+k}$ from origin t_0 is given by $\sum_{k=1}^{m} \alpha_k \hat{X}_{t_0}(k)$.

6.10 Show that forecasts, $\hat{X}_{t_0}(\ell)$, from the model $(1 - B)^2 (X_t - \mu) = a_t$ all lie on a straight line that passes through X_{t_0-1} and X_{t_0}.

6.11 Let $\hat{X}_{t_0}(\ell)$ denote the best forecast of $X_{t_0+\ell}$ given data to time t_0. Show that if X_{t_0+1} is observed, then the best forecast of $X_{t_0+\ell}$, i.e., $\hat{X}_{t_{0+1}}(\ell - 1)$, is given by the updating formula

$$\hat{X}_{t_{0+1}}(\ell - 1) = \hat{X}_{t_0}(\ell) + \psi_{\ell-1}(X_{t_{0+1}} - \hat{X}_{t_0}(1)),$$

where ψ_j denotes the jth ψ-weight from the ARMA model from which the forecasts are obtained.

6.12 Find the equation of the eventual forecast function for the ARMA(2,1) model and associated realization given on Example 6.2.

7

Parameter Estimation

7.1 Introduction

Suppose a realization $x_1, x_2, \dots, x_n$ is observed. The problem of fitting an ARMA(p,q) model to the data is the problem of estimating

$$\begin{aligned} &p, q \\ &\phi_1, \phi_2, \dots, \phi_p \\ &\theta_1, \theta_2, \dots, \theta_p \\ &\mu, \sigma_a^2. \end{aligned} \tag{7.1}$$

The estimation of p and q is called *order identification*, and this topic will be covered in detail in Chapter 8. In this chapter, we assume that p and q are known, and techniques will be discussed for estimating the additional parameters in (7.1). An analysis of a time series using an ARMA model will typically first involve order identification followed by the estimation procedures to be discussed in this chapter. However, the dichotomy between the order identification and parameter estimation steps is not clear-cut since the order identification techniques usually involve some parameter estimation as a means of selecting the appropriate model orders. Thus, for now, we discuss parameter estimation assuming known p and q.

Much work has been done recently by statisticians on the problem of obtaining maximum likelihood estimates of the parameters due to their well-known optimality properties. Maximum likelihood estimation of the ARMA parameters involves iterative procedures that require starting values. In this chapter, we first deal with the issue of finding computationally efficient preliminary estimates of the parameters that can be used as starting values for the maximum likelihood routines. It should be noted that these more computational efficient estimates have received considerable attention in the engineering and geophysical literature at least partially due to their need to either estimate parameters in real time, to obtain estimates for a large collection of time series, or for time series of several thousand points. In Section 7.2, we discuss techniques for obtaining preliminary estimates of the model parameters, and, then, in Section 7.3, we discuss maximum likelihood estimation. Section 7.4 discusses the topic of estimating σ_a^2 and describes backcasting. Section 7.5 gives asymptotic properties, Section 7.6 provides some example applications and comparisons of estimation methods, and Section 7.7 discusses ARMA spectral estimation.

7.2 Preliminary Estimates

In this section, we discuss techniques for obtaining preliminary estimates of the parameters of AR(p), MA(q), and ARMA(p,q) models.

7.2.1 Preliminary Estimates for AR(p) Models

We consider the problem of estimating the parameters of the AR(p) model

$$X_t - \mu - \phi_1(X_{t-1} - \mu) - \cdots - \phi_p(X_{t-p} - \mu) = a_t, \tag{7.2}$$

where p is assumed to be known. We focus here on Yule–Walker, least squares, and Burg estimates.

7.2.1.1 Yule–Walker Estimates

As the name suggests, the Yule–Walker estimates are based on the Yule–Walker equations in (3.39). In particular, recall that the autocorrelation, ρ_k, of a stationary AR(p) process satisfies the difference equation

$$\rho_k = \phi_1\rho_{k-1} - \cdots - \phi_p\rho_{k-p}, \tag{7.3}$$

for $k > 0$. Letting $k = 1, 2, \ldots, p$ in (7.3) yields the Yule–Walker equations:

$$\begin{aligned}
\rho_1 &= \phi_1 + \phi_2\rho_1 + \cdots + \phi_p\rho_{p-1} \\
\rho_2 &= \phi_1\rho_1 + \phi_2 + \cdots + \phi_p\rho_{p-2} \\
&\vdots \\
\rho_p &= \phi_1\rho_{p-1} + \phi_2\rho_{p-2} + \cdots + \phi_p.
\end{aligned} \tag{7.4}$$

In Chapter 3, these equations were seen to be useful for finding the ρ_k when $\phi_1, \phi_2, \ldots, \phi_p$ were known. If the ρ_k are known, then the Yule–Walker equations are p equations in the p unknowns $\phi_1, \ldots, \phi_p$ that can be solved for the ϕ_i. Yule–Walker estimates are found from a realization by replacing ρ_k, $k=1,\ldots,p$ in (7.4) with the sample autocorrelations, $\hat{\rho}_k$, discussed in Chapter 1. That is, $\hat{\rho}_k = \hat{\gamma}_k/\hat{\gamma}_0$, where $\hat{\gamma}_k$ is the biased estimator of γ_k given in (1.19). Specifically,

$$\begin{aligned}\hat{\gamma}_k &= \frac{1}{n}\sum_{t=1}^{n-|k|}(X_t-\overline{X})(X_{t+|k|}-\overline{X}), \quad k=0,\ \pm1,\ \pm2,\ldots,\ \pm(n-1)\\ &= 0, \quad k \geq n. \end{aligned} \tag{7.5}$$

If the $p \times p$ system of equations

$$\begin{aligned}\hat{\rho}_1 &= \hat{\phi}_1 + \hat{\phi}_2\hat{\rho}_1 + \cdots + \hat{\phi}_p\hat{\rho}_{p-1}\\ \hat{\rho}_2 &= \hat{\phi}_1\hat{\rho}_1 + \hat{\phi}_2 + \cdots + \hat{\phi}_p\hat{\rho}_{p-2}\\ &\vdots\\ \hat{\rho}_p &= \hat{\phi}_1\hat{\rho}_{p-1} + \hat{\phi}_2\hat{\rho}_{p-2} + \cdots + \hat{\phi}_p\end{aligned} \tag{7.6}$$

is solved for $\hat{\phi}_1,\ldots,\hat{\phi}_p$, the resulting estimates are referred to as the *Yule–Walker estimates* and will be denoted here as $\hat{\boldsymbol{\phi}}_{\text{YW}} = (\hat{\phi}_1,\ldots,\hat{\phi}_p)'$.

It is convenient to rewrite (7.6) as

$$\mathbf{P}\hat{\boldsymbol{\phi}} = \hat{\boldsymbol{\rho}}, \tag{7.7}$$

where

$$\mathbf{P} = \begin{bmatrix} 1 & \hat{\rho}_1 & \cdots & \hat{\rho}_{p-1}\\ \hat{\rho}_1 & 1 & \cdots & \hat{\rho}_{p-2}\\ \vdots & & \ddots & \vdots\\ \hat{\rho}_{p-1} & \hat{\rho}_{p-2} & \cdots & 1\end{bmatrix}, \tag{7.8}$$

$\hat{\boldsymbol{\phi}} = (\hat{\phi}_1,\hat{\phi}_2,\ldots,\hat{\phi}_p)'$, and $\hat{\boldsymbol{\rho}} = (\hat{\rho}_1,\hat{\rho}_2,\ldots,\hat{\rho}_p)'$, so that the Yule–Walker estimates can be written as $\hat{\phi}_{\text{YW}} = \mathbf{P}^{-1}\hat{\boldsymbol{\rho}}$.

Yule–Walker Estimation Using the Durbin–Levinson Algorithm

The matrix **P** in (7.8) is an example of a symmetric Toeplitz matrix (Graybill, 1983). Levinson (1947) and Durbin (1960) have shown that solutions of (7.7) can be obtained recursively without performing any matrix inversions. The recursion is as follows where $\hat{\phi}_{pj}$ denotes the estimate of the jth autoregressive coefficient for an AR(p) model:

$$\hat{\phi}_{11} = \hat{\rho}_1.$$

Then, for $k = 2, \ldots, p$,

$$\hat{\phi}_{kk} = \frac{\hat{\rho}_k - \sum_{i=1}^{k-1} \hat{\phi}_{k-1,i}\hat{\rho}_{k-i}}{1 - \sum_{i=1}^{k-1} \hat{\phi}_{k-1,i}\hat{\rho}_i}$$
$$\hat{\phi}_{kj} = \hat{\phi}_{k-1,j} - \hat{\phi}_{kk}\hat{\phi}_{k-1,k-j}, \tag{7.9}$$

$j = 1, 2, \ldots, k-1$. Of course, the analogous recursion for (7.4) provides an efficient technique for calculating the true parameters $\phi_1, \ldots, \phi_p$ when the autocorrelations are known.

An important property of the Yule–Walker estimates in the AR(p) setting is that the resulting model is stationary, that is, the roots of $1 - \hat{\phi}_1 z - \cdots - \hat{\phi}_p z^p = 0$ lie outside the unit circle (see Brockwell and Davis, 1991). It should be noted that unlike the case with $\hat{\gamma}_k$, if the estimates $\tilde{\gamma}_k$ given in (1.18) are used in the calculation of the sample autocorrelations, the resulting "Yule–Walker" estimates do not necessarily produce a stationary model. This is primarily due to the fact discussed in Section 1.4 that $\hat{\gamma}_k$ is positive semidefinite while $\tilde{\gamma}_k$ is not.

Before finishing this section, we note that the Yule–Walker estimates are method-of-moment (MM) estimates of $\phi_1, \ldots, \phi_p$ since they are obtained by equating sample autocorrelations to theoretical autocorrelations and solving the resulting equations. MM estimates of μ and σ_X^2 are given by $\hat{\mu} = \overline{X}$ and $\hat{\sigma}_X^2 = \hat{\gamma}_0$. For AR($p$) models, (3.43) shows that

$$\sigma_a^2 = \sigma_X^2(1 - \phi_1\rho_1 - \cdots - \phi_p\rho_p),$$

from which it follows that the MM estimator of σ_a^2 is

$$\hat{\sigma}_a^2 = \hat{\sigma}_X^2\left(1 - \hat{\phi}_1\hat{\rho}_1 - \cdots - \hat{\phi}_p\hat{\rho}_p\right). \tag{7.10}$$

In Section 7.4, we further discuss the estimation of σ_a^2.

7.2.1.2 Least Squares Estimation

Ordinary least squares (OLS) estimates are obtained by rewriting the AR(p) model of (7.2) as

$$X_t - \mu = \phi_1(X_{t-1} - \mu) + \cdots + \phi_p(X_{t-p} - \mu) + a_t, \tag{7.11}$$

and treating (7.11) as a regression model to estimate $\phi_1, \ldots, \phi_p$ and μ. Specifically, taking $t = p+1, p+2, \ldots, n$ in (7.11) gives the equations

$$\begin{aligned} X_{p+1} - \mu &= \phi_1(X_p - \mu) + \cdots + \phi_p(X_1 - \mu) + a_{p+1} \\ X_{p+2} - \mu &= \phi_1(X_{p+1} - \mu) + \cdots + \phi_p(X_2 - \mu) + a_{p+2} \\ &\vdots \\ X_n - \mu &= \phi_1(X_{n-1} - \mu) + \cdots + \phi_p(X_{n-p} - \mu) + a_n. \end{aligned} \tag{7.12}$$

OLS estimates are the values of $\hat{\mu}$, $\hat{\phi}_1, \ldots, \hat{\phi}_p$ that minimize the conditional sum of squares

$$\begin{aligned} S_c &= \sum_{t=p+1}^{n} \hat{a}_t^2 \\ &= \sum_{t=p+1}^{n} \left\{X_t - \hat{\mu} - \hat{\phi}_1(X_{t-1} - \hat{\mu}) - \cdots - \hat{\phi}_p(X_{t-p} - \hat{\mu})\right\}^2. \end{aligned} \tag{7.13}$$

It should be noted that the problem of finding the least squares solution of the parameters in (7.12) is different from that encountered in the classical regression setting in a number of ways. For example,

- X_t simultaneously plays the role of the response (dependent) and the predictor (independent) variables.
- μ is the mean of both the response and predictor variables.
- The predictor variables are correlated (and random).

Also notice that the equations for $t = 1, 2, \ldots, p$ are not used, since they involve unobservable data, and as a consequence, $X_1, X_2, \ldots, X_p$ play a less significant role than does, for example, X_{p+1}. Least squares estimates can still be obtained from (7.11), and Mann and Wald (1943) have shown that standard results about the variances and covariances of regression estimates hold asymptotically as $n \to \infty$ in this setting.

Minimizing S_c by differentiating with respect to $\mu, \phi_1, \ldots, \phi_p$ and setting these derivatives equal to zero produces normal equations that in general lead to solutions that are difficult to solve. As a consequence, least squares solutions obtained by minimizing (7.13) are found numerically.

Approximations have been proposed that allow the least squares type estimates to be obtained using a standard multiple regression package. The most common approximation is to take $\hat{\mu} = \overline{X}$ and minimize the sum-of-squares

$$S_c = \sum_{t=p+1}^{n} \left\{X_t - \overline{X} - \hat{\phi}_1(X_{t-1} - \overline{X}) - \cdots - \hat{\phi}_p(X_{t-p} - \overline{X})\right\}^2. \tag{7.14}$$

These approximate OLS estimates can be obtained directly using multiple regression packages designed to estimate the coefficients in the model

$$Y = b_0 + b_1 W_1 + \cdots + b_p W_p + e,$$

where

Y is the response variable
$W_1, \ldots, W_p$ are the predictor variables
$b_0, b_1, \ldots, b_p$ are the regression coefficients with b_0 being the intercept term

Approximate OLS estimates can be obtained by first calculating the differences $X_t^* = X_t - \overline{X}$, $t = 1, \ldots, n$, and defining the $n-p$ data vectors $(Y, W_1, \ldots, W_p)'$ to be $(X_t^*, X_{t-1}^*, \ldots, X_{t-p}^*)'$ for $t = p+1, p+2, \ldots, n$. The approximate OLS estimates $\hat{\phi}_i$, $i = 1, \ldots, p$ are the output $\hat{b}_i$, $i = 1, \ldots, p$ from the regression package when the regression is forced through the origin, that is, assuming $b_0 = 0$.

Estimating μ and σ_a^2
In the remainder of this chapter, we discuss various types of estimators for the parameters of an AR(p), MA(q), and ARMA(p,q) process. In many of these situations, μ and σ_a^2 can be estimated as a part of the estimation procedure. However, estimation is greatly simplified when we first estimate μ by $\overline{X}$, transform to the mean-corrected process $X_t - \overline{X}$, and then find ML estimates of the remaining parameters using the zero-mean process. We will follow this procedure throughout this book. The reader is referred to Priestley (1981) for other estimates of μ. We will discuss the estimation of σ_a^2 in Section 7.4 after the section on maximum likelihood estimation.

7.2.1.3 Burg Estimates

Recall that, when $\mu = 0$, the OLS estimators are based on the equations in (7.12) of the form

$$X_t = \phi_1 X_{t-1} + \cdots + \phi_p X_{t-p} + a_t. \tag{7.15}$$

Due to the fact that X_t, $t \leq 0$ are unobservable, we based the least squares estimates on only the $n-p$ equations in (7.12) with "response variables" X_{p+1} through X_n. The fact that use of these equations seems to make inefficient use of available information led Burg (1975), Ulrych and Clayton (1976), and Nuttal (1976b) to consider alternatives. They argued that for stationary processes, there is no "preference in the direction of time," since the correlation between X_t and X_{t+k} is the same as that between X_t and X_{t-k} for a stationary process. Cheng (1999) shows that stationary univariate Gaussian processes are always time reversible and that non-Gaussian general linear stationary processes are time reversible if the first two moments exist. Thus, for a given set of data for which the AR(p)

model in (7.15) is appropriate, we could just as well consider the "backward" or "reverse" AR(p) model

$$X_t = \phi_1 X_{t+1} + \cdots + \phi_p X_{t+p} + \delta_t, \tag{7.16}$$

where δ_t is white noise with zero mean and variance σ_a^2 and $t = 0, \pm 1, \ldots$. The standard AR(p) model specifies the outcome at time t in terms of the p preceding time values plus a random noise component. The autocorrelations between X_t and $X_{t-1}, \ldots, X_{t-p}$ are the same autocorrelations that exist between X_t and $X_{t+1}, \ldots, X_{t+p}$ in the model of (7.16).

From the model in (7.16), we obtain the $n-p$ "regression" equations

$$\begin{aligned} X_1 &= \phi_1 X_2 + \cdots + \phi_p X_{p+1} + \delta_1 \\ X_2 &= \phi_1 X_3 + \cdots + \phi_p X_{p+2} + \delta_2 \\ &\vdots \\ X_{n-p} &= \phi_1 X_{n-p+1} + \cdots + \phi_p X_n + \delta_{n-p}. \end{aligned} \tag{7.17}$$

Instead of minimizing the sum of squares $S_c = \sum_{t=p+1}^{n} a_t^2$, consider estimating the parameters of the model by minimizing the forward and backward sum of squares (FBSS)

$$\text{FBSS} = \sum_{t=p+1}^{n} a_t^2 + \sum_{t=1}^{n-p} \delta_t^2. \tag{7.18}$$

The Durbin–Levinson algorithm can be used to compute the $\hat{\phi}_i$, so that FBSS is minimized and at the same time guarantee that the resulting model is stationary. This result is due to Burg (1975), and the resulting estimates are referred to as the *maximum entropy* or Burg estimates. Explicitly, the Burg algorithm is as follows, where the input data (x_j in the algorithm) have been mean corrected.

Initialization:
Let

$$\begin{aligned} \text{den}^{(0)} &= 2\sum_{j=1}^{n} x_j^2 \\ f_j^{(0)} &= x_j, \quad j = 1, \ldots, n \\ b_j^{(0)} &= x_j, \quad j = 1, \ldots, n \\ a_0^{(0)} &= 0. \end{aligned}$$

For $k=1$ through $k=p$, the updating formulas for moving from $k-1$ to k are:

$$\begin{aligned}
\text{num}^{(k)} &= \sum_{j=k+1}^{n} f_j^{(k-1)} b_{j-1}^{(k-1)} \\
\text{den}^{(k)} &= \left[1 - \left(a_{k-1}^{(k-1)}\right)^2\right] \text{den}^{(k-1)} - \left(f_k^{(k-1)}\right)^2 - \left(b_n^{(k-1)}\right)^2 \\
a_k^{(k)} &= -2\text{num}^{(k)}/\text{den}^{(k)} \\
a_j^{(k)} &= a_j^{(k-1)} + a_k^{(k)} a_{k-j}^{(k-1)}, \quad j = 1, \ldots, k-1 \\
f_j^{(k)} &= f_j^{(k-1)} + a_k^{(k)} f_{j-1}^{(k-1)}, \quad j = k+1, \ldots, n \\
b_j^{(k)} &= b_{j-1}^{(k-1)} + a_k^{(k)} f_j^{(k-1)}, \quad j = k+1, \ldots, n.
\end{aligned}$$

The Burg estimates for the coefficients of an AR(p) model are $\hat{\phi}_j = -a_j^{(p)}$, $j=1,\ldots,p$. Burg estimates are computationally efficient and numerically stable, even in the presence of roots near the unit circle. We illustrate the numerical stability in Section 7.6.

Using GW-WINKS: Burg, Yule–Walker, and maximum likelihood (to be discussed later) estimates for AR(p) models can be obtained using the `Estimate Parameters` option under `Time Series Analysis`. The residuals and autoregressive spectral density estimates are also calculated and placed in data editor columns. Autoregressive spectral density estimates are discussed in Section 7.7. Analysis of residuals related to a model fit is discussed in Section 9.1.

Although, in many applications, it is desirable to obtain estimates of the ϕ_j that ensure a stationary model, such estimates are necessarily biased. Therefore, Nuttal (1976b) and Ulrych and Clayton (1976) considered forward–backward least squares estimation without imposing the stationarity constraint.

7.2.2 Preliminary Estimates for MA(q) Models

Estimation of the parameters of the MA(q) model

$$X_t = a_t - \theta_1 a_{t-1} - \cdots - \theta_q a_{t-q}, \tag{7.19}$$

is more difficult than the autoregressive estimation discussed earlier. Notice that, as before, we assume $\mu=0$, and *the estimation procedures discussed here are to be applied to the mean-corrected series* $X_t - \overline{X}$. In this section, we consider invertible MA(q) models, that is, models for which the roots of $\theta(B)=0$ are all outside the unit circle.

7.2.2.1 Method-of-Moment Estimation for an MA(q)

Method-of-moment (MM) estimation of the parameters $\theta_j, j=1,\ldots,q$ is based on the relationship between the autocorrelation function and parameters of an MA(q) model given in (3.4). Specifically, these estimates are found by solving the nonlinear equations

$$\hat{\rho}_k = \frac{-\hat{\theta}_k + \hat{\theta}_1\hat{\theta}_{k+1} + \hat{\theta}_2\hat{\theta}_{k+2} + \cdots + \hat{\theta}_{q-k}\hat{\theta}_q}{1+\hat{\theta}_1^2 + \cdots + \hat{\theta}_q^2}, \tag{7.20}$$

where $k=1,2,\ldots,q$ for $\hat{\theta}_1,\ldots,\hat{\theta}_q$. An MM estimate of σ_a^2 can be obtained from (3.2), that is,

$$\hat{\sigma}_a^2 = \frac{\hat{\gamma}_0}{1+\hat{\theta}_1^2 + \cdots + \hat{\theta}_q^2}. \tag{7.21}$$

Example 7.1 Method-of-Moments Estimation for an MA(1)

For an MA(1) model, (7.20) becomes

$$\hat{\rho}_1 = \frac{-\hat{\theta}_1}{1+\hat{\theta}_1^2}, \tag{7.22}$$

which can be written as the quadratic equation

$$\hat{\rho}_1\hat{\theta}_1^2 + \hat{\theta}_1 + \hat{\rho}_1 = 0. \tag{7.23}$$

The roots of (7.23) are

$$\hat{\theta}_1 = \frac{-1 \pm \sqrt{1-4\hat{\rho}_1^2}}{2\hat{\rho}_1}. \tag{7.24}$$

If $|\hat{\rho}_1| < 0.5$, then from (7.24), there are two distinct real solutions of (7.23). The invertible solution is given by

$$\hat{\theta}_1 = \frac{-1 + \sqrt{1-4\hat{\rho}_1^2}}{2\hat{\rho}_1}.$$

If $\hat{\rho}_1 = 0.5$, then (7.24) specifies two roots of $\hat{\theta}_1 = -1$, which does not result in an invertible model. Similarly, $\hat{\rho}_1 = -0.5$ results in the noninvertible model with $\hat{\theta}_1 = 1$. If $|\hat{\rho}_1| > 0.5$, no real solutions of (7.23) exist, and the MM estimates are said to not exist. Recall that for an MA(q) process, $|\rho_1| < 0.5$, so if $|\hat{\rho}_1|$ is much larger than 0.5, an MA(1) is a questionable model for the data.

In general, for $q > 1$, the equations in (7.20) must be solved numerically using a Newton–Raphson algorithm or other procedures. MM estimates do not exist (and numerical root finding techniques do not converge) whenever there are no real solutions to (7.20). Because of the frequent failure of the numerical routines to converge, we recommend the use of the innovations algorithm (discussed in the following) to find preliminary estimates of the moving average parameters.

7.2.2.2 MA(q) Estimation Using the Innovations Algorithm

Another approach to obtaining preliminary estimates of the MA parameters involves use of the innovations algorithm (see Brockwell and Davis, 1991). For a stationary process, X_t, let $\hat{X}_m(1)$ denote the best one-step ahead predictor of X_{m+1} given data $X_1, \ldots, X_m$. These predictors can be shown to follow the relationship

$$\begin{aligned} \hat{X}_n(1) &= 0, \quad \text{if } n = 0 \\ &= \sum_{j=1}^{n} \theta_{nj}\left(X_{n+1-j} - \hat{X}_{n-j}(1)\right), \quad \text{if } n = 1, 2, \ldots, \end{aligned} \tag{7.25}$$

for some constants θ_{nj}, $j = 1, 2, \ldots, n$. (see Brockwell and Davis, 1991, 2002). That is, the one-step ahead predictor is expressed as a linear combination of the prediction errors (innovations) $\hat{a}_{k+1} = X_{k+1} - \hat{X}_k(1)$. The estimated coefficients $\hat{\theta}_{m1}, \ldots, \hat{\theta}_{mm}$ and white noise variance $\hat{v}_m$, $m = 1, 2, \ldots$ are calculated recursively as follows:

$$\begin{aligned} \hat{v}_0 &= \hat{\gamma}_0 \\ \hat{\theta}_{m,m-k} &= \hat{v}_k^{-1}\left\{\sum_{j=0}^{k-1} \hat{\theta}_{m,m-j}\hat{\theta}_{k,k-j}\hat{v}_j - \hat{\gamma}_{m-k}\right\}, \end{aligned}$$

for $k = 0, 1, \ldots, m-1$ and

$$\hat{v}_m = \hat{\gamma}_0 - \sum_{j=0}^{m-1} \hat{\theta}_{m,m-j}^2 \hat{v}_j.$$

The innovations algorithm can be used to fit moving average models such as

$$X_t = a_t - \hat{\theta}_{m1}a_{t-1} - \cdots - \hat{\theta}_{mm}a_{t-m},$$

where $m = 1, 2, \ldots$. This technique is closely related to the use of the Durbin–Levinson algorithm given in (7.9) which produces Yule–Walker estimates of the autoregressive parameters by recursively solving the Yule–Walker equations. However, unlike the Yule–Walker estimates, the innovations estimates $\hat{\theta}_{q1}, \ldots, \hat{\theta}_{qq}$ are not consistent estimates of $\theta_1, \ldots, \theta_q$. In order to obtain consistent estimates, it is necessary to use $\hat{\theta}_{m1}, \ldots, \hat{\theta}_{mq}$, where m is substantially larger than q. In practice, m is typically increased until $\hat{\theta}_{m1}, \ldots, \hat{\theta}_{mq}$ stabilize, i.e., they do not change much as m continues to increase. For $100 \le n \le 500$, sufficient stability will often be obtained for $m \le n/4$ (see Brockwell and Davis, 1991).

Asymptotic properties of the estimation and resulting confidence intervals are based on the fact that if m is a sequence of positive integers, such that $m \to \infty$ and $m/n^{1/3} \to 0$ as $n \to \infty$, then the joint distribution of

$$n^{1/2}\left(\hat{\theta}_{m1} - \theta_1, \hat{\theta}_{m2} - \theta_2, \ldots, \hat{\theta}_{mk} - \theta_k\right)'$$

converges in distribution to a multivariate normal with mean $\mathbf{0}$ and covariance matrix $\mathbf{C} = (c_{ij})$, where $c_{ij} = \sum_{r=1}^{\min(i,j)} \theta_{i-r}\theta_{j-r}$.

Using GW-WINKS: The innovations algorithm is used to obtain preliminary estimates of moving average parameters. To obtain these estimates, select Burg/Tsay, Tiao estimates under the `Estimate Parameters` option in the `Time Series Analysis` menu.

Example 7.2 MA(2) Estimates Using the Innovations Algorithm

In order to illustrate the calculation of innovations estimates, a realization of length 400 was generated from the MA(2) model given by

$$X_t = a_t + 0.2a_{t-1} - 0.48a_{t-2}, \tag{7.26}$$

where $a_t \sim N(0, 1)$. The innovations algorithm is applied to the sample autocovariances, $\hat{\gamma}_k$, to recursively obtain $\hat{\theta}_{11}, \hat{\theta}_{22}, \hat{\theta}_{21}, \hat{\theta}_{33}, \hat{\theta}_{32}, \hat{\theta}_{31}, \ldots$, along with estimates $\hat{v}_m$, $m = 1, 2, 3, \ldots$ of the white noise variance. These estimates are shown in Table 7.1 for $1 \le m \le 10$, where it can be seen that the estimates $\hat{\theta}_{m,1}$ and $\hat{\theta}_{m,2}$ are quite stable for $m \ge 8$. An algorithm designed to produce innovations estimates of the parameters of an MA(q) process must make a determination concerning when "stability" has been reached.

7.2.3 Preliminary Estimates for ARMA(p,q) Models

We discuss procedures for finding preliminary estimates of the parameters of the ARMA model

$$X_t - \phi_1 X_{t-1} - \cdots - \phi_p X_{t-p} = a_t - \theta_1 a_{t-1} - \cdots - \theta_q a_{t-q}, \tag{7.27}$$

where again we assume $\mu = 0$.

TABLE 7.1

$\hat{\theta}_{mj}$, $m = 1, \ldots, 10$; $j = 1, \ldots, \min(m, 8)$ for a Realization of Length 400 from the MA(2) Model $X_t = a_t + 0.2a_{t-1} - 0.48a_{t-2}$ where $\sigma_a^2 = 1$

	$\hat{\theta}_{mj}$								
$m \backslash j$	1	2	3	4	5	6	7	8	$\hat{v}_m$
1	−0.10								1.22
2	−0.13	0.35							1.06
3	−0.14	0.35	0.05						1.05
4	−0.15	0.41	0.04	0.01					1.03
5	−0.16	0.41	0.05	0.01	−0.01				1.03
6	−0.17	0.43	0.04	0.04	−0.02	0.07			1.00
7	−0.18	0.44	0.03	0.04	−0.04	0.07	−0.03		0.99
8	−0.19	0.45	0.03	0.04	−0.04	0.06	−0.03	−0.06	0.98
9	−0.20	0.45	0.03	0.04	−0.04	0.06	−0.03	−0.06	0.98
10	−0.20	0.45	0.03	0.04	−0.04	0.06	−0.03	−0.07	0.98

7.2.3.1 Extended Yule–Walker Estimates of the Autoregressive Parameters

Recall that for an ARMA(p,q) model, the autocorrelation function satisfies the pth-order difference equation

$$\rho_k = \phi_1\rho_{k-1} + \cdots + \phi_p\rho_{k-p} \tag{7.28}$$

for $k > q$. If the true autocorrelations are known, then the p equations for $k = q+1, q+1, \ldots, q+p$ can be solved to obtain $\phi_1, \ldots, \phi_p$. These are called the *extended Yule–Walker equations*. The extended Yule–Walker estimates of the autoregressive parameters are obtained by replacing the true autocorrelations, ρ_k, in (7.28) with the estimates, $\hat{\rho}_k$, and solving the resulting system of p equations in p unknowns. The Yule–Walker estimates discussed earlier arise from the special case in which $q = 0$. However, unlike the case $q = 0$, when $q > 0$ the extended Yule–Walker estimates need not result in a stationary model. Choi (1986) gives an algorithm for solution of the extended Yule–Walker equations.

7.2.3.2 Tsay–Tiao (TT) Estimates of the Autoregressive Parameters

Tsay and Tiao (1984) obtain estimates of the autoregressive parameters of an ARMA(p,q) model motivated by iterated AR(k) regressions, and recursively adding MA type terms to an AR(k) model. If X_t follows an ARMA(p,q) model, then an AR(p) model

$$X_t = \sum_{j=1}^{p} \phi_{jp}^{(0)} X_{t-j} + e_{pt}^{(0)},$$

$t = p+1, \ldots, n$ is initially fit to the data. The least squares estimates of the model parameters are denoted by $\hat{\phi}_{jp}^{(0)}$. Based upon these estimates, the residuals

$$\hat{e}_{pt}^{(0)} = X_t - \sum_{j=1}^{p} \phi_{jp}^{(0)} X_{t-j},$$

are obtained. Note that if $q > 0$, then $\hat{e}_{pt}^{(0)}$ is not an estimator of white noise and will contain information about the model. The *first iterated regression* model

$$X_t = \sum_{j=1}^{p} \phi_{jp}^{(1)} X_{t-1} + \beta_{1p}^{(1)} e_{p,t-1}^{(0)} + e_{pt}^{(1)},$$

$t = p+2, \ldots, n$ is then formed. Let $\hat{\phi}_{jp}^{(1)}$ and $\hat{\beta}_{1p}^{(1)}$ denote the corresponding least squares estimates. If $q = 1$, then the $\hat{\phi}_{jp}^{(1)}$ are consistent for ϕ_j, $j = 1, \ldots, p$. If $q > 1$, the procedure is continued, by calculating residuals

$$\hat{e}_{pt}^{(1)} = X_t - \sum_{j=1}^{p} \phi_{jp}^{(1)} X_{t-1} - \hat{\beta}_{1p}^{(1)} \hat{e}_{p,t-1}^{(0)},$$

and forming a second iterated regression model. The TT estimates $\hat{\phi}_j$, $j = 1, \ldots, p$ for an ARMA(p,q) model are given by $\hat{\phi}_{jp}^{(q)}$, $j = 1, \ldots, p$.

Tsay and Tiao (1984) show that it is not necessary to go through this entire procedure, but note that the TT estimates $\hat{\phi}_{jp}^{(q)}$ can be obtained using the recursion given by

$$\hat{\phi}_{jk}^{(m)} = \hat{\phi}_{j,k+1}^{(m-1)} - \frac{\hat{\phi}_{j-1,k}^{(m-1)} \hat{\phi}_{k+1,k+1}^{(m-1)}}{\hat{\phi}_{kk}^{(m-1)}},$$

where $j = 1, \ldots, k$ with $k \geq 1$. The recursion is initialized with $\hat{\phi}_{0k}^{(m-1)} = -1$ and with estimates $\hat{\phi}_{jk}^{(0)}$ for the case $q = 0$ obtained using a method for estimating the parameters of an autoregressive process. Tsay and Tiao (1984) used least squares estimates to initialize this recursion. The TT estimates are asymptotically equivalent to the Yule–Walker estimates when X_t is stationary, and they are consistent for $\phi_1, \ldots, \phi_p$ even when the characteristic equation $\phi(z) = 0$ has roots on the unit circle. In this text and in the `GW-WINKS` software, we initialize the TT estimates using Burg estimates due to their stability properties. We refer to these as Burg/TT estimates in the text. `GW-WINKS` refers to these as `Burg/Tsay, Tiao` estimates.

7.2.3.3 *Estimating the Moving Average Parameters*

Consider the ARMA(p,q) model $\phi(B)X_t = \theta(B)a_t$. Since $Y_t = \phi(B)X_t$ satisfies the MA(q) model $Y_t = \theta(B)a_t$, preliminary estimates of the moving average parameters of an ARMA(p,q) process can be obtained by first estimating the autoregressive parameters (using a procedure such as Burg/TT) and then fitting an MA(q) model to $\hat{Y}_t = \hat{\phi}(B)X_t$. The implementation in `GW-WINKS` computes preliminary estimates of the moving average coefficients using the innovations algorithm applied to the transformed data, $\hat{Y}_t$ where $\phi(B)$ is based on Burg estimates.

7.3 Maximum Likelihood Estimation of ARMA(p,q) Parameters

The use of maximum likelihood (ML) estimation necessarily involves making distributional assumptions. For purposes of calculating the ML estimates of the parameters, we assume that the a_t are normally distributed and hence that $\mathbf{X} = (X_1, \ldots, X_n)'$ has a multivariate normal distribution with mean $\boldsymbol{\mu}$ and covariance matrix $\boldsymbol{\Sigma}_n$ where

$$\boldsymbol{\mu} = (\mu, \mu, \ldots, \mu)'$$

and

$$\boldsymbol{\Sigma}_n = \begin{pmatrix} \gamma_0 & \gamma_1 & \cdots & \gamma_{n-1} \\ \gamma_1 & \gamma_0 & \cdots & \gamma_{n-2} \\ \vdots & & \ddots & \\ \gamma_{n-1} & \gamma_{n-2} & \cdots & \gamma_0 \end{pmatrix}.$$

As with previous estimation procedures, we analyze $X_t - \overline{X}$ and assume $\mu = 0$.

7.3.1 Conditional and Unconditional Maximum Likelihood Estimation

Letting $\sigma_a^2 \mathbf{M}_n^{-1} = \boldsymbol{\Sigma}_n$, the joint probability density function of $\mathbf{X} = (X_1, \ldots, X_n)'$ is given by

$$f(x_1, \ldots, x_n) = \frac{1}{\left(2\pi\sigma_a^2\right)^{n/2}} \mid \mathbf{M}_n \mid^{1/2} \exp\left\{ -\frac{\mathbf{x}'\mathbf{M}_n\mathbf{x}}{2\sigma_a^2} \right\}. \tag{7.29}$$

The joint density function in (7.29) can be written

$$f(x_1, \ldots, x_n) = f(x_{p+1}, \ldots, x_n \mid x_1, \ldots, x_p) f(x_1, \ldots, x_p), \tag{7.30}$$

where $f(\cdot|\cdot)$ is the conditional probability density function. Since the a_t are uncorrelated $N(0, \sigma_a^2)$ random variables, it follows that

$$f(a_{p+1}, \ldots, a_n) = \frac{1}{\left(2\pi\sigma_a^2\right)^{(n-p)/2}} \exp\left\{ -\frac{1}{2\sigma_a^2} \sum_{t=p+1}^{n} a_t^2 \right\}.$$

Reordering the terms in (7.27), we obtain

$$a_t = X_t - \phi_1 X_{t-1} - \cdots - \phi_p X_{t-p} + \theta_1 a_{t-1} + \cdots + \theta_q a_{t-q}. \tag{7.31}$$

Conditional on $X_1 = x_1, \ldots, X_p = x_p$ and starting values $a_1 = a_1^*, \ldots, a_p = a_p^*$, it follows from (7.31) (assuming for convenience of notation that $q \leq p$) that for a choice of the ϕ_i's and θ_i's, the random variables $a_{p+1}, \ldots, a_n$ and $X_{p+1}, \ldots, X_n$ are related by the system of equations

$$\begin{aligned}
a_{p+1} &= X_{p+1} - \phi_1 X_p - \cdots - \phi_p X_1 + \theta_1 a_p + \cdots + \theta_q a_{p+1-q} \\
a_{p+2} &= X_{p+2} - \phi_1 X_{p+1} - \cdots - \phi_p X_2 + \theta_1 a_{p+1} + \cdots + \theta_q a_{p+2-q} \\
&\vdots \\
a_n &= X_n - \phi_1 X_{n-1} - \cdots - \phi_p X_{n-p} + \theta_1 a_{n-1} + \cdots + \theta_q a_{n-q}.
\end{aligned}$$

The values for $a_1, \ldots, a_p$ given earlier are often taken to be 0, their unconditional mean. Since the absolute value of the Jacobian of this transformation is one, it follows that

$$\begin{aligned}
&f(x_{p+1}, \ldots, x_n | x_1, \ldots, x_p) \\
&= \frac{1}{(2\pi\sigma_a^2)^{(n-p)/2}} \exp\left\{ -\frac{1}{2\sigma_a^2} \sum_{t=p+1}^{n} \left(x_t - \phi_1 x_{t-1} - \cdots - \phi_p x_{t-p} + \theta_1 a_{t-1} + \cdots + \theta_q a_{t-q}\right)^2 \right\} \\
&= \frac{1}{(2\pi\sigma_a^2)^{(n-p)/2}} \exp\left\{ -\frac{S_c}{2\sigma_a^2} \right\},
\end{aligned} \tag{7.32}$$

where S_c is given by

$$S_c = \sum_{t=p+1}^{n} (x_t - \phi_1 x_{t-1} - \cdots - \phi_p x_{t-p} + \theta_1 a_{t-1} + \cdots + \theta_q a_{t-q})^2.$$

Calculating $f(x_1,\ldots,x_n)$ using (7.30) involves $f(x_1,\ldots,x_p)$, which is given by

$$f(x_1,\ldots,x_p) = \frac{1}{\left(2\pi\sigma_a^2\right)^{p/2}}|\mathbf{M_p}|^{1/2}\exp\left\{-\frac{\mathbf{x}_p'\mathbf{M_p}\mathbf{x}_p}{2\sigma_a^2}\right\}, \quad (7.33)$$

where $\mathbf{x}_p = (x_1,\ldots,x_p)'$. Multiplying $f(x_{p+1},\ldots,x_n|x_1,\ldots,x_p)$ in (7.32) by $f(x_1,\ldots,x_p)$ in (7.33) yields

$$\begin{aligned} f(x_1,\ldots,x_n) &= f(x_{p+1},\ldots,x_n|x_1,\ldots,x_p)f(x_1,\ldots,x_p) \\ &= \frac{1}{\left(2\pi\sigma_a^2\right)^{(n-p)/2}}\exp\left\{-\frac{S_c}{2\sigma_a^2}\right\}\times\frac{1}{\left(2\pi\sigma_a^2\right)^{p/2}}|\mathbf{M_p}|^{1/2}\exp\left\{-\frac{\mathbf{x}_p'\mathbf{M_p}\mathbf{x}}{2\sigma_a^2}\right\} \\ &= \frac{1}{\left(2\pi\sigma_a^2\right)^{n/2}}|\mathbf{M_p}|^{1/2}\exp\left\{-\frac{S_u}{2\sigma_a^2}\right\}, \end{aligned}$$

where

$$S_u = S_c + \mathbf{x}_p'\mathbf{M_p}\mathbf{x}_p. \quad (7.34)$$

It is common terminology to refer to S_u as the *unconditional sum of squares* and to S_c as the *conditional sum of squares*. The log-likelihood function is given by

$$l\left(\phi_1,\ldots,\phi_p,\theta_1,\ldots,\theta_q,\sigma_a^2\right) = -\frac{n}{2}\ln\left(2\pi\sigma_a^2\right) + \frac{1}{2}\ln|\mathbf{M_p}| - \frac{S_u}{2\sigma_a^2}. \quad (7.35)$$

The maximum likelihood estimators are found by maximizing (7.35) with respect to the parameters $\phi_1,\ldots,\phi_p$, $\theta_1,\ldots,\theta_q$, and σ_a^2. The estimators so obtained are the *unconditional* or *exact maximum likelihood* estimates.

Example 7.3 Maximum Likelihood Estimation in the AR(1) Case

Consider an AR(1) model in which case

$$f(x_1,\ldots,x_n) = f(x_2,\ldots,x_n|x_1)f(x_1).$$

In this case, $f(x_1)$ is given by

$$\begin{aligned} f(x_1) &= \frac{1}{\left(2\pi\sigma_X^2\right)^{1/2}}\exp\left\{-\frac{x_1^2}{2\sigma_X^2}\right\} \\ &= \frac{1}{\left(2\pi\sigma_a^2\right)^{1/2}}\left(1-\phi_1^2\right)^{1/2}\exp\left\{-\frac{x_1^2\left(1-\phi_1^2\right)}{2\sigma_a^2}\right\}, \end{aligned}$$

since $\sigma_X^2 = \frac{\sigma_a^2}{1-\phi_1^2}$. Thus, in the notation of (7.33), $|\mathbf{M_p}| = 1 - \phi_1^2$ and

$$f(x_1, \ldots, x_n) = \left(\frac{1}{2\pi\sigma_a^2}\right)^{n/2} \left(1 - \phi_1^2\right)^{1/2} \times \exp\left\{ -\frac{1}{2\sigma_a^2} \left(\left(1 - \phi_1^2\right) x_1^2 + \sum_{t=2}^{n} (x_t - \phi_1 x_{t-1})^2 \right) \right\}.$$

The log-likelihood function is given by

$$l\left(\phi_1, \sigma_a^2\right) = -\frac{n}{2} \ln\left(2\pi\sigma_a^2\right) + \frac{1}{2} \ln\left(1 - \phi_1^2\right) - \frac{S_u}{2\sigma_a^2}, \tag{7.36}$$

where

$$S_u = \left(1 - \phi_1^2\right) x_1^2 + \sum_{t=2}^{n} (x_t - \phi_1 x_{t-1})^2.$$

The exact maximum likelihood estimates do not exist in closed form even for the AR(1) case of Example 7.3, and, in general, the derivatives of $\ln |\mathbf{M_p}|$ are complicated. Box et al. (2008) suggest two simplifications:

(1) For large n, they claim that $\ln |\mathbf{M_p}|$ is negligible compared to S_u. This suggests using the simplified log-likelihood function

$$l\left(\phi_1, \ldots, \phi_p, \theta_1, \ldots, \theta_q, \sigma_a^2\right) \approx -\frac{n}{2} \ln\left(2\pi\sigma_a^2\right) - \frac{S_u}{2\sigma_a^2}. \tag{7.37}$$

Approximations to the unconditional maximum likelihood estimates can be obtained by maximizing $l(\phi_1, \ldots, \phi_p, \theta_1, \ldots, \theta_q, \sigma_a^2)$ in (7.37). This maximization can be accomplished by minimizing the unconditional sum of squares S_u for finite, non-zero σ_a^2, and, consequently, these estimates are sometimes referred to as *unconditional least squares* estimates.

Box et al. (2008) suggest a practical method for approximating the unconditional sum of squares, S_u, based on a backforecasting (or *backcasting*) technique. They express the unconditional sum of squares as

$$S_u = \sum_{t=-\infty}^{n} \{E[a_t|\mathbf{X}]\}^2, \tag{7.38}$$

where $E[a_t|\mathbf{X}]$ denotes the conditional expectation of a_t given $X_1, \ldots, X_n$. The unconditional sum of squares in (7.38) can be approximated by with the sum

$$\hat{S}_u = \sum_{t=-K}^{n} \hat{a}_t^2, \tag{7.39}$$

for some positive integer K. This approach essentially involves "backcasting" $X_0, X_{-1}, \ldots,$ and using these backcasts and (7.31) to estimate $a_{-K}, a_{-K+1}, \ldots, a_n$. Backcasting will be discussed in Section 7.4. The estimates obtained by choosing the parameters that minimize (7.39) are sometimes called the *unconditional least squares* estimates, whereas the *unconditional or exact maximum likelihood estimates* are obtained by minimizing the log-likelihood in (7.35), where S_u is computed using (7.39). For additional discussion of conditional and unconditional ML estimation, see Priestley (1981).

(2) Another simplification is to use $f(x_{p+1}, \ldots, x_n|x_1, \ldots, x_p)$ as the likelihood function and maximize

$$l\left(\phi_1, \ldots, \phi_p, \theta_1, \ldots, \theta_q, \sigma_a^2\right) \approx -\frac{n-p}{2} \ln\left(2\pi\sigma_a^2\right) - \frac{S_c}{2\sigma_a^2}. \tag{7.40}$$

The estimates obtained by maximizing (7.40) are called the *conditional maximum likelihood estimates.*

Estimation of σ_a^2

The ML estimate of σ_a^2 (based on maximizing either (7.35) or (7.37)) can be obtained by differentiating $l(\phi, \theta, \sigma_a^2)$ with respect to σ_a^2, which yields (in either case)

$$\frac{\partial l}{\partial \sigma_a^2} = -\frac{n}{2}\left(\frac{1}{\sigma_a^2}\right) + \frac{S}{2\sigma_a^4},$$

where $S = S_u$ or S_c. Setting this partial derivative equal to zero and solving for σ_a^2 gives the estimate

$$\hat{\sigma}_a^2 = \frac{1}{n}\sum_{t=1}^{n} \hat{a}_t^2, \tag{7.41}$$

where the $\hat{a}_t$ s in (7.41) denote the estimates obtained via backcasting.

If backcasting is not used and the conditional sum-of-squares $\hat{S}_c = \sum_{t=p+1}^{n} \hat{a}_t^2$ is obtained (based on $a_1^* = a_2^* = \cdots = a_p^* = 0$), then, in this case, the conditional ML estimate of σ_a^2 is

$$\hat{\sigma}_a^2 = \frac{\hat{S}_c}{(n-p)}. \tag{7.42}$$

Here, $\hat{S}_c$ denotes the sum-of-squares function with conditional maximum likelihood estimates inserted for $\phi_1, \ldots, \phi_p, \theta_1, \ldots, \theta_q$.

7.3.2 ML Estimation Using the Innovations Algorithm

The exact likelihood function in the ARMA(p,q) case can be expressed in terms of the innovations algorithm (see Brockwell and Davis, 1991) for calculating one-step ahead forecasts. The log-likelihood function can be written as

$$l\left(\boldsymbol{\phi},\boldsymbol{\theta},\sigma_a^2\right) = -\frac{n}{2}\ln\left(2\pi\sigma_a^2\right) - \frac{1}{2}\sum_{t=0}^{n-1}\ln(r_t) - \frac{\sigma_a^2}{2}\sum_{t=1}^{n}(X_t - \hat{X}_{t-1}(1))^2/r_{t-1}, \tag{7.43}$$

where, using the notation of Chapter 6, $\hat{X}_{t-1}(1)$ is the one-step ahead forecast for X_t (using the ϕ_j's and θ_j's) and

$$r_t = \frac{E[X_{t+1} - \hat{X}_t(1)]}{\sigma_a^2}.$$

Taking the partial derivative of $l(\phi, \theta, \sigma_a^2)$ with respect to σ_a^2 results in

$$\hat{\sigma}_a^2 = \frac{1}{n}S_u(\hat{\phi},\hat{\theta}), \tag{7.44}$$

where

$$S_u(\hat{\boldsymbol{\phi}},\hat{\boldsymbol{\theta}}) = \sum_{t=1}^{n}(X_t - \hat{X}_t(1))^2/r_{t-1}$$

is evaluated at the ML estimates of the ϕ_j's and θ_j's. By inserting $\hat{\sigma}_a^2$ in (7.44) into (7.43), it can be seen that the ML estimates of the ϕ_j's and θ_i's are those values that minimize the reduced likelihood

$$l_r(\boldsymbol{\phi},\boldsymbol{\theta}) = \ln(n^{-1}S_u(\boldsymbol{\phi},\boldsymbol{\theta})) + n^{-1}\sum_{t=1}^{n}\ln r_{t-1}.$$

The least squares estimates can be obtained by simply finding the $\hat{\boldsymbol{\phi}}$ and $\hat{\boldsymbol{\theta}}$ that minimize

$$S_u(\boldsymbol{\phi},\boldsymbol{\theta}) = \sum_{t=1}^{n} \left(X_t - \hat{X}_t(1)\right)^2 / r_{t-1}.$$

These estimates are often very similar to the ML estimates.

A closed form of the unconditional or exact likelihood function for the ARMA(p,q) case has been given by Newbold (1974). Ansley (1979) and Ljung and Box (1979) discuss algorithms for finding the exact ML estimates.

Using `GW-WINKS`: Conditional maximum likelihood estimates can be obtained by selecting `MLE` estimates using the `Estimate Parameters` option under the `Time Series Analysis` menu.

7.4 Backcasting and Estimating σ_a^2

In this section, we discuss the use of backcasting to obtain the white noise estimates $\hat{a}_{-K}, \ldots, \hat{a}_n$ in (7.39) that are used to estimate the unconditional sum of squares given in (7.38). Not only does backcasting provide estimates, $\hat{a}_{-K}, \ldots, \hat{a}_0$, but it produces improved estimates of $\hat{a}_1, \ldots, \hat{a}_n$ compared to those obtained using the conditional approach (e.g., the white noise estimates in Table 6.1). Although a few estimation procedures (e.g., MM, Burg, and innovations) provide an estimate of the white noise variance, in practice, we recommend the use of the estimator given in (7.41) where the white noise estimates are obtained using backcasting. While the estimator in (7.41) is based on maximum likelihood estimation of the parameters, the procedure given there can be generalized.

The procedure used in `GW-WINKS` for backcasting is outlined as follows:

1. Obtain the parameter estimates $\hat{\phi}_1, \ldots, \hat{\phi}_p, \hat{\theta}_1, \ldots, \hat{\theta}_q$ using an estimation method.
2. Backcast using these parameter estimates to obtain estimates of X_0, $X_{-1}, \ldots, X_{-K}$ for $K = 50$ based on the (backward) model

$$X_t - \hat{\phi}_1 X_{t+1} - \cdots - \hat{\phi}_p X_{t+p} = \delta_t - \hat{\theta}_1 \delta_{t+1} - \cdots - \hat{\theta}_q \delta_{t+q}.$$

   Notice that $\delta_1, \ldots, \delta_q$ will need to be estimated whenever $q > 0$.
3. Use the backcast estimates of X_0, $X_{-1}, \ldots, X_{-K}$ to obtain the white noise estimates

$$\hat{a}_t = \hat{X}_t - \hat{\phi}_1 \hat{X}_{t-1} - \cdots - \hat{\phi}_p \hat{X}_{t-p} + \hat{\theta}_1 \hat{a}_{t-1} + \cdots + \hat{\theta}_q \hat{a}_{t-q}$$

$t = -K+p, \ldots, n$, where $\hat{X}_t$ is taken to be a value obtained using backcasting for $t = -K+p, \ldots, 0$, and $\hat{X}_t = X_t$, $t = 1, \ldots, n$. Also, $\hat{a}_t$ is taken to be 0 for $t < -K+p$.

4. Use the estimates $\hat{a}_t$, $t = 1, \ldots, n$ to obtain the white noise variance estimate

$$\hat{\sigma}_a^2 = \frac{1}{n}\sum_{t=1}^{n} \hat{a}_t^2. \tag{7.45}$$

Example 7.4 Using Backcasting to Estimate Residuals for Data in Table 6.1

In Example 6.2, we found forecasts based on the ARMA(2,1) model

$$(1 - 1.2B + 0.6B^2(X_t - 50) = (1 - 0.5B)a_t. \tag{7.46}$$

In that example, we found estimates of the residuals for purposes of calculating the forecasts. Actually, only $\hat{a}_{25}$ was needed in this example and $\hat{a}_{25}$ was obtained by the conditional implementation of setting $\hat{a}_1$ and $\hat{a}_2$ equal to zero. While this procedure is satisfactory for calculating $\hat{a}_{25}$, it does not provide estimates of a_1 and a_2, and the estimates of a_t for smaller values of t are poor because of their dependence on the starting values. Consequently, $\frac{1}{n-2}\sum_{t=3}^{n} \hat{a}_t^2$ may be a poor estimator of σ_a^2. We use the backcasting technique to obtain improved white noise variance estimates. We illustrate the backcasting procedure using the steps outlined earlier.

1. In this example, we simply use the true parameters as the estimated values for better comparison with the results in Table 6.1.
2. Backcasts from the model

$$X_t - 1.2X_{t+1} + 0.6X_{t+2} = \delta_t - 0.5\delta_{t+1}, \tag{7.47}$$

involve $\hat{\delta}_1$ and to calculate this quantity, we set $\hat{\delta}_{25} = \hat{\delta}_{24} = 0$ and recursively compute $\hat{\delta}_t, t = 23, 22, \ldots, 1$ using the formula

$$\hat{\delta}_t = X_t - 1.2X_{t+1} + 0.6X_{t+2} + 0.5\hat{\delta}_{t+1} - 50.04(0.4), \tag{7.48}$$

which is analogous to the formula used in Example 6.2 for calculating residuals based on the forward model in (7.46). For example,

$$\hat{\delta}_{23} = X_{23} - 1.2X_{24} + 0.6X_{25} + 0.5\hat{\delta}_{24} - 50.04(0.4) = 0.48,$$

TABLE 7.2

Realization of Length 25 and Associated Estimates of a_t Obtained Using Backcasting from the Model $(1 - 1.2B + 0.6B^2)(X_t - 50) = (1 - 0.5B)a_t$

t	$\hat{X}_t^B$	$\hat{a}_t$	t	X_t	$\hat{a}_t$	t	X_t	$\hat{a}_t$
−12	50.05	0.00	1	49.49	−0.21	14	50.00	−0.42
−11	50.02	−0.02	2	51.14	1.27	15	49.70	−0.61
−10	49.99	−0.03	3	49.97	−1.09	16	51.73	1.77
−9	49.98	−0.03	4	49.74	−0.10	17	51.88	0.49
−8	49.99	−0.02	5	50.35	0.58	18	50.88	−0.11
−7	50.05	0.02	6	49.57	−0.73	19	48.78	−1.22
−6	50.14	0.07	7	50.39	0.73	20	49.14	0.51
−5	50.22	0.10	8	51.04	0.66	21	48.10	−1.36
−4	50.24	0.09	9	49.93	−0.77	22	49.52	0.59
−3	50.13	0.01	10	49.78	0.09	23	50.06	−0.23
−2	49.90	−0.13	11	48.37	−1.38	24	50.66	0.17
−1	49.60	−0.28	12	48.85	−0.03	25	52.05	1.36
0	49.40	−0.34	13	49.85	0.22			

and continuing the calculation gives $\hat{\delta}_1 = -1.36$. We next backcast to obtain estimates of $X_0, \ldots, X_{-K}$ using the formula

$$\hat{X}_1^B(1) = 1.2\hat{X}_1^B(2) - 0.6\hat{X}_1^B(3) - 0.5\hat{\delta}_1 + 50.04(0.4),$$

and for $\ell > 1$

$$\hat{X}_1^B(\ell) = 1.2\hat{X}_1^B(\ell+1) - 0.6\hat{X}_1^B(\ell+2) + 50.04(0.4), \tag{7.49}$$

where $\hat{X}_1^B(\ell)$ is the backcast of $X_{1-\ell}$ given data from $X_1, X_2, \ldots$. In `GW-WINKS`, we use $K=50$, but in Table 7.2, we only show the backcasts for $X_0, X_{-1}, \ldots, X_{-12}$.

3. Set $\hat{a}_{-50} = \hat{a}_{-49} = 0$ and calculate the residuals

$$\hat{a}_t = \hat{X}_t^B - 1.2\hat{X}_{t-1}^B + 0.6\hat{X}_{t-2}^B + 0.5\hat{a}_{t-1} - 50.04(0.4),$$

where $\hat{X}_t^B$ is defined in Step 2. These values are shown for $t = -12, -11, \ldots, 25$ in Table 7.2, where it can be seen that for $t \geq 11$ the residuals thus obtained agree with those shown in Table 6.1 obtained using the conditional approach.

4. Estimate σ_a^2 using $\hat{\sigma}_a^2 = \frac{1}{25}\sum_{t=1}^{n} \hat{a}_t^2 = 0.68$.

Using `GW-WINKS`: Residuals are calculated (using backcasting) each time parameter are calculated estimates using `Estimate Parameters`. The residuals, $\hat{a}_t$, $t = 1, \ldots, n$, are placed in a column in the data editor, and $\hat{\sigma}_a^2$ is printed in the written `Report`.

7.5 Asymptotic Properties of Estimators

In this section, we state some asymptotic results concerning the estimators discussed in this chapter.

7.5.1 Autoregressive Case

First, note that for an AR(p) process, the Yule–Walker, Burg, and maximum likelihood estimates (either conditional or unconditional) have the same asymptotic distribution. Specifically, if $\boldsymbol{\phi}$ denotes the vector of autoregressive parameter estimators obtained using any of these techniques, then

$$n^{1/2}(\hat{\boldsymbol{\phi}} - \boldsymbol{\phi}) \Rightarrow N(0, \sigma_a^2 \boldsymbol{\Gamma}^{-1}), \tag{7.50}$$

where $\Rightarrow$ denotes convergence in distribution, and $\boldsymbol{\Gamma}$ is the $p \times p$ matrix whose ijth element is γ_{i-j}, that is,

$$\boldsymbol{\Gamma} = \begin{pmatrix} \gamma_0 & \gamma_1 & \cdots & \gamma_{p-1} \\ \gamma_1 & \gamma_0 & \cdots & \gamma_{p-2} \\ \vdots & \vdots & \ddots & \vdots \\ \gamma_{p-1} & \gamma_{p-2} & \cdots & \gamma_0 \end{pmatrix} \tag{7.51}$$

(see Brockwell and Davis, 1991).

Even though the estimators are asymptotically equivalent in the AR(p) case, it is worth noting that the YW estimates can be quite poor in practice when some roots of the characteristic equation are very near the unit circle or when there are repeated roots. Consider, for example, the AR(3) model

$$(1 - 0.995B)(1 - 1.2B + 0.8B^2)X_t = a_t \tag{7.52}$$

in (5.12) which has a positive real root very close to the unit circle $\left(|r_i^{-1}| = 0.995\right)$ and complex conjugate roots that are further from the unit circle $\left(|r_i^{-1}| = \sqrt{0.8} = 0.89\right)$. Figure 5.2c shows that the autocorrelation function for this model approximates the damped exponential behavior of a first-order process, and, in Section 5.2, we discussed the fact that the (limiting) autocorrelations satisfy a first-order difference equation. As a consequence, the third-order theoretical Yule–Walker equations associated with (7.52) are a near singular system, and the use of sample autocorrelations in these equations will tend to reflect the weakened effect of the second-order behavior. In Tables 7.6 and 7.7, we will see that the Yule–Walker estimates based on a realization from this model are quite poor while Burg estimates are much more stable. This is a common pattern for such models, and, for this reason, it is preferable to use the ML or Burg estimates if they are available. In Example 7.5, we use (7.50) to find the asymptotic distributions for estimators of an AR(1) and an AR(2) process.

Example 7.5 Asymptotic Distributions for AR(1) and AR(2) Models

(i) *AR(1)*

In the case of an AR(1) model, $\mathbf{\Gamma}$ is the 1×1 matrix whose element is γ_0, which from (3.35) is given by $\gamma_0 = \sigma_a^2/(1-\phi_1^2)$. Thus, it follows that

$$\sigma_a^2 \mathbf{\Gamma}^{-1} = 1 - \phi_1^2, \tag{7.53}$$

so that $\hat{\phi}_1$ is asymptotically normal with mean ϕ_1 and variance $(1-\phi_1^2)/n$.

(ii) *AR(2)*

The matrix $\mathbf{\Gamma}$, for an AR(2), is given by

$$\mathbf{\Gamma} = \begin{bmatrix} \gamma_0 & \gamma_1 \\ \gamma_1 & \gamma_0 \end{bmatrix}, \tag{7.54}$$

where γ_0 and γ_1 are given by (3.113) and (3.114) (with $\theta_1 = 0$) as

$$\begin{aligned} \gamma_0 &= \frac{(1-\phi_2)\sigma_a^2}{(1+\phi_2)((1-\phi_2)^2 - \phi_1^2)} \\ \gamma_1 &= \frac{\phi_1 \sigma_a^2}{(1+\phi_2)((1-\phi_2)^2 - \phi_1^2)}. \end{aligned} \tag{7.55}$$

After some algebra, it follows that

$$\sigma_a^2 \mathbf{\Gamma}^{-1} = \begin{bmatrix} 1-\phi_2^2 & -\phi_1(1+\phi_2) \\ -\phi_1(1+\phi_2) & 1-\phi_2^2 \end{bmatrix}. \tag{7.56}$$

From (7.50) and the diagonal elements of (7.56), it can be seen that, asymptotically, $\hat{\phi}_1$ is distributed as a normal random variable with mean ϕ_1 and variance $(1-\phi_2^2)/n$, while $\hat{\phi}_2$ is asymptotically normal with mean ϕ_2 and the same asymptotic variance as $\hat{\phi}_1$.

7.5.1.1 Confidence Intervals: Autoregressive Case

A common use for the result in (7.50) is to compute approximate large sample confidence intervals for the parameters. From (7.50), it follows that

$$(\hat{\boldsymbol{\phi}} - \boldsymbol{\phi})' \frac{n\hat{\mathbf{\Gamma}}}{\hat{\sigma}_a^2} (\hat{\boldsymbol{\phi}} - \boldsymbol{\phi}) \Rightarrow \chi^2(p), \tag{7.57}$$

where $\chi^2(p)$ denotes a chi-square distribution with p degrees of freedom, and where $\hat{\sigma}_a^2$ and $\hat{\mathbf{\Gamma}}$ denote data-based estimates of σ_a^2 and $\mathbf{\Gamma}$ (see Brockwell and Davis, 1991). From (7.57), an approximate $(1-\alpha)\times 100\%$ confidence ellipsoid for $\boldsymbol{\phi}$ is given by

$$(\hat{\boldsymbol{\phi}} - \boldsymbol{\phi})'\hat{\mathbf{\Gamma}}(\hat{\boldsymbol{\phi}} - \boldsymbol{\phi}) \le \frac{\hat{\sigma}_a^2}{n}\chi^2_{1-a}(p), \tag{7.58}$$

where $\chi^2_\beta(p)$ denotes the $\beta \times 100$th percentile of the chi-square distribution with p degrees of freedom. If confidence intervals are desired for an individual parameter, ϕ_j, then (7.58) can be modified to yield

$$\hat{\phi}_j \pm z_{1-a/2}\frac{\hat{\sigma}_a\hat{v}_{jj}^{1/2}}{\sqrt{n}}, \tag{7.59}$$

where z_β denotes the $\beta \times 100$th percentile of the standard normal distribution, and $\hat{v}_{jj}$ denotes the jth diagonal element in $\hat{\mathbf{\Gamma}}^{-1}$.

Before continuing to the ARMA(p,q) case, a few comments can be made concerning the estimation of σ_a^2 and $\mathbf{\Gamma}$. First, for the AR(1) and AR(2) cases, $\sigma_a^2\mathbf{\Gamma}^{-1}$ can be estimated directly by inserting parameter estimates into (7.53) and (7.56), respectively. For higher order models, σ_a^2 can be estimated by $\hat{\sigma}_a^2$, the estimated white noise variance associated with the particular estimation procedure employed using backcasting as discussed in Section 7.4. Then, the matrix $\hat{\mathbf{\Gamma}}$ can be obtained by estimating the elements γ_k using $\hat{\gamma}_k$ in (7.5).

In Example 7.8a and d, we compute confidence intervals for model parameters based on the YW, Burg, and ML estimates for realizations from known autoregressive models.

7.5.2 ARMA(p,q) Case

In this section, we briefly discuss the asymptotic distribution of $\hat{\boldsymbol{\beta}} = (\hat{\phi}_1, \ldots, \hat{\phi}_p, \hat{\theta}_1, \ldots, \hat{\theta}_q)'$ for the more general ARMA(p,q) process,

$$\phi(B)X_t = \theta(B)a_t, \tag{7.60}$$

where the parameter estimates are obtained using maximum likelihood or least squares (conditional or unconditional). This asymptotic distribution is given by

$$n^{1/2}(\hat{\boldsymbol{\beta}} - \boldsymbol{\beta}) \Rightarrow N\left(\mathbf{0}, \sigma_a^2\mathbf{\Gamma}^{-1}\right), \tag{7.61}$$

where $\boldsymbol{\beta}$ denotes the vector of true parameter values. In order to obtain the matrix $\boldsymbol{\Gamma}$ in this case, we consider the two autoregressive processes U_t and V_t defined by

$$\phi(B)U_t = a_t \tag{7.62}$$

$$\theta(B)V_t = a_t, \tag{7.63}$$

where $\phi(B)$ and $\theta(B)$ are as in (7.60). We use the notation $\gamma_U(k) = E[U_t U_{t+k}]$, $\gamma_V(k) = E[V_t V_{t+k}]$ and $\gamma_{UV}(k) = E[U_t V_{t+k}]$. Then, $\boldsymbol{\Gamma}$ is the $(p+q) \times (p+q)$ matrix defined by

$$\boldsymbol{\Gamma} = \begin{bmatrix} \boldsymbol{\Gamma}_{pp} & \boldsymbol{\Gamma}_{pq} \\ \boldsymbol{\Gamma}_{qp} & \boldsymbol{\Gamma}_{qq} \end{bmatrix}, \tag{7.64}$$

where
$\boldsymbol{\Gamma}_{pp}$ is the $p \times p$ matrix whose ijth element is $\gamma_U(i-j)$
$\boldsymbol{\Gamma}_{qq}$ is the $q \times q$ matrix with ijth element $\gamma_V(i-j)$

The matrix $\boldsymbol{\Gamma}_{pq}$ is the $p \times q$ matrix whose ijth element is $\gamma_{UV}(i-j)$, while $\boldsymbol{\Gamma}_{qp}$ is the $q \times p$ matrix with ijth element given by $\gamma_{UV}(j-i)$. It follows from (7.64) that the $\boldsymbol{\Gamma}$ matrix for the MA(q) model, $X_t = \theta(B)a_t$, is identical to that for the AR(q) model $\theta(B)X_t = a_t$. Using this fact, in Example 7.6, we find the asymptotic distribution associated with an MA(1) and an MA(2) model, and, in Example 7.7, we find the asymptotic distribution for an ARMA(1,1). For the derivation of the asymptotic distribution in the ARMA(p,q) case, see Box et al. (2008) and Brockwell and Davis (1991).

Example 7.6 Asymptotic Distributions for MA(1) and MA(2) Models

(i) *MA(1)*

It follows from the earlier discussion that for an MA(1) model, $\boldsymbol{\Gamma}$ is the 1×1 matrix whose element is given by $\gamma_0 = \sigma_a^2/(1-\theta_1^2)$. Thus, $\sigma_a^2\boldsymbol{\Gamma}^{-1} = 1-\theta_1^2$, and $\hat{\theta}_1$ is asymptotically normal with mean θ_1 and variance $(1-\theta_1^2)/n$.

(ii) *MA(2)*

It follows from (7.56) that

$$\sigma_a^2\boldsymbol{\Gamma}^{-1} = \begin{bmatrix} 1-\theta_2^2 & -\theta_1(1+\theta_2) \\ -\theta_1(1+\theta_2) & 1-\theta_2^2 \end{bmatrix}. \tag{7.65}$$

As in the AR(2) case, it can be seen from the diagonal elements in (7.65) that, asymptotically, $\hat{\theta}_1$ is distributed as a normal random variable with mean θ_1 and variance $(1-\theta_2^2)/n$ while $\hat{\theta}_2$ has mean θ_2 and the same asymptotic variance as $\hat{\theta}_1$.

Example 7.7 Asymptotic Distributions for an ARMA(1,1) Model

Consider the ARMA(1,1) model

$$(1 - \phi_1 B)X_t = (1 - \theta_1 B)a_t. \tag{7.66}$$

In this case, each of the submatrices in (7.64) is 1×1, and $\mathbf{\Gamma}$ is given by

$$\mathbf{\Gamma} = \begin{bmatrix} \gamma_U(0) & \gamma_{UV}(0) \\ \gamma_{UV}(0) & \gamma_V(0) \end{bmatrix},$$

where $\gamma_U = \sigma_a^2/\left(1 - \phi_1^2\right)$ and $\gamma_V(0) = \sigma_a^2/\left(1 - \theta_1^2\right)$. Now,

$$\begin{aligned}\gamma_{UV}(0) &= E[U_t V_t] \\ &= E[(\phi_1 U_{t-1} + a_t)(\theta_1 V_{t-1} + a_t)] \\ &= \phi_1\theta_1 E[U_{t-1}V_{t-1}] + \theta_1 E[V_{t-1}a_t] + \phi_1 E[U_{t-1}a_t] + \sigma_a^2,\end{aligned}$$

so that

$$\gamma_{UV}(0) = \phi_1\theta_1\gamma_{UV}(0) + \sigma_a^2,$$

and $\gamma_{UV}(0)$ is given by

$$\gamma_{UV}(0) = \frac{\sigma_a^2}{1 - \phi_1\theta_1}.$$

Thus, we have

$$\mathbf{\Gamma} = \sigma_a^2 \begin{bmatrix} \dfrac{1}{1-\phi_1^2} & \dfrac{1}{1-\phi_1\theta_1} \\ \dfrac{1}{1-\phi_1\theta_1} & \dfrac{1}{1-\theta_1^2} \end{bmatrix},$$

so that

$$\sigma_a^2\mathbf{\Gamma}^{-1} = \frac{1-\phi_1\theta_1}{(\phi_1-\theta_1)^2}\begin{pmatrix} \left(1-\phi_1^2\right)(1-\phi_1\theta_1) & -\left(1-\phi_1^2\right)\left(1-\theta_1^2\right) \\ -\left(1-\phi_1^2\right)\left(1-\theta_1^2\right) & \left(1-\theta_1^2\right)(1-\phi_1\theta_1) \end{pmatrix}. \tag{7.67}$$

Asymptotically, $\hat{\phi}_1$ has a normal distribution with mean ϕ_1 and variance

$$\text{var}(\hat{\phi}_1) = \frac{\left(1-\phi_1^2\right)(1-\phi_1\theta_1)^2}{n(\phi_1-\theta_1)^2},$$

while $\hat{\theta}_1$ is normal with mean θ_1 and variance

$$\text{var}(\hat{\theta}_1) = \frac{\left(1-\theta_1^2\right)(1-\phi_1\theta_1)^2}{n(\phi_1-\theta_1)^2}.$$

7.5.2.1 Confidence Intervals for ARMA(p,q) Parameters

The result in (7.61) provides large sample confidence intervals for the model parameters of an ARMA(p,q) model. A confidence ellipsoid corresponding to (7.57) and individual confidence intervals corresponding to (7.58) are given by

$$(\hat{\boldsymbol{\beta}}-\boldsymbol{\beta})'\hat{\boldsymbol{\Gamma}}(\hat{\boldsymbol{\beta}}-\boldsymbol{\beta}) \le \frac{\hat{\sigma}_a^2}{n}\chi^2_{1-\alpha}(p+q), \tag{7.68}$$

and

$$\hat{\beta}_j \pm z_{1-\alpha/2}\frac{\hat{\sigma}_a\hat{v}_{jj}^{1/2}}{\sqrt{n}}. \tag{7.69}$$

In practice, the results in Examples 7.6 and 7.7 can be used for the special cases MA(1), MA(2), and ARMA(1,1) by substituting estimated quantities for their theoretical values as in the autoregressive case. The estimation of the elements of $\boldsymbol{\Gamma}$ for more complex ARMA(p,q) models, however, is not as straightforward as it was in the autoregressive case, since the processes U_t and V_t are not observed. In order to estimate the components of $\boldsymbol{\Gamma}$, we modify the alternative technique suggested for the autoregressive case. That is, the elements of $\boldsymbol{\Gamma}_{pp}$ can be estimated using the Yule–Walker equations to find the theoretical autocovariances associated with the AR(p) model

$$U_t - \hat{\phi}_1 U_{t-1} - \cdots - \hat{\phi}_p U_{t-p} = a_t, \tag{7.70}$$

while the components of $\boldsymbol{\Gamma}_{qq}$ can be estimated from the autocovariances for the AR(q) model

$$V_t - \hat{\theta}_1 V_{t-1} - \cdots - \hat{\theta}_q V_{t-q} = a_t. \tag{7.71}$$

The elements of $\boldsymbol{\Gamma}_{pq}$ and $\boldsymbol{\Gamma}_{qp}$ are estimated by noting that since U_t and V_t are autoregressive processes based on the same white noise process, they can be expressed as

$$Y_t = \sum_{j=0}^{\infty} c_j a_{t-j}$$

and

$$Z_t = \sum_{m=0}^{\infty} d_m a_{t-m},$$

where the weights c_j and d_m are the ψ-weights discussed in Chapter 3. Thus, $\gamma_{YZ}(k)$ can be expressed in linear process form as

$$\gamma_{YZ}(k) = \sigma_a^2 \sum_{j=0}^{\infty} c_j d_{j+k}. \tag{7.72}$$

The cross covariances $\gamma_{YZ}(k)$ can be approximated by summing (7.72) until the terms become "small enough." Thus, the cross covariances $\gamma_{UV}(k)$ can be estimated using the estimated AR models in (7.70) and (7.71), finding the appropriate weights for each model and approximating the corresponding infinite series in (7.72).

7.5.3 Asymptotic Comparisons of Estimators for an MA(1)

Although the various estimation techniques discussed in this chapter for estimating the parameters of an autoregressive process produce asymptotically equivalent estimators, this is not true for estimators of the parameters of a moving average process. To see this, we examine the MA(1) process, $X_t = a_t - \theta_1 a_{t-1}$. In Example 7.6, we found that the MLE of $\hat{\theta}_1$ is asymptotically normal with asymptotic variance given by $(1-\theta_1^2)/n$. The innovations and MM estimators are also asymptotically normal, and the asymptotic variances for the MM, innovations, and ML estimators are given in Table 7.3 (see Brockwell and Davis, 1991). From the table, it is seen that for an MA(1)

$$\text{Asyvar}\left\{\hat{\theta}_1^{(MM)}\right\} > \text{Asyvar}\left\{\hat{\theta}_1^{(1)}\right\} > \text{Asyvar}\left\{\hat{\theta}_1^{(ML)}\right\}.$$

In Table 7.4, these asymptotic variances (multiplied by n) are shown for various values of θ_1.

TABLE 7.3

Asymptotic Variances for Estimators of θ_1

Estimator	Asymptotic Variance
Methods of moments: $\hat{\theta}_1^{MM}$	$(1+\theta_1^2+\theta_1^4+\theta_1^6+\theta_1^8)/(n(1-\theta_1^2))$
Innovations: $\hat{\theta}_1^{(I)}$	$1/n$
Maximum likelihood: $\hat{\theta}_1^{ML}$	$(1-\theta_1^2)/n$

TABLE 7.4

Asymptotic Variance and Efficiency Ratios for Estimates of θ_1 for Various Values of $|\theta_1|$

	$n \times$ Asymptotic Variance			Efficiency	
$\|\theta_1\|$	MM	I	ML	E_1	E_2
0.1	1.031	1.000	0.990	1.041	1.010
0.3	1.356	1.000	0.910	1.490	1.099
0.5	2.701	1.000	0.750	3.601	1.333
0.7	10.095	1.000	0.510	19.794	1.961
0.9	149.482	1.000	0.190	784.211	5.263

Also shown in the table are the efficiency ratios:

$$E_1 = \frac{\text{Aysvar}\left\{\hat{\theta}_1^{(MM)}\right\}}{\text{Aysvar}\left\{\hat{\theta}_1^{(ML)}\right\}}$$

$$E_2 = \frac{\text{Aysvar}\left\{\hat{\theta}_1^{(I)}\right\}}{\text{Aysvar}\left\{\hat{\theta}_1^{(ML)}\right\}}.$$

From the table, it is seen that the three estimators have quite similar asymptotic variances for small values of θ_1. However, although the asymptotic variance of the innovations estimator does not depend on θ_1, as $|\theta_1|$ increases, the MM estimators deteriorate while the ML estimates become less variable. Thus, there are substantial differences among the estimators when $|\theta_1|$ is near one. Because of these results coupled with the fact that the moment equations often do not yield real solutions, in practice, we do not recommend the use of MM estimates for moving average parameters.

7.6 Estimation Examples Using Data

In this section, we will use realizations discussed in Chapters 3 and 5 from known models to illustrate some of the concepts of this chapter.

Using `GW-WINKS`: To estimate ARMA model parameters use the `Estimate Parameters` option on the `Time Series Analysis` menu. You will need to specify the column in the data editor containing the data, p, q, and the type of estimator you want. For AR(p) models, the options are Burg, conditional

MLE (default), and Yule-Walker. For models in which $q > 0$, the options are Burg/Tsay, Tiao, and conditional ML (default). The residuals (using back-casting) are computed and placed in a data editor column with a variable name beginning with `RESIDUALS`. Additionally, the Report file contains parameter estimates (including $\hat{\sigma}_a^2$) and a factor table of the estimated model. Also note that the spectral density of the estimated model (i.e., the ARMA spectral density estimator) is placed in a data editor column with a name beginning with `SPECTRUM`. ARMA spectral estimation will be discussed in Section 7.7.

Example 7.8 Estimating Parameters in MA(q) Models

We consider the realization of length $n = 100$ in Figure 3.1a from the MA(1) model $X_t = (1 - 0.99B)a_t$. In this case $\theta_1 = 0.99$ and $\sigma_a^2 = 1$. The MA characteristic equation has its root (1/0.99) very close to the noninvertible region. Using the innovations algorithm, the estimate is $\hat{\theta}_1 = 0.8$ with $\hat{\sigma}_a^2 = 0.97$ while the MLE is $\hat{\theta}_1 = 0.89$ with $\hat{\sigma}_a^2 = 0.94$. We see that in each case, the parameter is underestimated. It should be noted that the MM estimate does not exist because $\hat{\rho}_1 = -0.54$. See Example 7.1. As previously mentioned, we do not recommend the MM estimates for MA(q) models.

We revisit a few of the AR(p) examples considered in Chapters 3 and 5. As mentioned earlier, the methods (YW, Burg, and MLE) all behave similarly asymptotically, but for AR(p) models with roots close to the nonstationary region, the YW estimates do not perform as well.

Example 7.9 Parameter Estimation for Selected AR Models

(a) *AR(1)*
Figure 3.5a shows a realization of length $n = 100$ from the AR(1) model $(1 - 0.95B)X_t = a_t$ when $\sigma_a^2 = 1$. For these data, the YW, Burg, and ML estimates of $\phi_1 = 0.95$ are 0.93, 0.94, and 0.95, respectively, and, in all cases, $\hat{\sigma}_a^2 = 0.93$. Thus, each of these estimates of ϕ_1 are quite similar in this case. From Example 7.5i, the asymptotic standard deviation is

$$SD(\hat{\phi}_1) = \sqrt{(1 - 0.95^2)/100} = 0.03,$$

from which it follows that all of these estimates are within one SE of the true value.

(b) *AR(4)*
In this case, we consider the AR(4) model

$$X_t - 1.15X_{t-1} + 0.19X_{t-2} + 0.64X_{t-3} - 0.61X_{t-4} = a_t,$$

previously discussed in Example 3.10. A realization of length $n = 200$ from this model is shown in Figure 3.16a. In Table 7.5, we show the

TABLE 7.5

Parameter Estimates for the Realization in Figure 3.16a from the AR(4) Model in (3.81)

	ϕ_1	ϕ_2	ϕ_3	ϕ_4	σ_a^2
True	1.15	−0.19	−0.64	0.61	1.0
YW	1.11	−0.22	−0.59	0.57	1.08
Burg	1.11	−0.21	−0.61	0.58	1.08
MLE	1.11	−0.22	−0.61	0.59	1.08

four autoregressive model parameters along with the estimates using the YW, Burg, and MLE methods. Clearly, all of the techniques produce similar estimates that are quite close to the model parameters. The factor table for this model is shown in Table 3.3 where it is seen that there are no roots extremely close to the unit circle and no dominating factors since the absolute reciprocals of the roots of the characteristic equation range from 0.8 for the negative real root to 0.95 for the positive real root.

(c) *AR(3) with a dominant positive real root*
Figure 5.3c shows a realization of length $n = 200$ from the AR(3) model

$$X_t - 2.195X_{t-1} + 1.994X_{t-2} - 0.796X_{t-3} = a_t,$$

which can be written in factored form as

$$(1 - 0.995B)(1 - 1.2B + 0.8B^2)X_t = a_t.$$

This model has a positive real root very close to the unit circle and complex conjugate roots further removed. As noted in the discussion following (7.52), the Yule–Walker system of equations is near singular, and, as a consequence, the Yule–Walker estimates of the model parameters were predicted to be poor. The parameter estimates using the YW, Burg, and MLE procedures are shown in Table 7.6

TABLE 7.6

Parameter Estimates for the Realization in Figure 5.3c from the AR(3) Model $X_t - 2.195X_{t-1} + 1.994X_{t-2} - 0.796X_{t-3} = a_t$

	ϕ_1	ϕ_2	ϕ_3	σ_a^2
True	2.195	−1.994	0.796	1.0
YW	1.73	−1.11	0.35	1.24
Burg	2.15	−1.89	0.74	0.87
MLE	2.14	−1.89	0.74	0.87

TABLE 7.7

Factor Tables Associated with Parameter Estimates for the Realization in Figure 5.3c from the AR(3) Model $(1 - 2.195B + 1.994B^2 - 0.796B^3)X_t = a_t$

AR Factors	Roots (r_j)	$\lvert r_j^{-1}\rvert$	f_{0j}
(a) True parameter values			
$1 - 0.995B$	1.005	0.995	0.0
$1 - 1.2B + 0.8B^2$	$0.75 \pm 0.83i$	0.89	0.13
(b) Yule–Walker estimates			
$1 - 0.96B$	1.05	0.96	0.0
$1 - 0.77B + 0.37B^2$	$1.04 \pm 1.27i$	0.61	0.14
(c) Burg estimates			
$1 - 0.98B$	1.02	0.98	0.0
$1 - 1.16B + 0.75B^2$	$0.78 \pm 0.86i$	0.86	0.13
(d) ML estimates			
$1 - 0.98B$	1.02	0.98	0.0
$1 - 1.16B + 0.75B^2$	$0.77 \pm 0.86i$	0.87	0.13

where it is seen that the Burg and ML estimates are quite good while the YW estimates are poor.

Table 7.7 shows the factor tables associated with the fitted models, where it is seen that all three fitted models have a positive real root close to the unit circle and complex conjugate roots associated with a system frequency of about $f = 0.13$. The poor performance of the Yule–Walker estimates can be traced to the fact that the second-order factor is estimated to be much weaker (further removed from the unit circle) than it actually is. Again, this can be attributed to the fact that the dominant positive real root tends to diminish the influence of the second-order factor in the sample autocorrelations. In Chapter 9, we will discuss the use of differencing such realizations as part of the model building procedure.

(d) *AR(2) with repeated roots*

In Figure 3.9d, we show a realization of length $n = 100$ from the AR(2) model $X_t - 1.9X_{t-1} + 0.9025X_{t-2} = a_t$, which can be written in factored form $(1 - 0.95B)(1 - 0.95B)X_t = a_t$. We see that this model has two real roots of 1.05 ($= 1/0.95$). Table 7.8 shows the parameter estimates using the YW, Burg, and MLE procedures where it is seen that the Burg and ML estimates are quite good, while, as in the previous example, the YW estimates are very poor. For the data in Figure 3.9d, we calculate that $\hat{\gamma}_0 = 910.15$ and $\hat{\gamma}_1 = 882.85$, so that

$$\boldsymbol{\Gamma} = \begin{pmatrix} 910.15 & 882.85 \\ 882.85 & 910.15 \end{pmatrix},$$

TABLE 7.8

Parameter Estimates for the Realization in Figure 3.9d from the AR(2) Model $(1 - 0.9B + 0.9025B^2)X_t = a_t$

	ϕ_1	ϕ_2	σ_a^2	Half Width
True	1.9	−0.9025	1.0	
YW	1.04	−0.07	5.9	0.065
Burg	1.88	−0.89	0.94	0.026
MLE	1.89	−0.89	0.93	0.026

and

$$\boldsymbol{\Gamma}^{-1} = \begin{pmatrix} 0.0186 & -0.018 \\ -0.018 & 0.0186 \end{pmatrix}.$$

Thus, the 95% confidence intervals in (7.59) based on any of the three estimation methods are given by

$$\hat{\phi}_j \pm 1.96\frac{\hat{\sigma}_a\sqrt{0.0186}}{\sqrt{100}}, \tag{7.73}$$

for $j = 1, 2$, where $\hat{\sigma}_a$ is the square root of the estimated white noise variance for the parameter estimation technique employed. The last column of Table 7.8 shows the half widths $\pm 1.96\hat{\sigma}_a\sqrt{0.0186}/\sqrt{100} = \pm 1.96\hat{\sigma}_a(0.0136)$. In the table, it can be seen that the Burg- and MLE-based confidence intervals cover the true values while even though the YW-based confidence intervals are much wider, they still do not cover the true values.

Table 7.9 shows factor tables for the estimated models to better illustrate their features. Table 7.9a is the factor table associated with the true model, which shows the repeated real roots fairly close to the unit circle. The Yule–Walker estimates in Table 7.8 are clearly quite poor, and, in the factor table in Table 7.9b, we see that these estimates "saw" only one root close to 1.05. Based on the Yule–Walker estimates for an AR(2) fit, one would (in this case incorrectly) conclude that an AR(1) model is more appropriate. The tendency to approximate first-order behavior can be attributed to the fact that if there are two roots on the unit circle, and these roots are repeated positive real roots (i.e., have multiplicity two), then the limiting autocorrelations (ρ_k^*) will tend to satisfy a first-order difference equation (see Theorem 5.1).

The factor table for the Burg estimates makes another important point. That is, although the Burg parameter estimates in Table 7.8 are close to the true values, the factor table in Table 7.9c does not at

TABLE 7.9

Factor Tables Associated with Parameter Estimates for the Realization in Figure 3.9d from the AR(2) Model $(1 - 0.9B + 0.9025B^2)X_t = a_t$

AR Factors	Roots (r_j)	$\lvert r_j^{-1} \rvert$	f_{0j}
(a) True parameter values			
$1 - 0.95B$	1.05	0.95	0.0
$1 - 0.95B$	1.05	0.95	0.0
(b) Yule–Walker estimates			
$1 - 0.97B$	1.03	0.97	0.0
$1 - 0.07B$	13.82	0.07	0.0
(c) Burg estimates			
$1 - 1.88B + 0.89B^2$	$1.06 \pm 0.09i$	0.94	0.013
(d) ML estimates			
$1 - B$	1.0	1.0	0.0
$1 - 0.89B$	1.12	0.89	0.0

first glance look anything like the factor tables in Table 7.9a, since the characteristic equation for the Burg model has complex roots. However, examination of the roots shows that the complex part ($0.09i$) is quite small and that the real part (1.06) is close to the true value.

Additionally, the system frequency associated with the complex conjugate roots is $f = 0.013$, which is close to zero. In many cases, a pair of positive real roots in a theoretical model will be manifested in the estimated model as complex conjugate roots associated with a very small system frequency. The ML estimates shown in Table 7.9d show two positive real roots, with one of the roots on the unit circle. That is, the fitted model is nonstationary. The decision concerning whether to use the Burg or the ML estimates may be dictated by the physical situation. For example, if the physical process is known to have a long cycle (of about 1/0.013), then clearly the Burg fit is the preferred one. However, if a long cycle does not make sense in the particular setting, then the ML model would be preferred.

Tables 7.7 and 7.9 illustrate the important fact that more information can often be learned from the factor tables of the fitted models than is obtained by examining parameter estimates along with the standard errors or confidence intervals. For example, while it is true that the 95% confidence intervals (not shown here) on the ϕ_i's for the Yule–Walker estimates in Tables 7.6 and 7.8 do not cover the true parameter values, a much better understanding of the nature of and differences among the fitted models is achieved by examining the factor tables in Tables 7.7 and 7.9.

TABLE 7.10

Parameter Estimates for the Realization in Figure 3.22a from the ARMA(2,1) Model $(1 - 1.6B + 0.9B^2)\, X_t = (1 - 0.8B)a_t$

	ϕ_1	ϕ_2	θ_1	σ_a^2
True	1.6	−0.9	0.8	1.0
Burg/TT	1.54	−0.81	0.86	1.05
MLE	1.55	−0.83	0.86	1.05

Example 7.10 Estimating Parameters in an ARMA(2,1) Model

We consider the use of Tsay–Tiao estimates initialized using Burg estimates (Burg/TT) and ML estimates (initialized by Burg/TT) to estimate the parameters of an ARMA(2,1) model fit to the realization in Figure 3.22a of length $n = 200$ from the model $(1 - 1.6B + 0.9B^2)X_t = (1 - 0.8B)a_t$. Table 7.10 shows the parameter estimates, and it can be seen that both techniques provide reasonably good estimates. Factor tables (not shown) reveal that in each case (i.e., for the true model and the two fitted models), the autoregressive characteristic function has complex roots with a system frequency of $f = 0.09$. However, the complex roots associated with the two fitted models are somewhat further from the unit circle than are those in the true model.

7.7 ARMA Spectral Estimation

In (1.25) and (1.26), we defined the power spectrum and the spectral density of a stationary process, and in Section 1.6, we discussed the periodogram, sample spectrum, and smoothed versions of these quantities as methods for estimating the spectrum and spectral density. In this section, we discuss the method of *ARMA spectral estimation*, which is based on the spectral density of an ARMA(p,q) process shown in the following [and shown earlier in (3.92)] given by

$$S_X(f) = \frac{\sigma_a^2}{\sigma_X^2} \frac{\left|1 - \theta_1 e^{-2\pi if} - \theta_2 e^{-4\pi if} - \cdots - \theta_q e^{-2q\pi if}\right|^2}{\left|1 - \phi_1 e^{-2\pi if} - \phi_2 e^{-4\pi if} - \cdots - \phi_p e^{-2p\pi if}\right|^2}.$$

The ARMA spectral density estimator based on an ARMA(p,q) fit to a set of data (using any parameter estimation technique) is simply *the spectral density associated with the fitted model*. That is, the ARMA spectral density estimator is given by

$$\hat{S}_X(f) = \frac{\hat{\sigma}_a^2}{\hat{\sigma}_X^2} \frac{\left|1 - \hat{\theta}_1 e^{-2\pi i f} - \hat{\theta}_2 e^{-4\pi i f} - \cdots - \hat{\theta}_q e^{-2q\pi i f}\right|^2}{\left|1 - \hat{\phi}_1 e^{-2\pi i f} - \hat{\phi}_2 e^{-4\pi i f} - \cdots - \hat{\phi}_p e^{-2p\pi i f}\right|^2}. \tag{7.74}$$

We will use the generic term ARMA spectral estimator to indicate either the spectrum or spectral density based on the fitted model. Clearly, the ARMA spectral estimator for a set of data will only be as good as the parameter estimates, and thus care must be taken in parameter estimation. We note that ARMA spectral estimators based on AR(p) fits are often called *autoregressive spectral estimators*.

The ARMA spectral density estimator has an inherent smoothness. By increasing the orders p and q of a model fit, the corresponding spectral density becomes less smooth similar to the manner in which increasing the truncation point M in the spectral estimates in Chapter 1 produces spectral estimators that are less smooth. Thus, choosing orders p and q that are too low may produce estimators that lose important detail, while using orders that are too high can produce spectral estimators with spurious peaks. Thus, the choice of p and q (to be discussed in Chapter 8) is an important step in the modeling procedure.

Using `GW-WINKS`: As a part of the output from the `Estimate Parameters` option, the `GW-WINKS` program places the (log) ARMA spectral density estimator for the fitted model in a column of the Data Editor with a name indicating that it is a spectral file (the default name is `SPECTRUM`). This data column contains 251 data values that should be considered to be the 251 values of $\hat{S}_X(f)$, $f = 0, 0.002, 0.004, \ldots, 0.5$. The ARMA spectral density estimator can be plotted using the `Graphs/Chart` menu item and then selecting `Line/Time Series Plots` followed by `Multiple (matrix) of line plots`, and specifying the variable containing the spectral estimates. The name `SPECTRUM` causes the default plot mode to be a spectral plot. If you change the variable name in the data editor, you may need to edit the plot to create a spectral plot.

Example 7.11 ARMA Spectral Estimation

(a) *An AR(4) model*

In Figure 7.1a, we show the true spectral density, previously shown in Figure 3.16c for the AR(4) that was previously discussed in Example 7.9b. In this case, all of the estimation routines performed equally well, and Figure 7.1b shows the AR spectral estimator based on the ML estimates. In the figure, it is seen that the spectral density in Figure 7.1a is well estimated. Figure 7.1c and d shows spectral estimators

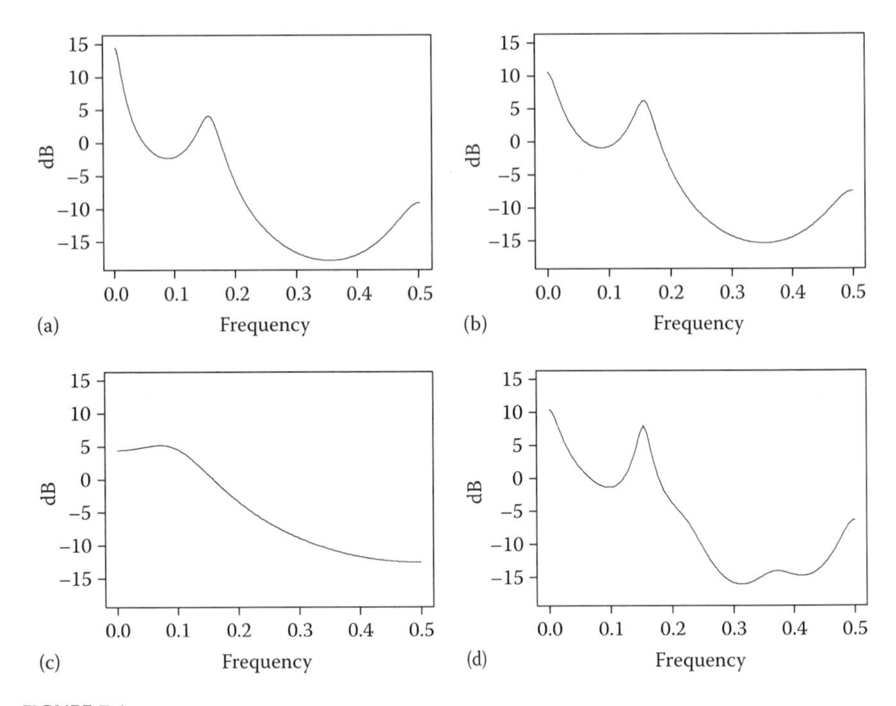

FIGURE 7.1
(a) True spectral density for the AR(4) model in Example 7.9b. (b) AR(4) spectral density estimator. (c) AR(2) spectral density estimator. (d) AR(12) spectral density estimator.

based on AR(2) and AR(12) fits to the data using ML estimates. Figure 7.1c shows that an AR(2) fit to the data is inappropriate and that the frequency behavior associated with system frequencies $f=0$ and $f=0.16$ (see Table 3.3) is smeared together. The AR(12) spectral estimator in Figure 7.1d is much better than the one based on the AR(2) fit, but we see a small peak at around 0.38 that is in fact a spurious peak. This example illustrates the fact that a high-order fit will tend to show the dominant peaks along with spurious peaks that must be considered. One technique for ARMA spectral estimation is to fit a high-order autoregressive model to data with the understanding that this may create spurious peaks.

(b) *An ARMA(2,1) model*
Figure 7.2a shows the true spectral density for the ARMA(2,1) model discussed in Example 7.10. Table 7.10 shows estimation results based on the realization from this model that was shown in Figure 3.22a. All of these results are similar, and, in Figure 7.2b, we show the ARMA(2,1) spectral density estimator based on ML estimates. It is

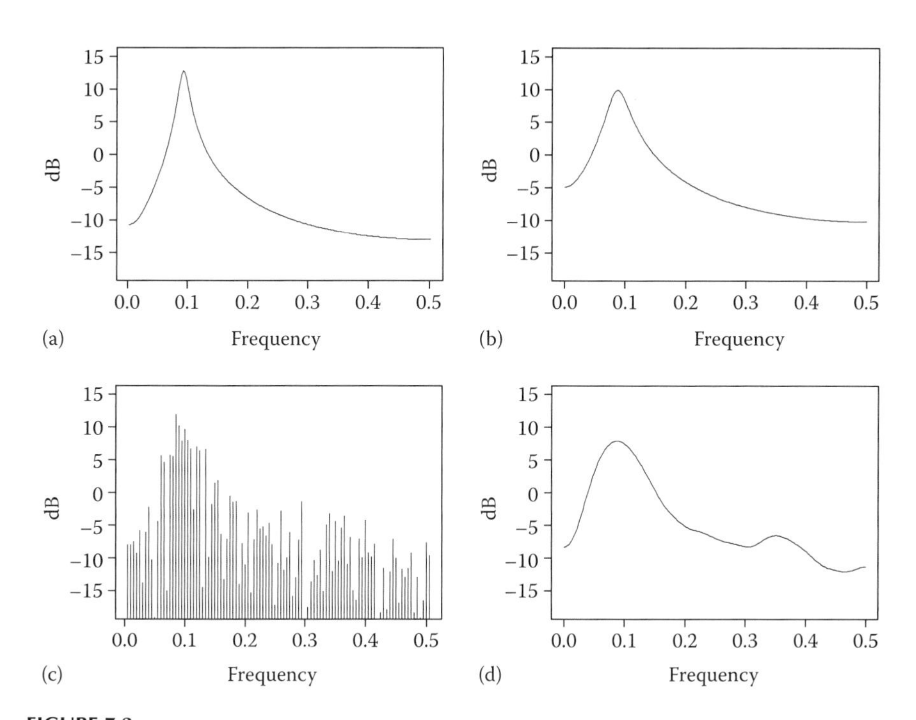

FIGURE 7.2
(a) True spectral density for the ARMA(2,1) model in Example 7.10. (b) ARMA(2,1) spectral density estimator. (c) Periodogram. (d) Parzen windowed spectral estimator.

seen that the estimated spectral density is very similar to the true spectral density in Figure 7.2a in that it has a single peak at about $f = 0.1$ along with a dip at $f = 0$ associated with the moving average component. Figure 7.2c shows the periodogram and Figure 7.2d shows the Parzen spectral density estimator based on a truncation point $M = 28(=2\sqrt{200})$. These estimates have the general appearance of the true spectral density in Figure 7.2a. As usual, the periodogram is extremely variable, and, in this case, the smoothed version shows a spurious peak at about $f = 0.37$.

(c) *An AR(3) with estimation difficulties*
Figure 7.3a shows the true spectral density associated with the AR(3) in Example 7.9c. This model has a positive real root close to the unit circle that causes the Yule–Walker estimates to be poor. Figure 7.3b and c shows AR(3) spectral density estimates associated with ML and Yule–Walker estimates, respectively, while Figure 7.3d shows Parzen windowed spectral estimate using $M = 28$. In the figure, it is seen that the ML estimates produce an excellent spectral density

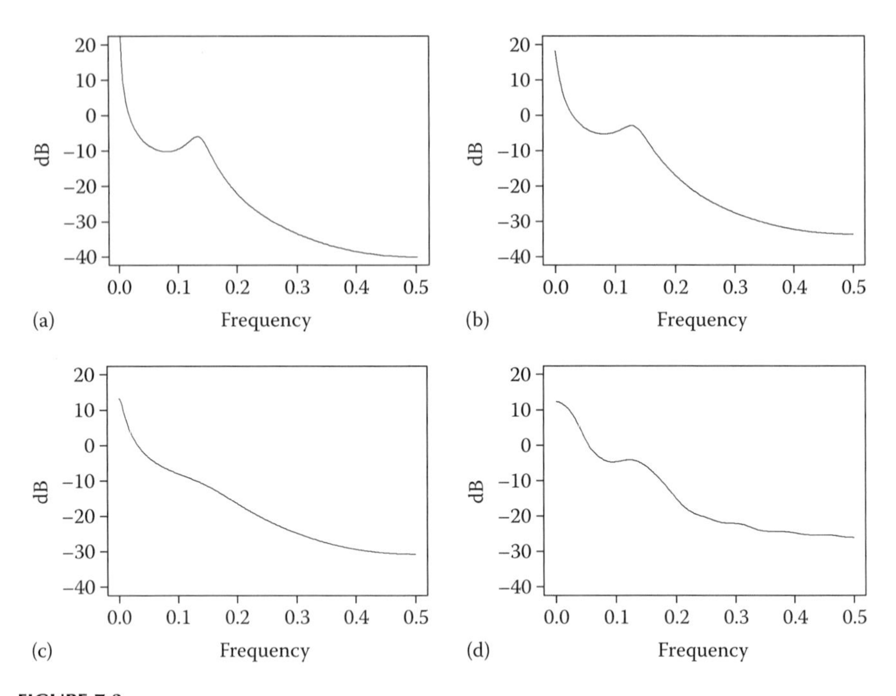

FIGURE 7.3
(a) True spectral density for the AR(3) model in Example 7.9c. (b) AR(3) ML spectral density estimator. (c) AR(3) YW spectral density estimator. (d) Parzen windowed spectral estimator.

estimate, while the Yule–Walker-based spectral estimate loses the smaller peak at about $f = 0.13$ because of the parameter estimation problems associated with the Yule–Walker fit. Note that this peak shows up in the Parzen spectral estimator. If a fitted model disagrees substantially from nonparametric spectral estimates as in this case, it is advisable to examine the model fit to the data.

7.8 ARUMA Spectral Estimation

ARUMA(p,d,q) models such as

$$(1 - B)(1 + B^2)(1 - B + 0.8B^2)X_t = a, \tag{7.75}$$

have spectral densities with infinite spikes, in this case, at $f = 0$ and $f = 0.25$. In practice, if an ARUMA model such as (7.75) is fit to a set of data, then we recommend plotting the ARUMA spectral density estimates by plotting

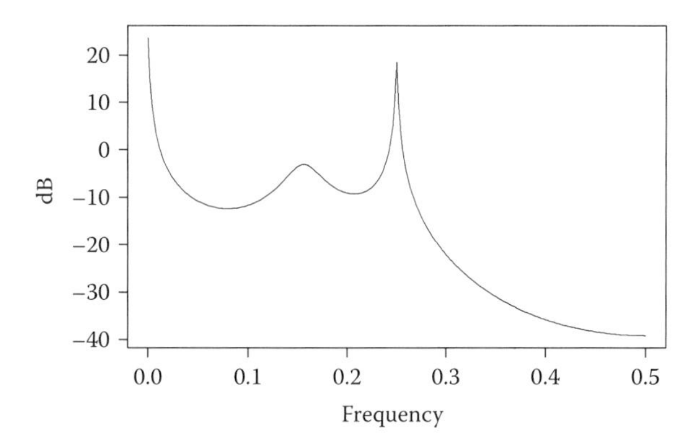

FIGURE 7.4
ARUMA spectral density estimate associated with the ARUMA(2,3,0) model in (7.75).

ARMA spectral density estimates of related near-nonstationary models. In this case, we could plot the true spectral density associated with

$$(1 - 0.995B)(1 + 0.99B^2)(1 - B + 0.8B^2)X_t = a_t. \tag{7.76}$$

Note that the coefficients were selected, so that all three near nonstationary roots are equally close to the unit circle. In this case, the roots are 1.005 and $\pm 1.005i$. Figure 7.4 shows an approximate ARUMA(2,3,0) spectral density estimate associated with the model fit in (7.75). The sharp peaks at $f = 0$ and $f = 0.25$ represent the infinite peaks at those frequencies.

NOTE: Techniques for fitting ARUMA(p,d,q) models to data will be discussed in Section 8.3.

Using `GW-WINKS`: The approximate ARUMA(2,3,0) spectral density estimate shown in Figure 7.4, which is based on the AR(5) model in (7.76), can be obtained in `GW-WINKS` by using the `Generate Realization` option on the `Time Series Analysis` menu and specifying the factors of the AR(5) model in (7.76). (The white noise variance and sample size are irrelevant.) The data in the column `TRUE_SPECTRUM` contain the desired "approximate" ARUMA(4,1,0) spectral density estimate associated with the fitted model.

Exercises

If you use `GW-WINKS` to work the following problems, you will find the `Estimate Parameters`, `Generate Realization`, and `Spectrum` options on the `Time series Analysis` menu to be useful.

7.1 Suppose an AR(4) model is to be fit to a time series realization, and the sample autocorrelations, $\hat{\rho}_k$, $k=0, 1, \ldots, 6$ were calculated to be 1, 0.69, 0.22, 0.22, 0.61, 0.77, and 0.47.

(a) Use the Durbin–Levinson algorithm to solve the Yule–Walker equations for $\hat{\phi}_k, k = 1, \ldots, 4$.

(b) Find 95% confidence intervals on ϕ_k, $k=1, \ldots, 4$.

(c) Based on these estimates and confidence intervals, does an AR(4) seem to be the best model for these data? What other model would you suggest and why?

7.2 The following realization of length $n=10$ has been observed:

t	X_t	t	X_t
1	26.8	9	31.7
2	27.8	10	31.5
3	30.0	11	31.9
4	31.6	12	31.5
5	33.0	13	30.0
6	34.2	14	29.4
7	34.1	15	29.9
8	33.2		

Find the approximate OLS estimates (i.e., those obtained by minimizing (7.14)) for the following:

(a) The parameter ϕ_1 assuming the data are from an AR(1) model

(b) The parameters ϕ_1 and ϕ_2 assuming the data are from an AR(2) model

NOTE: You can either find these by hand (calculator), write a program to find the estimates, or use a regression program.

7.3 Find the Yule–Walker, Burg, and ML estimates of the model parameters in the following:

(a) SUN.XLS assuming an AR(2) and an AR(8) model

(b) The data in Problem 7.2 assuming an AR(1) model

(c) The data in Problem 7.2 assuming an AR(2) model

(d) Using the asymptotic distribution results in Example 7.5, calculate 95% confidence intervals for the parameters in the AR(1) and AR(2) models in (b) and (c)

7.4 For each of the following models, perform a "mini simulation" studying by generating 3 different realizations of length $n=100$ and finding the Yule–Walker, Burg, and ML estimates for each realization.

(A) $(1 - 1.3B + 0.6B^2)(X_t + 20) = a_t$

(B) $(1 - 3.1B + 4.1B^2 - 2.63B^3 + 0.72B^4)(X_t - 100) = a_t$

(C) $(1 - 0.99B + 0.03B^2 - 0.297B^3)(X_t - 250) = a_t$

(a) Find Factor Tables for each of these models.
(b) Compare the estimator performance on the three models. It may be useful to factor the estimated models.

7.5 Using the sunspot data in SUN.XLS, calculate and plot the following:
(a) The Parzen window spectral density estimate using the default truncation point
(b) The AR(2) spectral density estimate based on ML estimates of the parameters
(c) The AR(8) spectral density estimate based on ML estimates of the parameters

Compare and contrast the three spectral densities. What feature of the AR(8) and Parzen spectral density estimate is missing from the AR(2) spectral density estimate. Compare to Problem 3.12. Comparison of spectral density estimates is yet another method for comparing models.

7.6 Using one realization from each of the models in Problem 7.4, find and plot the following:
(a) The Parzen spectral density
(b) The AR(p) spectral density estimate based on the fitted model
  (i) Using the Yule–Walker estimates
  (ii) Using the Burg estimates
  (iii) Using the ML estimates

Compare and contrast these spectral density estimates. How does this comparison relate to the findings in Problem 7.4b?

8

Model Identification

In Chapter 7, we discussed the problem of estimating the parameters of a given ARMA(p,q) process assuming that the model orders p and q are known. The procedure for finding an appropriate model actually begins with the intriguing and sometimes challenging problem of determining the number of parameters that should be estimated. In this chapter, we consider the problem of identifying the model orders p and q of a stationary ARMA(p,q) model as well as p, d, and q for the nonstationary ARUMA(p,d,q) model, and the model orders associated with a multiplicative seasonal model. Since there is realistically not a "true ARMA model" for a real process, our goal will be to find a model that adequately describes the process but does not contain unnecessary parameters. Tukey (1961) refers to such a model as being *parsimonious*.

8.1 Preliminary Check for White Noise

In searching for a parsimonious model, the investigation should begin with a decision concerning whether or not the process can be adequately modeled using the simplest of all ARMA models, white noise. Unnecessarily complicated models can be obtained by the careless analyst for processes that should have been modeled as white noise in the first place. Several techniques are available for testing for white noise, and we will address this problem further in Section 9.1 when we discuss the examination of residuals from a fitted model (which should essentially be white noise if the model is appropriate).

At this point, we focus on checking for white noise by examining the realizations and sample autocorrelations. The first advice we give in this context is that *you should always first examine a plot of the data*! The failure to realize that a set of time series data should be modeled as white noise or to notice other very basic characteristics is often the result of having failed to plot the data. In Figures 1.3b, 1.4b, and 1.9b, we showed a realization, the true autocorrelations, and the sample autocorrelations from a white noise model. There is no noticeable cyclic or trending behavior in the data, and while the true autocorrelations for a white noise process have the property that $\rho_k = 0$ for $k \neq 0$, the sample autocorrelations will not be identically zero as is illustrated in Figure 1.9b. Each of the sample autocorrelations $\hat{\rho}_k$, $k > 1$ in that figure estimates a quantity that is zero. In order to test $H_0 : \rho_k = 0$, we first note that $\hat{\rho}_k$ is asymptotically unbiased for ρ_k since $E[\hat{\rho}_k] \approx 0$ for $k \neq 0$ and realization length n "large." Additionally, $\hat{\rho}_k$ is asymptotically normal (Brockwell and Davis, 1991). Recall also that if X_t is Gaussian white noise, then Bartlett's approximations in (1.21) and (1.22) imply that

$$\text{Cov}(\hat{\rho}_k, \hat{\rho}_{k+r}) \approx 0, \quad r \neq 0$$

and

$$\text{Var}(\hat{\rho}_k) \approx \frac{1}{n}, \quad k \neq 0.$$

Thus, if a given realization is from a white noise process, then the sample autocorrelations should be small and approximately uncorrelated at different lags. For a given k, it then follows that

$$|\hat{\rho}_k| > 2\left(\frac{1}{\sqrt{n}}\right) \tag{8.1}$$

provides evidence against the null hypothesis of white noise at approximately the $\alpha = 0.05$ level of significance (where we have rounded $z_{0.95} = 1.96$ to 2). It should be noted that the significance level $\alpha = 0.05$ applies separately for each k, and it would not be unusual for about 5% of the $\hat{\rho}_k$'s to exceed $2(1/\sqrt{n})$ when the data are white noise. As a result, this "test" is actually more of an informal guide for determining whether or not the data should be modeled as white noise. It should be noted, however, that many alternatives of interest, that is, an MA(q) model or an AR(p) model with roots relatively far removed from the unit circle, have nonzero autocorrelations at low lags followed by zero or near zero autocorrelations at higher lags. Thus, if the "nonzero" $\hat{\rho}_k$'s occur for small k, then other models should be considered. This will be discussed in Example 8.A.2.

Example 8.1 A White Noise Process

Figure 8.1a displays the white noise realization of length $n = 200$ from Figure 1.3b, and Figure 8.1b shows the sample autocorrelations, $\hat{\rho}_k, k = 0, \ldots, 44$, from Figure 1.9b, along with horizontal lines at $\pm 2(1/\sqrt{n}) = \pm 0.14$. The sample autocorrelations at lags 10 and 36 marginally exceed the 95% limits with all others being within the lines. This behavior is typical of sample autocorrelations from white noise data. To draw the conclusion that there is some pertinent information for modeling purposes due to the fact that the lag 10 and lag 36 sample autocorrelation exceed the 95% limits would be a mistake. As mentioned earlier, one should expect about 5% of the $\hat{\rho}_k$ values to exceed the 95% limits if the data are white noise.

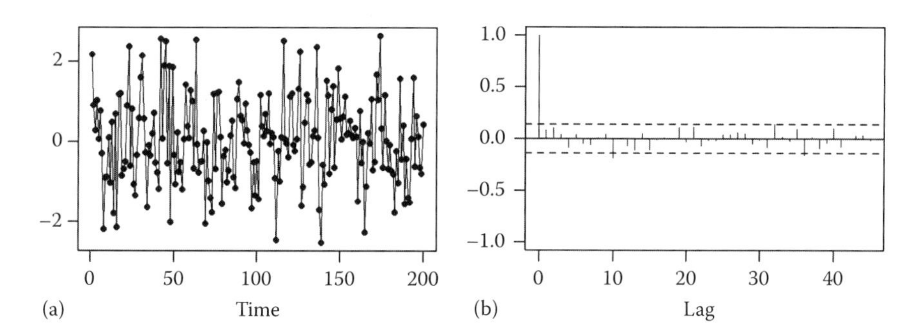

FIGURE 8.1
(a) Realization from a white noise process. (b) Sample autocorrelations from (a) along with 95% limits.

Example 8.2 Checking for White Noise

In Figure 8.2a, we show sample autocorrelations (with 95% limits) based on the realization of length $n = 190$ in Figure 8.2b. As in Figure 8.1b, more than 95% of the sample autocorrelations stay within the 95% limit lines. Consequently, examination of the sample autocorrelation suggests white noise. However, examination of Figure 8.2b shows that the realization has

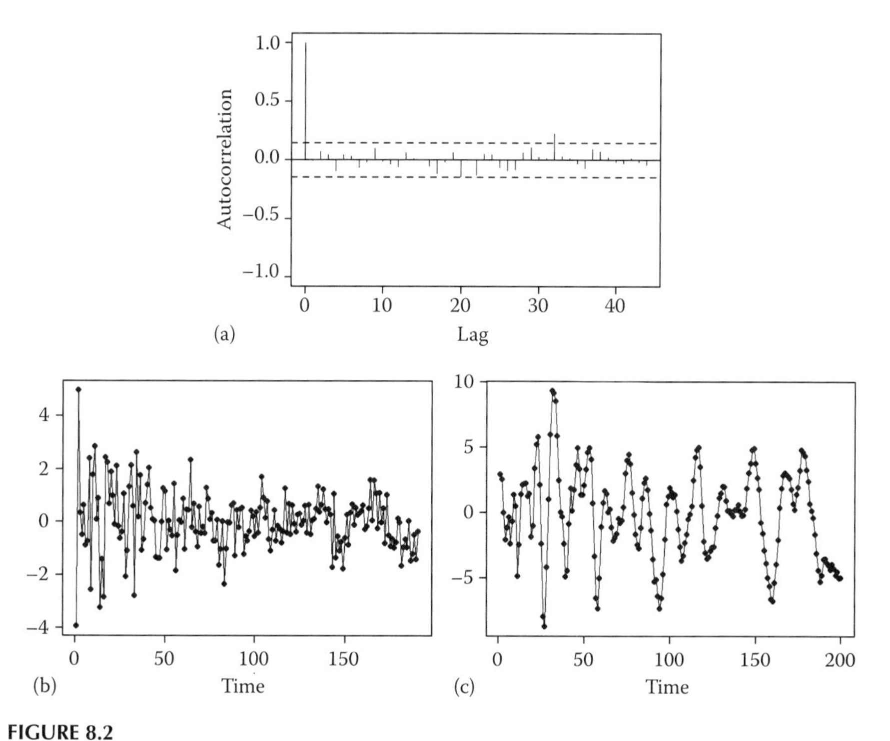

FIGURE 8.2
(a) Sample autocorrelations along with 95% limit lines for the realization in (b). The data shown in (b) are the residuals from an AR(10) model (incorrectly) fit to the TVF data set in (c).

characteristics inconsistent with white noise. For example, the variability of the data seems larger early in the series, and there appears to be a certain amount of piecewise trending behavior after $t = 100$. While local irregularities can occur in white noise data, the total picture of this realization is that, although the sample autocorrelations are small, the realization does not appear to be white noise. The realization in Figure 8.2b consists of the residuals from an AR(10) fit to the data set in Figure 8.2c. Note that this realization displays frequency behavior that decreases with time similar to the bat echolocation signal in Figure 1.26a. An ARMA(p,q) model cannot appropriately be fit to such data because of its nonstationary behavior, and it is not surprising that the residuals from such a fit are not white.

In Chapters 12 and 13, we will discuss the analysis of TVF data sets such as the one shown in Figure 8.2c. In Section 9.1, we return to the topic of white noise testing when we discuss portmanteau tests.

8.2 Model Identification for Stationary ARMA Models

In this section, we discuss procedures for identifying the orders p and q of an ARMA(p,q) model. Model identification for nonstationary or near-nonstationarymodels will be discussed in Section 8.3. A considerable amount of research has gone into the study of how to best select the orders p and q of an ARMA(p,q) fit to a set of data. Box and Jenkins popularized the use of ARMA models for purposes of modeling time series data in the 1970 edition of the current Box et al. (2008) book. They proposed a general procedure for determining the orders p and q of an ARMA(p,q) model based on the examination of plots of the autocorrelations and partial autocorrelations, which will be defined in Appendix 8.A. Extensions of this approach were given by Gray et al. (1978), Beguin et al. (1980), Woodward and Gray (1981), Tsay and Tiao (1984), among others. These and other similar approaches can be characterized as pattern recognition approaches in that they are based on identifying patterns of expected behavior that would occur if the underlying model were an ARMA(p,q). While some of these methods have been automated, their successful use tends to be dependent on the skill of the analyst in identifying the pertinent patterns. These methods will briefly be discussed in Appendix 8.A.

In this section, we discuss automated techniques based on Akaike's Information Criterion (AIC) (Akaike, 1973). After years of analyzing time series data (and after having developed some of the pattern recognition approaches), the authors of this book have begun to depend nearly exclusively on the AIC-based approaches, especially when several time series realizations are to be analyzed. In some cases, careful analysis involves using any and all available tools including the pattern-recognition approaches discussed in Appendix 8.A.

8.2.1 Model Identification Based on AIC and Related Measures

AIC is a general criterion for statistical model identification with a wide range of applications (including model selection in multivariate regression and, in our case, identifying p and q in an ARMA(p,q) model). Maximum likelihood (ML) estimation is a common tool that applies to a model or class of distributions with a fixed set of parameters, say $\alpha_1, \ldots, \alpha_k$. In this setting, the ML estimators are the estimators, $\hat{\alpha}_1, \ldots, \hat{\alpha}_k$ that maximize the likelihood function. However, simply maximizing the likelihood is not an appropriate strategy when parameters are being added to an existing model. For example, suppose ML methods are used to choose between a model with k parameters (e.g., an AR(k)) and a model from the same class with $k+1$ parameters, for example, an AR($k+1$). Consider the strategy of maximizing the likelihood under each scenario and selecting the model with the maximized likelihood. This strategy would always select the AR($k+1$) model, since

$$L_{MAX}(\hat{\alpha}_{k1}, \ldots, \hat{\alpha}_{kk}) \leq L_{MAX}(\hat{\alpha}_{k+1,1}, \ldots, \hat{\alpha}_{k+1,k+1}), \quad (8.2)$$

where $\hat{\alpha}_{mj}$ denotes the maximum likelihood estimator of α_j for a model with m parameters. To see why (8.2) is true, note that when there are $k+1$ parameters, one choice for $\hat{\alpha}_{k+1,k+1}$ is $\hat{\alpha}_{k+1,k+1} = 0$. The maximized likelihood on the left-hand side of (8.2) is thus based on the constraint that $\hat{\alpha}_{k+1,k+1} = 0$ while the right-hand side is maximized without this constraint. AIC addresses this problem in the ARMA case by imposing a penalty for adding parameters. The method is based on selecting the p and q that minimize the expected Kullback–Leibler distance between the true underlying model of the data and an ARMA(p,q) model fit to the data where $0 \leq p \leq P$ and $0 \leq q \leq Q$. See Akaike (1976). Essentially, AIC imposes a penalty for adding terms to a model, and the resulting procedure is to select the model that minimizes

$$\text{AIC} = -2[\text{In (maximized likelihood)}] + 2(\text{\# of free parameters}). \quad (8.3)$$

In the setting of ARMA(p,q) modeling, the AIC criterion becomes

$$\text{AIC} = -2[\ln \text{(maximized likelihood)}] + 2(p+q+1), \quad (8.4)$$

since there are $p+q+1$ parameters including the constant term. It can be shown that AIC in (8.4) is approximately equal to

$$\text{AIC} = \ln(\hat{\sigma}_a^2) + 2(p+q+1)/n \quad (8.5)$$

where $\hat{\sigma}_a^2$ is the estimated white noise variance associated with the maximum likelihood fit of an ARMA(p,q) model (see Box et al., 2008; Shumway and Stoffer, 2006). We will subsequently apply AIC by finding the p and q ($0 \leq p \leq P$ and $0 \leq q \leq Q$) that minimize AIC as defined in (8.5).

A problem with the use of AIC is that as the realization length increases, AIC tends to overestimate the model orders (i.e., select models with too many parameters). Several authors have considered modifications to AIC to adjust for this problem. Two of the popular alternatives are AICC (Hurvich and Tsai, 1989) and BIC (sometimes called the Schwarz Information Criterion (SIC)) (Schwarz, 1978). These are given by

$$\text{AICC} = \ln\left(\hat{\sigma}_a^2\right) + \frac{n+p+q+1}{n-p-q-3} \tag{8.6}$$

and

$$\text{BIC} = \ln\left(\hat{\sigma}_a^2\right) + (p+q+1)\frac{\ln(n)}{n}. \tag{8.7}$$

Simulation studies have shown BIC to perform best for large realization lengths, n, while AICC generally performs better for smaller n when the number of parameters is relatively large (see McQuarrie and Tsai, 1998). If computing time related to the use of maximum likelihood estimation is an issue, then the quantities in (8.5) through (8.7) can be approximated using alternative estimation techniques. For example, if Burg/TT estimates (discussed in Section 7.2) are used, then (8.5) through (8.7) can be calculated using $\hat{\sigma}_a^2$ obtained based on Burg/TT estimates of the model parameters.

Other automated model identification techniques have been proposed in the literature. For example, Akaike (1969) suggested the use of final prediction error (FPE) for identifying the order of an autoregressive model. Another early technique for identifying the order of an autoregressive model is the criterion autoregressive transfer (CAT) function proposed by Parzen (1974).

Using `GW-WINKS`: AIC, AICC, and BIC can be calculated by selecting `AIC` on the `Time Series Analysis menu` and selecting the desired model identification measure. You then select the data set to be analyzed and the ranges of p and q to be considered. The program successively fits the models within the selected range using MLE estimates (default setting) or Burg/TT and reports the models in AIC (AICC or BIC) order, showing the model with minimum AIC (AICC or BIC) as the first model.

NOTE: In this book, we use AIC as the model identification criterion. This is not meant as an endorsement of this criterion over the others, but is simply done for consistency of presentation.

Example 8.3 Model Identification Using AIC

In this section, we reconsider some of the simulated data sets (for which we know the true model orders) discussed in earlier chapters and illustrate the use of AIC.

(a) In Example 3.10, we considered the AR(4) model

$$X_t - 1.15X_{t-1} + 0.19X_{t-2} + 0.64X_{t-3} - 0.61X_{t-4} = a_t, \tag{8.8}$$

and, in Figure 3.16a, we showed a realization of length $n = 200$ from this model with $\mu = 0$ and $\sigma_a^2 = 1$. The factor table is shown in Table 3.3. AIC with $P = 8$ and $Q = 4$ correctly selects $p = 4$ and $q = 0$ for this realization. The ML estimates for an AR(4) fit are shown in Table 7.5 (along with Yule–Walker and Burg estimates).

(b) In Example 3.14, we discussed the features of the ARMA(2,1) model

$$X_t - 1.6X_{t-1} + 0.9X_{t-2} = a_t - 0.8a_{t-1}. \tag{8.9}$$

Figure 3.22a shows a realization of length $n = 200$ from this model, and Table 7.10 shows ML and Burg/TT estimates that are quite similar to each other and reasonably close to the true values. AIC (using MLE) with $P = 5$ and $Q = 3$ correctly selects $p = 2$ and $q = 1$ for this realization. It should be noted that if we allow $P = 8$ and $Q = 4$, then AIC selects an ARMA(6,1) model whose factor table shows a factor very near to $1 - 1.6B + 0.9B^2$.

(c) Figure 5.3c shows a realization of length $n = 200$ generated from the AR(3) model

$$X_t - 2.195X_{t-1} + 1.994X_{t-2} - 0.796X_{t-3} = a_t, \tag{8.10}$$

which can be written as $(1 - 0.995B)(1 - 1.2B + 0.8B^2)\ X_t = a_t$. In Table 7.7, we show the factor table for the ML, Burg, and YW estimates of the parameters. The ML and Burg estimates are quite similar to each other and reasonably close to the true values while the YW estimates are poor. AIC with $P = 8$ and $Q = 4$ correctly selects $p = 3$ and $q = 0$. It should be noted that if Burg estimates are used in the AIC procedure, an AR(3) model is also selected while if estimates are based on YW estimates, then AIC picks AR(4), ARMA(3,2), and AR(5) as the top three model choices. The message is to avoid using Yule–Walker estimates.

8.3 Model Identification for Nonstationary ARUMA(p,d,q) Models

In the previous section, we consider a variety of methods for identifying the orders of ARMA(p,q) processes. However, it will often be necessary to fit a nonstationary model to a given set of data. In this section, we discuss procedures for fitting ARUMA(p,d,q) models. Recall that an ARUMA(p,d,q) model is specified by

$$\phi(B)\lambda(B)(X_t - \mu) = \theta(B)a_t \tag{8.11}$$

where

$\phi(B)$ and $\theta(B)$ are pth and qth order operators, respectively
All the roots of $\phi(z)=0$ and $\theta(z)=0$ are outside the unit circle
All roots of $\lambda(z)=1-\lambda_1 z-\cdots-\lambda_d z^d=0$ lie on the unit circle

To begin this discussion, we recall the Findley–Quinn theorem (Theorem 5.1) that says that if $p>0$, the extended autocorrelations, ρ_k^*, satisfy a lower order difference equation than $p+d$. The order of this difference equation is based on the roots on the unit circle of highest multiplicity. It was also seen in Example 3.12 that the stationary components are often "hidden" by near-nonstationary components, and, in Example 7.9c, we noted that roots near the unit circle can cause parameter estimation problems.

As a consequence of the comments in the previous paragraph, it follows that model identification in the presence of nonstationarities or near-nonstationarities must be done carefully. We make the following general comments in this regard:

General comments:

1. ML (and Burg/TT) estimates are better than YW (or extended YW) in the nonstationary or nearly nonstationary case. (See the estimation results shown in Tables 7.6 and 7.7.)
2. Model ID techniques (such as the Box–Jenkins procedure) that are based on YW-type estimation can be misleading in the presence of nonstationary or near-nonstationary models. AIC using ML (or Burg/TT) estimates will not be as susceptible to nonstationarity or near-nonstationarity.
3. The decision whether to fit a nonstationary model or a stationary model that has roots close to the unit circle will often depend on the application, that is, the physical problem being considered.
4. YW and Burg estimates of the parameters of an autoregressive model will always produce stationary models.

5. Indications of nonstationarity or near-nonstationarity include
   a. Slowly damping autocorrelations
   b. Realizations exhibiting wandering or strong cyclic behavior.

Because of items (1) and (2), we will use ML (and sometimes Burg/TT) in our examples. We will further discuss these ideas in the following material.

8.3.1 Including a Nonstationary Factor in the Model

In the following, we discuss methods that aid in the decision concerning whether to use a nonstationary component in the model. We first discuss the general procedure for analysis if the decision is made to include a nonstationary component, such as $1-B$, in the model for X_t. In this case, the model is $(1-B)\phi(B)(X_t-\mu_X)=\theta(B)a_t$. The next step is to transform the data to stationarity. In this example, we calculate $Y_t=(1-B)X_t=X_t-X_{t-1}$. If $1-B$ is the only nonstationary component in the model, then Y_t satisfies the ARMA(p,q) model $\phi(B)(Y_t-\mu_Y)=\theta(B)a_t$. The final step is to identify the orders p and q and estimate the parameters of the stationary model for Y_t using techniques discussed in Section 8.2. To summarize, we outline the steps in the following:

Steps for obtaining an ARUMA(p,d,q) model

(a) Identify the nonstationary component(s) of the model.
(b) Transform the data using the nonstationary factor(s) identified in (a).
(c) Estimate the parameters of the "stationarized data."

It is important to realize that the decision concerning whether to include a factor $1-B$ in a model cannot be answered simply by examining the parameter estimates, since a factor $1-B$ will generally not appear exactly in the factored form of the estimated model. That is, if there is a factor $1-B$ in the true model, one would expect the estimated model to contain a factor that approximates $1-B$, but is not exactly equal to it. As mentioned in General comment 4, the YW and Burg/TT estimates will always produce a stationary AR model. Studies have also shown that ML estimates of ϕ_1 in the AR(1) model $(1-\phi_1 B)X_t=a_t$, where $\phi_1 \approx 1$ tend to have a negative bias (Kang, 1992). Thus, the fact that $|\hat{\phi}_1|<1$ is not sufficient evidence of a stationary AR(1) model, nor does the fact that all roots of $1-\hat{\phi}_1 z-\cdots-\hat{\phi}_p z^p=0$ fall outside the unit circle imply that the underlying model is stationary.

8.3.2 Identifying Nonstationary Component(s) in a Model

General comment 5 has been demonstrated in Examples 3.12, 5.3, and 5.4 where we noticed that time series models with near unit roots have realizations whose characteristics are dominated by this nonstationary (or near nonstationary) behavior. For example, in Example 5.3, we considered the model $(1-\alpha_1 B)(1-1.2B+0.8B^2)X_t=a_t$, where $\alpha_1=0.8$, 0.95, and 0.995. We noticed that as $\alpha_1 \rightarrow 1$ (i.e., as the associated root of the characteristic equation approached +1), the realizations were characterized by approximately linear wandering behavior with no attraction to a mean level and theoretical autocorrelations that tend to damp slowly. We note that sample autocorrelations will tend to have a similar behavior but will often damp more quickly than theoretical autocorrelations. In that example, we also noted that stationary features of the model may not be apparent because of the dominance of the near unit root. Similarly, in Example 5.4, we examined an AR(3) model $(1-0.8B)(1-1.34B+0.995B^2)X_t=a_t$ with a pair of complex roots close to the unit circle and noted that the realization and autocorrelations had a cyclic behavior and that the autocorrelations damped slowly. Additionally, the realization and autocorrelations showed very little evidence of the $1-0.8B$ factor. The behavior of the autocorrelations is consistent with the results of the Findley–Quinn theorem (Theorem 5.1).

It is common practice to make a decision concerning inclusion of a factor of $1-B$ based on examination of realizations and sample autocorrelations. Realizations with wandering or piecewise trending behavior along with slowly damping sample autocorrelations suggest that a unit root of +1, that is, that a nonstationary factor of $1-B$, should be included in the model (e.g., see Box et al., 2008, p. 197). The classical "Box–Jenkins" procedure is then to stationarize the data by differencing. If the differenced data indicate continued evidence of a unit root of +1, then the Box–Jenkins procedure is to difference again, etc., until stationarity is achieved. We present an easy method for inspecting a model for unit roots or near unit roots anywhere on the unit circle. We refer to this method as *overfitting*, and it is based on the following result given by Tiao and Tsay (1983).

Theorem 8.1 (Tiao–Tsay Theorem)

If X_t is ARUMA(p,d,q), that is, $\phi_s(B)\lambda(B)(X_t-\mu)=\theta(B)a_t$, then the model obtained using ordinary least squares (OLS) estimates of an AR($d+j$), $j>0$ fit to the data can be factored as $\hat{\phi}_{d+j}(B)=\hat{\phi}_d(B)\hat{\phi}_j(B)$, where $\hat{\phi}_d(B)$ is a consistent estimator of $\lambda(B)$.

The implication of the Tiao–Tsay Theorem (Theorem 8.1) is that if a "high-order" autoregressive model is fit to data from an ARUMA(p,d,q) model, then the features of the model associated with the roots on the unit circle will be well estimated in the "overfit" model. Gray and Woodward (1986) demonstrated the application of Theorem 8.1 for identifying nonstationarities using Burg estimates instead of OLS estimates. The procedure recommended here is to successively overfit data with AR(k) models. The factored forms of the resulting models are examined for near-nonstationary factors that tend to show in each of the overfit AR models. This procedure is illustrated in Example 8.4.

Example 8.4 Identifying Nonstationarities Using Overfitting

(a) Figure 8.3a shows a realization of length $n = 200$ from the near-nonstationary AR(3) model

$$(1 - 0.995B)(1 - 1.2B + 0.8B^2)X_t = a_t. \tag{8.12}$$

This realization was previously shown in Figure 5.3c. In Figure 8.3b we show the associated sample autocorrelations. The sample autocorrelations in Figure 8.3b are slowly damping, but as is often true in practice, especially for small realization lengths, they do not damp as slowly as the true autocorrelations in Figure 5.2c. The wandering or piecewise trending behavior in the realization and the slowly damping nature of the sample autocorrelations suggest nonstationarity or near nonstationarity and, in particular, suggest a factor $1 - \alpha_1 B$ for positive α_1, where $\alpha_1 \leq 1$ and $\alpha_1 \approx 1$. Table 8.1 shows factor tables for "overfit" AR(k), $k = 6$, 8, and 10 models fit to the realization in Figure 8.3a. It should be noted that, although the true model order is AR(3), in each of the overfit factor tables, we see a factor $1 - \alpha_1 B$, where α_1 is close to one. This illustrates the result in Theorem 8.1 and suggests the inclusion of a factor $1 - B$ in the model. The fact that a factor close to $1 - B$ shows up only once in each of the overfit models suggests that only one factor of $1 - B$ is appropriate. That is, the data should be differenced once to achieve stationarity.

NOTE: In this book, we show overfit factor tables based on Burg estimates. Final models use ML estimates.

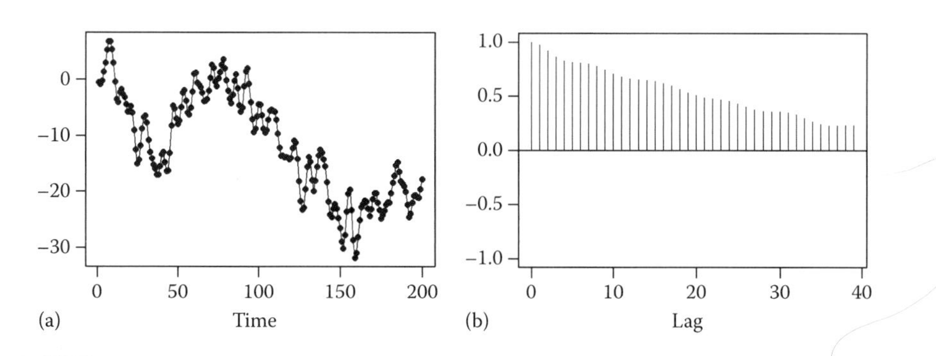

FIGURE 8.3
(a) Realization and (b) sample autocorrelations from the near nonstationary AR(3) model in model in (8.12).

TABLE 8.1

Factor Table for Overfit AR(k) Models Fit to the AR(3) Realization in Figure 8.3a

AR Factors	Roots (r_j)	$\lvert r_j^{-1} \rvert$	f_{0j}
$k = 6$			
$1 - 0.97B$	1.04	0.97	0
$1 - 0.98B + 0.63B^2$	$0.78 \pm 0.99i$	0.80	0.14
$1 + 0.58B + 0.21B^2$	$-1.40 \pm 1.71i$	0.46	0.36
$1 - 0.42B$	2.36	0.42	0
$k = 8$			
$1 - 0.95B$	1.05	0.95	0
$1 - 1.08B + 0.73B^2$	$0.74 \pm 0.91i$	0.85	0.14
$1 - 0.70B$	1.42	0.70	0
$1 + 0.23B + 0.34B^2$	$-0.35 \pm 1.67i$	0.59	0.28
$1 + 0.40B$	-2.49	0.40	0.5
$1 + 0.30B$	-3.28	0.30	0.5
$k = 10$			
$1 - 0.96B$	1.04	0.96	0
$1 - 1.06B + 0.74B^2$	$0.71 \pm 0.92i$	0.86	0.14
$1 + 0.26B + 0.47B^2$	$-0.27 \pm 1.44i$	0.68	0.28
$1 + 1.08B + 0.36B^2$	$-1.51 \pm 0.72i$	0.60	0.43
$1 - 1.13B + 0.33B^2$	$1.69 \pm 0.38i$	0.58	0.03
$1 + 0.014B$	-70.26	0.014	0.50

Using `GW-WINKS`: The "overfit" factor tables in Table 8.1 can be obtained by using the `Factoring Routines` option on the `Time Series Analysis` menu, requesting `Calculate Burg estimates`, and successively choosing the order p to be 6, 8, and 10.

(b) We next consider the near-nonstationary AR(3) model

$$(1 - 0.8B)(1 - 1.6B + 0.995B^2)X_t = a_t, \tag{8.13}$$

where $\sigma_a^2 = 1$. Model (8.13) is nearly nonstationary because of two complex roots close to the unit circle (associated with the factor $1 - 1.6B + 0.995B^2$). In Figure 8.4a, we show a realization of length $n = 200$ from this model. The sample autocorrelations in Figure 8.4b show a slowly damping sinusoidal behavior.

The strong cyclic behavior in the realization and the slowly damping sinusoidal sample autocorrelations are indications of nonstationarity or near-nonstationarity due to complex roots close to the unit circle. Table 8.2 shows factor tables for overfit autoregressive models of orders $k = 6$, 8,

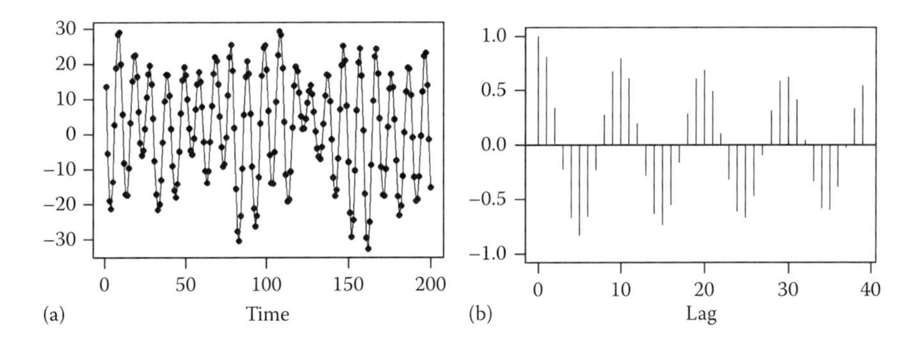

FIGURE 8.4
(a) Realization and (b) sample autocorrelations from the near nonstationary AR(3) model in model in (8.13).

TABLE 8.2

Factor Table for Overfit AR(k) Models Fit to the AR(3) Realization in Figure 8.4a

AR Factors	Roots (r_j)	$\lvert r_j^{-1} \rvert$	f_{0j}
$k = 6$			
$1 - 1.58 + 0.98B^2$	$0.81 \pm 0.61i$	0.99	0.10
$1 - 0.80B$	1.25	0.80	0
$1 + 0.37B$	−2.68	0.37	0.50
$1 - 0.33B$	3.01	0.33	0
$1 - 0.03B$	29.45	0.03	0
$k = 8$			
$1 - 1.58B + 0.98B^2$	$0.81 \pm 0.61i$	0.99	0.10
$1 - 0.80B$	1.24	0.80	0
$1 - 0.37B$	2.69	0.37	0
$1 + 0.62B + 0.11B^2$	$-2.94 \pm 0.94i$	0.32	0.45
$1 - 0.24B + 0.06B^2$	$2.07 \pm 3.62i$	0.24	0.17
$k = 10$			
$1 - 1.58B + 0.98B^2$	$0.81 \pm 0.61i$	0.99	0.10
$1 - 0.81B$	1.23	0.81	0
$1 + 1.21B + 0.44B^2$	$-1.37 \pm 0.63i$	0.66	0.43
$1 + 0.20B + 0.36B^2$	$-0.28 \pm 1.64i$	0.60	0.27
$1 - 0.85B + 0.34B^2$	$1.26 \pm 1.18i$	0.58	0.12
$1 - 0.55B$	1.82	0.55	0

and 10 fit to the realization in Figure 8.4a. It should be noted that, although the true model order is AR(3), in each of the overfit factor tables, we see a factor of $1 - 1.58B + 0.98B^2$. This again illustrates the result in Theorem 8.1 and suggests that a factor of about $1 - 1.58B + 0.98B^2$ should be included in the model.

NOTE: If it is concluded that a nonstationary factor is appropriate (i.e., $\phi_2 = -1$), then ϕ_1 should not be taken to be 1.58 but instead should be selected, so that the corresponding system frequency is $f = 0.10$. That is, using (3.63) with $\phi_2 = -1$, we select ϕ_1, so that

$$f = 0.1 = \frac{1}{2\pi} \cos^{-1}\left(\frac{\phi_1}{2}\right).$$

In this case, $\phi_1 = 2 \cos (0.2\pi) = 1.62$, so that the nonstationary factor is

$$1 - 1.62B + B^2 \tag{8.14}$$

in order to retain the system frequency $f = 0.1$ (i.e., cycles of length 10).

To summarize what we have learned from Example 8.4(a) and (b), the point is that the overfitting procedure applied to the realization from (8.12) identifies the possible need for differencing (one time) that is also suggested by the classical "Box–Jenkins" procedure. However, for data from model (8.13), the overfitting procedure also identifies analogous nonstationary-like behavior associated with complex roots on the unit circle.

8.3.3 Decision between a Stationary or a Nonstationary Model

For the two models considered in the previous example, it is important to note that AIC based on ML (and Burg/TT) estimates selects the correct order ($p = 3$) in both cases. However, the resulting model is stationary. *In order to select a nonstationary model, you must make a conscious decision that (a) such a model is indicated and (b) that such a model makes sense.* For example, if stock prices have a wandering behavior and other indicators suggest a possible root of $+1$, your decision whether to use of a factor $1 - B$ should be based on your assessment concerning whether there is or is not some sort of attraction to a mean level, where such an attraction would suggest the use of a stationary (near-nonstationary) model. However, it is also often the case that the model $X_t = X_{t-1} + a_t$ is physically more appealing than, for example, $X_t = 0.98X_{t-1} + a_t$.

8.3.4 Deriving a Final ARUMA Model

After we have been determined that one or more nonstationary factors should be included in the model, we are ready to perform the final two steps mentioned in the ARUMA modeling outline. Specifically, we transform the data to stationarity (Step (b)) and model the stationarized data with an ARMA(p,q) model (Step (c)) using model identification using techniques of Section 8.2 and model estimation discussed in Chapter 7. These two steps will be illustrated in Example 8.5 using the two data sets considered in Example 8.4.

Example 8.5 Final Models for the Data in Figures 8.3a and 8.4a

Suppose that for each of the realizations in Example 8.4, the decision is made to include a nonstationary factor in the model, that is, a factor of $1 - B$ for the realization in Figure 8.3a and a factor of $1 - 1.62B + B^2$ for the data in Figure 8.4a. We illustrate steps (b) and (c) of the ARUMA modeling outline.

(a) *Final Model for Data in Figure 8.3a*
Since we have concluded that a factor of $1 - B$ should be included in the model, the next step is to difference the data (i.e., transform using the operator $1 - B$) to obtain stationary data. Figure 8.5a shows this differenced data. That is, Figure 8.5a is a plot of length $n - 1 = 199$ of the data $y_t = x_t - x_{t-1}$, where x_t represents the data in Figure 8.3a. In Figure 8.5a, it should be noted that the data show the cyclic behavior produced by the stationary factor $1 - 1.2B + 0.8B^2$ much more prominently than it was seen in Figure 8.3a. The cyclic behavior was previously somewhat hidden by the wandering behavior induced by the root very near +1. Also, the sample autocorrelations in Figure 8.5b damp quite quickly suggesting a stationary model. That is, the

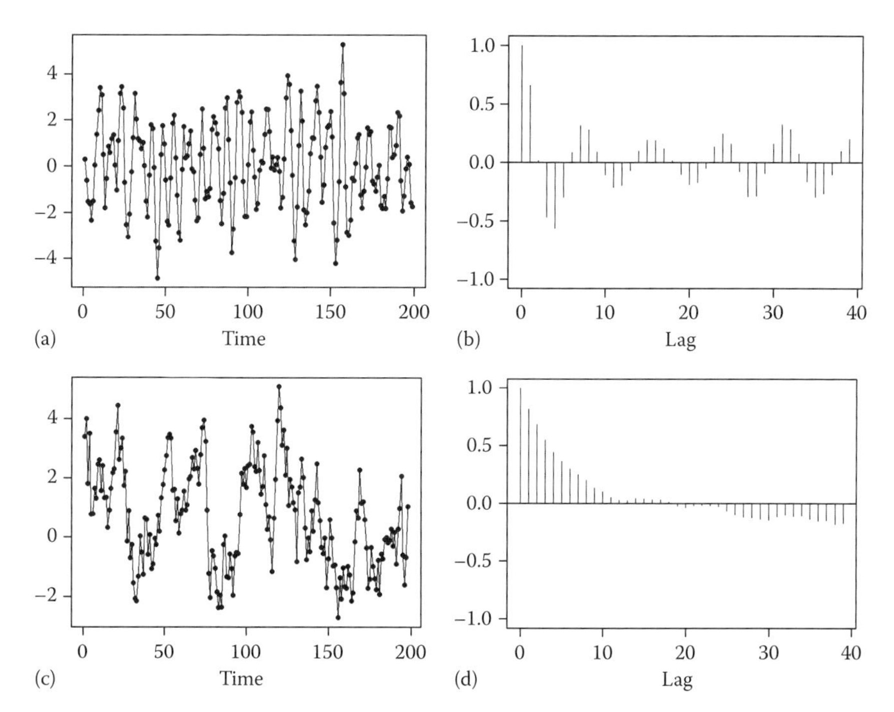

FIGURE 8.5
(a) Data in Figure 8.3a differenced. (b) Sample autocorrelations for the differenced data in (a). (c) Data in Figure 8.4a transformed by $1 - 1.62B + B^2$. (d) Sample autocorrelations for the transformed data in (c).

sample autocorrelation suggests that one difference is sufficient to produce stationarity.

Using AIC with $P=8$ and $Q=4$, an AR(2) is identified for the differenced data in Figure 8.5a. The AR(2) model fit to these data is given by

$$(1 - 1.16B + 0.75B^2)(Y_t - \overline{Y}) = a_t, \tag{8.15}$$

where $\hat{\sigma}_a^2 = 0.88$ and $\overline{Y} = -0.09$. Thus, a final model for the data in Figure 8.3a is given by

$$(1 - B)(1 - 1.16B + 0.75B^2)(X_t - \overline{X}) = a_t, \tag{8.16}$$

where $\hat{\sigma}_a^2 = 0.88$ and $\overline{X} = -11.6$. The model in (8.16) is quite close to the model from which the data were generated, namely

$$(1 - 0.995B)(1 - 1.2B + 0.8B^2)X_t = a_t, \tag{8.17}$$

with $\sigma_a^2 = 1$. Interestingly, the system frequency associated with the true second-order factor, $1 - 1.2B + 0.8B^2$, and with the estimated factor, $1 - 1.16B + 0.75B^2$, is $f = 0.13$ in both cases.

(b) *Final Model for Data in Figure 8.4a*

As we commented earlier, transforming the data in Figure 8.4a by the factor $1 - 1.62B + B^2$ is analogous to the classical "Box–Jenkins" procedure of differencing the data in Figure 8.3a. In Figure 8.5c, we plot a realization of length $n - 2 = 198$ of the data $y_t = x_t - 1.62x_{t-1} + x_{t-2}$, where x_t represents the data in Figure 8.4a. The data in Figure 8.5c display wandering behavior consistent with the factor $1 - 0.8B$ in the model that was previously barely visible in Figure 8.4a. The sample autocorrelations display damped exponential behavior for which the damping is quite rapid, suggesting stationary data.

Using AIC with $P=8$, and $Q=4$, we select an AR(1) model for the transformed data. The model for Y_t is

$$(1 - 0.82B)(Y_t - \overline{Y}) = a_t, \tag{8.18}$$

where $\overline{Y} = 0.5$, and the resulting model for X_t is

$$(1 - 1.62B + B^2)(1 - 0.82B)(X_t - \overline{X}) = a_t, \tag{8.19}$$

where $\overline{X} = 0.71$ and $\hat{\sigma}_a^2 = 0.93$. Again, this model is very similar to the model

$$(1 - 1.6B + 0.995B^2)(1 - 0.8B)X_t = a_t, \tag{8.20}$$

with $\sigma_a^2 = 1$ from which the data were generated.

Using GW-WINKS: The transformed data sets shown in Figure 8.5a and c can be obtained using the `Edit` menu in `GW-WINKS`. To obtain differenced data, select `New Variable, Common Transformations` on the `Edit` menu, drag down to the `by Differencing series` selection, select the data set to difference, and enter 1 in the entry box. `GW-WINKS` will create a new variable and place it in a new column in the data editor. To obtain the data in Figure 8.5c, select `New Variable, Common Transformations` on the `Edit` menu, drag down to `by Autoregressive Transformation`, select the data set to transform, and enter 1.62, −1 (include comma) in the entry box. Note that the differenced data in Figure 8.5a could also be obtained by entering 1 in the `Autoregressive Transformation` entry box. The final models in (8.16) and (8.19) are obtained by first differencing or transforming (using the `Edit` menu) and then analyzing the newly created series using standard techniques (i.e., using `AIC` to identify p and q and estimating the parameters of the resulting model). The estimates $\hat{\sigma}_a^2$ reported earlier are those that appear in the output from the parameter estimation on the transformed data sets.

8.3.5 More on the Identification of Nonstationary Components

In the preceding discussion concerning identification of nonstationary components in the model, we focused on nonstationary factors such as $1 - B$ or $1 - \phi_1 B + B^2$. In practice, we may encounter variations of these two types of nonstationarities. For example, the model may need to include one or more nonstationary factors of each type. We discuss techniques for modeling non stationary data in the following.

8.3.5.1 Including a Factor $(1 - B)^d$ *in the Model*

In Figure 8.6a, we show a realization from the model

$$(1 - B)^2(1 - 1.2B + 0.6B^2)X_t = a_t, \tag{8.21}$$

for which the characteristic equation has two roots of +1 and a pair of complex roots outside the unit circle associated with a system frequency of $f = 0.1$. The extreme wandering (random trending behavior) in Figure 8.6a and the slowly damping sample autocorrelations in Figure 8.6b indicate the need for differencing the data. The differenced data are shown in Figure 8.6c. The wandering behavior in Figure 8.6c and the slowly damping sample autocorrelations in Figure 8.6d again suggest that the differenced data should be differenced. In Figure 8.6e, we show the twice differenced data, that is, $(1 - B)^2 X_t$, along with the sample autocorrelations. In Figure 8.6e, the cyclic behavior associated with the factor $1 - 1.2B + 0.6B^2$ is visible, and the sample autocorrelations in Figure 8.6f have a damped sinusoidal behavior. It is

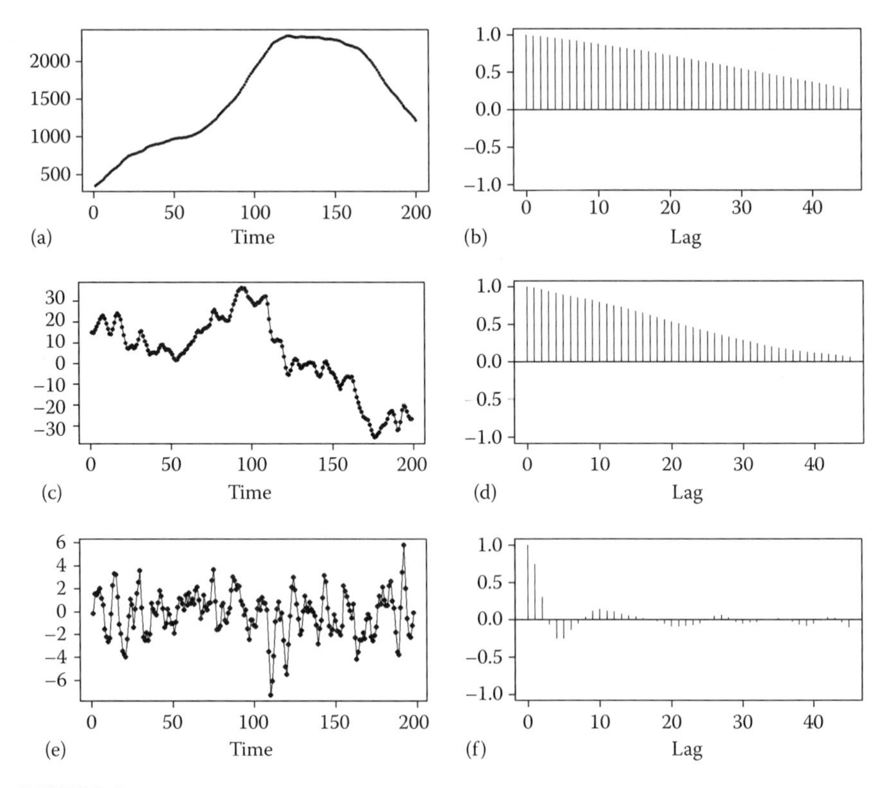

FIGURE 8.6
(a) Realization from model (8.21). (b) Sample autocorrelations for the realization in (a). (c) Data in (a) differenced. (d) Sample autocorrelation for data in (c). (e) Data in (a) twice differenced. (f) Sample autocorrelation for data in (e).

interesting to note that the cyclic behavior is not visible at all in the original data or the sample autocorrelations. In the once differenced data, the cyclic behavior begins to appear in the data but not in the sample autocorrelations. Also note that the sample autocorrelations of the original data and the once-differenced data have a similar (first-order) appearance. This is to be expected since according to the Findley–Quinn Theorem (Theorem 5.1), the limiting autocorrelations, ρ_k^*, for model (8.21), satisfy the same difference equation ($\rho_k^* - \rho_{k-1}^* = 1$) in each case.

Table 8.3 shows overfit AR(8) and AR(10) model fits for the data in Figure 8.6a. These tables suggest the inclusion of a factor $(1 - B)^2$ in the model, but this is not obvious at first glance. Notice in both factor tables in Table 8.3 that the first factor is associated with pair of complex root very close to the unit circle ($|r_i^{-1}| \approx 0.995$) and a system frequency of about 0.005 (and associated period of $1/0.005 = 200$). What should also be noted is that the two complex roots are very close to $+1$ (real part is close to $+1$ and imaginary part is quite small). Since the period associated with the system frequency is as long as the realization

TABLE 8.3

Factor Table for Overfit AR(k) Models Fit to the AR(4) Realization in Figure 8.6a

AR Factors	Roots (r_j)	$\lvert r_j^{-1} \rvert$	f_{0j}
$k = 8$			
$1 - 1.988B + 0.989B^2$	$1.005 \pm 0.03i$	0.995	0.005
$1 - 1.19B + 0.63B^2$	$0.95 \pm 0.83i$	0.79	0.11
$1 - 0.56B$	1.78	0.56	0
$1 + 0.48B$	−2.10	0.48	0.5
$1 + 0.15B + 0.22B^2$	$-0.33 \pm 2.09i$	0.47	0.28
$k = 10$			
$1 - 1.986B + 0.987B^2$	$1.006 \pm 0.03i$	0.994	0.005
$1 - 1.37B + 0.72B^2$	$0.94 \pm 0.70i$	0.85	0.10
$1 + 0.82B$	−1.23	0.82	0.5
$1 + 0.80B + 0.62B^2$	$-0.64 \pm 1.09i$	0.79	0.33
$1 - 0.62B + 0.60B^2$	$0.52 \pm 1.19i$	0.77	0.18
$1 - 0.77B$	1.30	0.77	0

length and since no periodicity is noticeable in the data, an appropriate interpretation is that the associated factor, for example ($1 - 1.988B + 0.989B^2$ for $k = 8$), is an estimate of $1 - 2B + B^2 = (1 - B)^2$. These factor tables are fundamentally different from the ones in Table 8.1, which consistently showed only one factor close to $1 - B$. Thus, the overfit factor tables in Table 8.3 immediately suggest the need for a factor $(1 - B)^2$ without first needing to difference the data to check whether nonstationary behavior remains.

Using `GW-WINKS`: The transformed data sets shown in Figure 8.6c and e can be obtained using the `Edit` menu in `GW-WINKS`. To obtain the differenced data in Figure 8.6a, select `New Variable, Common Transformations` on the `Edit` menu, drag down to the `by Differencing series` selection, select the data set to difference (i.e., the data shown in Figure 8.6a), and enter 1 in the entry box. `GW-WINKS` will create a new variable (differenced data) and place it in a new column in the data editor. To obtain the data in Figure 8.6e, repeat the preceding procedure to difference the new differenced data variable. Note that differencing twice as we have done here, that is, applying the transformation $(1 - B)^2$, is not the same as applying a second-order seasonal filter, by entering 2 in the selection box for differencing (which would transform the data by $1 - B^2$). The two differences resulting in $(1 - B)^2$ could be obtained in one step by transforming the series using $1 - 2B + B^2$, that is, by selecting `New Variable, Common Transformations` on the `Edit` menu, then choosing `by Autoregressive Transformation`, and entering the coefficients "`2, −1`" (without quotes).

8.3.5.2 Testing for a Unit Root

Before leaving this topic, we comment that formal tests are available (and are quite popular, especially in the econometrics literature) for testing for the existence of unit roots, i.e., factors of $(1-B)^d$. Consider the model

$$X_t - \mu = \phi_1(X_{t-1} - \mu) + \cdots + \phi_p(X_{t-p} - \mu) + a_t, \tag{8.22}$$

where $1 - \phi_1 z - \cdots - \phi_p z^p = 0$ may have one or more unit roots (+1). The model in (8.22) can be written as

$$\nabla X_t = \beta_0 + \tau X_{t-1} + \beta_1 \nabla X_{t-1} + \cdots + \beta_{p-1} \nabla X_{t-p+1} + a_t, \tag{8.23}$$

where $\nabla X_{t-j} = (1-B)X_{t-j} = X_{t-j} - X_{t-j-1}$, $\beta_0 = \mu(1 - \phi_1 - \cdots - \phi_p)$, $\tau = -(1 - \phi_1 - \cdots - \phi_p)$, and $\beta_j = -(\phi_{j+1} + \cdots + \phi_p)$, for $j = 1, \ldots, p-1$. The form in (8.23) is called the *error correction form*. (See Problem 8.10.) Note that there are $p-1$ terms involving the ∇X_{t-j} in (8.23). These are sometimes called the *augmenting* terms. Note that if $1 - \phi_1 z - \cdots - \phi_p z^p = 0$ has a unit root, then $1 - \phi_1 - \cdots - \phi_p = 0$, and also $\tau = 0$. As a consequence, tests for unit roots have been developed that test the hypothesis $H_0: \tau = 0$ vs. $H_a: \tau < 0$ in (8.23). The augmented Dickey–Fuller (tau) test (see Dickey and Fuller, 1979) is based on considering (8.23) to be a regression equation with dependent variable ΔX_t and independent variables $X_{t-1}, \nabla X_{t-1}, \ldots, \nabla X_{t-p+1}$. Using standard ordinary least squares regression (including an intercept term), the test statistic $t_{DF} = \hat{\tau}/se(\hat{\tau})$ is calculated, and $H_0: \tau = 0$ is rejected if $t_{DF} < C$ for a critical value C. Dickey and Fuller (1979) show that under the null hypothesis $H_0: \tau = 0$, the sampling distribution of the test statistic, t_{DF}, does not follow the usual t-distribution and that its limiting distribution is skewed to the left. The α-level critical values, $t^*_{DF}(\alpha)$, based on the limiting distribution are −2.57, −2.86, and −3.43 for $\alpha = 0.1$, 0.05, and 0.01, respectively (Dickey, 1976). It should be noted that in practice, the value of p is chosen sufficiently large, so that the residuals, $\hat{a}_t$, appear to be uncorrelated.

Testing for a factor $(1-B)^d$, that is, d unit roots, requires application of the procedure d times. For example, if the null hypothesis of a unit root is not rejected, then a second unit root test could be performed by letting $Y_t = (1-B)X_t$, considering the model

$$\nabla Y_t = \alpha_0 + \tau Y_{t-1} + \beta_1 \nabla Y_{t-1} + \cdots + \beta_{p-1} \nabla Y_{t-p+1} + a_t,$$

and testing $H_0: \tau = 0$ in this new model.

Example 8.6 Testing the Data in Figure 8.6 for Unit Roots

We apply the augmented Dickey–Fuller (tau) test to the realization from (8.21) shown in Figure 8.6a. Note that this model has two unit roots and a second-order stationary component. Using SAS `PROC ARIMA`, we apply the augmented Dickey–Fuller test (letting $p=4$) and obtain $t_{DF}=-2.03$, which has a p-value 0.27. Thus, the null hypothesis is not rejected, and we conclude that the model has a unit root. Based on Figure 8.6c, it is reasonable to consider a second unit root. Applying the augmented Dickey–Fuller test to the differenced data (this time using $p=3$), we obtain $t_{DF}=-0.50$ with p-value 0.89. Thus, again, we do not reject the null hypothesis of a unit root. Taking a second difference, we apply the Dickey–Fuller test to the data in Figure 8.6e using $p=2$. This realization shows no indication of a unit root, and, in this case $t_{DF}=-7.96$ with p-value <0.0001, which strongly rejects a third unit root.

It should be noted that SAS `PROC ARIMA` gives three versions of the ADF test. The "zero mean" version is based on model (8.23) assuming $\beta_0=0$, by applying least squares regression is applied in the "zero intercept" form. The version of the test described earlier is the "single mean" version, which assumes a constant mean that may be nonzero. The third form of the test is based on a model of the form (8.23) but with β_0 replaced by $\beta_0+\beta_1 t$, allowing for the possibility that some of the trending behavior in the data is caused by an underlying trend line.

A final comment should be made about unit root tests. While these tests give the investigator a "yes-or-no" answer regarding a unit root, there are areas of concern regarding their use. Note that the decision to retain a unit root is based on failure to reject a null hypothesis, and depending on the power of the test, the conclusion that the model has a unit root may not be one that can be made with much confidence. Unit root tests can be useful tools to assist in the decision, but the final decision concerning the inclusion of nonstationary factors rests with the investigator. The point is that finite realizations from the nonstationary model $(1-B)X_t=a_t$ and, for example, the stationary (but nearly nonstationary) model $(1-0.97B)X_t=a_t$ will be very similar, and tests such as the augmented Dickey–Fuller test will often conclude the existence of a unit root when applied to realizations from the aforementioned nearly nonstationary model. The final decision concerning whether to include a factor $1-B$ should be based on such considerations as whether there is any reason to believe that there is an attraction to the mean, and a decay in the autocorrelations at lag lengths of interest (in which case a stationary model is appropriate). A unit root model is often used to model the behavior of stock market prices, where such an attraction to the mean does not seem to exist.

8.3.5.3 Including a Seasonal Factor $(1-B^s)$ *in the Model*

In Section 5.3, we discussed the multiplicative seasonal model. Here, we consider the special case

$$(1-B^s)\phi(B)(X_t-\mu)=\theta(B)a_t. \tag{8.24}$$

In these models, the s roots of $1-z^s=0$ all lie on the unit circle, and the seasonal factor, $1-B^s$, models a tendency in the data for the patterns observed at times

$t=1,2,\ldots,s$ to be similar to the patterns for observations $t=s+1,s+2,\ldots,2s$, etc. For example, quarterly sales data may exhibit a seasonal behavior if the general pattern of sales in each quarter (e.g., relatively lower sales in first quarter, higher sales in fourth quarter, etc.) tends to be repeated from year to year. As mentioned in Section 5.3, it is typical to use a seasonal model with $s=4$ for quarterly data exhibiting such behavior or $s=12$ for monthly data with this type of pattern. In Example 5.7, we discussed the "airline model"

$$(1-B)(1-B^s)\phi(B)(X_t-\mu)=\theta(B)a_t, \tag{8.25}$$

as a model that allows for seasonal and trending behavior. However, a seasonal component may or may not be appropriate for data collected monthly, quarterly, etc. In the following examples, we discuss techniques that help you determine whether a seasonal component is appropriate.

Example 8.7 Simulated Seasonal Data

Figure 8.7 shows a realization and sample autocorrelations from the seasonal model

$$(1-B^{12})(1-1.25B+0.9B^2)(X_t-50)=a_t, \tag{8.26}$$

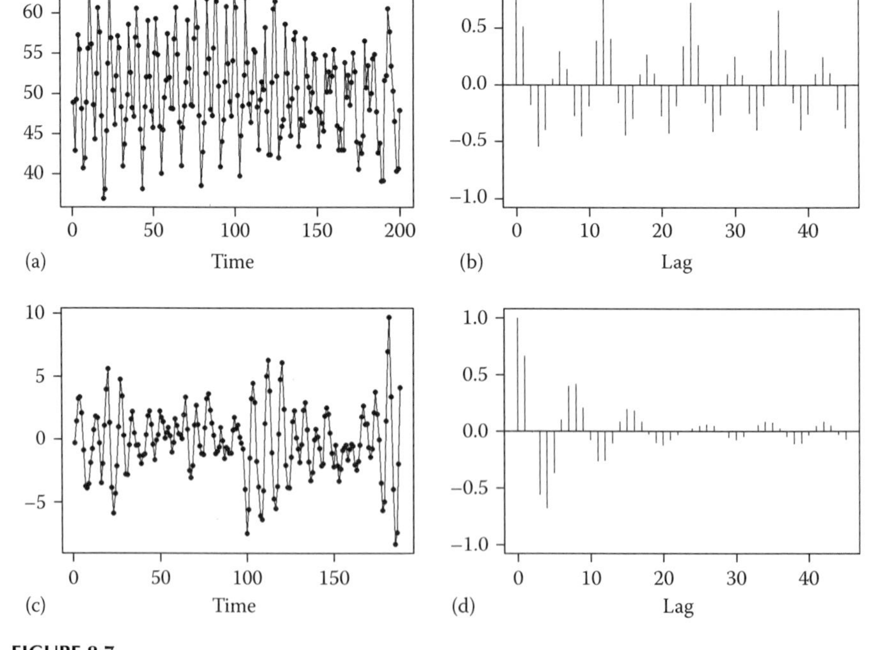

FIGURE 8.7
(a) Seasonal data from model in (8.26). (b) Sample autocorrelations of the data in (a). (c) Plot of (a) transformed by $1-B^{12}$. (d) Sample autocorrelations of transformed data in (c).

TABLE 8.4

Factor Table for AR(16) Model Fit to the Realization in Figure 8.7a

AR Factors	Roots (r_j)	$\lvert r_j^{-1} \rvert$	f_{0j}
$1 + 1.73B + 0.998B^2$	$-0.87 \pm 0.50i$	0.999	0.417
$1 + 1.01B + 0.997B^2$	$-0.50 \pm 0.87i$	0.998	0.334
$1 + 0.998B$	-1.002	0.998	0.5
$1 - 0.01B + 0.995B^2$	$0.01 \pm 1.00i$	0.997	0.249
$1 - B + 0.993B^2$	$0.50 \pm 0.87i$	0.997	0.167
$1 - 1.71B + 0.984B^2$	$0.87 \pm 0.51i$	0.992	0.084
$1 - 0.988B$	1.012	0.988	0
$1 - 1.27B + 0.855B^2$	$0.74 \pm 0.79i$	0.925	0.130
$1 + 0.164B$	-6.109	0.164	0.5
$1 - 0.152B$	6.583	0.152	0

with white noise variance $\sigma_a^2 = 1$. Inspection of the realization shows that a 12-unit seasonal pattern seems to be replicated, and the sample autocorrelations show spikes at multiples of 12. To further examine the need for the nonstationary seasonal component, $1 - B^{12}$, in the fitted model, we use the overfitting procedure used previously. In order to decide whether to include a seasonal factor, $1 - B^{12}$, we overfit using AR(p) models with $p = 16, 18$, and 20. Table 8.4 shows the factor table for $p = 16$, which is similar to those for the other two overfit orders considered. This factor table should be compared to the factors of $1 - B^{12}$ shown in Table 5.5.

By examining the factors associated with the first 7 rows of the factor table in Table 8.4, it can be seen that these factors are very similar to those of $1 - B^{12}$ in Table 5.5, and the associated roots are all very close to the unit circle. Because of the solid indication in Table 8.4 that a factor of $1 - B^{12}$ is appropriate, the next step is to transform the data by $1 - B^{12}$, i.e., calculate $Y_t = X_t - X_{t-12}$, and find an appropriate model for the transformed data. The transformed data are shown in Figure 8.7c which, according to the model in (8.26), should have cyclic behavior with a frequency of about $f = 0.14$ associated with the remaining non-seasonal second order factor, $1 - 1.25B + 0.9B^2$. Examination of Figure 8.7c and the associated sample autocorrelations in Figure 8.7d shows the expected behaviors, i.e., pseudo periodic behavior in the data and damped sinusoidal sample autocorrelations, both with cycle lengths about 7 (=1/0.14).

AIC with $P = 8$ and $Q = 4$ applied to the transformed data in Figure 8.7c selects an AR(2). Estimating the parameters of this AR(2) model using ML estimates we obtain the AR(2) model $(1 - 1.26B + 0.86B^2)(Y_t - \overline{Y}) = a_t$ with $\hat{\sigma}_a^2 = 1.21$. Thus, the final model is

$$(1 - B^{12})(1 - 1.26B + 0.86B^2)(X_t - 50.3) = a_t, \qquad (8.27)$$

with $\hat{\sigma}_a^2 = 1.21$, which is very similar to the model in (8.26).

Using GW-WINKS: The seasonally differenced data in Figure 8.7c were obtained via the `Edit` menu by choosing `New Variable, Common Transformations`, dragging down to the `by Differencing series` selection, choosing the data set to difference (i.e., the data shown in Figure 8.7a), and entering `12` in the entry box. `GW-WINKS` will create a new variable (the seasonally differenced data showing in Figure 8.7c), and place it in a new column in the data editor.

Example 8.8 Airline Data

Figure 8.8 shows the log airline data (number of international airline passengers) previously shown in Figure 1.25a and discussed in Examples 5.7 and 6.7. It has been noted that these data have a seasonal pattern with high airline travel in November and December, low in January and February, etc., and that the number of passengers has increased through time. In Examples 5.7 and 6.7 we have modeled these data using the airline model.

$$(1 - B)(1 - B^{12})\phi(B)(X_t - \mu) = \theta(B)a_t. \tag{8.28}$$

The model in (8.28) has two factors of $1 - B$ which model the trending behavior in the data and the forecasts. However, instead of using a model

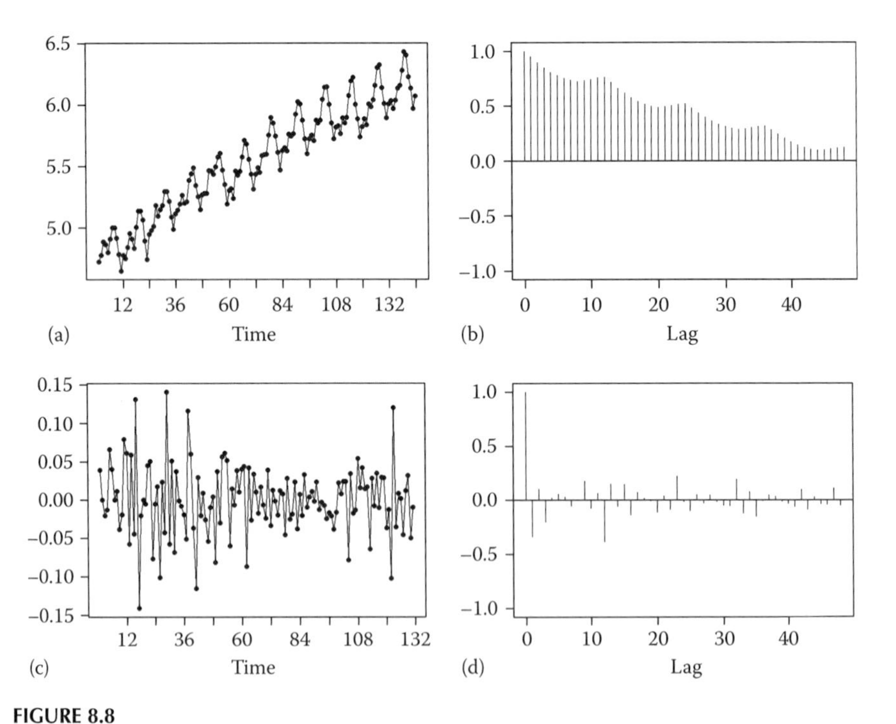

FIGURE 8.8
(a) Log airline data. (b) Sample autocorrelations for data in (a). (c) Log airline data transformed by $(1 - B)(1 - B^{12})$. (d) Sample autocorrelations for data in (c).

TABLE 8.5

Factor Table for AR(16) Model Fit to the Log Airline Data

AR Factors	Roots (r_j)	$\lvert r_j^{-1}\rvert$	f_{0j}
$1 - 1.73B + 0.996B^2$	$0.87 \pm 0.50i$	0.998	0.084
$1 - 0.99B + 0.993B^2$	$0.50 \pm 0.87i$	0.996	0.168
$1 + 0.99B + 0.989B^2$	$-0.50 \pm 0.87i$	0.995	0.333
$1 - 1.98B + 0.981B^2$	$1.01 \pm 0.02i$	0.990	0.004
$1 - 0.03B + 0.974B^2$	$0.01 \pm 1.01i$	0.987	0.248
$1 + 1.69B + 0.962B^2$	$-0.88 \pm 0.52i$	0.981	0.416
$1 + 1.70B + 0.745B^2$	$-1.14 \pm 0.20i$	0.863	0.472
$1 - 0.37B + 0.307B^2$	$0.61 \pm 1.70i$	0.554	0.196

such as (8.28) just because it seems reasonable, we examine factor tables of overfit AR models in order to determine whether the factors $(1 - B)(1 - B^{12})$ are present in the data. In Table 8.5 we show the factor table for an AR(16) fit to the data. Comparison of the factors in Table 8.4 with those in Table 5.5 shows that the AR(16) model has factors very close to $1 - \sqrt{3}B + B^2$, $1 - B + B^2$, $1 + B + B^2$, $1 + B^2$, and $1 + \sqrt{3}B + B^2$. The two factors of $1 - B$ are represented by the factor $1 - 1.98B + 0.981B^2$ which is very close to $(1 - B)^2 = 1 - 2B + B^2$. Note in Table 8.5 that this factor is associated with complex conjugate roots with real part close to +1 and complex part close to zero. Additionally, the system frequency associated with this factor is very small (0.004). The only factor in Table 5.5 that does not seem to appear in the AR(16) overfit model is $1 + B$. Table 8.5 has a factor of $1 + 1.70B + 0.745B^2$ which has high frequency behavior (0.472) and complex conjugate roots with real part close to −1. We conclude that $(1 - B)(1 - B^{12})$ is appropriate as the nonstationary component of the final ARUMA(p,d,q) model fit to the data.

The next step is to transform the data by $(1 - B)(1 - B^{12})$ which can be accomplished by differencing the data to obtain $Y_t = X_t - X_{t-1}$ and then applying the transformation $Z_t = Y_t - Y_{t-12}$ to the differenced data. This data set, Z_t, (which is shown in Figure 8.8c) has $144 - 13 = 131$ data values and should then be modeled as a stationary process. The autocorrelations of Z_t are shown in Figure 8.8d, and these are consistent with the autocorrelations of a stationary process. Using AIC with $P = 16$ and $Q = 4$, an ARMA(12,1) and an AR(13) are the top two models. Using the ARMA(12,1) fit to the transformed data we obtain the model

$$(1 - B)(1 - B^{12})\phi(B)(X_t - 5.54) = \theta(B)a_t, \tag{8.29}$$

where

$$\phi(B) = 1 + 0.05B - 0.07B^2 + 0.11B^3 + 0.04B^4 - 0.08B^5 - 0.07B^6 + 0.04B^7 - 0.04B^8 - 0.14B^9 + 0.05B^{10} + 0.12B^{11} + 0.40B^{12},$$

$\theta(B) = 1 - 0.42B$, and $\hat{\sigma}_a^2 = 0.0013$. A more parsimonious model might be of the form $\phi(B) = 1 - \phi_{12}B^{12}$, but this will not be pursued here. The model using the AR(13) is given by Gray and Woodward (1981).

Using `GW-WINKS`: The data in Figure 8.8c were obtained via the `Edit` menu by first obtaining a 12th order seasonal difference using the procedure outlined in Example 8.7. Then apply the transformation $1 - B$ to the new 12th order seasonally differenced data using the differencing procedure described in Example 8.5. This will create a new variable of length $n - 12 - 1 = 131$ containing the data plotted in Figure 8.8c. The stationary ARMA(12,1) model was obtained via an analysis of this new data set. Note that the order of the differencing operations is not important. That is, you could first difference the log airline data and then take a 12th order seasonal difference of the differenced data.

Example 8.9 Modeling the Pennsylvania Monthly Temperature Data

In Figure 8.9a, we show the average monthly temperatures (°F) for the state of Pennsylvania from January, 1990 through December, 2004 that were

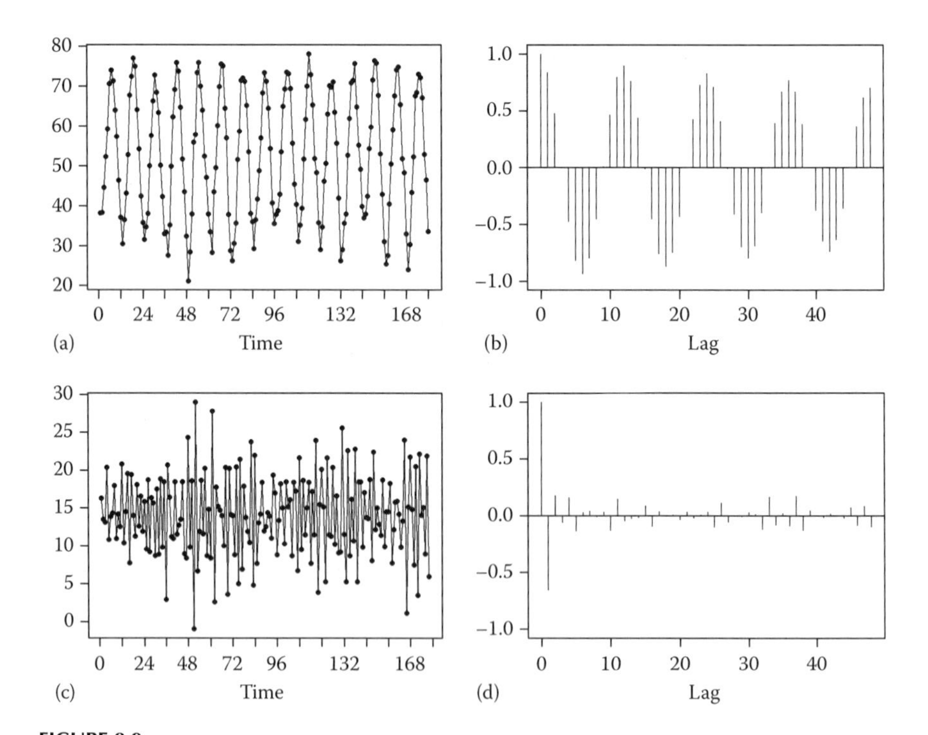

FIGURE 8.9
(a) Pennsylvania monthly temperature data (January 1990–December 2004). (b) sample autocorrelations for the realization in (a). (c) Data in (a) transformed by $1 - 1.732B + B^2$. (d) Sample autocorrelation of data in (c).

TABLE 8.6

Factor Table for AR(16) Model Fit to Pennsylvania Temperature Data

AR Factors	Roots (r_j)	$\lvert r_j^{-1} \rvert$	f_{0j}
$1 - 1.73B + 0.998B^2$	$0.87 \pm 0.50i$	0.999	0.084
$1 + 0.27B + 0.856B^2$	$-0.16 \pm 1.07i$	0.925	0.273
$1 + 1.01B + 0.808B^2$	$-0.62 \pm 0.92i$	0.899	0.345
$1 - 1.02B + 0.788B^2$	$0.65 \pm 0.92i$	0.888	0.152
$1 + 1.73B + 0.785B^2$	$-1.10 \pm 0.24i$	0.886	0.467
$1 - 1.72B + 0.754B^2$	$1.14 \pm 0.15i$	0.868	0.021
$1 - 0.39B + 0.687B^2$	$0.28 \pm 1.17i$	0.829	0.212
$1 + 1.41B + 0.683B^2$	$-1.03 \pm 0.64i$	0.827	0.412

previously plotted in Figure 1.1b. In Figure 8.9b, we show the sample autocorrelations. Since these data are monthly and (as expected) show a distinct seasonal pattern, we consider a model of the form

$$(1 - B^{12})\phi(B)(X_t - \mu) = \theta(B)a_t.$$

As with the other potential seasonal data sets, we examine factor tables associated with high order autoregressive fits to the data (with $p > 12$). Table 8.6 shows a factor table associated with an AR(16) fit to these data. The features of this table are similar to factor tables for other values of p.

In Table 8.6, it can be seen that only the pair roots associated with $f_0 = 0.084$ $(=1/12)$ is close to the unit circle. That is, the data are not seasonal but instead have a cyclic nature with a period of 12. Examining the data we see that the temperature readings move from winter to summer and back to winter in a fairly sinusoidal pattern. Consequently, the seasonal behavior can be modeled using an ARUMA(p,d,q) model with one nonstationary factor, $1 - \sqrt{3}B + B^2$. Note that the seasonal behavior in the airline data follows an irregular pattern. For example, air travel is highest in November and December followed by a drop to an annual low in March with an increased travel in the summer, slight drop in early fall, etc. Such a pattern is not sinusoidal and is best modeled including $1 - B^{12}$ since, ignoring the steady increase from year to year, the dominant factor influencing air travel in a given month is the air travel in the previous year during that month. Note that the factor $1 - \sqrt{3}B + B^2$ is one of the factors of $1 - B^{12}$, but for the Pennsylvania temperature data it is the only factor needed to model the periodic behavior.

Completing the analysis we transform the temperature data to obtain $Y_t = X_t - 1.732X_{t-1} + X_{t-2}$. The transformed realization is shown in Figure 8.9c and the sample autocorrelations are shown in Figure 8.9d. These data and sample autocorrelations suggest a stationary model, so we used AIC to fit an ARMA(p,q) model to Y_t. Using AIC with $P = 15$ and $Q = 5$, AIC selects an ARMA(10,4) as the first choice with an MA(4) being a close second. For the sake of parsimony, we select the MA(4) model. The final model is

$$\begin{aligned}(1 - 1.732B + B^2)(X_t - 52.63) &= a_t - 1.23a_{t-1} + 0.52a_{t-2} + 0.05a_{t-3} + 0.13a_{t-4} \\ &= (1 - 1.55B + 0.87B^2)(1 + 0.32B + 0.15B^2)a_t \end{aligned} \tag{8.30}$$

with $\hat{\sigma}_a^2 = 9.42$. One point of interest is that the first factor of the MA component in (8.30) has a system frequency, $f_0 = 0.094$, which is close to the dominant system frequency of the AR part, i.e., $f_0 = 1/12$. The discussion in Example 4.1 suggests that a harmonic component model is a possible model for the temperature data. In fact, the harmonic model may in fact be more appropriate due to the stability of the 12-month period in this type of temperature data.

Example 8.10 ARUMA Spectral Estimation and Hidden Peaks

Consider the ARUMA(4,1,0) model

$$(1 - B)(1 - 1.79B + 1.75B^2 - 1.61B^3 + 0.765B^4)X_t = a_t. \tag{8.31}$$

The factor table for this model is given in Table 8.7 where it can be seen that the model has a root of +1 and two pairs of complex conjugate roots outside the unit circle associated with system frequencies $f = 0.25$ and 0.04.

Figure 8.10a shows a realization from this model and Figure 8.10b shows the sample autocorrelations. We consider the problem of modeling the realization in Figure 8.10a (not knowing the true model) and finding the associated spectral density estimator. The wandering nature of the realization and the slow exponential damping in the sample autocorrelations suggest a model with a unit root. Overfitting AR(p) models of order $p = 10$, 15, and 20 shows a consistent root close to +1. Consequently, we continue the modeling process by differencing the data. The differenced data, Y_t, are shown in Figure 8.11a and the associated sample autocorrelations are shown in Figure 8.11b.

TABLE 8.7

Factor Table for the ARUMA(4,1,0) Model in (8.31)

AR Factors	Roots (r_j)	$\lvert r_j^{-1} \rvert$	f_{0j}
$1 - B$	1	1	0.0
$1 + 0.9B^2$	$\pm 1.05i$	0.95	0.25
$1 - 1.79B + 0.85B^2$	$1.05 \pm 0.26i$	0.92	0.04

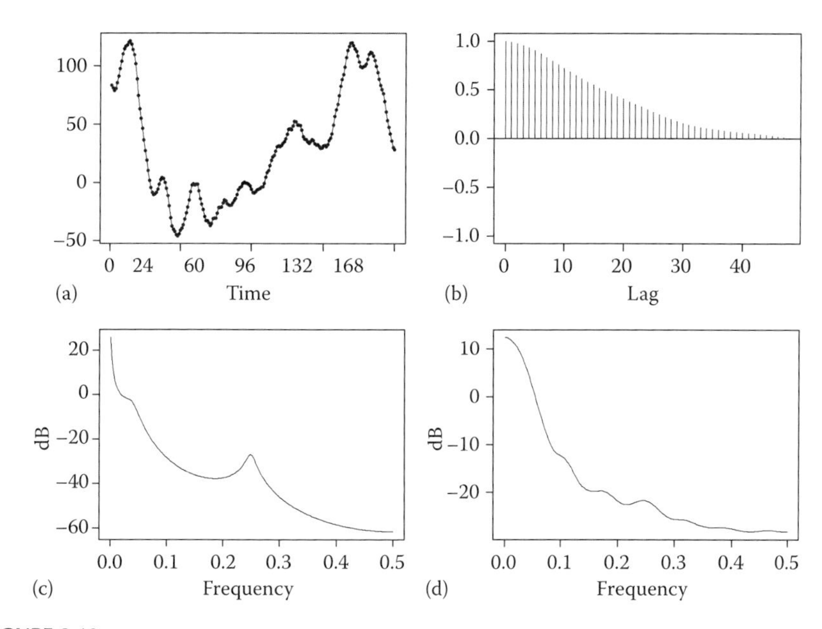

FIGURE 8.10
(a) Realization from the ARUMA(4,1,0) model in (8.31). (b) Sample autocorrelations for the realization in (a). (c) ARUMA(4,1,0) spectral density estimate for (a). (d) Parzen spectral density estimate for (a).

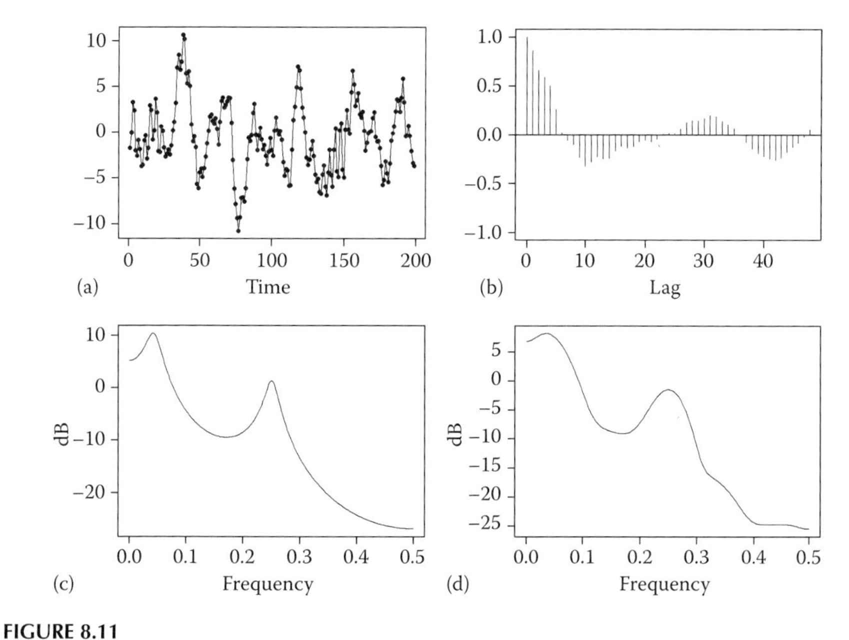

FIGURE 8.11
(a) Realization in Figure 8.10a differenced. (b) Sample autocorrelations for the differenced data in (a). (c) AR(4) spectral density estimate for (a). (d) Parzen spectral density estimate for (a).

TABLE 8.8

Factor Table for the ARUMA(4,1,0) Model in (8.32)

AR Factors	Roots (r_j)	$\lvert r_j^{-1} \rvert$	f_{0j}
$1 - B$	1	1	0.0
$1 + 0.02B + 0.874B^2$	$-0.01 \pm 1.07i$	0.94	0.25
$1 - 1.79B + 0.847B^2$	$1.04 \pm 0.30i$	0.92	0.045

Both plots suggest a stationary model. Using AIC with $P = 8$ and $Q = 4$, an AR (4) model is selected and the fitted model is

$$(1 - B)(1 - 1.75B + 1.69B^2 - 1.53B^3 + 0.74B^4)X_t = a_t. \tag{8.32}$$

The factor table associated with the fitted model is given in (8.32) is given in Table 8.8.

As suggested in Section 7.8, the associated ARUMA(4,1,0) spectral density estimate is plotted using the ARMA(5,0) spectral density estimate associated with a nearly nonstationary model such as

$$(1 - 0.995B)(1 + 0.9B^2)(1 - 1.79B + 0.85B^2)X_t = a_t. \tag{8.33}$$

The resulting ARUMA spectral density estimate is shown in Figure 8.10c, and in Figure 8.10d we show the corresponding Parzen spectral density estimator. Notice that in the ARUMA spectral density estimate there is only a somewhat weak indication of the system frequency at $f = 0.04$ while in the Parzen spectral density estimate there is no indication at all. This illustrates the important fact that frequency information is often more visible in the factor table than it is in the spectral density. See Gray and Woodward (1986, 1987).

The problem is that the peak in the spectral density at the frequency associated with the nonstationary factor (in this case $f = 0$) dominates the contribution in the spectral density associated with frequencies nearby (in this case $f = 0.04$). One technique to more clearly see the more subtle features of the spectral density is to transform the data to remove the nonstationarity, and find the spectral density of the transformed data. Figure 8.11c shows the AR(4) spectral density estimate for the stationary model $(1 - 1.75B + 1.69B^2 - 1.53B^3 + 0.74B^4)Y_t = a_t$ fit to the differenced data. Figure 8.11d shows the associated Parzen spectral density estimate. It can be seen that the AR(4) spectral density and the Parzen spectral density both show a peak at about $f = 0.04$. We repeat the fact that the "hidden peak" at $f = 0.04$ shows up clearly in the factor table of the fitted model shown in Table 8.8.

Using `GW-WINKS`: The approximate ARUMA(4,1,0) spectral density estimate shown in Figure 8.10c (which is based on the near nonstationary AR(5) model in (8.33)) can be obtained in `GW-WINKS` by using the `Generate`

`Realization` option on the `Time Series Analysis` menu and specifying the factors of the model in (8.33). (The white noise variance and sample size are irrelevant.) The data in the column `TRUE_SPECTRUM` contain the desired "approximate" ARUMA(4,1,0) spectral density estimate associated with the fitted model.

8.A Appendix: Model Identification Based on Pattern Recognition

In this appendix, we discuss techniques for identifying model orders on the basis of pattern recognition. We have already discussed one example of this technique for white noise checking. In Section 8.1, we examined sample autocorrelations with respect to whether they behaved consistently with the theoretical behavior for a white noise process, i.e., $\rho_k = 0$, $k \neq 0$. In this appendix, we will discuss the classical Box–Jenkins procedure of looking for patterns in autocorrelations and partial autocorrelations. We then discuss the generalized partial autocorrelation introduced by Woodward and Gray (1981) which is a natural generalization of the Box–Jenkins approach.

8.A.1 Patterns in the Autocorrelations

Consider the autocorrelations of an MA(q) process which from (3.4) have the form

$$\begin{aligned} \rho_k &= 0 \quad \text{for } k > q \\ &\neq 0 \quad \text{for } k = q. \end{aligned} \tag{8.A.1}$$

The behavior described in (8.A.1) uniquely determines the order q of an MA(q) process if the true autocorrelations are known. Thus, sample autocorrelations that are near zero for $k > q$ while $\hat{\rho}_q \neq 0$ suggest an MA(q) as a possible model. From Bartlett's (1946) approximation (1.22) it follows that if X_t is an MA(q) process, then

$$\text{Var}(\hat{\rho}_k) \approx \frac{1}{n}\left\{1 + 2\sum_{\nu=1}^{q} \rho_\nu^2\right\}$$

for $k > q$. Further, since $E[\hat{\rho}_k] \approx 0$ for $k > q$, then a rough guide is that if

$$|\hat{\rho}_k| > 2\sqrt{\frac{1 + 2\sum_{\nu=1}^{q} \hat{\rho}_\nu^2}{n}}, \tag{8.A.2}$$

for some $k > q$, then this is evidence against an MA(q) at the $\alpha = 0.05$ level of significance. Thus, if an MA(q) is to be considered as the model, then $\hat{\rho}_k$ should essentially remain within the limits specified by (8.A.2) for $k > q$. However, as with the test for white noise, the quoted significance level applies separately for each k so that it will not be unusual to observe a small percentage of $\hat{\rho}_k$'s, $k \geq q$, outside the specified limits when the model is an MA(q). It should be noted that when X_t is white noise, i.e., X_t is MA(q) with $q = 0$, then the summation in (8.A.2) is taken to be zero, and (8.A.2) reduces to (8.1).

Example 8.A.1 Model Identification Based on Sample Autocorrelations

Figure 8.A.1a shows a realization of length $n = 100$ from the MA(3) process

$$X_t = a_t - 0.7a_{t-1} - 0.54a_{t-2} + 0.81a_{t-3}. \tag{8.A.3}$$

Figure 8.A.1b is a plot of the true autocorrelations, and Figure 8.A.1c shows the sample autocorrelations with limits obtained based on (8.A.2). There it can be seen that $\hat{\rho}_k$, $k = 1, 2, 3$ are similar to the theoretical quantities, and $\hat{\rho}_k$, $k > 3$ are near zero. The limits associated with lag k are plotted at

$$\pm 2\sqrt{\frac{1 + 2\sum_{\nu=1}^{k} \hat{\rho}_\nu^2}{n}}, \tag{8.A.4}$$

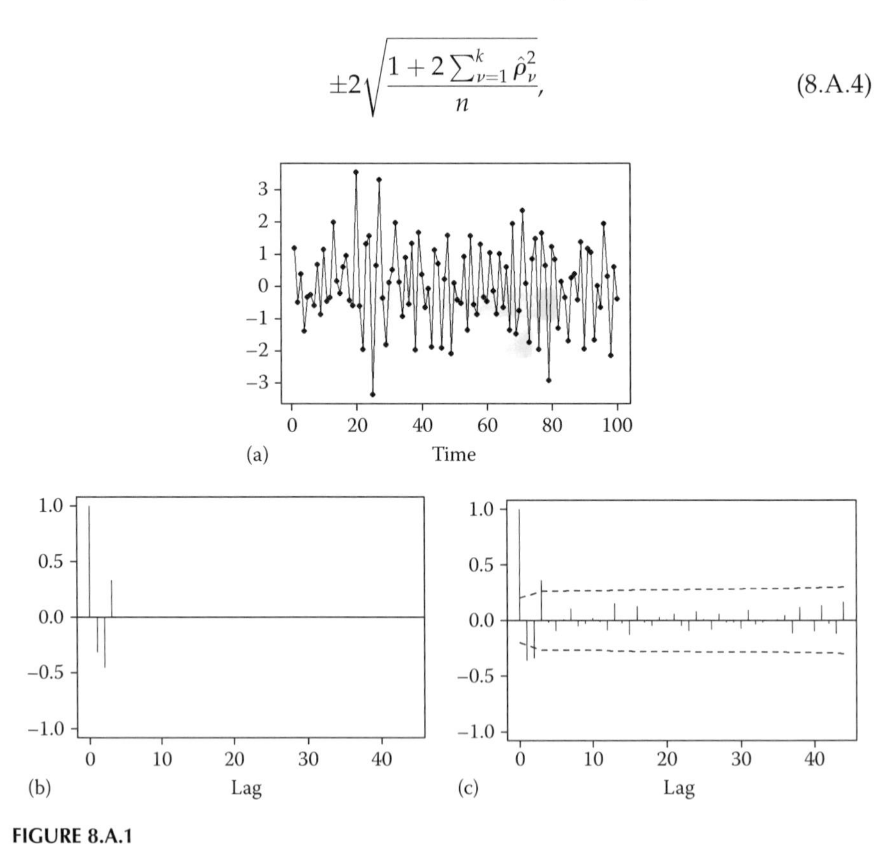

FIGURE 8.A.1
(a) Realization from the MA(3) model in (8.A.3), (b) true autocorrelations from the model in (8.A.3), and (c) sample autocorrelations with MA(q) limits.

i.e., these are the limits based on (8.A.2) where $q = k$. Thus, at each lag k, the limits shown are those for checking $\rho_j = 0$ for a given $j > k$. To test for an MA(k) model, one would essentially compare $\hat{\rho}_j$, $j > k$, to the limits in (8.A.4) associated with an MA(k). We see that the sample autocorrelations, $\hat{\rho}_j$, stay within these limits for $j > 3$, and $\hat{\rho}_3$ is outside the limits. Thus, an MA(3) seems to be an appropriate model.

Example 8.A.2 The Need for Caution When Using Autocorrelations for Model Identification

Care needs to be exercised before an MA(q) model is selected in the manner suggested above. In Figure 8.A.2c we plot sample autocorrelations, and it should be noted that they fall within the limit lines for $j > 2$ suggesting an MA(2). However, the autocorrelations are actually based on the realization of length $n = 150$ shown in Figure 8.A.2a from the AR(1) model

$$X_t - 0.65X_{t-1} = a_t. \tag{8.A.5}$$

Examination of Figure 8.A.2c reveals that the sample autocorrelations tend to damp exponentially toward zero as would be expected for an

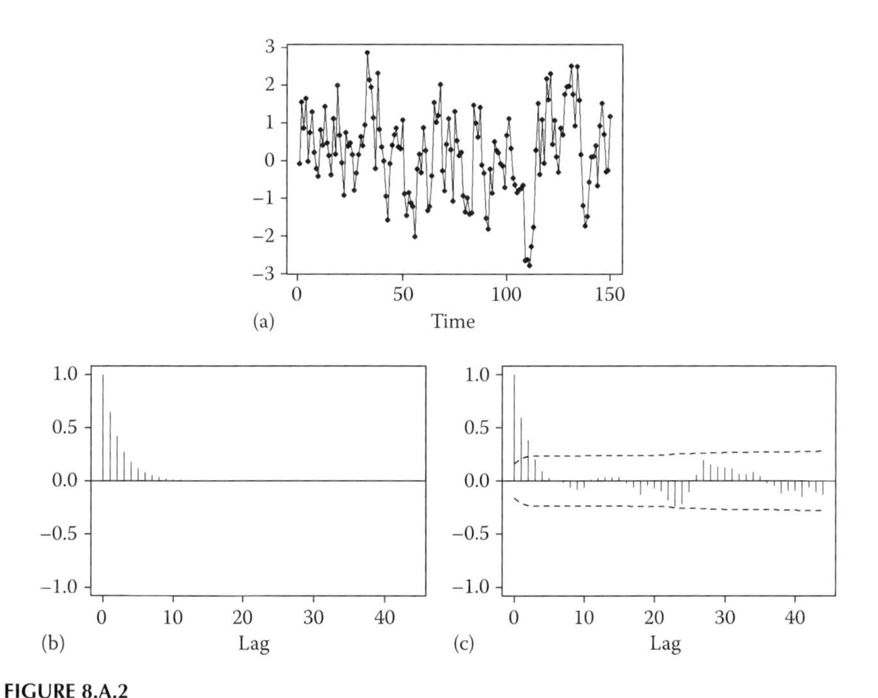

FIGURE 8.A.2
(a) Realization from the AR(1) model in (8.A.5), (b) true autocorrelations, and (c) sample autocorrelations with MA(q) limits.

AR(1) with positive ϕ_1. The corresponding true autocorrelations are shown in Figure 8.A.2b. Because of this damping, the sample autocorrelations eventually tend to remain within the limits associated with some MA(k), which in this case is $k = 3$. This illustrates the fact that one must make a judgment concerning whether:

(a) The autocorrelations have "died out" like mixtures of damped exponentials and/or sinusoids as would be consistent with many AR(p) and ARMA(p,q) models with $p > 0$,

or

(b) The decay to zero is abrupt as in Figure 8.A.1 which is typical of MA(q) processes.

The use of partial autocorrelations helps with the above decision.

8.A.2 Patterns in Partial Autocorrelations

It was shown in Section 3.2 that the autocorrelations of an AR(p) process behave like a mixture of damped exponentials and/or sinusoids depending on the roots of $\phi(z) = 0$. As a consequence, examination of the autocorrelations does not provide sufficient information to identify model orders even if the true autocorrelations are available. For this reason the partial autocorrelation function has been widely used as a supplement to the autocorrelations for purposes of identifying the order of AR(p) models.

Suppose that X_t is a stationary process with autocorrelations ρ_j, $j = 0, 1, 2, \ldots,$ and consider the following system of equations:

$$\begin{aligned}
\rho_1 &= \phi_{k1}\rho_0 + \phi_{k2}\rho_1 + \cdots + \phi_{k,k-1}\rho_{k-2} + \phi_{kk}\rho_{k-1} \\
\rho_2 &= \phi_{k1}\rho_1 + \phi_{k2}\rho_0 + \cdots + \phi_{k,k-1}\rho_{k-3} + \phi_{kk}\rho_{k-2} \\
&\vdots \\
\rho_k &= \phi_{k1}\rho_{k-1} + \phi_{k2}\rho_{k-2} + \cdots + \phi_{k,k-1}\rho_1 + \phi_{kk}\rho_0,
\end{aligned} \tag{8.A.6}$$

where ϕ_{ki} denotes the ith coefficient associated with a $k \times k$ system. The partial autocorrelation function is defined to be ϕ_{kk}, $k = 1, 2, \ldots,$ and an expression for it can be obtained by using Cramer's rule to solve (8.A.6) for ϕ_{kk}. That is,

$$\phi_{11} = \rho_1,$$

$$\phi_{22} = \frac{\begin{vmatrix} \rho_0 & \rho_1 \\ \rho_1 & \rho_2 \end{vmatrix}}{\begin{vmatrix} \rho_0 & \rho_1 \\ \rho_1 & \rho_0 \end{vmatrix}}$$

and for $k \geq 2$,

$$\phi_{kk} = \frac{\begin{vmatrix} \rho_0 & \cdots & \rho_{k-2} & \rho_1 \\ \rho_1 & \cdots & \rho_{k-3} & \rho_2 \\ \vdots & & & \\ \rho_{k-1} & \cdots & \rho_1 & \rho_k \end{vmatrix}}{\begin{vmatrix} \rho_0 & \cdots & \rho_{k-2} & \rho_{k-1} \\ \rho_1 & \cdots & \rho_{k-3} & \rho_{k-2} \\ \vdots & & & \\ \rho_{k-1} & \cdots & \rho_1 & \rho_0 \end{vmatrix}}. \tag{8.A.7}$$

Note that by Theorem 1.1 the denominator in ϕ_{kk} is nonzero for autoregressive processes since for these processes $\gamma_0 > 0$ and by (3.47) $\rho_k \to 0$ as $k \to \infty$. Examination of (8.A.6) shows that if X_t is an AR(k) process, these equations are simply the associated Yule–Walker equations. Thus ϕ_{kk} is the solution of the Yule–Walker equations for the kth autoregressive coefficient based on "assuming" X_t is AR(k).

The partial autocorrelation function is useful for model identification because of the fact that it has an identifiable pattern when X_t is an AR(p) process. For a stationary AR(p) process, it follows that $\phi_{pp} = \phi_p$, which is nonzero. However, when $k > p$, the last column of the numerator of (8.A.7) is a linear combination of the first p columns, and as a result $\phi_{kk} = 0$. This is consistent with the intuition that, for example, if an AR($p+1$) is fit to a process that is actually an AR(p), then $\hat{\phi}_k$ should be near zero. In summary, if X_t is AR(p), then

$$\begin{aligned} \phi_{kk} &= 0 && \text{for } k > p \\ &= \phi_p(\neq 0) && \text{for } k = p, \end{aligned} \tag{8.A.8}$$

which is analogous to the property in (8.A.1) satisfied by the autocorrelations for an MA(q) process.

Before proceeding we note that there is another definition of the partial autocorrelation from which it derives its name. That is, ϕ_{kk} is the correlation between X_t and X_{t+k} in the bivariate distribution of X_t and X_{t+k} conditional on $X_{t+1}, X_{t+2}, \ldots, X_{t+k-1}$. For more discussion, see Brockwell and Davis (1991).

In practical situations, identification of p will be based on estimates of $\hat{\phi}_{kk}$. These estimates can be obtained via any estimation procedures (e.g., Yule–Walker, Burg, and MLE) for estimating the parameters of an AR(k) model. It is common practice to plot partial autocorrelation estimates based on Yule–Walker estimates which can be easily obtained using the Durbin–Levinson recursion in (7.9). It was shown by Quenouille (1949) that for an AR(p)

process, the $\hat{\phi}_{kk}$, $k=p+1$, $p+2,\ldots$ based on Yule–Walker estimates are asymptotically normal and approximately independent with zero mean and Var $(\hat{\phi}_{kk}) \approx 1/n$. Thus, the interval $\pm 2(1/\sqrt{n})$ is often used as a guide for testing $\phi_{kk}=0$. We illustrate the use of the partial autocorrelation function for identification of the order of an autoregressive process in the following example.

Example 8.A.3 Model Identification Based on Sample Partial Autocorrelations

In Figure 8.A.3a, we show the true autocorrelations, and in Figure 8.A.3c we show the true partial autocorrelations for the AR(3) model

$$X_t - 0.7X_{t-1} - 0.54X_{t-2} + 0.81X_{t-3} = a_t. \tag{8.A.9}$$

In Figure 8.A.3b and d, we show the sample autocorrelations and sample partial autocorrelations respectively from a realization (not shown) of length $n=200$ from the model in (8.A.9). Examination of the true and

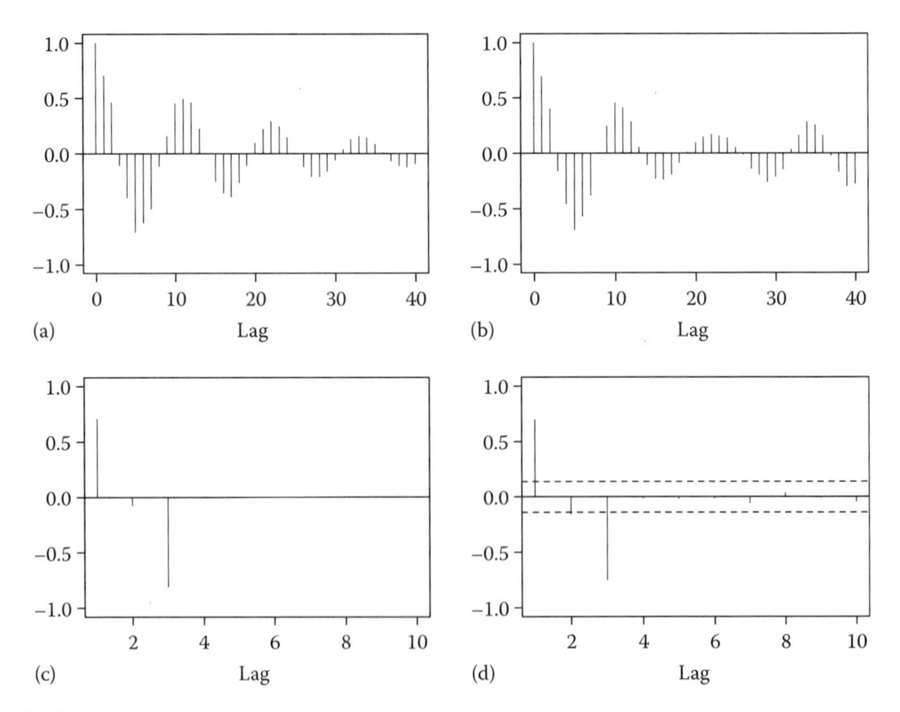

FIGURE 8.A.3
(a) True autocorrelations, (b) sample autocorrelations, (c) true partial autocorrelations, and (d) sample autocorrelations based on the AR(3) model in (8.A.9).

sample autocorrelations in Figure 8.A.3a and b gives us very little information concerning the fact that the model is an AR(3). In fact the damped sinusoidal nature is similar to that seen in several examples of AR(2) models in Chapter 3. However, in Figure 8.A.3c it is clear that the associated model is an AR(3) since the true partial autocorrelation at $k=3$ is nonzero while those for $k>3$ are zero. The sample partial autocorrelations in Figure 8.A.3d clearly show the AR(3) behavior, and the model identification of the process is as an AR(3).

8.A.3 Box-Jenkins Procedure for ARMA Model Identification

The classical "Box-Jenkins procedure" (see Box et al., 2008) for identifying the orders p and q of an ARMA(p,q) process involves identifying patterns exhibited by autocorrelations and partial autocorrelations. While these functions provide perfect identification for MA(q) and AR(p) models when true quantities are known, the identification of ARMA(p,q) processes is more problematic. The Box–Jenkins identification patterns are shown in Table 8.A.1, and their use involves recognizing behavior exhibited by solutions to difference equations. For example, the autocorrelations, ρ_k, for an ARMA(p,q) process satisfy a linear homogeneous difference equation with constant coefficients for $k>q$. The Box–Jenkins procedure is based on the fact that the functional form of the solution is satisfied by ρ_k for $k>p-q$. Unfortunately these types of patterns do not provide perfect identification even when true quantities are known, and model identification is even more difficult when only the sample autocorrelations and sample partial autocorrelations are available.

TABLE 8.A.1

Patterns Used for Box–Jenkins Model Identification

AR(p)
$\phi_{kk}=0,\ k>p \quad (\phi_{pp}=\phi_p\neq 0)$
ρ_k is a damped $\begin{cases}\text{exponential } (p=1)\\ \text{sinusoid and/or exponential } (p>1)\end{cases}$
MA(q)
$\rho_k=0,\ k>q \quad (\rho_q\neq 0)$
ϕ_{kk} is dominated by damped $\begin{cases}\text{exponential } (q=1)\\ \text{sinusoid and/or exponential } (q>1)\end{cases}$
ARMA(p,q)
ρ_k is a damped exponential or sinusoid after first $q-p$ lags
ϕ_{kk} is dominated by damped exponential or sinusoid after first $p-q$ lags

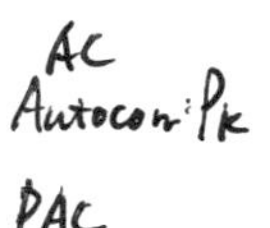

Example 8.A.4 Box–Jenkins Model Identification for ARMA(1,2) and ARMA(4,3) Models

(a) In Figure 8.A.4a–d, we show true autocorrelations, sample autocorrelations, true partial autocorrelations, and sample partial autocorrelations for the ARMA(1,2) model

$$X_t - 0.9X_{t-1} = a_t - 1.6a_{t-1} + 0.9a_{t-2}. \quad (8.A.10)$$

In the figure it can be seen that the true autocorrelations have the appearance of a damped exponential for $k > 1$, while the sample autocorrelations approximate this behavior. The true and sample partial autocorrelations exhibit a consistent behavior from the beginning of the plot. Based on these behaviors we conclude that $q - p = 1$, and due to the simple damped exponential behavior of the autocorrelations, a reasonable conclusion is that $p = 1$ and thus that $q = 2$.

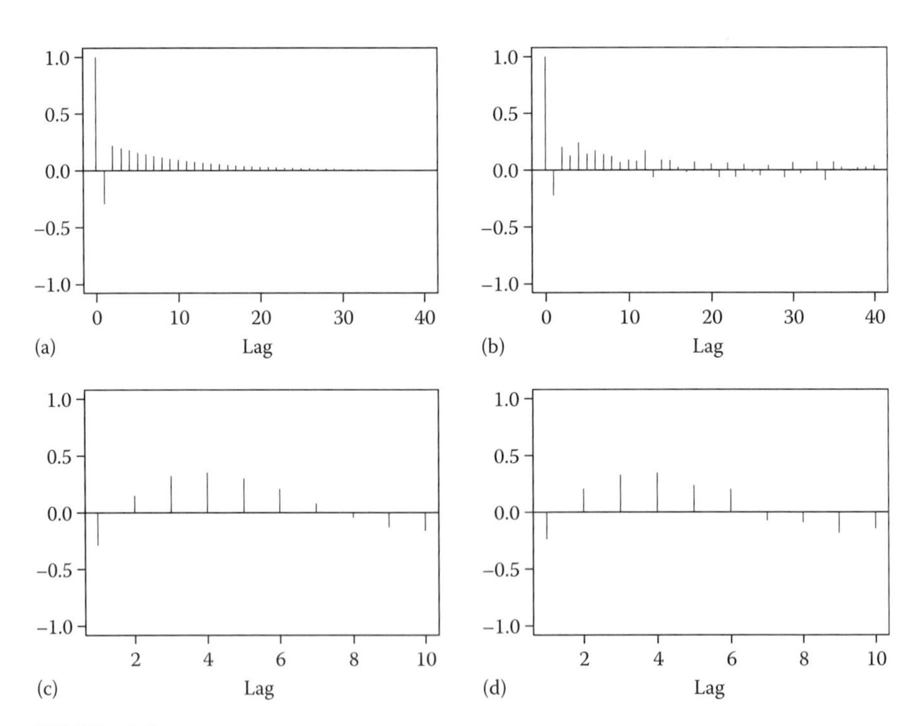

FIGURE 8.A.4
(a) True autocorrelations, (b) sample autocorrelations, (c) true partial autocorrelations, and (d) sample autocorrelations. Plots (a) and (c) are based on the ARMA(1,2) model in (8.A.10).

(b) We next consider the ARMA(4,3) model

$$X_t - 0.3X_{t-1} - 0.9X_{t-2} - 0.1X_{t-3} + 0.8X_{t-4} = a_t + 0.9a_{t-1} + 0.8a_{t-2} + 0.72a_{t-3}, \tag{8.A.11}$$

shown previously in (3.94) with the factored form shown in (3.95). In Figure 3.23, we showed a realization of length $n=200$, the true autocorrelations, and spectral density for this model. In Figure 8.A.5 we show true autocorrelations, sample autocorrelations, true partial autocorrelations, and sample partial autocorrelations for this ARMA (4,3) model. In the figure it can be seen that the true and sample autocorrelations have a damped sinusoidal behavior. There is some suggestion in the partial autocorrelations that a pattern begins at $k=2$ but this is not absolutely clear. Based on these observations one might conclude that $p-q=1$, but due to the fact that the autocorrelations appear to have second order behavior, an ARMA(2,1) selection is a possible choice. If you examine a sample spectral density estimator and conclude that there are two peaks in the spectrum between $f=0$ and $f=0.5$, then the correct ARMA(4,3) identification is plausible.

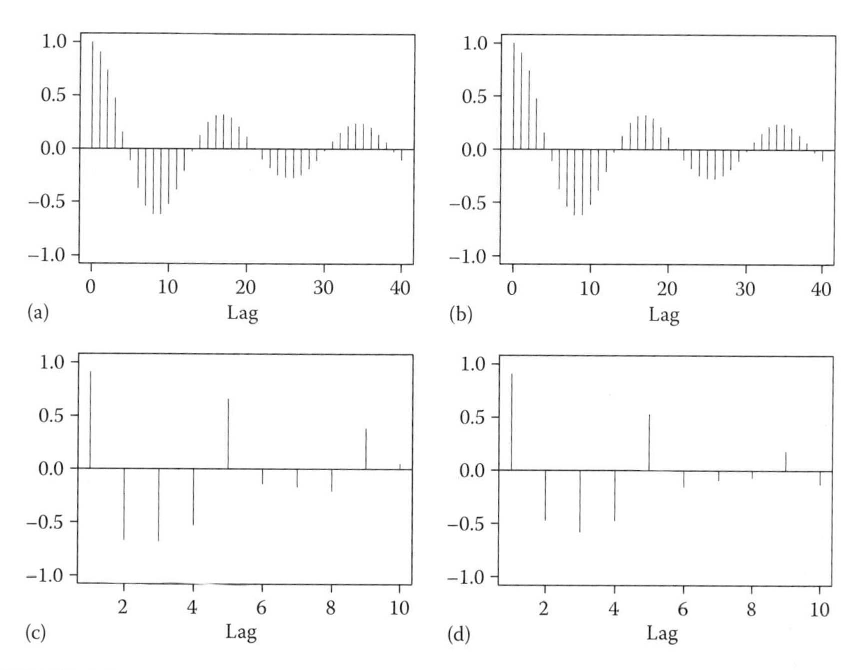

FIGURE 8.A.5
(a) True autocorrelations, (b) sample autocorrelations, (c) true partial autocorrelations, and (d) sample partial autocorrelations. Plots (a) and (c) are based on the ARMA(4,3) model in (8.A.11).

8.A.4 Generalized Partial Autocorrelations

One major concern with the Box–Jenkins approach in the case of an ARMA(p,q) with both $p>0$ and $q>0$ is that there is uncertainty concerning model selection *even when examining the true autocorrelations and partial autocorrelations*. The failure of the true autocorrelation and partial autocorrelation functions to uniquely identify p and q in a mixed ARMA(p,q) model as discussed above led Woodward and Gray (1981) to define the generalized partial autocorrelation in order to better handle the mixed process. Letting X_t be a stationary process with autocorrelations ρ_j, $j=0,\pm 1,\pm 2,\ldots$, and consider the following $k \times k$ system of equations:

$$\begin{aligned}
\rho_{j+1} &= \phi_{k1}^{(j)}\rho_j + \phi_{k2}^{(j)}\rho_{j-1} + \cdots + \phi_{k,k-1}^{(j)}\rho_{j-k+2} + \phi_{kk}^{(j)}\rho_{j-k+1} \\
\rho_{j+2} &= \phi_{k1}^{(j)}\rho_{j+1} + \phi_{k2}^{(j)}\rho_j + \cdots + \phi_{k,k-1}^{(j)}\rho_{j-k+3} + \phi_{kk}^{(j)}\rho_{j-k+2} \\
&\vdots \\
\rho_{j+k} &= \phi_{k1}^{(j)}\rho_{j+k-1} + \phi_{k2}^{(j)}\rho_{j+k-2} + \cdots + \phi_{k,k-1}^{(j)}\rho_{j+1} + \phi_{kk}^{(j)}\rho_j.
\end{aligned} \tag{8.A.12}$$

Notice that these equations are of the same form as (8.A.6) except for the fact that they begin with ρ_{j+1} instead of ρ_1 on the left-hand side of the first equation. In (8.A.12), $\phi_{ki}^{(j)}$ denotes the ith coefficient associated with the $k \times k$ system in which ρ_{j+1} is on the left-hand side of the first equation. The *generalized partial autocorrelation function* is defined to be $\phi_{kk}^{(j)}$, $j=0,1,\ldots$. Similar to the use of (8.A.7) to solve (8.A.6), in (8.A.13), Cramer's rule can be used to solve (8.A.12) yielding

$$\phi_{kk}^{(j)} = \frac{\begin{vmatrix} \rho_j & \cdots & \rho_{j-k+2} & \rho_{j+1} \\ \rho_{j+1} & \cdots & \rho_{j-k+3} & \rho_{j+2} \\ \vdots & & & \\ \rho_{j+k-1} & \cdots & \rho_{j+1} & \rho_{j+k} \end{vmatrix}}{\begin{vmatrix} \rho_j & \cdots & \rho_{j-k+2} & \rho_{j-k+1} \\ \rho_{j+1} & \cdots & \rho_{j-k+3} & \rho_{j-k+2} \\ \vdots & & & \\ \rho_{j+k-1} & \cdots & \rho_{j+1} & \rho_j \end{vmatrix}}. \tag{8.A.13}$$

The motivation for the definition of $\phi_{kk}^{(j)}$ comes from the fact that if X_t is an ARMA (k,j) process, then the equations in (8.A.12) are the extended Yule–Walker equations discussed in Section 7.2.3. Thus $\phi_{kk}^{(j)}$ is the solution of the extended Yule–Walker equations for the last autoregressive coefficient, "assuming" the process is ARMA(k,j). Therefore, $\phi_{kk}^{(j)}$ is a generalization of the partial autocorrelation, and in fact $\phi_{kk}^{(0)} = \phi_{kk}$. It should be noted, however, that the generalized partial autocorrelation function does not have a correlation interpretation, and in fact it may assume values larger than one in absolute value.

The generalized partial autocorrelations as defined in the preceding paragraph satisfy properties that uniquely determine p and q for an ARMA(p,q) process when the true autocorrelations are known. For an ARMA(p,q) process it is clear that $\phi_{pp}^{(q)} = \phi_p$ which is nonzero. Also, as with the partial autocorrelation function, it can be shown that if $k > p$ then $\phi_{kk}^{(q)} = 0$. See Woodward and Gray (1981). Thus, the generalized partial autocorrelation provides a unique identification of p and q for an ARMA(p,q) model in much the same way that the partial autocorrelation does for identification of p for an AR(p). In fact (8.A.8) is simply the special case of this result for $q = 0$.

The generalized partial autocorrelation function has another useful property that enhances model identification. Notice that if a process is ARMA(p,q), then

$$\rho_j = \phi_1 \rho_{j-1} + \phi_2 \rho_{j-2} + \cdots + \phi_p \rho_{j-p}, \tag{8.A.14}$$

for $j > q$. This gives rise to the extended Yule–Walker equations

$$\begin{aligned} \rho_{q+1} &= \phi_1 \rho_q + \phi_2 \rho_{q-1} + \cdots + \phi_p \rho_{q-p+1} \\ \rho_{q+2} &= \phi_1 \rho_{q+1} + \phi_2 \rho_q + \cdots + \phi_p \rho_{q-p+2} \\ &\vdots \\ \rho_{q+p} &= \phi_1 \rho_{q+p-1} + \phi_2 \rho_{q+p-2} + \cdots + \phi_p \rho_q, \end{aligned} \tag{8.A.15}$$

that can be solved for the autoregressive parameters if the autocorrelations are known. However, since (8.A.14) is satisfied for $j > q$, then if h is a non-negative integer it follows that the p equations associated with $j = q + h$, $q + h + 1, \ldots, q + h + p - 1$ can also be used to solve for the autoregressive parameters. Thus it follows that if X_t is ARMA(p,q), then $\phi_{pp}^{(q+h)} = \phi_p$ for $h = 0, 1, \ldots$. Theorem 8.A.1 summarizes the above discussion.

Theorem 8.A.1

Let X_t be an ARMA process with autoregressive order greater than zero. Then

(a) X_t is ARMA(p,q) if and only if $\phi_{kk}^{(q)} = 0$, $k > p$ and $\phi_{pp}^{(q)} \neq 0$.
(b) If X_t ARMA(p,q) then $\phi_{pp}^{(q+h)} = \phi_p$, $h \geq 0$.

Proof:

See Woodward and Gray (1981).

8.A.5 The GPAC Array

Woodward and Gray (1981) recommend examining the generalized partial autocorrelations by calculating $\phi_{kk}^{(j)}$, $k = 1, 2, \ldots, P$ and $j = 1, 2, \ldots, Q$ for some P and Q and then placing these values in a generalized partial autocorrelation array (GPAC array) as in Table 8.A.2. Note that the first row of the GPAC array contains the partial autocorrelations. The results of Theorem 8.A.1 indicate that the GPAC array will display specific patterns when X_t is an ARMA(p,q) process, and this pattern is shown in Table 8.A.3. In order to identify p and q, the GPAC array is examined in order to find a row in which zeros begin occurring beyond a certain point. This row is the qth row, and the zeros begin in the $p+1$st column. Also values in the pth column

TABLE 8.A.2

GPAC Array

		Autoregressive Order			
	j/k	1	2	…	P
	0	$\phi_{11}^{(0)}$	$\phi_{22}^{(0)}$	…	$\phi_{PP}^{(0)}$
Moving average order	1	$\phi_{11}^{(1)}$	$\phi_{22}^{(1)}$	…	$\phi_{PP}^{(1)}$
	2	$\phi_{11}^{(2)}$	$\phi_{22}^{(2)}$	…	$\phi_{PP}^{(2)}$
	⋮	⋮	⋮		⋮
	Q	$\phi_{11}^{(Q)}$	$\phi_{22}^{(Q)}$	…	$\phi_{PP}^{(Q)}$

TABLE 8.A.3

GPAC Array for an ARMA(p,q) Process

		Autoregressive Order						
		1	2	…	p	$p+1$	…	
	0	$\phi_{11}^{(0)}$	$\phi_{22}^{(0)}$	…	$\phi_{pp}^{(0)}$	$\phi_{p+1,p+1}^{(0)}$	…	
	1	$\phi_{11}^{(1)}$	$\phi_{22}^{(1)}$	…	$\phi_{pp}^{(1)}$	$\phi_{p+1,p+1}^{(1)}$	…	
Moving average order	⋮	⋮						
	$q-1$	$\phi_{11}^{(q-1)}$	$\phi_{22}^{(q-1)}$	…	$\phi_{pp}^{(q-1)}$	$\phi_{p+1,p+1}^{(q-1)}$	…	
	q	$\phi_{11}^{(q)}$	$\phi_{22}^{(q)}$	…	ϕ_p	0	0	…
	$q+1$	$\phi_{11}^{(q+1)}$	$\phi_{22}^{(q+1)}$	…	ϕ_p	U[a]	U	…
	⋮				ϕ_p	U	U	…

[a] U = undefined.

are constant from the qth row and below. This constant is $\phi_p \neq 0$. Given the true autocorrelations for an ARMA(p,q) process, the patterns in the GPAC array uniquely determine the model orders if P and Q are chosen sufficiently large. In addition to the results in Theorem 8.A.1, it can also be shown that if $k > p$ and $j > q$, then $\phi_{kk}^{(j)}$ is undefined in the sense that $\phi_{kk}^{(j)}$ as calculated by (8.A.13) is %. See Woodward and Gray (1981).

In practice the GPAC array must be estimated using the observed realization. The ability of the GPAC array to identify p and q ultimately depends on how identifiable the patterns are when the components of the GPAC array are estimated. We refer to an estimated GPAC array based on a realization as a "sample GPAC array." As with the partial autocorrelation function, the most obvious method of estimating $\phi_{kk}^{(j)}$ is to use the defining relation in (8.A.13) with ρ_k replaced by the usual sample autocorrelation, $\hat{\rho}_k$. When this is done, the resulting estimate of $\phi_{kk}^{(j)}$ is simply the extended Yule–Walker estimate of ϕ_k based on assuming the process is ARMA(k,j). However, as noted in Chapter 7, extended Yule–Walker estimators have undesirable properties, especially when some roots of the characteristic equation are close to the unit circle. Our implementation of the sample GPAC arrays is based on Burg/TT estimates. Gray et al. (1978) and Woodward and Gray (1981) have shown that "so called" R and S recursion can be used to solve for the kth parameter in the extended $k \times k$ system. The implementation of the GPAC array in `GW-WINKS` is based on an efficient algorithm using Burg/TT estimate along with R and S recursion. While $\phi_{kk}^{(j)}$ could be obtained as the ML estimate of the fitted ARMA(k,j) model, this procedure is overly time consuming because of the multitude of models necessary for estimating an entire GPAC array.

Example 8.A.5 Model Identification Using the GPAC Array

In Figure 3.23a, we showed the realization of length $n = 200$ from the ARMA(4,3) model in (8.A.11) for which the theoretical and sample autocorrelations and partial autocorrelations are shown in Figure 8.A.5. In Table 8.A.4, we show the theoretical GPAC array for this model using $P = 7$ and $Q = 6$. The string of zeros in row $j = 3$ beginning with the $k = 5$ column along with the fact that the $k = 4$ column becomes constant (−0.80) beginning in the $j = 3$ row uniquely identifies X_t as an ARMA(4,3) process. In Table 8.A.5, we show the sample GPAC array for the realization in Figure 3.23a. In this table, the identification as an ARMA(4,3) is strongly suggested by the fact that the elements in row $j = 3$ are small beginning in the $k = 5$ column while the elements in the $k = 4$ column are approximately constant (−0.70) beginning in row $j = 3$. That is $\phi_{kk}^{(3)} \approx 0$ for $k > 4$ and $\phi_{33}^{(2+h)} \approx$ constant for $h \geq 0$. Examination of the entire sample GPAC array indicates that no other "GPAC identification pattern" is as strong as that for an ARMA(4,3) with an ARMA(4,4) being the next strongest pattern.

TABLE 8.A.4

Theoretical GPAC Array for ARMA(4,3) Model in (8.A.11)

		Autoregressive Order						
		1	2	3	4	5	6	7
	0	0.91	−0.66	−0.68	−0.52	0.64	−0.16	−0.16
	1	0.79	−1.89	−0.41	−1.58	0.59	−0.82	−0.04
Moving	2	0.58	−0.97	0.92	−0.90	0.42	−0.69	7.18
average	3	0.15	−0.87	−2.15	−0.80	0	0	0
order	4	−4.23	−0.91	−1.07	−0.80	*U*	*U*	*U*
	5	2.11	−1.12	−0.73	−0.80	*U*	*U*	*U*
	6	1.32	−0.80	−0.34	−0.80	*U*	*U*	*U*

TABLE 8.A.5

Sample GPAC Array for the Realization in Figure 3.23a from Model (8.A.11)

		Autoregressive Order						
		1	2	3	4	5	6	7
	0	0.91	−0.49	−0.61	−0.52	0.60	−0.07	−0.16
	1	0.81	−1.35	−0.35	−1.03	0.56	−1.43	−0.10
Moving	2	0.64	−0.98	1.37	−0.85	0.38	−0.68	2.37
average	3	0.31	−0.74	−1.63	−0.74	−0.12	−0.01	−0.02
order	4	−0.89	−0.91	−0.89	−0.70	−0.08	0.41	0.00
	5	3.03	−1.11	−0.57	−0.70	−1.20	0.42	12.41
	6	1.42	−0.77	−0.08	−0.69	2.09	−0.59	−0.88

8.A.5.1 Measuring Patterns in the GPAC Array

One measure of the strength of the GPAC identification pattern is the W-statistic. For a given P and Q, then for each $j \le P$ and $k \le Q$, the W-statistic measures the strength of the GPAC pattern associated with an ARMA(j,k) model. Using the notation in Tables 8.A.2 and 8.A.3, then the W-statistic measures the extent to which $\phi_{kk}^{(j)}, \phi_{kk}^{(j+1)}, \phi_{kk}^{(j+2)}, \phi_{kk}^{(j+3)}$ are "constant" and the values $\phi_{k+1,k+1}^{(j)}, \phi_{k+2,k+2}^{(j)}, \phi_{k+2,k+2}^{(j)}$ are near zero. That is, only four values are used to assess the constant behavior and only three to measure the zero behavior since these behaviors become more variable as you move away from the true p,q location. For this reason, we also weight the constants and zeros. For example, implementation in `GW-WINKS` uses the weighted average of the constants

$$A = \frac{wc_0\phi_{kk}^{(j)} + wc_1\phi_{kk}^{(j+1)} + wc_2\phi_{kk}^{(j+2)} + wc_3\phi_{kk}^{(j+3)}}{wc_0 + wc_1 + wc_2 + wc_3}.$$

Similarly, we weight the three "zeros" using the weights wz_1, wz_2, wz_3. The W-statistic is then calculated as $W = (C + Z)/|A|$ where C is the (weighted) sample standard deviation of the four "constants" and Z is the square root of the (weighted) average sum of squares of the "zeros". The W-statistic uses $wc_0 = wc_1 = 1$, $wc_2 = 0.8$, $wc_3 = 0.6$, and $wz_1 = 1$, $wz_2 = 0.8$, $wz_3 = 0.6$. Clearly, smaller values of W suggest a stronger pattern. For the sample GPAC array in Table 8.A.5, the top three W-statistic choices are:

```
p = 4   q = 3   W-STAT = 0.139
p = 4   q = 4   W-STAT = 0.357
p = 4   q = 0   W-STAT = 0.756
```

Consequently, the W-statistic correctly identifies the realization as coming from an ARMA(4,3) model.

Using `GW-WINKS`: To compute a sample GPAC for a realization in `GW-WINKS`, select the `GPAC-Array` option on the `Time Series Analysis` menu. You will be asked to specify P and Q. The GPAC-array, such as that in Figure 8.A.5, will be shown along with the top three order choices based on the W-statistic.

8.A.6 A Final Note about Pattern Recognition Methods

Other pattern recognition methods have been developed for ARMA(p,q) model identification based on arrays, but these will not be discussed here. Other such techniques include the corner method (Beguin et al., 1980), R- and S-arrays (Gray et al., 1978), and extended sample autocorrelation function (ESACF) arrays (Tsay and Tiao, 1984). We reiterate the fact that we recommend the use of AIC-based approaches for most situations. However, pattern recognition methods can provide useful supplemental information in cases where special care in the modeling process is called for.

Exercises

If you use `GW-WINKS` to work problems below, you will find the `AIC`, `Estimate Parameters`, `Factoring Routines`, `GPAC`, and `Generate Realization` options on the `Time series Analysis` menu to be useful.

8.1 Data file `PROB8.1.XLS` has four time series realizations (A–D) which might be white noise. Using procedures discussed in this chapter, what is your conclusion concerning whether each data set is white noise?

8.2 Consider the three ARMA models:
(A) $(1 - 1.5B + 1.21B^2 - 0.455B^3)(X_t - 50) = a_t$
(B) $(1 - 1.6B + 1.7B^2 - 1.28B^3 + 0.72B^4)(X_t - 200) = a_t$
(C) $(1 - B + 1.7B^2 - 0.8B^3 + 0.72B^4)(X_t + 10) = (1 - 0.9B)a_t$
For each of these models
(a) Compute a factor table
(b) Generate a realization of length $n = 200$
(c) Estimate the model orders for the realization generated in (b) using AIC with $p \leq 8$ and $q \leq 3$.
(d) Using ML estimates, fit a model of the order identified in (c). Completely specify your model, including your estimates for μ and σ_a^2.
(e) Examine the factors of the fitted model and compare with those of the true model.

NOTE: Even when the model orders obtained by AIC are not the true orders, by factoring you should be able to identify common characteristics.

8.3 Consider the three time series data sets:
(A) West Texas intermediate crude oil prices: Figure 1.1a, File: `WTCRUDE.XLS`
(B) 30-year conventional mortgage rates, 1/1/1991–4/1/2010: Figure 10.1c, File: `ECODATA.XLS`
(C) Global Temperature Data: Figure 1.24a, File: `HADLEY.XLS`
(a) Fit an ARMA model to each data set.
(i) Use AIC to identify model orders
(ii) Estimate the parameters using model orders obtained in (i). Completely specify your model, including your estimates for μ and σ_a^2.
(b) Examine the factors of the fitted model and discuss the sense in which dominant factors in the model relate to characteristics of the data.

8.4 (a) Suppose you have used the `Factoring Routines` procedure in `GW-WINKS` and have found the factor $(1 - 1.5B + 0.97B^2)$ as one of the model factors. Since these roots are near the unit circle, it is decided to transform the time series using the associated nonstationary factor associated with roots on the unit circle and with the same frequency content as this factor. What is the associated nonstationary factor that you should use?
(b) In general, given a near nonstationary irreducible second order polynomial, $1 - \alpha_1 B - \alpha_2 B^2$, associated with roots close to the unit circle, what are the coefficients of the corresponding nonstationary factor with the same frequency content?
(c) For a first order factor, $1 - \alpha_1 B$, associated with roots near the unit circle, what is the associated nonstationary factor with the same frequency content?

8.5 Simulate a realization of length $n = 200$ from the model

$$(1 - B^6)(1 - 0.8B)(X_t - 25) = a_t.$$

(a) Using the overfit procedures discussed in this chapter, fit an ARUMA model to the simulated realization. Specifically:
   (i) Overfit to identify nonstationary factors in the model
   (ii) Transform the data by the identified nonstationary factors
   (iii) Using AIC with ML estimates, fit an ARMA model to the stationarized data.

(b) How does the final fitted model compare with the true model?

NOTE: In GW-WINKS you will need to replace the factor $(1 - B^6)$ with a factor such as $(1 - 0.999B^6)$ to generate the realization.

8.6 Find appropriate ARUMA models for the following data sets using the procedures discussed in this chapter. Discuss features of your model including seasonality, etc. Use the overfitting procedure outlined in Problem 8.5(a).

(a) `FREIGHT.XLS`—9 years of monthly freight shipment data
(b) `FREEZE.XLS`—Each data value represents the minimum temperature over a 10-day period at a location in South America.
(c) `LAVON.XLS`—Gives lake levels of Lake Lavon (a reservoir northeast of Dallas). The data values are feet above sea level and are obtained at the end of each quarter (i.e., March 31, June 30, September 30, and December 31).

8.7 If X_t is AR(p), show that
(a) $\phi_{kk} \neq 0$ for $k = p$.
(b) $\phi_{kk} = 0$ for $k > p$.

8.8 Use GPAC with the W-statistic to fit ARMA models to data sets (A)–(C) in Problem 8.2.

8.9 Use GPAC with the W-statistic to fit ARMA models to data sets (A)–(C) in Problem 8.3.

8.10 Consider the model

$$X_t - \mu = \phi_1(X_{t-1} - \mu) + \cdots + \phi_p(X_{t-p} - \mu) + a_t.$$

Show that this model can be written as

$$\nabla X_t = \beta_0 + \tau X_{t-1} + \beta_1 \nabla X_{t-1} + \cdots + \beta_{p-1} \nabla X_{t-p+1} + a_t,$$

where $\nabla X_{t-j} = (1 - B)X_{t-j} = X_{t-j} - X_{t-j-1}, \beta_0 = \mu(1 - \phi_1 - \cdots - \phi_p)$, $\tau = -(1 - \phi_1 - \cdots - \phi_p)$, and $\beta_j = -(\phi_{j+1} + \cdots + \phi_p)$, for $j = 1, \ldots, p - 1$.

NOTE: This alternative expression is known as the *error correction form*.

9

Model Building

In Chapter 8, we discussed techniques for fitting a model to a time series realization. After a model is fit to a realization, you must take steps to assure that the model is suitable. The first step in this process is examine the residuals. This topic will be discussed in Section 9.1. Suppose now that you have obtained a model by using a model identification technique (such as AIC), estimated the parameters, and examined the residuals to satisfy yourself that they are sufficiently white (i.e., uncorrelated). Having obtained such a model does not necessarily indicate that the model is the most suitable one for your situation. After finding a model that seems appropriate, you should make a decision concerning whether the model *makes sense*. For example, the following questions are among those that should be asked in this regard.

1. Does the stationarity or nonstationarity, seasonality, or nonseasonality, of the fitted model make physical or practical sense?
2. Is a signal-plus-noise model more appropriate than a strictly correlation-based model?
3. Are characteristics of the fitted model consistent with those of the observed data?
   a. Do forecasts and spectral estimates based on the fitted model reflect what is known about the physical setting?
   b. Do realizations and their characteristics behave like the data?

We discuss these topics in this chapter and illustrate the ideas through examples.

9.1 Residual Analysis

Calculation of residuals from a fitted model, was discussed in Section 7.4. For the ARMA(p,q) model

$$\phi(B)(X_t - \mu) = \theta(B)a_t,$$

we can write

$$a_t = X_t - \phi_1 X_{t-1} - \cdots - \phi_p X_{t-p} - \mu(1 - \phi_1 - \cdots - \phi_p) + \theta_1 a_{t-1} + \cdots + \theta_q a_{t-q}. \quad (9.1)$$

Based on a fitted model, we calculate the residuals

$$\hat{a}_t = X_t - \hat{\phi}_1 X_{t-1} - \cdots - \hat{\phi}_p X_{t-p} - \overline{X}(1 - \hat{\phi}_1 - \cdots - \hat{\phi}_p) + \hat{\theta}_1 \hat{a}_{t-1} + \cdots + \hat{\theta}_q \hat{a}_{t-q}. \quad (9.2)$$

In Section 7.4, we discussed the use of backcasting to compute necessary starting values, so that the residuals $\hat{a}_1, \hat{a}_2, \ldots, \hat{a}_p$ can be effectively calculated. A key point of the current section is that if the ARMA(p,q) model fit to the data is appropriate, then the residuals should be well modeled as white noise. If the residuals of the fitted model contain significant autocorrelation, then the model has not suitably accounted for the correlation structure in the data. Another assessment that may be of interest is whether the residuals appear to be normally distributed, an assumption of the MLE procedures.

We begin by discussing methods for testing residuals for white noise.

9.1.1 Check Sample Autocorrelations of Residuals versus 95% Limit Lines

One method of testing residuals for white noise has already been discussed in Section 8.1. Specifically, for white noise, the autocorrelations satisfy $\rho_k = 0$, $k \neq 0$, and in (8.1) we gave 95% limits $(\pm 2(1/\sqrt{n}))$ for testing $H_0 : \rho_k = 0$. We pointed out that if substantially more than 5% of the sample autocorrelations fall outside these lines, then this is evidence against white noise. See Section 8.1 for a more complete discussion of the use of these limit lines for white noise checking.

9.1.2 Ljung–Box Test

The problem with checking the sample autocorrelations against, say, the 95% limit lines, is that the 5% chance of exceeding these lines applies to the autocorrelations at each lag separately. In order to provide a single portmanteau test that simultaneously tests the autocorrelations at several lags, Box and Pierce (1970) and Ljung and Box (1978) developed tests of the null hypothesis that the data (in our application the residuals) are white noise. The test developed by Ljung and Box (1978) is widely used for this purpose. The Ljung–Box test statistic given by

$$L = n(n+2) \sum_{k=1}^{K} \frac{\hat{\rho}_k^2}{n-k}, \quad (9.3)$$

is used for testing the hypotheses

$$\begin{aligned} &H_0: \rho_1 = \rho_2 = \cdots \rho_K = 0 \\ &H_a: \text{at least one } \rho_k \neq 0 \quad \text{for } 1 \leq k \leq K. \end{aligned} \tag{9.4}$$

Ljung and Box (1978) show that L in (9.3) is approximately distributed as χ^2 with K degrees of freedom when the data are white noise. When the data to be tested are residuals from a fitted ARMA(p,q) model, then the test statistic is approximately χ^2 with $K - p - q$ degrees of freedom. Examination of (9.3) shows that the test statistic measures the size of the first K sample autocorrelations as a group, so that large values of L suggest that the data are not white noise. Thus, the null hypothesis of white noise residuals is rejected if $L > \chi^2_{1-\alpha}(K - p - q)$ where $\chi^2_{1-\alpha}(m)$ denotes the $(1 - \alpha) \times 100\%$ percentile of the χ^2 distribution with m degrees of freedom. In the examples that follow and `GW-WINKS`, in the program we use the Ljung–Box test with $K = 24$ and $K = 48$, which is consistent with the value of K used by Box et al. (2008). For small sample sizes ($n \leq 100$), Box et al. (2008) and Ljung (1986) recommend the use of smaller values of K.

Using `GW-WINKS`: The Ljung–Box test can be applied by selecting `Ljung-Box Test` on the `Time Series Analysis` menu. You will be asked to select a data set on which to apply the test and then to specify p and q. If you are applying a Ljung–Box test as the initial step in the analysis of a data set, then you should enter 0 for p and q. If you are testing the residuals from a fitted model (`GW-WINKS` creates a column containing the residuals from a model obtained using the `Estimate Parameters` option on the `Time Series Analysis` menu), then you should enter p and q from the estimated model. `GW-WINKS` reports the Ljung–Box test using $K = 24$ and $k = 48$. A small p-value supports rejection of the null hypothesis in (9.4).

9.1.3 Other Tests for Randomness

There are many other tests for white noise that could be used. Testing for white noise is essentially a test for randomness, so that any such test (e.g., runs tests) could be used in this regard. See Davis and Jones (1968) for a test based on the log periodogram. Li (1988) and Chan (1994) have developed tests designed to be more robust than the Ljung–Box test to non-normal white noise. Peña and Rodriguez (2002) have developed a portmanteau test that is claimed to be a more powerful alternative to the Ljung–Box test in the presence of normal noise.

Example 9.1 Testing Residuals for White Noise

In this example, we examine the residuals for several of the data sets we modeled in Chapter 8 examples.

(a) In Example 8.3, we considered realizations from an AR(4), ARMA (2,1), and AR(3) model and showed that in each case, AIC selected the correct model orders. In this section, we return to these examples and examine the residuals associated with the first two of these model fits based on maximum likelihood (ML) estimates.

(i) The realization shown in Figure 3.16a is from the AR(4) model

$$X_t - 1.15X_{t-1} + 0.19X_{t-2} + 0.64X_{t-3} - 0.61X_{t-4} = a_t. \tag{9.5}$$

In Figure 9.1a, we show the residuals from the fitted AR(4) model obtained using ML estimates, and, in Figure 9.1b, we show the sample autocorrelations of the residuals in Figure 9.1a along with the 95% limit lines. The residuals in Figure 9.1a look white (there is one unusually large (negative) residual at about 130), and the sample autocorrelations stay well within the limit lines. The Ljung–Box test using $K = 24$ (i.e., with $K - p - q = 24 - 4 - 0 = 20$ degrees of freedom) has a p-value of 0.99. For $K = 48$, GW-WINKS reports a p-value greater than 0.99. Thus, on the basis of the residuals, the AR(4) model using ML estimates appears to be an adequate model.

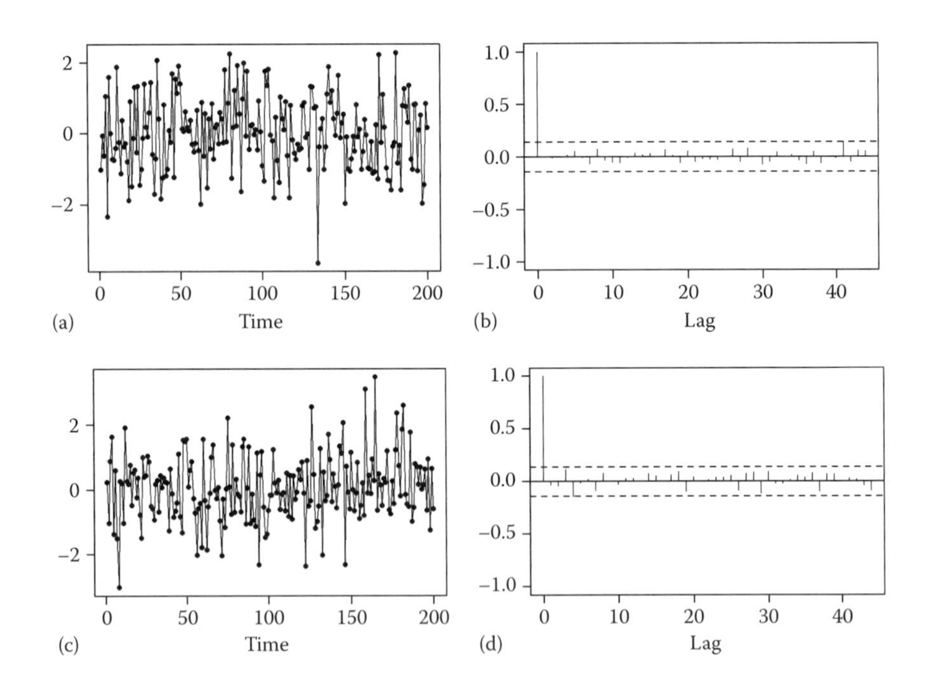

FIGURE 9.1
(a) Residuals from an AR(4) MLE fit to the data in Figure 3.16a. (b) Sample autocorrelations for the residuals in (a). (c) Residuals from an ARMA(2,1) MLE fit to the data in Figure 3.22. (d) Sample autocorrelations for the residuals in (c).

(ii) Figure 3.22a shows a realization of length $n = 200$ from the ARMA(2,1) model

$$X_t - 1.6X_{t-1} + 0.9X_{t-2} = a_t - 0.8a_{t-1}, \tag{9.6}$$

and, in Figure 9.1c, we show the residuals from the fitted ARMA (2,1) model based on ML estimates shown in Table 7.10. The sample autocorrelations of the residuals in Figure 9.1c are shown in Figure 9.1d. The residuals appear to be random, and the sample autocorrelations except for $k = 4$ stay within the limit lines. Using $K = 24$, the degrees of freedom are $K - p - q = 24 - 2 - 1 = 21$, and the Ljung–Box test reports a p-value of 0.58. For $K = 48$, the p-value is 0.80. Consequently, an analysis of the residuals does not provide any suggestion of an inadequate model fit.

(b) We next consider the fits to the airline data and Pennsylvania monthly temperature data discussed in Examples 8.8 and 8.9.

Airline data

In Figure 9.2a, we show the residuals from the ARUMA(12,1,0)× $(0,1,0)_{12}$ model in (6.53) fit to the log airline data, and in Figure 9.2b

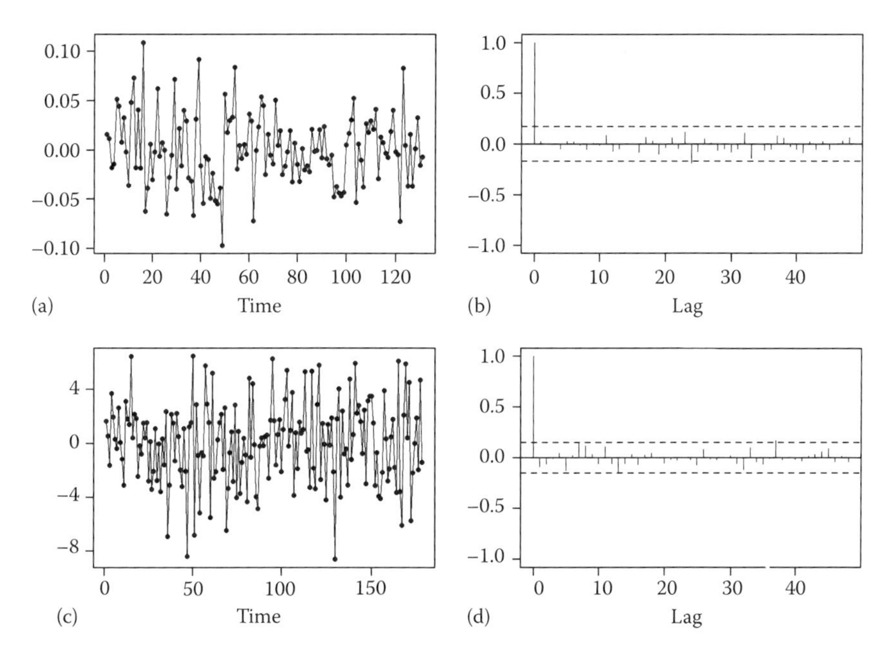

FIGURE 9.2
(a) Residuals from an ARUMA(12,13,1) fit to the log airline data. (b) Sample autocorrelations for the residuals in (a). (c) Residuals from an ARUMA(0,2,4) fit to the Pennsylvania temperature data. (d) Sample autocorrelations for the residuals in (c).

we show the corresponding sample autocorrelations. The residuals look fairly white with some wandering behavior beginning at about $t = 45$. The sample autocorrelations stay within the limit lines, and the Ljung–Box statistics at $K = 24$ and 48 have p-values of 0.13 and 0.74, respectively. Thus, while the residual analysis causes some concern, it does not provide substantive indication of an inadequate fit. Note that we apply the Ljung-Box test to the transformed data, $(1 - B)(1 - B^{12})X_t$, so that the values of p and q used to calculate the degrees of freedom of the Ljung–Box statistic are 12 and 0 in this example.

Pennsylvania temperature data
In Figure 9.2c and d, we show the residuals and their sample autocorrelations from the ARUMA(0,2,4) fit in (8.30) to the Pennsylvania temperature data in Figure 8.9a. The sample autocorrelations in Figure 9.2d tend to stay within the limit lines, while the Ljung–Box test reports p-values of 0.27 and 0.65 for $k = 24$ and 48, respectively. As in the case of the airline model, we conclude that the residuals do not provide significant evidence to reject the model.

9.1.4 Testing Residuals for Normality

Another issue that may need to be checked concerning residuals is normality. Normality is an assumption upon which ML estimates are based. Additionally, if uncorrelated residuals are non-normal, then ARCH-type behavior is a possibility. (See Section 4.2.) Methods for checking normality include the use of Q–Q plots or formal tests for normality such as the Anderson–Darling test (see Anderson and Darling, 1952) and the Shapiro–Wilk test (see Shapiro and Wilk, 1965). In Figure 9.3, we show Q–Q plots of the (standardized) residuals in Figure 9.2. The normality of the residuals for the airline model seems like an appropriate assumption while the normality of the PA

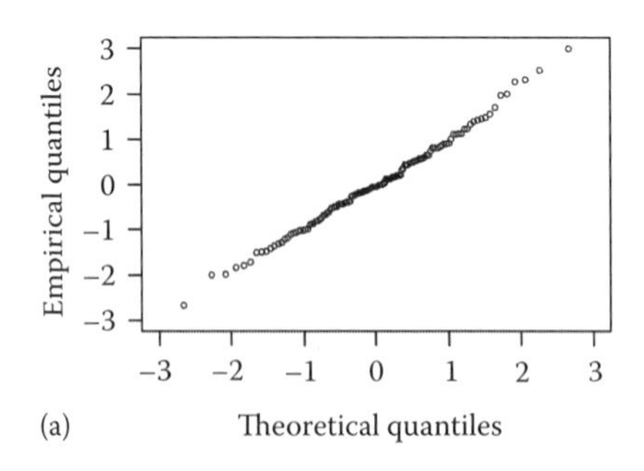

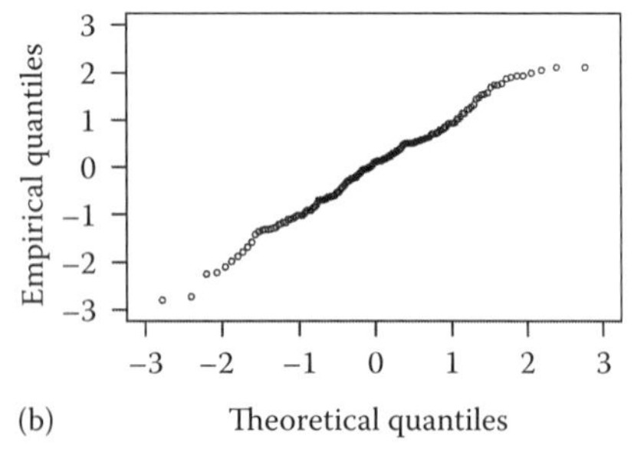

FIGURE 9.3
Q–Q plots for the residuals in Figure 9.2. (a) Histogram of airline residuals. (b) Q–Q plot for PA temperature residuals

temperature residuals seems more questionable. Neither the Anderson–Darling nor the Shapiro–Wilk test reject normality for either set of residuals at $\alpha = 0.05$.

Using `GW-WINKS`: Normality checks are available in GW-WINKS by selecting `Analyze` on the top menu. Then by selecting `Descriptives`, `Tests for Normality`, and then by selecting the variable on the Data Screen to be tested. A written report containing results from the Anderson–Darling and Lilliefors test appears on the `Output Viewer`. By selecting the `Graph` on the top left menu and scrolling through the graph option you are shown, among other things, plots of the histogram and normal probability plot.

9.2 Stationarity versus Nonstationarity

As discussed in Section 8.3, the decision concerning whether or not to include a nonstationary component in the model (e.g., a factor of $1 - B$, a seasonal factor $1 - B^s$, and so forth) is one that will need to be made by the investigator. For example, AIC will not make such a decision for you. It was noted in Section 8.3 that there are tests for a unit root (i.e., the need for a factor $1 - B$ in the model). However, as was noted there, the decision to retain a unit root is the result of failure to reject a null hypothesis, and as such the unit root tests do not have sufficient power in general to distinguish between models with near unit roots and those with unit roots. The final decision concerning whether to include a factor $1 - B$ should be based on such considerations as whether there is any physical reason to believe that there is an attraction to the mean.

Another important issue is that realizations from the model

$$(1 - \delta_1 B)\phi(B)(X_t - \mu) = \theta(B)a_t, \tag{9.7}$$

where δ_1 is close to or equal to 1 (but not greater than 1) may show a piecewise trending behavior. In Example 5.1, we discussed the problem of determining whether the upward trend in the global temperature data is due to a deterministic (e.g., man-made) signal or is simply random trending behavior. This is further discussed in the following example.

Example 9.2 Modeling the Global Temperature Data

(a) *A stationary model*

In Figure 1.24a, we showed the annual average global temperature for the years 1850–2009 and noted the upward trending behavior

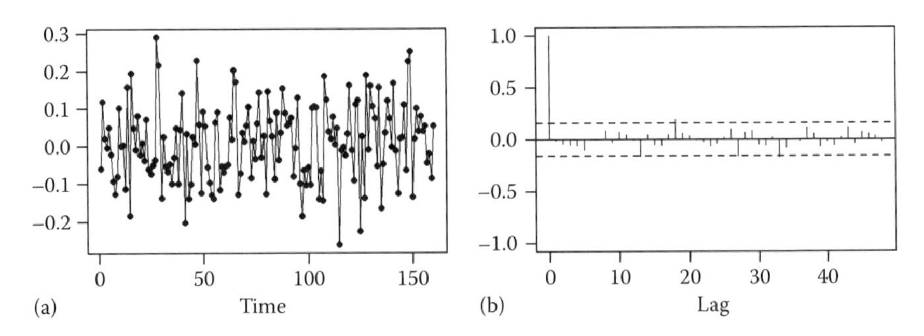

FIGURE 9.4
(a) Residuals from an AR(4) fit to the global temperature data. (b) Sample autocorrelations for the residuals in (a).

(especially in the last 25 years) that has concerned scientists. To fit an ARMA(p,q) model to the data in Figure 1.24a, we use AIC with $P = 10$ and $Q = 5$, which gives $p = 4$ and $q = 0$. The resulting model was previously shown in (5.3) (in its zero-mean form). The factored form of (5.3), including the mean, is given by

$$(1 - 0.99B)(1 - 0.2B + 0.455B^2)(1 + 0.535B)(X_t + 0.17) = a_t, \qquad (9.8)$$

where $\hat{\sigma}_a^2 = 0.0107$. In Figure 9.4a and b, we show the residuals from the model in (9.8) and their sample autocorrelations, respectively. The residuals in Figure 9.4a appear to be uncorrelated, and the sample autocorrelations in Figure 9.4b stay within the 95% limit lines. The Ljung–Box statistic reports p-values of 0.38 and 0.47 for $K = 24$ and 48, respectively. Accordingly, on the basis of a residual analysis, the AR(4) fit in (9.8) seems to be adequate.

(b) *A model with a unit root*
The factor table for the AR(4) model in (9.8) has been previously given in Table 5.1, where it can be seen that model is stationary with a positive real root ($1/0.975 = 1.025$) close to the unit circle. The sample autocorrelations shown in Figure 1.24b damp quite slowly, and the wandering behavior of the realization makes the inclusion of a unit root (i.e., a factor of $1 - B$) plausible. Therefore, we consider modeling the temperature data by differencing the data to calculate $Y_t = X_t - X_{t-1} = (1 - B)X_t$, and then modeling Y_t as an AR(p) process. The differenced data are shown in Figure 9.5a along with the sample autocorrelations of the differenced data in Figure 9.5b. It can be seen that the difference removed the trending behavior in the data as expected. While the differenced data appear fairly white, the first two sample autocorrelations exceed the 95% limit lines so we continue with the modeling process. To find the best-fitting AR model

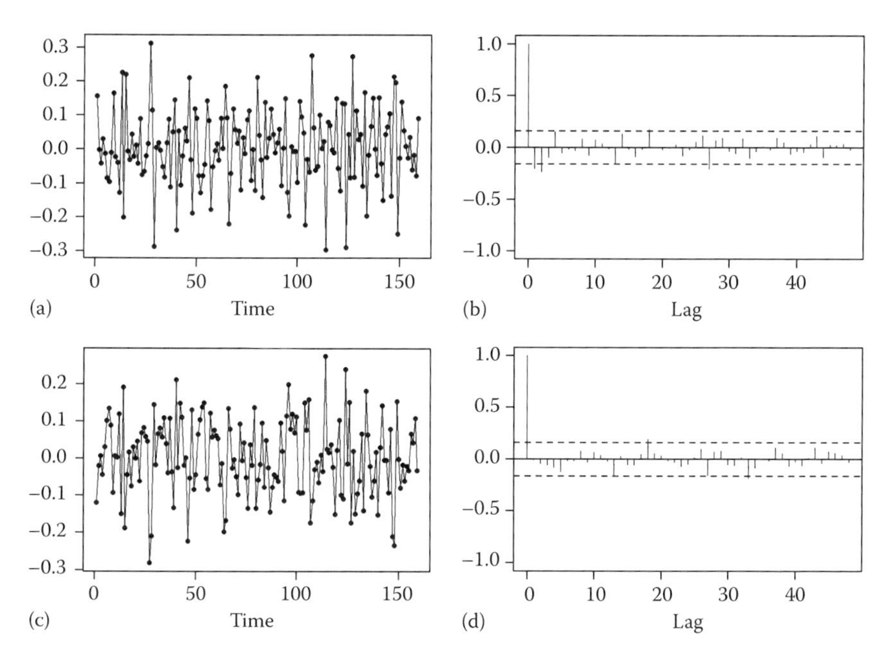

FIGURE 9.5
(a) Differenced global temperature data in Figure 1.24a. (b) Sample autocorrelations for the data in (a). (c) Residuals from an AR(3) MLE fit to the differenced data in (a). (d) Sample autocorrelations for the residuals in (c).

for the differenced data, we use $P = 8$ and $Q = 0$, and AIC selects an AR(3). Estimating the parameters of the AR(3) model gives

$$Y_t + 0.34Y_{t-1} + 0.36Y_{t-2} + 0.26Y_{t-3} = a_t.$$

The final model is

$$(1 - B)(1 + 0.34B + 0.36B^2 + 0.26B^3)(X_t + 0.17) = a_t, \tag{9.9}$$

with $\hat{\sigma}_a^2 = 0.0107$. The factor table for the AR(3) model fit to Y_t is given in Table 9.1. Residuals from the model in (9.9) are given in Figure 9.5c along with their sample autocorrelations in

TABLE 9.1

Factor Table for AR(3) Model Fit to Differenced Temperature Data

AR Factors	Roots (r_j)	$\lvert r_j^{-1}\rvert$	f_{0j}
$1 - 0.21B + 0.47B^2$	$0.22 \pm 1.44i$	0.69	0.23
$1 + 0.55B$	-1.82	0.55	0.5

Figure 9.5d. Again, the residuals appear white and the sample autocorrelations stay sufficiently within the limit lines. The Ljung–Box test (based on $p=3$ and $q=0$) reports p-values of 0.40 and 0.49 at $k=24$ and 48, respectively, for residuals from the model in (9.9).

NOTE: Using AIC to select to an ARMA(p, q) model for the differenced data using $P=8$ and $Q=4$, gives an MA(2) model. However, since the residuals from that model did not pass the Ljung–Box white noise test, we chose not to use the MA(2) model.

(c) *Forecasts*

Thermometer-based temperature readings for the past 160 or so years, such as the data shown in Figure 1.24a, indicate that temperatures have generally been increasing over this time period.

However, it is well known using proxy data such as tree rings, ocean sediment data, etc., that temperatures have not remained static over time leading up to the rise over the past 160 years, but have tended to vary. Although the data are insufficient to draw strong conclusions, there is some evidence that, at least in much of the northern hemisphere, the temperatures during the medieval warm period (950–1100) may have been as high as they are today. The medieval warm period was followed by a period known as the Little Ice Age. While not a true ice age, this was a period during which available temperature data indicate there was cooling that lasted into the nineteenth century.

As we examine possible models and their forecasts for a set of data, such as the global temperature data, it is important to understand the properties of the models that are obtained. The models given in Equations 9.8 and 9.9 are two competing models for the global temperature data. Model (9.8) is a stationary model, and, thus, under this model, the global temperatures are in an equilibrium around some mean temperature. The recent concern about global warming centers around fear that man-made greenhouse gases have moved us out of such an equilibrium. The model in (9.9) is not stationary and, therefore, under this model, the temperatures are not in an equilibrium about a mean level.

In Figure 9.6a and b, we show forecasts for the next 25 years based on the models in (9.8) and (9.9). The forecasts in Figure 9.6a tend to the sample mean of the observed data while the forecasts in Figure 9.6b from the unit root model eventually stay constant at a value similar to the most recent observation. These forecasts are consistent with the properties of the models. That is, the forecasts in Figure 9.6a predict that temperatures are at or near a relative maximum that will eventually abate and tend back toward a mean level, best estimated by the sample mean. The forecasts in Figure 9.6b suggest that there is

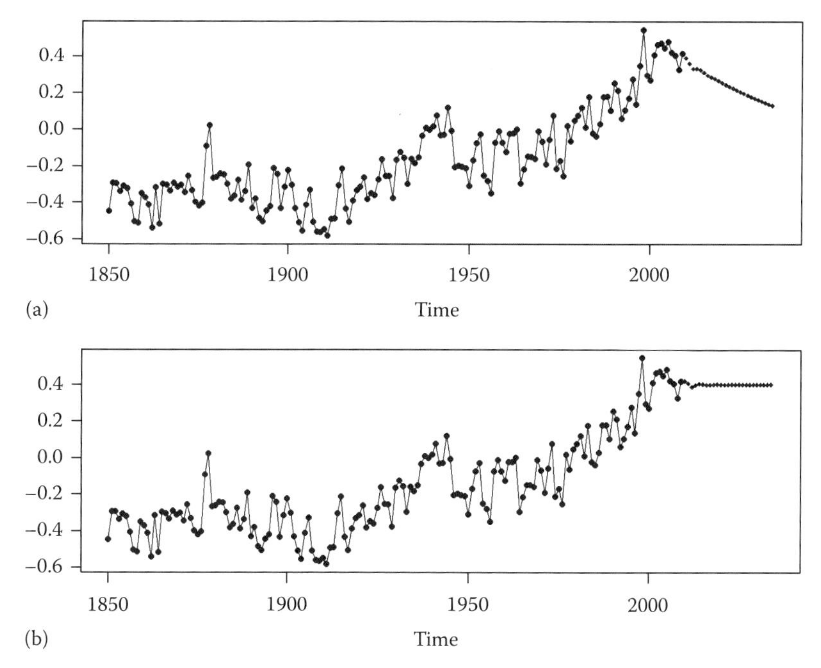

FIGURE 9.6
Forecasts for global temperature for the years 2010–2034 (a) using the model in (9.8) and (b) using the model in (9.9).

no attraction to a mean level, and the temperatures from 2010 forward are as likely to increase as they are to decrease. Neither of these models predicts the current trend to continue. Recall that in Figure 6.9, it was seen that the trend in the airline data was predicted to continue using the model $(1 - B)(1 - B^{12})\phi_1(B)X_t = a_t$, where $\phi_1(B)$ is a stationary 12th order factor given in (6.54). As was mentioned in Example 6.7, the trending behavior in the airline data forecasts is caused by the fact that $1 - B$ is a factor of $1 - B^{12}$, and, thus, the "airline model" has two factors of $1 - B$. As seen in Chapter 6, this produces forecasts that tend to follow the trending behavior immediately preceding the forecasts. However, a second factor of $1 - B$ is not suggested for the temperature data as can be seen by examining factor tables associated with AR(k) fits to the temperature data for $k = 10, 12, 16$ (not shown here).

Do these models provide conclusive evidence that the increasing trend should not be predicted to continue? The answer to this questions is a resounding "No." What we can conclude is that if the stationary AR(4) model in (9.8) that we found for the available past data remains valid into the future, then the best predictor forecasts the trend to abate. However, if the nonstationary model in (9.9) is a valid representation of future behavior, then the best predictor is "silent" regarding whether temperatures will increase or decrease.

Using `GW-WINKS`: The forecasts in Figure 9.6a are obtained by first estimating the parameters of an AR(4) fit to the data using the `Estimate Parameters` option on the `Time Series Analysis` menu. By next selecting the `Forecasting` option, you can enter the parameters of the AR(4) model. If you have moved directly from estimation to forecasting, these values should be available as the default options. In order to obtain the forecasts in Figure 9.6b, you will need to difference the temperature data using the `New Variable, Common Transformations` option on the `Edit` menu and selecting `by Differencing`. Then fit an AR(3) model to the differenced data. You then next select the `Forecasting` option and specify that the data to be forecast are the original temperature series (not the differenced data). To obtain forecasts from this nonstationary model, enter 1 in the `Difference` entry box on the `Forecasting` entry screen. You then specify $p = 3, q = 0$. As above you next enter the autoregressive parameters noting that if you have moved directly from estimation to forecasting, these values should be available as the default options.

Note that a signal-plus-noise model, assuming for simplicity a linear trend model, could be fit to the temperature data. Such a model would predict an increasing trend if the slope of the trend line is significantly different from zero. Thus, it is clear that the forecasts are only as good as the model being used, and it is also clear that forecasts can be obtained to support an unscrupulous investigator's position by selecting the model to be used for forecasting to meet his or her needs or preconceptions. At the same time, erroneous conclusions may be unintentionally drawn from various analyses due to lack of understanding by the researcher. Thus, it is important to be aware of underlying properties associated with any forecasting method being used.

9.3 Signal-plus-Noise versus Purely Autocorrelation-Driven Models

Based on the preceding discussion, a question of importance is "how do we test for trend in time series data?" In Section 5.1, we considered the signal-plus-noise model.

$$X_t = s_t + Z_t, \tag{9.10}$$

where

s_t is a deterministic signal

Z_t is a zero-mean stationary noise component that is usually modeled using an ARMA(p,q) model

For purposes here, we consider the case in which $s_t = a + bt$. That is, the signal-plus-noise (regression) model is given by

$$X_t = a + bt + Z_t. \tag{9.11}$$

Suppose we want to decide whether the observed data follow an ARMA(p,q) model, which could be denoted

$$X_t = \mu + Z_t, \tag{9.12}$$

or a signal-plus-noise model of the form (9.11). Noting that (9.11) is the more general model and simplifies to (9.12) if $b = 0$, then the natural approach would be to fit a model of the form (9.11) and test $H_0 : b = 0$. If this hypothesis is rejected, then the decision would be to include a deterministic component in the model. For simplicity in the following, we consider the case in which Z_t satisfies an AR(p) model. Since the residuals in the regression model in (9.11) may be correlated, the test for $H_0 : b = 0$ using standard regression-based procedures assuming uncorrelated residuals may not be appropriate. Testing $H_0 : b = 0$ in (9.11) when the residuals are correlated is a difficult problem, primarily due to the fact, as discussed in Section 5.1, that realizations from stationary AR(p) models such as (9.12) can produce realizations with trending behavior. Several authors have addressed this problem, and we discuss a few approaches here.

9.3.1 Cochrane Orcutt, ML, and Frequency Domain Method

Consider the case in which $(1 - \phi_1 B)Z_t = a_t$. In this case, it follows from (9.11) that

$$\begin{aligned} Y_t &= (1 - \phi_1 B)X_t \\ &= (1 - \phi_1 B)(a + bt + Z_t) \\ &= a(1 - \phi_1) + b(1 - \phi_1 B)t + (1 - \phi_1 B)Z_t \\ &= c + bt_{\phi_1} + a_t, \end{aligned} \tag{9.13}$$

where (9.13) is a regression equation with the same slope coefficient, b, as in (9.11) with a new independent variable $t_{\phi_1} = (1 - \phi_1 B)t = t - \phi_1(t - 1)$ and uncorrelated residuals. In practice, however, ϕ_1, a, and b will not be known and must be estimated from the data. Cochrane and Orcutt (1949) suggested fitting a regression line to the data using least-squares estimates and fitting an AR(1) model to the *residuals* $\hat{Z}_t = X_t - \hat{a} - \hat{b}t$. Letting $\hat{\phi}_1$ denote the estimated coefficient, they suggested transforming to obtain $\hat{Y}_t = X_t - \hat{\phi}_1 X_{t-1}, t = 2, \ldots, n$. Thus, the resulting series, $\hat{Y}_t$, is of length $n - 1$ and satisfies

$$\begin{aligned} \hat{Y}_t &= (1 - \hat{\phi}_1 B)X_t \\ &= a(1 - \hat{\phi}_1) + b(1 - \hat{\phi}_1 B)t + (1 - \hat{\phi}_1 B)Z_t \\ &= c + bt_{\hat{\phi}_1} + g_t, \end{aligned} \tag{9.14}$$

where $t_{\hat{\phi}_1} = t - \hat{\phi}_1(t-1)$. Noting that $Z_t = (1 - \phi_1 B)^{-1} a_t$, we see that $g_t = (1 - \hat{\phi}_1 B)(1 - \phi_1 B)^{-1} a_t$, which is not white noise but should be almost uncorrelated. The Cochrane–Orcutt (CO) procedure is then to test the null hypothesis $H_0 : b = 0$ in (9.14) where the independent variable is $t_{\hat{\phi}_1}$ using standard least-squares methods. That is, the CO test involves calculating the test statistic

$$t_{\text{CO}} = \frac{\hat{b}_{\text{CO}}}{\widehat{SE}(\hat{b}_{\text{CO}})}, \tag{9.15}$$

where $\hat{b}_{\text{CO}}$ is the usual least squares estimate of the slope b in (9.14) and $\widehat{SE}(t_{\text{CO}})$ is its usual standard error. Then, $H_0 : b = 0$ is tested assuming that the statistic t_{CO} in (9.15) has a Students t distribution. Beach and MacKinnon (1978) developed a ML approach that also involves calculation of a test statistic that has a t distribution under the null. While the CO and ML approaches assume that the residuals are AR(1), extensions allow for residuals that are AR(p) or ARMA(p,q). Another test developed by Bloomfield and Nychka (1992) (BN) involves a frequency-domain expression for the standard error of $\hat{b}$.

Although the CO, ML, and BN tests seem reasonable, it has been shown that they tend to have significance levels higher than their nominal levels (see Park and Mitchell, 1980; Woodward and Gray, 1993; Woodward et al., 1997). While these tests have the desired asymptotic properties, Woodward et al. (1997) demonstrate that this asymptotic behavior is often not approached by realizations of moderate length (e.g., $n \approx 100$) in the presence of even moderate correlation structure in the data. For example, Woodward et al. (1997) show that when Z_t is an AR(1) with $\phi_1 = 0.9$ or higher and $b = 0$, realizations of length $n = 1000$ or more may be required in order for the CO, ML, or BN tests to have observed significance levels close to the nominal levels. The simulations performed by Woodward et al. (1997) showed that if the true model is $(1 - 0.9B)X_t = a_t$ (i.e., a stationary model with no deterministic signal), then the CO method, run at the nominal 5% level of significance, (incorrectly) detects a significant trend in 20% of 1000 realizations of length $n = 100$ while Beach and McKinnon's (1978) maximum likelihood method detects a trend in 17% of these realizations. For realizations of length $n = 1000$, the CO and ML procedures detected significant trends in about 8% of the realizations, which are still significantly high.

9.3.2 A Bootstrapping Approach

Woodward et al. (1997) suggested a bootstrap method to estimate the actual distribution of the test statistics obtained using the CO and MLE procedures. In the following, we will consider the application of the bootstrap to t_{CO} in (9.15). First, the t_{CO} statistic is calculated as described previously on the original (observed) data set. To simulate realizations under H_0, an autoregressive model is fit to the observed data. This procedure implicitly assumes under H_0 that any trending behavior in the series is due to the correlation structure alone. Then, K realizations are generated from the autoregressive model fit to the original data. Woodward et al. (1997) used a parametric bootstrap procedure that assumes normal errors and generated the a_t's used in the generated realizations as random normal deviates. If normality of the errors is questionable, then a nonparametric bootstrap could be used in which the residual series, $\hat{a}_t, t = 1, \ldots, K$, from the original AR($p$) fit to the data are resampled with replacement for use as the error terms in the simulated realizations. For the bth realizations, $b = 1, \ldots, K$, we calculate $t_{CO}^{(b)}$ using the CO procedure. For a two-sided test, $H_0 : b = 0$ is rejected at significance level α if $t_{CO} > t^*_{1-\alpha/2}$ or $t_{CO} < t^*_{\alpha/2}$, where t^*_β is the βth empirical quantile of $\left\{t_{CO}^{(b)}\right\}_{b=1}^{K}$ with standard adjustments made for one-sided tests.

9.3.3 Other Methods for Trend Testing

Other researchers have developed tests for slope in the presence of correlated residuals for which the observed significance level is closer to the nominal levels. Canjels and Watson (1997) and Vogelsang (1998) propose tests that are appropriate whether or not the model for Z_t has unit roots, while Sun and Pantula (1999) describe several alternative methods. Brillinger (1989) proposed a test for detecting a monotonic trend, that is, for the case in which s_t is a monotonic function, and Woodward and Gray (1993) demonstrated that this test has inflated significance levels (similar to CO and ML) for moderate realization lengths. Sanders (2009) investigated bootstrap tests for monotonic and other nonparametric trends.

Needless to say, the problem of detecting a trend in the presence of correlated data is a very challenging one. It is worth making special note of the fact that if a significant trend is detected and a nonzero slope coefficient is included in the model, then the implicit decision is made that the trending behavior is predicted to continue. Such a model should be used with caution especially if more than simply short-term forecasts are needed. See Woodward (2006).

9.4 Checking Realization Characteristics

Even though a model is selected by a particular model identification procedure or has residuals that seem appropriate and the model makes some sense physically, it may in fact not be an appropriate model for forecasting, spectral estimation, etc. One useful diagnostic concerning the appropriateness of a model fit to a set of time series data is to determine whether realizations generated from the model have characteristics consistent with those of the observed data. For example, do realizations from the model have spectral densities, sample autocorrelations, and so forth, similar to those of the original data set?

Example 9.3 Checking Realization Characteristics for Models Fit to the Sunspot Data

We consider the annual sunspot numbers shown in Figure 1.23a that, for convenience, are plotted in Figure 9.7a where the horizontal axis is the sequence number (1-261). The 10–11 year cycle previously mentioned along with a tendency for there to be more variability in the heights of the cycles than in the troughs is evident in Figure 9.7a. The sample autocorrelations in Figure 9.7b have a damped sinusoidal pattern with a period of 10–11 years.

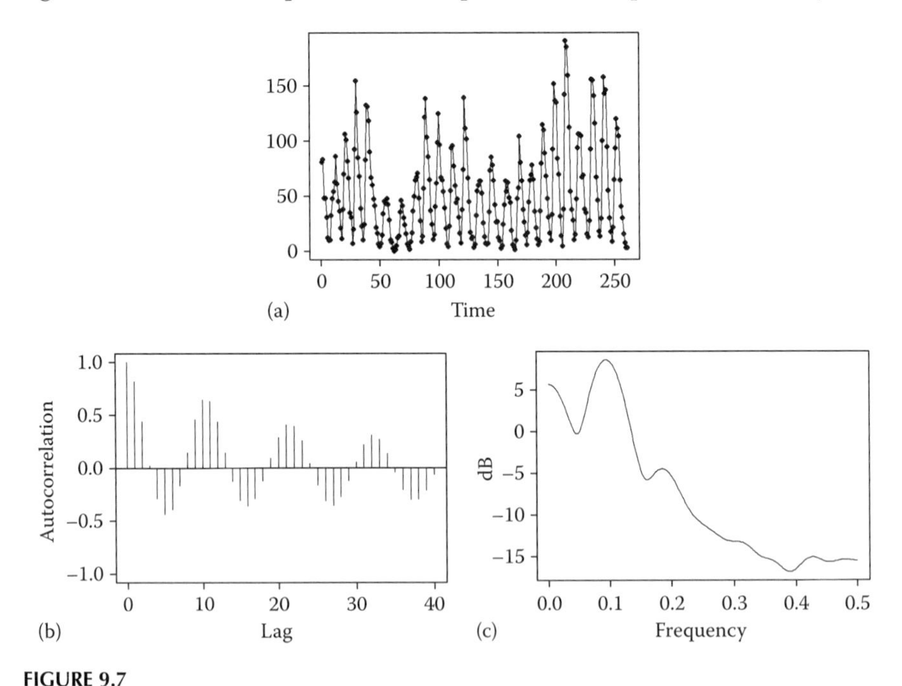

FIGURE 9.7
(a) Sunspot data, (b) sample autocorrelations, and (c) Parzen spectral density estimator using $M = 32$.

The smoothed sample spectral density estimator based on a Parzen window with $M = 32$ is shown in Figure 9.7c, where, as expected, we see a peak at about $f = .1$ associated with the 10–11 year cycle. We also see a peak at $f = 0$ associated with the wandering or piecewise trending behavior of the cycle crests.

A subset of these data (called Series C) was considered by Box et al. (2008), and one of the models they fit to the data set was an AR(2). This model was suggested by the fact that the partial autocorrelations are quite small after the first two partial autocorrelations, which are about 0.8 and –0.7. This same pattern of partial autocorrelations can be found using the updated series in Figure 9.7a. Thus, we fit an AR(2) model to the sunspot data in Figure 9.7a using ML estimates and obtain the model

$$(1 - 1.39B + 0.69B^2)(X_t - 52.2) = a_t, \tag{9.16}$$

with $\hat{\sigma}_a^2 = 293$. Applying AIC with $P = 8$ and $Q = 4$, we select the AR(9) model

$$\begin{aligned}(1 &- 1.165B + 0.40B^2 + 0.18B^3 - 0.155B^4 + 0.10B^5 - 0.025B^6 - 0.024B^7 \\ &+ 0.06B^8 - 0.24B^9)(X_t - 52.2) = a_t\end{aligned} \tag{9.17}$$

with $\hat{\sigma}_a^2 = 238$.

When comparing models (9.16) and (9.17), we first note that (9.16) is a much more parsimonious model and should be considered to be preferable if there are no substantive advantages found for using the higher order model. In Figures 9.8 and 9.9, we show four realizations from models (9.16) and model (9.17), respectively. Clearly, neither model does a good job of generating realizations that have more variability in the peaks of the cycles than the troughs, and the realizations do not satisfy the physical constraint that sunspot numbers are non-negative. For this reason, you might conclude that neither model is adequate. Refer to Tong (1990) for nonlinear models for the sunspot data.

Regardless of the issue regarding the need for a nonlinear model, suppose that we want to fit an ARMA model to the sunspot data and have decided to choose between the AR(2) and AR(9) models given above. It appears that the realizations in Figure 9.9 from the AR(9) model tend to have more consistent 10–11 year cycles (similar to the sunspot series), but only the realization in Figure 9.9d is reasonably close in appearance to the sunspot data. In Figure 9.10a, we show sample autocorrelations for the sunspot data (bold) and for the four realizations in Figure 9.8. It can be seen that the sample autocorrelations for the generated realizations damp out more quickly than those for the sunspot data, and the first damped cycle in the sample autocorrelations often differs from 10–11 lags. In Figure 9.9b, we show the sample autocorrelations for the realizations from the AR(9) model in (9.17), and there it can be seen

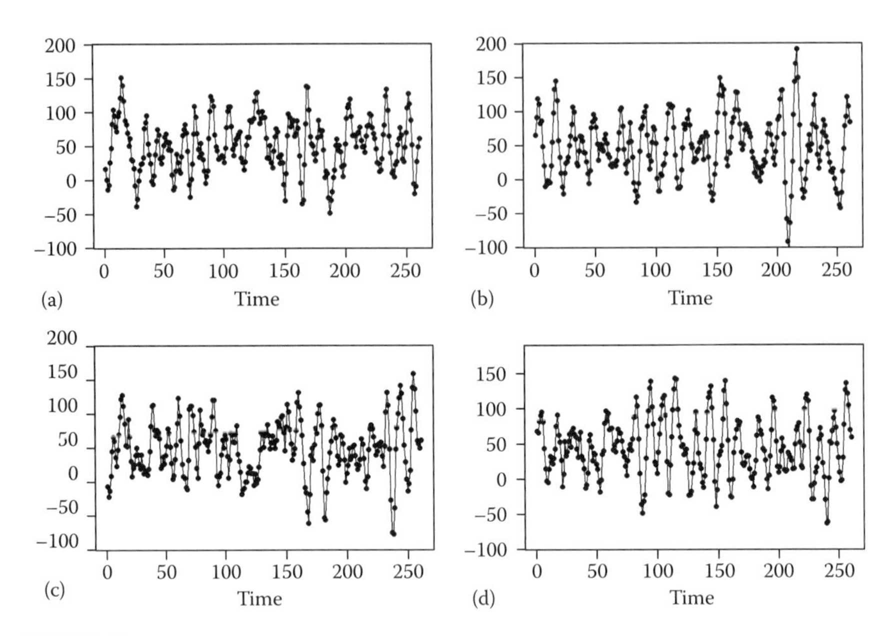

FIGURE 9.8
Four realizations (a–d) from the AR(2) model in (9.16) fit to the sunspot data.

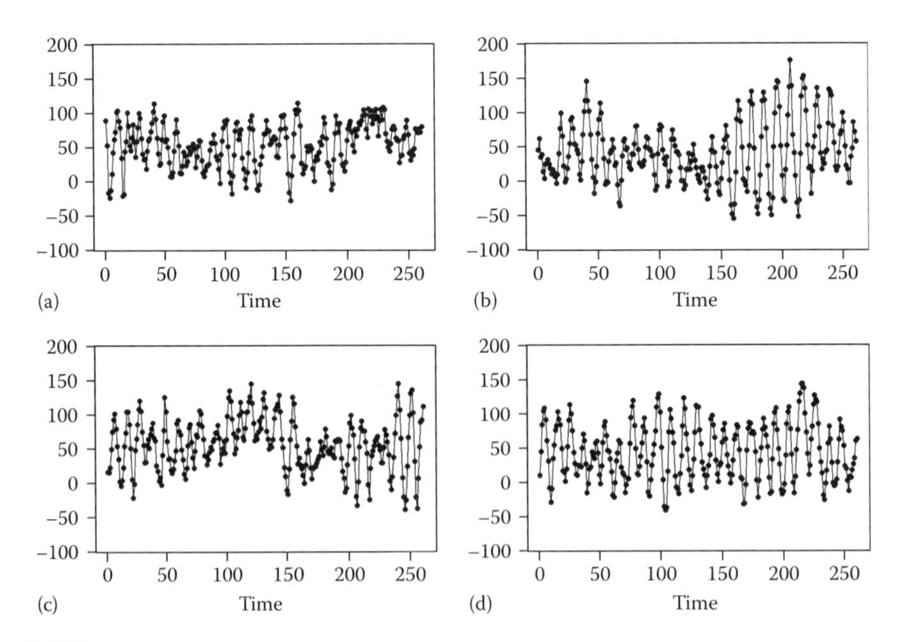

FIGURE 9.9
Four realizations (a–d) from the AR(9) model in (9.17) fit to the sunspot data.

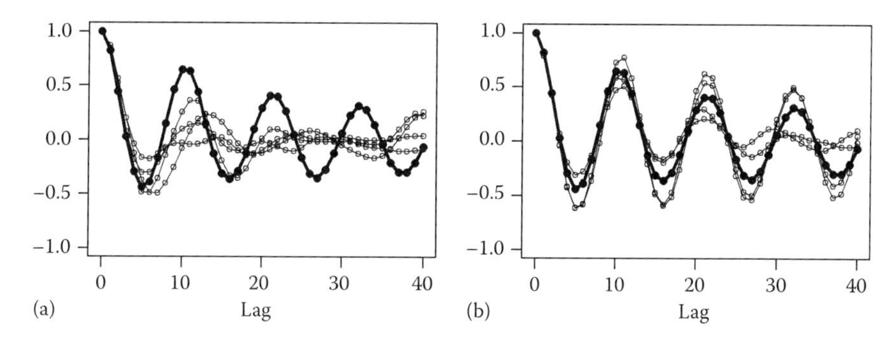

FIGURE 9.10
(a) Sample autocorrelations for sunspot data (bold) and for the four realizations in Figure 9.8. (b) Sample autocorrelations for sunspot data (bold) and for the four realizations in Figure 9.9.

that sample autocorrelations from these realizations are very similar in appearance to those of the sunspot data. Figure 9.11 shows spectral density estimates using a Parzen window with $M = 32$ for the sunspot data (bold) and for the realizations in Figures 9.8 and 9.9 from models (9.16) and (9.17), respectively. The peaks in spectral densities in Figure 9.10a for the realizations from the AR (2) model tend to not be as sharp as the peaks for the sunspot data. Also, there is no consistent tendency for the spectral densities to have a peak at $f = 0$. In Figure 9.11b, the spectral density estimates for the generated realizations from the AR(9) model are quite consistent with the spectral density for the sunspot data typically showing distinct peaks at $f = 0$ and $f = 0.1$. As a consequence, based on examination of the sample autocorrelations and spectral densities, the AR(9) model in (9.17) is the preferred autoregressive model of the two considered. See Woodward and Gray (1978) for a related discussion.

Before proceeding, we note that much can be learned about the AR(2) and AR(9) fits to the sunspot data using the corresponding factor tables given in

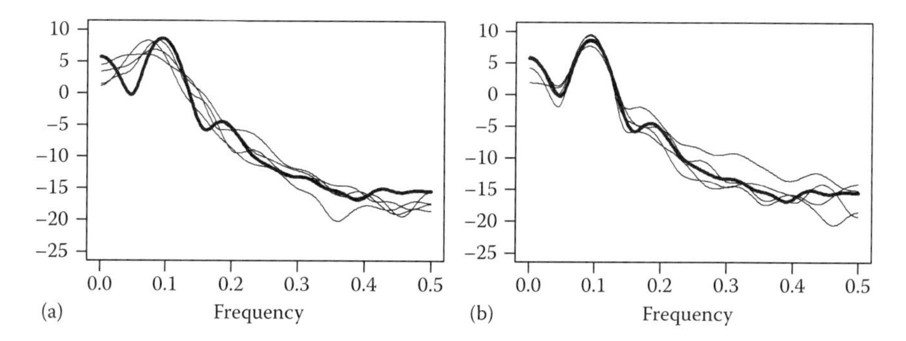

FIGURE 9.11
Smoothed sample spectral density estimates using a Parzen window with $M = 32$ (a) for sunspot data (bold) and for the four realizations in Figure 9.8; (b) for the sunspot data (bold) and for the four realizations in Figure 9.9.

TABLE 9.2

Factor Table for the AR(2) Model in (9.16)

AR Factors	Roots (r_j)	$\lvert r_j^{-1} \rvert$	f_{0j}
$1 - 1.39B + 0.69B^2$	$1.01 \pm 0.66i$	0.83	0.092

TABLE 9.3

Factor Table for the AR(9) Model in (9.17)

AR Factors	Roots (r_j)	$\lvert r_j^{-1} \rvert$	f_{0j}
$1 - 1.62B + 0.96B^2$	$0.85 \pm 0.57i$	0.98	0.095
$1 - 0.94B$	1.06	0.94	0
$1 - 0.62B + 0.74B^2$	$0.42 \pm 1.09i$	0.86	0.19
$1 + 1.47B + 0.61B^2$	$-1.20 \pm 0.44i$	0.78	0.445
$1 + 0.55B + 0.59B^2$	$-0.47 \pm 1.21i$	0.77	0.309

Tables 9.2 and 9.3. Not surprisingly, the AR(2) model has a pair of complex roots associated with a system frequency $f = 0.092$. It is clear that no AR(2) model with a system frequency near $f = 0.1$ can also have another system frequency of $f = 0$. Notice that the spectral density estimates in Figure 9.11a are consistent with this observation. Table 9.3 shows that the roots closest to the unit circle in the AR(9) model are a pair of complex conjugate roots with a system frequency $f = 0.095$, and the second nearest root to the unit circle is a positive real root. The roots associated with the system frequency $f = 0.095$ have absolute reciprocal 0.98. Thus, the associated spectral estimates from realizations from the AR(9) model will have sharper peaks than those from the AR(2) model where the absolute reciprocal of the corresponding roots is 0.83. Additionally, the sample autocorrelations from the AR(9) model would be expected to damp more slowly than those for the AR(2) model as is seen in Figure 9.10. The positive real root in the AR(9) model has absolute reciprocal 0.94, and other factors are associated with roots much further removed from the unit circle. As a consequence, if it is determined that the peak at $f = 0$ in the spectral density estimator in Figure 9.7c for the sunspot data is real, then the AR(9) model is the clear choice.

NOTE: We comment that residuals from the AR(2) model in (9.16) do not pass the white noise tests. However, the AR(2) model considered by Box, et al. (2008) for a shorter sunspot realization did produce residuals that appear to be white. Their AR(2) model produces similar realizations to those seen for the AR(2) model in (9.16). The point is that the estimated model can be inadequate, even if the residuals pass white noise tests.

Tsay (1992) formalizes the procedure discussed in Example 9.3 for assessing the performance of a model fit by examining realizations from the fitted model and assessing how characteristics (e.g., the autocorrelations density) calculated from these realizations compare with the corresponding quantities for the actual data. Consider, for example, the spectral density estimates shown in Figure 9.11. Instead of simply overlaying several spectral density plots as we did here, Tsay (1992) proposes a parametric bootstrap procedure in which many realizations are generated from the fitted model, and for each frequency the empirical $100(1-\alpha/2)\%$ and $100\alpha/2\%$ quantiles of the corresponding sample estimates are obtained. This forms an acceptance envelope for each frequency based on the simulated values. The procedure is to assess the extent to which the spectral density estimates for the observed data remain within these envelopes.

Rather than comparing the entire spectral density with the acceptance envelope, another possible strategy is to compute a few variables of particular interest. For example, for the sunspot models one might calculate $(\hat{S}^{(b)}(0), \hat{S}^{(b)}(0.1))$, i.e., the spectral density estimates at $f = 0$ and $f = 0.1$ for each of the K bootstrap realizations, and compare the observed values of $(\hat{S}(0), \hat{S}(0.1))$ calculated on the original data with this empirical bootstrap distribution. Given a time series realization, Woodward and Gray (1995) introduce a procedure for generating bootstrap realizations from each of two candidate models and using discriminant analysis to ascertain which model generates realizations that best match the characteristics of the observed realization.

9.5 Comprehensive Analysis of Time Series Data: A Summary

In this chapter, we have discussed issues involved in an analysis of time series data. To summarize, given a realization to be modeled, a comprehensive analysis should involve:

1. Examination
   a. Of data
   b. Of sample autocorrelations
2. Obtaining a model
   a. Correlation-based or signal-plus-noise
   b. Identifying p and q
   c. Estimating parameters
3. Checking for model appropriateness
4. Obtaining forecasts, spectral estimates, etc., as dictated by the situation.

We have discussed the issue of model building based on the types of models discussed to this point, that is, ARMA(p,q) models, ARUMA(p,d,q)

models, multiplicative seasonal models, and signal-plus-noise models. It is important to understand that these are not the only models and methods that are available for the modeling and analysis of time series data. In this text, we will discuss the following models and time-series analysis methods.

- Multivariate and state-space models (Chapter 10)
- Long memory models (Chapter 11)
- Wavelet analysis (Chapter 12)
- Models for data with time-varying frequencies (Chapter 13)

Exercises

Applied Problems

9.1 For each of the realizations in Problem 8.2, examine the appropriateness of the fitted models:

(a) Check the whiteness of the residuals.

(b) Plot the associated ARMA spectral density estimator and compare it with a window-based estimator.

(c) Examine the appropriateness of the model in terms of forecasting performance by determining how well the model forecasts the last ℓ steps for whatever value or values you decide ℓ should be.

(d) Generate realizations from the fitted model to determine whether realizations have a similar appearance and characteristics as the original realization.

9.2 For each of the fitted models found in Problem 8.3, examine the model appropriateness using the outline in Problem 9.1.

9.3 For the fitted model found in Problem 8.5, examine the model appropriateness using the outline in Problem 9.1.

9.4 For each of the fitted models found in Problem 8.6, examine the model appropriateness using the outline in Problem 9.1.

9.5 (a) Fit a model to the monthly temperature data in `PATEMP.XLS` of the form

$$X_t = \beta_0 + \beta_1 \cos(2\pi t/12) + \beta_2 \sin(2\pi t/12) + Z_t$$

where Z_t satisfies a stationary ARMA(p,q) process. Estimate $\beta_0, \beta_1, \beta_2, p, q, \phi_1, \ldots, \phi_p, \theta_1, \ldots, \theta_q$, and σ_a^2.

(b) In Example 8.9, an ARUMA(0,2,4) model was found for the data in `PATEMP. XLS`. Compare and contrast these two models.
   (i) How do forecasts from the two models differ?
   (ii) Which model creates realizations most similar to the temperature data?
   (iii) Which model for the temperature data do you prefer?

9.6 Consider a realization, $x_t, t = 1, \ldots, n$, which is known to be from one of the following models:

Model A

$$X_t = a_0 + X_{t-1} + a_t, \quad t = 1, 2, \ldots,$$

where $a_0 \neq 0, X_0 = 0$ and a_t is zero-mean white noise.

Model B

$$X_t = \beta_0 + \beta_1 t + e_t, \quad t = 0, \pm 1, \pm 2, \ldots,$$

where e_t is zero-mean white noise.

(a) For each model:
   (i) Find $E[X_t]$.
   (ii) Find $\text{Var}[X_t]$.
   (iii) For an α_0, β_0, and β_1 that you select, plot a realization you have generated from each model, and describe the appearance of typical realizations.

(b) Consider the transformation $Z_t = X_t - X_{t-1}$
   (i) Discuss the properties of Z_t (stationarity, what model is satisfied by Z_t, etc.) if X_t actually follows Model A. Also discuss the properties of Z_t if X_t satisfies Model B.
   (ii) Discuss how one could use the properties obtained above to distinguish between the two types of models.

(c) Consider the data in file `PROB9.6.XLS`. These data were generated using either Model A or Model B. Based on what you learned in (a) and (b), decide which of these two models was used to generate the data.

9.7 Using the 30 year mortgage-rate data in `ECODATA.XLS`,

(a) Fit an ARMA/ARUMA model (you may have already done this in Problem 8.3).

(b) Use the CO procedure to fit a line + noise model to the data and test the hypothesis $H_0 : b = 0$.

(c) For each model:
   (i) Generate and plot 4 realizations.
   (ii) Compute forecasts of the next 10 years.

(d) Which model do you prefer? Explain based on the results in (c) and any features of the realizations (spectral estimates, etc.) that you believe might provide useful information for discriminating between the two models.

10

Vector-Valued (Multivariate) Time Series

In Chapters 1 through 9, we have considered time series "one at a time." In some situations, it is important to understand and model not only the autocorrelation structure down the time axis but also the relationships among a collection of time series along the same time axis. For example, in Figure 10.1a–c, we show monthly data from January 1, 1991 through April 1, 2010, on interest rates in the United States for 6-month CDs, Moody's seasoned Aaa corporate bonds, and 30-year conventional mortgages, respectively. The understanding of such data may be improved by considering not only the autocorrelation structure of the univariate interest rate series but also the interrelationships among the three as they respond to market and economic conditions across time.

In this chapter, we extend the definitions and techniques presented earlier in the univariate setting to multivariate time series. Among the topics covered in this chapter are basic terminology and stationarity conditions for multivariate time series, multivariate autoregressive-moving average models (focusing on autoregressive models), and their applications. We also discuss the problem of assessing whether two time series are correlated along with state-space models.

10.1 Multivariate Time Series Basics

Loosely speaking, a multivariate time series can be thought of as a collection of vector-valued observations made sequentially in time. To be more formal, we give the following definition that extends Definition 1.1 to the multivariate case.

Definition 10.1 A multivariate stochastic process $\{\mathbf{X}_t; t \in T\}$ is a collection of vector-valued random variables

$$\mathbf{X}_t = \begin{pmatrix} X_{t1} \\ X_{t2} \\ \vdots \\ X_{tm} \end{pmatrix},$$

where T is an index set for which the random variables, $\mathbf{X}_t, t \in T$, are defined on the same sample space. When T represents time, we refer to $\{\mathbf{X}_t; t \in T\}$ as a multivariate time series.

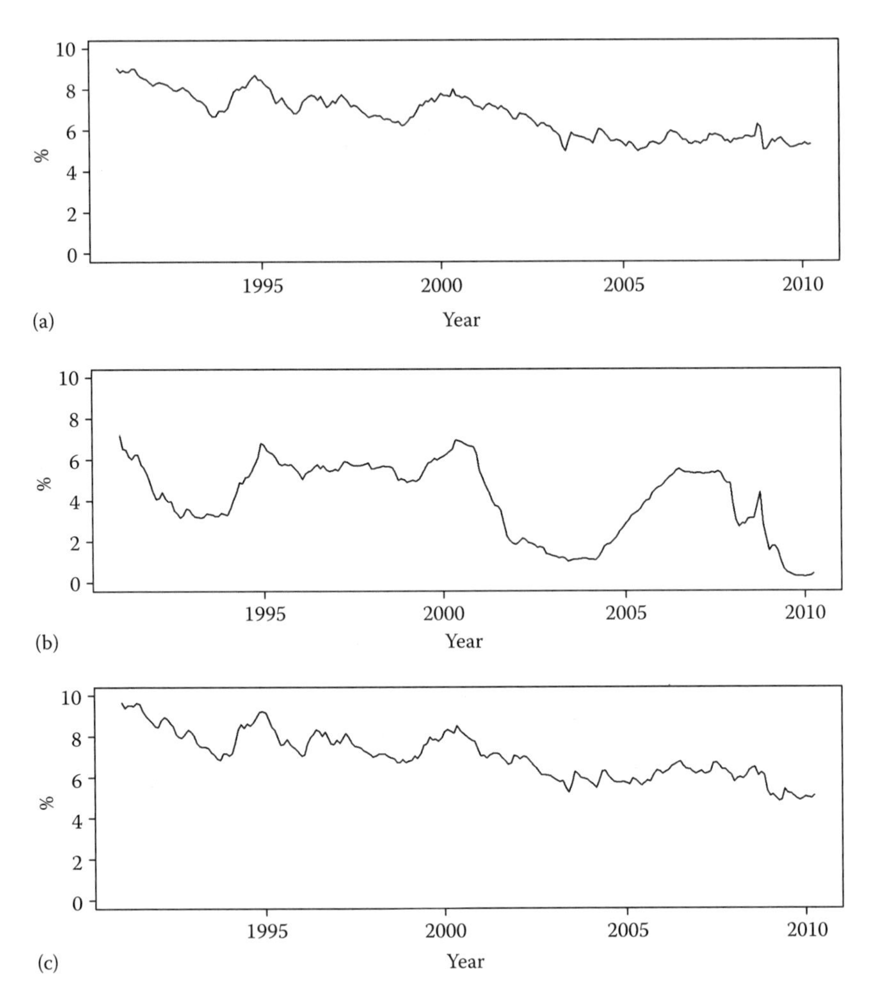

FIGURE 10.1
Monthly interest rates in the US for (a) 6 month CDs, (b) Moody's Seasoned Aaa Corporate Bonds, and (c) 30-year conventional mortgage rates.

NOTES:

1. The individual (univariate) time series, $\{X_{ti}; t \in T\}, i = 1, \ldots, m$ are called *components* of the multivariate time series $\{\mathbf{X}_t; t \in T\}$.
2. Since the components of the multivariate time series discussed in this chapter are discrete parameter time series, we will use the subscript notation for time.
3. As in the univariate case, when no confusion will arise, we shorten the notation $\{\mathbf{X}_t; t \in T\}$ to $\{\mathbf{X}_t\}$ or even $\mathbf{X}_t$.

4. A finite realization, $\mathbf{x}_t, t = 1, \ldots, n$, of a multivariate time series is a collection of the univariate realizations $x_{ti}, t = 1, \ldots, n$, from the component time series, where $i = 1, \ldots, m$. For the interest rate data in Figure 10.1, we let X_{t1} denote rates on 6-month CDs, X_{t2} stands for rates of Moody's seasoned Aaa corporate bonds, and X_{t3} is the 30-year conventional mortgage rate. These three univariate realizations can be considered to be a single realization, $\mathbf{x}_t = (x_{t1}, x_{t2}, x_{t3})'$, from a trivariate time series, $\mathbf{X}_t = (X_{t1}, X_{t2}, X_{t3})'$.
5. The mean of a multivariate time series is given by

$$\boldsymbol{\mu}_t = E[\mathbf{X}_t] = \begin{pmatrix} \mu_{t1} \\ \mu_{t2} \\ \vdots \\ \mu_{tm} \end{pmatrix},$$

where $\mu_{ti} = E[X_{ti}], i = 1, \ldots, m$.

6. The covariance matrix is given by

$$\boldsymbol{\Gamma}(t_1, t_2) = E[(\mathbf{X}_{t_1} - \boldsymbol{\mu}_{t_1})(\mathbf{X}_{t_2} - \boldsymbol{\mu}_{t_2})'] = \begin{pmatrix} \gamma_{11}(t_1, t_2) & \cdots & \gamma_{1m}(t_1, t_2) \\ \vdots & \ddots & \vdots \\ \gamma_{m1}(t_1, t_2) & \cdots & \gamma_{mm}(t_1, t_2) \end{pmatrix},$$

where $\gamma_{ij}(t_1, t_2)$ is the covariance between the univariate random variables $X_{t_1 i}$ and $X_{t_2 j}$, i.e., $\gamma_{ij}(t_1, t_2) = E[(X_{t_1 i} - \mu_{t_1 i})(X_{t_2 j} - \mu_{t_2 j})]$, where $t_1 \in T$ and $t_2 \in T$.

10.2 Stationary Multivariate Time Series

In this section, we discuss the concept of covariance stationarity for multivariate time series. The basic definition is given in Definition 10.2.

Definition 10.2 The multivariate time series $\{\mathbf{X}_t; t \in T\}$ is said to be covariance stationary if

1. $E(\mathbf{X}_t) = \boldsymbol{\mu}$ (constant across t), i.e.,

$$\boldsymbol{\mu} = \begin{pmatrix} \mu_1 \\ \mu_2 \\ \vdots \\ \mu_m \end{pmatrix}.$$

2. $\mathbf{\Gamma}(0) = E[(\mathbf{X}_t - \boldsymbol{\mu})(\mathbf{X}_t - \boldsymbol{\mu})']$ exists and is constant across time t
3. $\mathbf{\Gamma}(t_1, t_2)$ is a function only of $t_2 - t_1$, i.e., $\text{cov}(X_{t_1 i}, X_{t_2 j})$ is a function only of $t_2 - t_1$ for each i and j. For stationary multivariate time series, we will use the notation

$$\mathbf{\Gamma}(k) = \begin{pmatrix} \gamma_{11}(k) & \cdots & \gamma_{1m}(k) \\ \vdots & \ddots & \vdots \\ \gamma_{m1}(k) & \cdots & \gamma_{mm}(k) \end{pmatrix}, \tag{10.1}$$

where $\gamma_{ij}(k) = E[(X_{ti} - \mu_i)(X_{t+k,j} - \mu_j)]$, $k = 0, \pm 1, \ldots$.

Note that if $i \neq j$, then $\gamma_{ij}(k)$ is called the *cross covariance* at lag k. Note also that $\gamma_{ij}(k) = \gamma_{ji}(-k)$ and $\mathbf{\Gamma}(-k) = \mathbf{\Gamma}'(k)$, where A$'$ is the transpose of the matrix A. The jth diagonal element of $\mathbf{\Gamma}(0)$ is $\text{Var}[X_{tj}]$.

The *cross-correlation*, $\rho_{ij}(k)$, is defined by

$$\rho_{ij}(k) = \frac{\gamma_{ij}(k)}{\sqrt{\gamma_{ii}(0)\gamma_{jj}(0)}}, \tag{10.2}$$

and, correspondingly, the correlation matrix $\boldsymbol{\rho}(k)$, $k = 0, 1, \ldots$ is given by

$$\boldsymbol{\rho}(k) = \begin{pmatrix} \rho_{11}(k) & \cdots & \rho_{1m}(k) \\ \vdots & \ddots & \vdots \\ \rho_{m1}(k) & \cdots & \rho_{mm}(k) \end{pmatrix}. \tag{10.3}$$

Note the following:

(a) The autocorrelations for variable X_{tj} are the jth diagonal elements of the matrices $\boldsymbol{\rho}(k)$, $k = 0, \pm 1, \ldots$.

(b) The values $\rho_{ij}(0)$ of $\boldsymbol{\rho}(0)$ measure what is referred to as the concurrent or contemporaneous linear relationship between X_{ti} and X_{tj}. Note that $\boldsymbol{\rho}(0)$ has 1's along the diagonal.

(c) A useful expression for $\boldsymbol{\rho}(k)$ is

$$\boldsymbol{\rho}(k) = \begin{pmatrix} \gamma_{11}^{-1/2} & 0 & \cdots & 0 \\ 0 & \gamma_{22}^{-1/2} & 0 & \vdots \\ \vdots & 0 & \ddots & 0 \\ 0 & \cdots & 0 & \gamma_{mm}^{-1/2} \end{pmatrix} \mathbf{\Gamma}(k) \begin{pmatrix} \gamma_{11}^{-1/2} & 0 & \cdots & 0 \\ 0 & \gamma_{22}^{-1/2} & 0 & \vdots \\ \vdots & 0 & \ddots & 0 \\ 0 & \cdots & 0 & \gamma_{mm}^{-1/2} \end{pmatrix}. \tag{10.4}$$

Definition 10.3 extends the concept of white noise to the multivariate setting.

Definition 10.3 The multivariate time series $\{\mathbf{a}_t; t \in T\}$ is said to be multivariate white noise if

1. $E(\mathbf{a}_t) = \mathbf{0}$ where $\mathbf{0}$ is an m-component vector of 0's.
2. $\boldsymbol{\rho}_{\mathbf{a}}(k) = \mathbf{0}, k \neq 0$, where in this case $\mathbf{0}$ denotes an $m \times m$ matrix of 0's.

It should be noted that Definition 10.3 imposes no restrictions on $\boldsymbol{\rho}_{\mathbf{a}}(0)$. That is, a_{ti} and a_{tj} may be contemporaneously correlated. That is, it may be that $\rho_{ij}(0) \neq 0$ for at least some i, j with $i \neq j$. We use the notation $\mathbf{a}_t \sim \text{MWN}(\mathbf{0}, \boldsymbol{\Gamma}_{\mathbf{a}})$ to specify that $\mathbf{a}_t$ is a multivariate white noise process with $\boldsymbol{\Gamma}(0) = \boldsymbol{\Gamma}_{\mathbf{a}}$. At first glance, it may seem contradictory for a multivariate white noise process to be contemporaneously correlated. However, in Figure 10.2, we show realizations of length $n = 75$ from a bivariate white noise time series $\mathbf{a}_t = (a_{t1}, a_{t2})'$ in which it can be seen that each realization appears to be uncorrelated down the time axis, but close inspection shows that the two realizations are positively correlated. In this case, $\boldsymbol{\Gamma}(0) = \boldsymbol{\Gamma}_{\mathbf{a}}$ is given by

$$\boldsymbol{\Gamma}_{\mathbf{a}} = \begin{pmatrix} 1 & 1.78 \\ 1.78 & 4 \end{pmatrix},$$

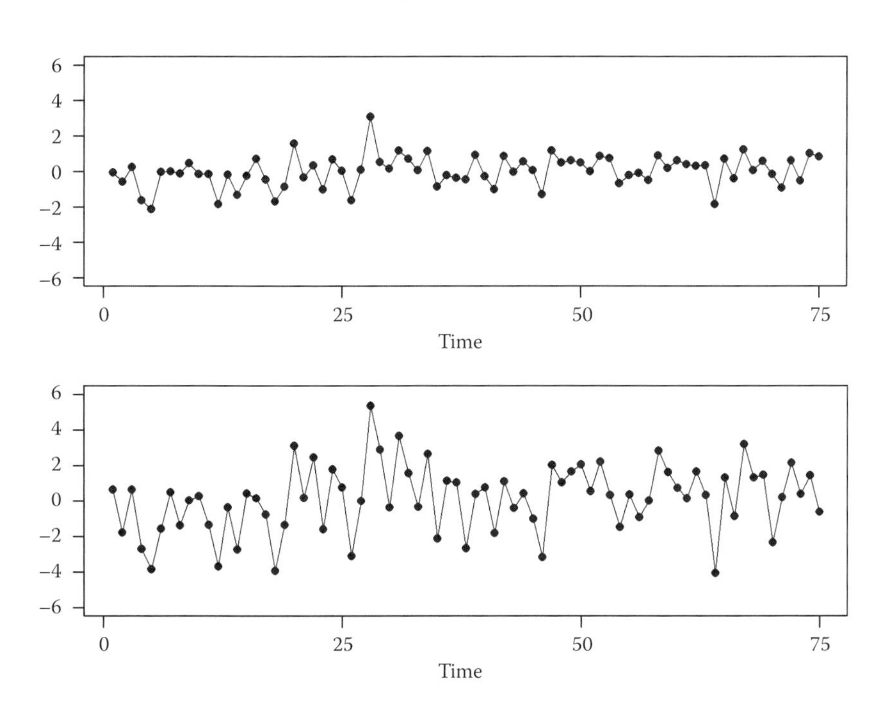

FIGURE 10.2
Realization from a bivariate white noise process with positive contemporaneous correlation between the two white noise processes, i.e., $\rho_{12}(0) \neq 0$.

and

$$\boldsymbol{\rho}_{\mathbf{a}}(0) = \begin{pmatrix} 1 & 0.89 \\ 0.89 & 1 \end{pmatrix}. \tag{10.5}$$

Definition 10.4 extends the concept of a general linear process defined in Definition 2.1 to the multivariate case, and Definition 10.5 defines the spectrum of a multivariate time series.

Definition 10.4 A multivariate time series $\{\mathbf{X}_t; t = 0, \pm 1, \pm 2, \ldots\}$ is said to be a causal and stationary multivariate general linear process if it is given by

$$\mathbf{X}_t - \boldsymbol{\mu} = \sum_{j=0}^{\infty} \boldsymbol{\Psi}_j \mathbf{a}_{t-j}, \tag{10.6}$$

where $\mathbf{a}_t$ is multivariate white noise and the $\boldsymbol{\Psi}_j$'s are $m \times m$ matrices, such that $\boldsymbol{\Psi}_0 = \mathbf{I}$ (the $m \times m$ identity matrix) and whose components are absolutely summable. That is, letting $\psi_j(k, l)$ denote the element in the kth row and lth column of $\boldsymbol{\Psi}_j$, then $\sum_{j=0}^{\infty} |\psi_j(k, l)| < \infty$ for $1 \le k, l \le m$.

Definition 10.5 Let $\{\mathbf{X}_t; t = 0, \pm 1, \pm 2, \ldots\}$ be a causal stationary multivariate process with covariance matrix $\boldsymbol{\Gamma}(k)$ given in (10.1) and where $\sum_{k=-\infty}^{\infty} |\gamma_{jh}(k)| < \infty$ for $1 \le j, h \le m$. For $0 \le f \le 0.5$, the power spectrum of $\mathbf{X}_t$ is given by

$$\mathbf{P}_{\mathbf{X}}(f) = \sum_{k=-\infty}^{\infty} \mathrm{e}^{-2\pi i f k} \boldsymbol{\Gamma}(k). \tag{10.7}$$

Note that in the multivariate case, $\mathbf{P}_{\mathbf{X}}(f)$ in (10.7) is often called the *spectral density*. For consistency of notation with the univariate case, we will refer to it as the spectrum. Note also that $\mathbf{P}_{\mathbf{X}}(f)$ is an $m \times m$ matrix for each f. Analogous to (1.28) for the univariate case, it can be shown that

$$\boldsymbol{\Gamma}(k) = \int_{-0.5}^{0.5} \mathrm{e}^{2\pi i f k} \mathbf{P}_{\mathbf{X}}(f) df, \tag{10.8}$$

where $i = \sqrt{-1}$. The jhth element of $\mathbf{P}_{\mathbf{X}}(f)$ is given by

$$P_{jh}(f) = \sum_{k=-\infty}^{\infty} \mathrm{e}^{-2\pi i f k} \gamma_{jh}(k),$$

where $|f| \le 0.5$, and the $\gamma_{jh}(k)$'s are defined following (10.1). Note that $P_{jj}(f)$ is the usual spectrum of the univariate time series X_{tj}, while if $j \ne h$ then

$P_{jh}(f)$ is called the *cross spectrum* of X_{tj} and X_{th}, which can be complex valued with the property that $P_{hj}(f)$ is the complex conjugate of $P_{jh}(f)$. Denoting $P_{jh}(f) = c_{jh}(f) - iq_{jh}(f)$, then $c_{jh}(f)$ is called the *cospectrum* of X_{tj} and X_{th}, and $q_{jh}(f)$ is called the *quadrature spectrum*. Expressing $P_{jh}(f)$ in polar form, we have $P_{jh}(f) = \alpha_{jh}(f)e^{i\varphi_{jh}(f)}$, where $\alpha_{jh}(f) = \sqrt{c_{jh}^2(f) + q_{jh}^2(f)}$ is referred to as the *amplitude spectrum* and $\varphi_{jh}(f) = \tan^{-1}(-q_{jh}(f)/c_{jh}(f))$ is the *phase spectrum*.

The *squared coherency*, $\kappa_{jh}^2(f)$, of the two univariate time series X_{tj} and X_{th} is defined by

$$\kappa_{jh}^2(f) = \frac{|P_{jh}(f)|^2}{P_{jj}(f)P_{hh}(f)}. \tag{10.9}$$

It should be noted that $\kappa_{jh}^2(f) \leq 1$ and can be interpreted as a squared correlation-type measure of the linear relationship between the random components of X_{tj} and X_{th} at frequency f.

10.2.1 Estimating the Mean and Covariance for Stationary Multivariate Processes

We briefly discuss the estimation of the mean vector $\boldsymbol{\mu}$ and the covariance matrix $\boldsymbol{\Gamma}(k)$ based on a realization of length n from the stationary multivariate process $\{\mathbf{X}_t; t = 0, \pm 1, \pm 2, \ldots\}$. These estimation procedures are entirely analogous to the corresponding procedures for univariate processes.

10.2.1.1 Estimating $\boldsymbol{\mu}$

The natural unbiased estimator of the m-component vector $\boldsymbol{\mu}$ is the m-component vector $\overline{\mathbf{X}}$ given by

$$\overline{\mathbf{X}} = \begin{pmatrix} \overline{X}_1 \\ \overline{X}_2 \\ \vdots \\ \overline{X}_m \end{pmatrix}, \tag{10.10}$$

where $\overline{X}_j = \frac{1}{n}\sum_{t=1}^{n} X_{tj}$. We say that $\overline{\mathbf{X}}$ is ergodic for $\boldsymbol{\mu}$ if $\overline{\mathbf{X}}$ converges in the mean-square sense as $n \to \infty$, i.e., if $E\left[\left(\overline{\mathbf{X}} - \boldsymbol{\mu}\right)\left(\overline{\mathbf{X}} - \boldsymbol{\mu}\right)'\right] \to \mathbf{0}$. Corollary 1.1 gives a sufficient condition for ergodicity in the univariate case, and by applying that result to each component of $\overline{\mathbf{X}}$, it follows that $\overline{\mathbf{X}}$ is ergodic for $\boldsymbol{\mu}$ if $\lim_{k\to\infty} \gamma_{jj}(k) = 0$ for each j, $1 \leq j \leq m$.

10.2.1.2 *Estimating* $\mathbf{\Gamma}(k)$

It is common practice to estimate the covariance matrix $\mathbf{\Gamma}(k)$ by the $m \times m$ matrix $\hat{\mathbf{\Gamma}}(k)$ given by

$$\hat{\mathbf{\Gamma}}(k) = \begin{pmatrix} \hat{\gamma}_{11}(k) & \cdots & \hat{\gamma}_{1m}(k) \\ \vdots & \ddots & \vdots \\ \hat{\gamma}_{m1}(k) & \cdots & \hat{\gamma}_{mm}(k) \end{pmatrix}, \tag{10.11}$$

where $\hat{\gamma}_{ij}(k) = \frac{1}{n}\sum_{t=1}^{n-k}(X_{ti} - \overline{X}_i)(X_{t+k,j} - \overline{X}_j)$. Note that the estimator in (10.11) is analogous to the use of (1.19) to estimate γ_k. For discussion of the properties of the estimators, see Brockwell and Davis (1991, 2002). The cross-correlation $\rho_{ij}(k)$ between X_{ti} and X_{tj} at lag k is estimated by

$$\hat{\rho}_{ij}(k) = \frac{\hat{\gamma}_{ij}(k)}{\sqrt{\hat{\gamma}_{ii}(0)\hat{\gamma}_{jj}(0)}}. \tag{10.12}$$

Note that $\hat{\rho}_{ij}(0)$ is the usual Pearson product-moment correlation calculated on the basis of the n observations on X_{ti} and X_{tj}.

10.3 Multivariate (Vector) ARMA Processes

The AR(p), MA(q), ARMA(p,q), ARUMA(p,d,q), and corresponding seasonal models can be extended to the multivariate case. We note that our focus will be almost exclusively on the multivariate AR(p) process. We begin this discussion with Example 10.1.

Example 10.1 A Bivariate Process

Consider the bivariate process $\mathbf{X}_t = \begin{pmatrix} X_{t1} \\ X_{t2} \end{pmatrix}$, $t = 0, \pm 1, \pm 2, \ldots$ specified by

$$\begin{aligned} X_{t1} &= \phi_{11}X_{t-1,1} + \phi_{12}X_{t-1,2} + a_{t1} \\ X_{t2} &= \phi_{21}X_{t-1,1} + \phi_{22}X_{t-1,2} + a_{t2}, \end{aligned} \tag{10.13}$$

where and $\mathbf{a}_t = (a_{t1}, a_{t2})'$ is a bivariate white noise processes. The equations in (10.13) can be written in matrix form

$$\mathbf{X}_t = \mathbf{\Phi}_1\mathbf{X}_{t-1} + \mathbf{a}_t, \tag{10.14}$$

where $\mathbf{\Phi}_1 = \begin{pmatrix} \phi_{11} & \phi_{12} \\ \phi_{21} & \phi_{22} \end{pmatrix}$. The bivariate process in (10.13) is an example of a multivariate (or vector) AR(1) process. The vector AR(p) process is defined in Definition 10.6 that follows.

Definition 10.6 Let $\mathbf{X}_t$ be a causal, stationary multivariate process that satisfies

$$\mathbf{X}_t - \mathbf{\Phi}_1\mathbf{X}_{t-1} - \cdots - \mathbf{\Phi}_p\mathbf{X}_{t-p} = \boldsymbol{\alpha} + \mathbf{a}_t, \tag{10.15}$$

where

(a) $\mathbf{X}_t = (X_{t1}, \ldots, X_{tm})'$ is an $m \times 1$ vector
(b) $\mathbf{\Phi}_k$ is a real-valued $m \times m$ matrix for each $k = 1, \ldots, p$
(c) $\mathbf{a}_t$ is multivariate white noise with covariance matrix $E\left[\mathbf{a}_t\mathbf{a}_t'\right] = \mathbf{\Gamma}_\mathbf{a}$
(d) $\boldsymbol{\alpha} = (\mathbf{I} - \mathbf{\Phi}_1 - \cdots - \mathbf{\Phi}_p)\boldsymbol{\mu}$, where $\boldsymbol{\mu} = E(\mathbf{X}_t)$.

Then $\mathbf{X}_t$ is called a VAR(p) process, i.e., a vector autoregressive process of order p.

Note that (10.15) can be written in multivariate operator notation $\mathbf{\Phi}(B)(\mathbf{X}_t - \boldsymbol{\mu}) = \mathbf{a}_t$, where

$$\mathbf{\Phi}(B) = \mathbf{I} - \mathbf{\Phi}_1 B - \cdots - \mathbf{\Phi}_p B^p, \tag{10.16}$$

and $B^k\mathbf{X}_t = \mathbf{X}_{t-k}$. Theorem 10.1 below gives stationarity conditions for VAR(p) models that are analogous to those in Theorem 3.2 for univariate AR(p) models.

Theorem 10.1

A multivariate process $\mathbf{X}_t$ satisfying the difference equation in (10.15) is a stationary and causal VAR(p) process if and only if the roots of the determinantal equation,

$$|\mathbf{\Phi}(z)| = \left|\mathbf{I} - \mathbf{\Phi}_1 z - \cdots - \mathbf{\Phi}_p z^p\right| = 0,$$

lie outside the unit circle.

Proof:

See Brockwell and Davis (1991) and Reinsel (1997).

Note that a VAR(p) model $\mathbf{\Phi}(B)(\mathbf{X}_t - \boldsymbol{\mu}) = \mathbf{a}_t$ satisfying Theorem 10.1 can be written in the multivariate general linear process form

$$\begin{aligned}\mathbf{X}_t - \boldsymbol{\mu} &= \mathbf{\Psi}(B)\mathbf{a}_t \\ &= \sum_{j=0}^{\infty} \mathbf{\Psi}_j \mathbf{a}_{t-j},\end{aligned} \tag{10.17}$$

where $\mathbf{\Psi}(B) = \mathbf{\Phi}^{-1}(B)$.

The multivariate Yule–Walker equations are given by

$$\mathbf{\Gamma}(k) = \mathbf{\Gamma}(k-1)\mathbf{\Phi}_1' + \cdots + \mathbf{\Gamma}(k-p)\mathbf{\Phi}_p', \quad k = 1, \ldots, p \tag{10.18}$$

where

$$\mathbf{\Gamma}(0) = \mathbf{\Gamma}(-1)\mathbf{\Phi}_1' + \cdots + \mathbf{\Gamma}(-p)\mathbf{\Phi}_p' + \mathbf{\Gamma}_\mathbf{a}, \tag{10.19}$$

which is analogous to (3.42) in the univariate case.

In the VAR(1) case (10.18) and (10.19) simplify to $\mathbf{\Gamma}(k) = \mathbf{\Gamma}(k-1)\mathbf{\Phi}_1', k > 0$ and $\mathbf{\Gamma}(0) = \mathbf{\Gamma}(-1)\mathbf{\Phi}_1' + \mathbf{\Gamma}_\mathbf{a}$, respectively. Since $\mathbf{\Gamma}(1) = \mathbf{\Gamma}(0)\mathbf{\Phi}_1'$ for a VAR(1) and since $\mathbf{\Gamma}(-k) = \mathbf{\Gamma}'(k)$, for any causal stationary multivariate process, it follows that $\mathbf{\Gamma}(0) = \mathbf{\Phi}_1\mathbf{\Gamma}(0)\mathbf{\Phi}_1' + \mathbf{\Gamma}_\mathbf{a}$. These expressions illustrate the fact that knowing $\mathbf{\Phi}_1$ (and $\mathbf{\Gamma}_\mathbf{a}$) for a VAR(1) allows for straightforward computation of the covariance matrices $\mathbf{\Gamma}(k)$, $k = 0, \pm 1, \ldots$. Since $\mathbf{\Phi}_1' = \mathbf{\Gamma}^{-1}(0)\mathbf{\Gamma}(1)$, it follows that $\mathbf{\Phi}_1$ can also be computed if $\mathbf{\Gamma}(0)$ and $\mathbf{\Gamma}(1)$ are known. These relationships are analogous to properties previously observed for the univariate AR(1) case. See for example (3.34) and (3.35).

Example 10.2 A Bivariate VAR(1) Process

Consider the zero-mean VAR(1) process $\mathbf{\Phi}(B)\mathbf{X}_t = (\mathbf{I} - \mathbf{\Phi}_1 B)\mathbf{X}_t = \mathbf{a}_t$, where

$$\mathbf{\Phi}_1 = \begin{pmatrix} 0.9 & 0.6 \\ -0.3 & 0.7 \end{pmatrix} \tag{10.20}$$

and

$$\mathbf{\Gamma}_\mathbf{a} = \begin{pmatrix} 1 & 1.78 \\ 1.78 & 4 \end{pmatrix} \tag{10.21}$$

In this example $\mathbf{a}_t$ has the same $\mathbf{\Sigma}_\mathbf{a}$ as that for the bivariate white noise realization shown in Figure 10.2. Equations (10.13) in this case are

$$\begin{aligned} X_{t1} &= 0.9X_{t-1,1} + 0.6X_{t-1,2} + a_{t1} \\ X_{t2} &= -0.3X_{t-1,1} + 0.7X_{t-1,2} + a_{t2}, \end{aligned} \tag{10.22}$$

and the equation $|\mathbf{I} - \mathbf{\Phi}_1 z| = \begin{vmatrix} 1 - \phi_{11}z & -\phi_{12}z \\ -\phi_{21}z & 1 - \phi_{22}z \end{vmatrix} = 0$ is given by $(\phi_{11}\phi_{22} - \phi_{12}\phi_{21})z^2 - (\phi_{11} + \phi_{22})z + 1 = 0$. For this example, the equation is $1 - 1.6z + 0.81z^2 = 0$, which has roots r and r^* given by $0.99 \pm 0.51i$ and $|r^{-1}| = 0.90$. Thus, the roots are outside the unit circle.

NOTE: The equation, $1 - 1.6z + 0.81z^2 = 0$, is the same form as that of a characteristic equation for a univariate ARMA(p,q) model. Consequently the roots (and associated assessment concerning stationarity) can be obtained from the `Factoring Routines` in the `Time Series Analysis` menu of `GW-WINKS` and entering 1.6, −0.81 as the coefficients.

A realization of length $n = 75$ from the VAR(1) process in (10.22) is shown in Figure 10.3 where it can be seen that even the VAR(1) model can exhibit pseudo-cyclic behavior in the bivariate case. Using the fact that $\boldsymbol{\Gamma}(0) - \boldsymbol{\Phi}_1\boldsymbol{\Gamma}(0)\boldsymbol{\Phi}_1' = \boldsymbol{\Gamma}_{\mathbf{a}}$, and then solving for $\boldsymbol{\Gamma}(0)$, it follows that $\boldsymbol{\Gamma}(0) = \begin{pmatrix} 27.74 & -0.40 \\ -0.40 & 13.07 \end{pmatrix}$. This is consistent with Figure 10.3, where it can be seen that X_{t1} (upper plot) has larger variance than X_{t2} (lower plot). Using the property that $\boldsymbol{\Gamma}(k) = \boldsymbol{\Gamma}(k-1)\boldsymbol{\Phi}_1'$, $k > 0$, it can be shown that $\boldsymbol{\Gamma}(1) = \begin{pmatrix} 27.73 & -8.60 \\ 7.48 & 9.37 \end{pmatrix}$, $\boldsymbol{\Gamma}(2) = \begin{pmatrix} 17.09 & -13.34 \\ 12.30 & 4.24 \end{pmatrix}, \ldots$.

To find the correlation matrices $\boldsymbol{\rho}(k)$, we use the bivariate version of (10.4) to yield $\boldsymbol{\rho}(k) = \begin{pmatrix} \gamma_{11}^{-1/2} & 0 \\ 0 & \gamma_{22}^{-1/2} \end{pmatrix} \boldsymbol{\Gamma}(k) \begin{pmatrix} \gamma_{11}^{-1/2} & 0 \\ 0 & \gamma_{22}^{-1/2} \end{pmatrix}$, from which it follows that

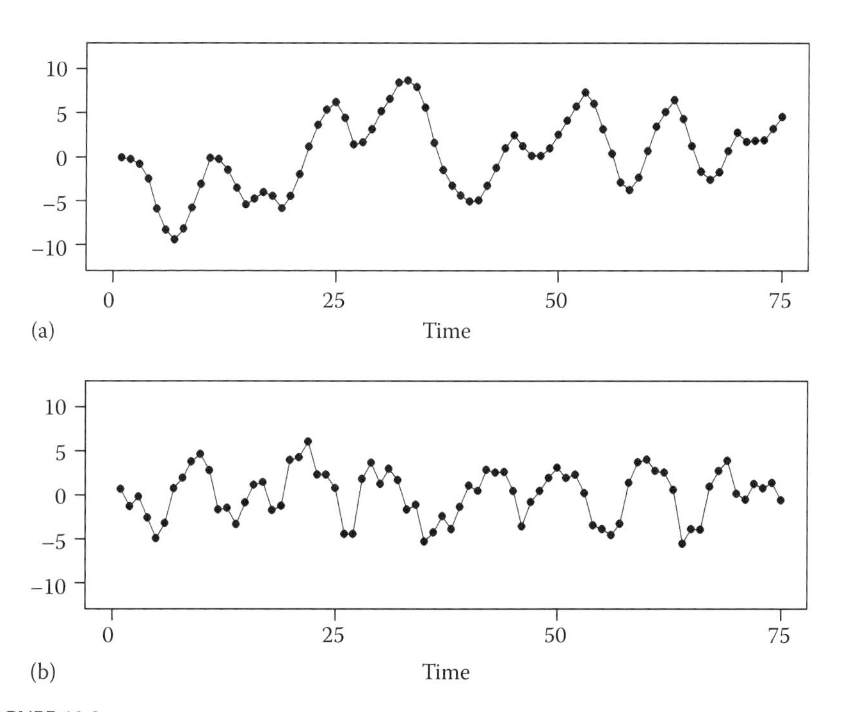

FIGURE 10.3
Realizations from the bivariate VAR(1) process in Example 10.2. (a) Realization from X_{t1}. (b) Realization from X_{t2}.

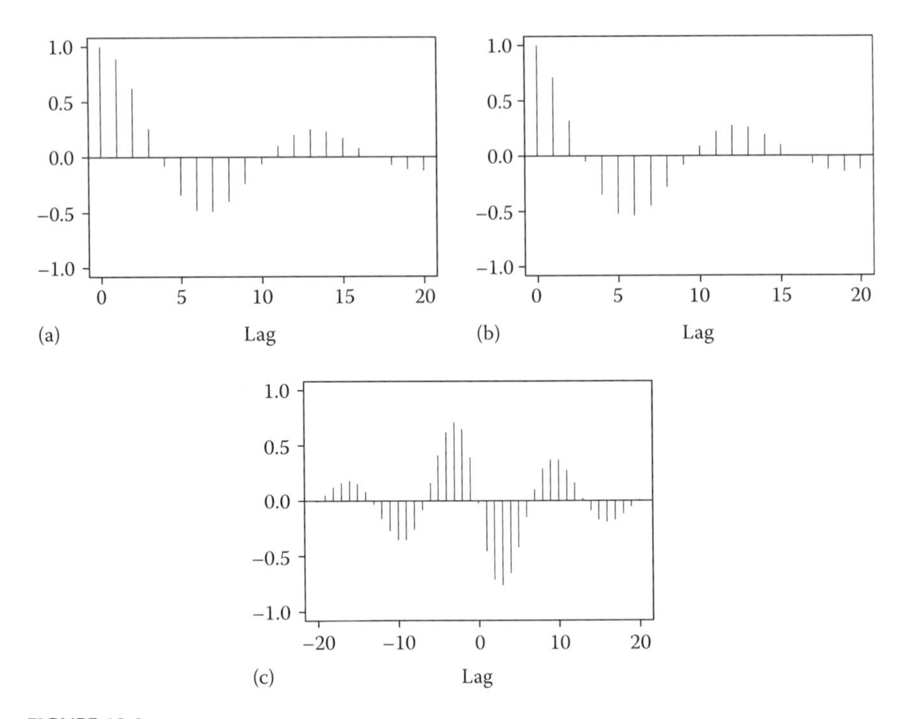

FIGURE 10.4
Autocorrelations, $\rho_{11}(k)$ and $\rho_{22}(k)$, along with cross correlations, $\rho_{12}(k)$ for the bivariate AR(1) process in Example 10.2. (a) ρ_{11} for realization in Figure 10.2a. (b) ρ_{22} for realization in Figure 10.2b. (c) ρ_{12} for bivariate realization.

$\boldsymbol{\rho}(0) = \begin{pmatrix} 1 & -0.02 \\ -0.02 & 1 \end{pmatrix}$, $\boldsymbol{\rho}(1) = \begin{pmatrix} 0.89 & -0.45 \\ 0.39 & 0.71 \end{pmatrix}$, $\boldsymbol{\rho}(2) = \begin{pmatrix} 0.62 & -0.71 \\ 0.65 & 0.32 \end{pmatrix}$, $\boldsymbol{\rho}(3) = \begin{pmatrix} 0.26 & -0.76 \\ 0.72 & -0.06 \end{pmatrix}, \ldots$. From these matrices, it can be seen that while the contemporaneous cross-correlation between X_{t1} and X_{t2} is small ($\rho_{12}(0) = -0.02$), there is substantial cross correlation at lags 1, 2, and 3. Figure 10.4a and b show the autocorrelations for X_{t1} and X_{t2}, respectively, where it can be seen that the autocorrelations reflect a damped sinusoidal behavior with period about 14 for both X_{t1} and X_{t2}. The cross-correlations in Figure 10.4c show that while there is very little contemporaneous cross-correlation, the cross correlations, $\gamma_{12}(k)$, for $k = 1$ through 5 are more substantial (and negative) while for $k = -1$ to -5, the cross correlations are positive.

This relationship between X_{t1} and X_{t2} is not easy to visualize from examination of the realizations in Figure 10.3. In Figure 10.5a, we show the two realizations in Figure 10.3 overlaid, and it can be seen that there is very little contemporaneous correlation between X_{t1} and X_{t2}, which is consistent with the fact that $\rho_{12}(0) = -0.02$. From Figure 10.4c, we see that the maximum positive value for $\rho_{12}(k)$ occurs at $k = -3$. Figure 10.5b shows overlays of

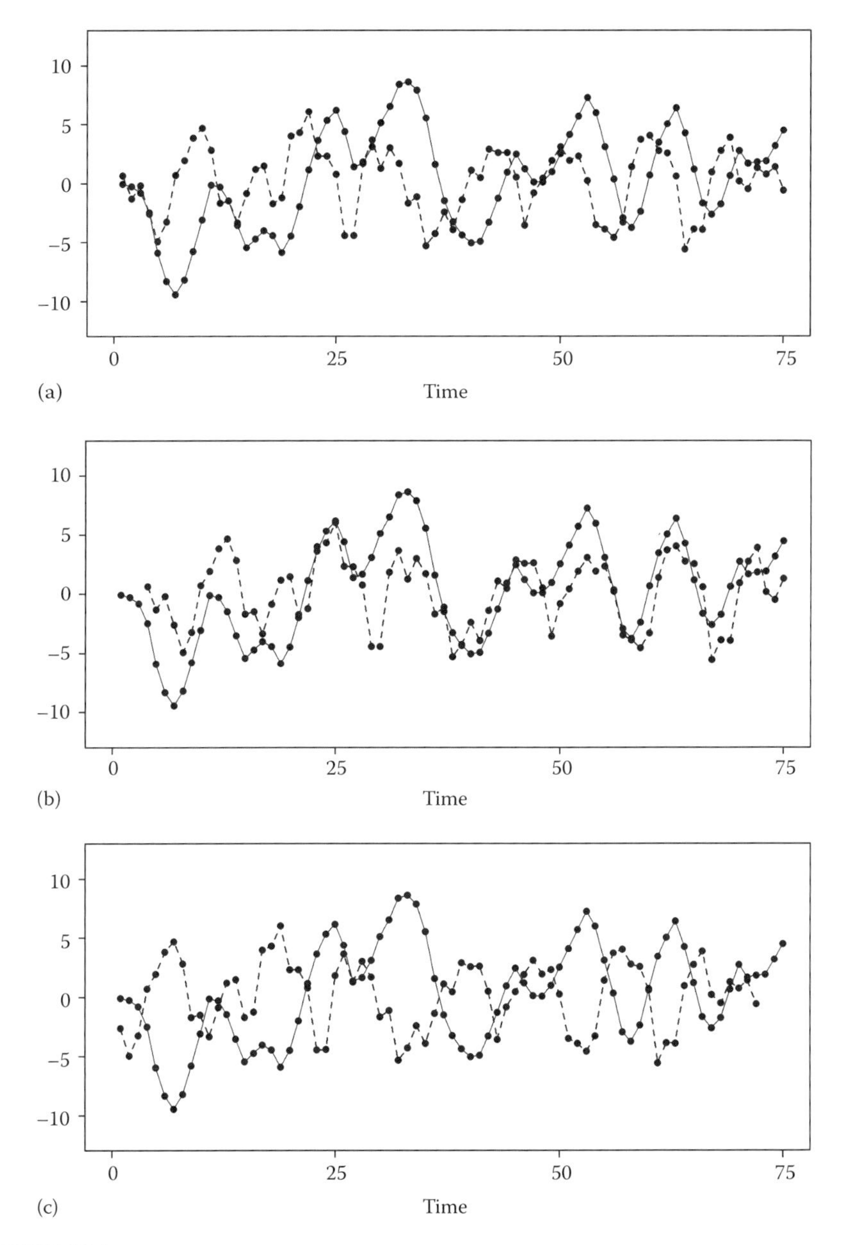

FIGURE 10.5
Overlays based on bivariate realization shown in Figure 10.3 showing (a) X_{t1} and X_{t2}, (b) X_{t1} and $X_{t-3,2}$, and (c) X_{t1} and $X_{t+3,2}$.

X_{t1} and $X_{t-3,2}$, where it can be seen that the resulting curves are positively correlated, i.e., they tend to reach their cyclic peaks and troughs together. Correspondingly, in Figure 10.5c, we show overlays of X_{t1} and $X_{t+3,2}$, which reveal a negative correlation that is consistent with the fact that $\rho_{12}(3) = -0.76$.

Definition 10.7 Let $\mathbf{X}_t$ be a causal, stationary multivariate process that satisfies

$$\mathbf{X}_t - \mathbf{\Phi}_1\mathbf{X}_{t-1} - \cdots - \mathbf{\Phi}_p\mathbf{X}_{t-p} = \boldsymbol{\alpha} + \mathbf{a}_t - \mathbf{\Theta}_1\mathbf{a}_{t-1} - \cdots - \mathbf{\Theta}_q\mathbf{a}_{t-q},$$

where the $\mathbf{\Phi}_k$'s and $\mathbf{\Theta}_k$'s are real-valued $m \times m$ matrices, and $\mathbf{a}_t$ is multivariate white noise. Then, $\mathbf{X}_t$ is a VARMA(p,q) process.

NOTES:

1. A VMA(q) process is a VARMA(p,q) process with $p = 0$.
2. Recall that Definition 3.5, which defines the univariate ARMA(p,q) process, includes the restriction that the operators $\phi(B)$ and $\theta(B)$ have no common factors. In the case of the VARMA(p,q) process, there are additional related restrictions not listed in Definition 10.7. One of the problems with multivariate ARMA(p,q) processes is that of identifiability. For example, there are cases in which
   a. A bivariate AR(1) model is identical to a bivariate MA(1) model
   b. Two bivariate ARMA(1,1) models have different parameters but have identical properties.

Because of these issues, we restrict our attention in this chapter to VAR(p) processes. For more discussion of VARMA(p,q) processes and the identifiability problem, see Box et al. (2008), Tsay (2002), and Shumway and Stoffer (2006).

10.3.1 Forecasting Using VAR(p) Models

Given a VAR(p) model, Box–Jenkins style forecasts, $\hat{\mathbf{X}}_{t_0}(\ell)$, for $\mathbf{X}_{t_0+\ell}$ "assuming" the infinite past $\mathbf{X}_t$, $t = t_0, t_0 - 1, t_0 - 2, \ldots$ can be obtained in a manner analogous to that given in Section 6.2 in the univariate case. Specifically, these forecasts can be expressed in terms of the difference equation, giving

$$\hat{\mathbf{X}}_{t_0}(\ell) = \sum_{j=1}^{p} \mathbf{\Phi}_j\hat{\mathbf{X}}_{t_0}(\ell - j) + (\mathbf{I} - \mathbf{\Phi}_1 - \cdots - \mathbf{\Phi}_p)\boldsymbol{\mu}, \tag{10.23}$$

which corresponds to (6.21). As in Chapter 6, we calculate forecasts based on the assumption that the model is known. Since $\boldsymbol{\mu}$ must be estimated from data, we express the basic formula for calculating forecasts from a VAR(p) model as

$$\hat{\mathbf{X}}_{t_0}(\ell) = \sum_{j=1}^{p} \mathbf{\Phi}_j \hat{\mathbf{X}}_{t_0}(\ell - j) + (\mathbf{I} - \mathbf{\Phi}_1 - \cdots - \mathbf{\Phi}_p)\overline{\mathbf{X}}, \tag{10.24}$$

which corresponds to (6.24) in the univariate case. In practice, the model is estimated from the data, and the $\mathbf{\Phi}_j'$s in (10.24) are replaced by their estimated values $\hat{\mathbf{\Phi}}_j$. Parameter estimation in VAR(p) models will be discussed shortly.

Example 10.3 Forecasts Based on the Bivariate VAR(1) Process in Example 10.2

For the bivariate VAR(1) process in Example 10.2, the forecasts in (10.24) become

$$\begin{aligned}\hat{\mathbf{X}}_{t_0}(\ell) &= \mathbf{\Phi}_1 \hat{\mathbf{X}}_{t_0}(\ell - 1) + (\mathbf{I} - \mathbf{\Phi}_1)\overline{\mathbf{X}} \\ &= \begin{pmatrix} 0.9 & 0.6 \\ -0.3 & 0.7 \end{pmatrix} \hat{\mathbf{X}}_{t_0}(\ell - 1) + \begin{pmatrix} 0.1 & -0.6 \\ 0.3 & 0.3 \end{pmatrix} \overline{\mathbf{X}},\end{aligned} \tag{10.25}$$

which can be written as

$$\begin{aligned}\hat{X}_{t_0 1}(\ell) &= 0.9\hat{X}_{t_0 1}(\ell - 1) + 0.6\hat{X}_{t_0 2}(\ell - 1) + 0.1\overline{X}_1 - 0.6\overline{X}_2 \\ \hat{X}_{t_0 2}(\ell) &= -0.3\hat{X}_{t_0 1}(\ell - 1) + 0.7\hat{X}_{t_0 2}(\ell - 1) + 0.3\overline{X}_1 + 0.3\overline{X}_2.\end{aligned} \tag{10.26}$$

From (10.26), it can be seen that forecasts for X_{t1} involve previous values and forecasts from both X_{t1} and X_{t2}. For the bivariate realization in Figure 10.3, we calculate $\overline{X}_1 = 0.31$ and $\overline{X}_2 = 0.12$, i.e., $\overline{\mathbf{X}} = \begin{pmatrix} 0.31 \\ 0.12 \end{pmatrix}$, and (10.26) becomes

$$\begin{aligned}\hat{X}_{t_0 1}(\ell) &= 0.9\hat{X}_{t_0 1}(\ell - 1) + 0.6\hat{X}_{t_0 2}(\ell - 1) - 0.041 \\ \hat{X}_{t_0 2}(\ell) &= -0.3\hat{X}_{t_0 1}(\ell - 1) + 0.7\hat{X}_{t_0 2}(\ell - 1) + 0.129.\end{aligned} \tag{10.27}$$

Using the fact that $\hat{X}_{75,1}(0) = X_{75,1} = 4.51$ and $\hat{X}_{75,2}(0) = X_{75,2} = -0.59$, it follows that

$$\hat{X}_{75,1}(1) = 0.9(4.51) + 0.6(-0.59) - 0.041 = 3.664$$

$$\hat{X}_{75,2}(1) = -0.3(4.51) + 0.7(-0.59) + 0.129 = -1.637,$$

$$\hat{X}_{75,1}(2) = 0.9(3.664) + 0.6(-1.637) - 0.041 = 2.274$$

$$\hat{X}_{75,2}(2) = -0.3(3.664) + 0.7(-1.637) + 0.129 = -2.116,$$

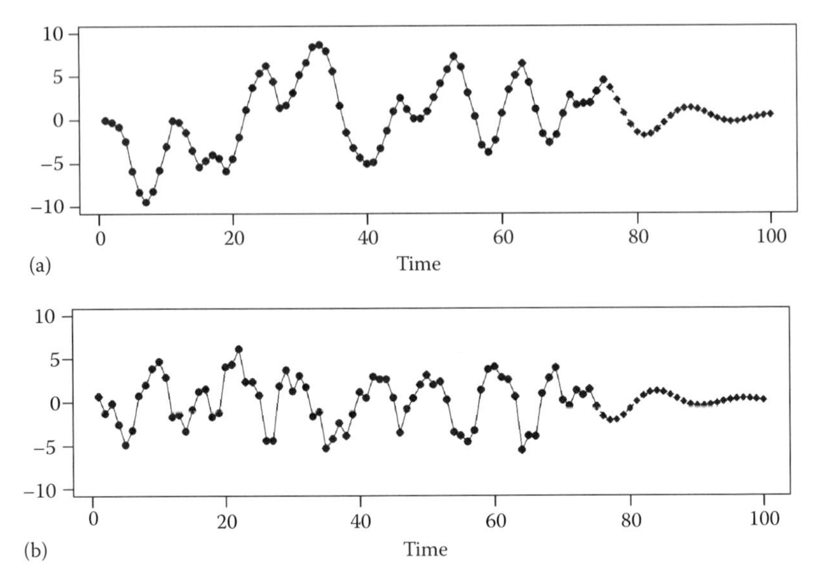

FIGURE 10.6
Realization shown in Figure 10.3 from the bivariate VAR(1) model in Example 10.2 along with forecasts of the next 25 values. (a) Realization and forecasts from X_{t1}. (b) Realization and forecasts from X_{t2}.

etc. Forecasts $\hat{X}_{75,1}(\ell)$ and $\hat{X}_{75,2}(\ell)$, $\ell = 1, \ldots, 25$ are shown in Figure 10.6 where it can be seen that the forecasts follow the general pattern of the data and eventually damp as in the stationary AR(p) case.

As in the univariate case, the forecast $\hat{\mathbf{X}}_{t_0}(\ell)$ in (10.23) can be expressed in terms of the multivariate general linear process form in (10.17). Specifically, $\hat{\mathbf{X}}_{t_0}(\ell)$ can be expressed as

$$\hat{\mathbf{X}}_{t_0}(\ell) = \boldsymbol{\mu} + \sum_{j=\ell}^{\infty} \boldsymbol{\Psi}_j \mathbf{a}_{t_0+\ell-j}, \tag{10.28}$$

which corresponds to (10.6) and (6.9). Properties of the forecasts are best derived from (10.28). Among these properties is the fact that $\hat{\mathbf{X}}_{t_0}(\ell)$ is the minimum mean square error matrix predictor of $\mathbf{X}_{t_0+\ell}$ given $\mathbf{X}_{t_0}, \mathbf{X}_{t_0-1}, \ldots$ Using the general linear process form in (10.17) and the forecasts in (10.28), the ℓ-step ahead forecasts errors, $\mathbf{e}_{t_0}(\ell)$, are given by

$$\begin{aligned} \mathbf{e}_{t_0}(\ell) &= \mathbf{X}_{t_0+\ell} - \hat{\mathbf{X}}_{t_0}(\ell) \\ &= \sum_{j=0}^{\ell-1} \boldsymbol{\psi}_j \mathbf{a}_{t_0+\ell-j}, \end{aligned} \tag{10.29}$$

from which it follows that $\mathbf{e}_{t_0}(\ell)$ has expected value $\mathbf{0}$ and covariance matrix, $\mathbf{\Gamma}_{\mathbf{e}}$, given by

$$\mathbf{\Gamma}_{\mathbf{e}} = E[\mathbf{e}_{t_0}(\ell)\mathbf{e}_{t_0}(\ell)'] = \sum_{j=0}^{\ell-1} \mathbf{\Psi}_j \mathbf{\Gamma}_{\mathbf{a}} \mathbf{\Psi}_j'. \tag{10.30}$$

In the case of normal errors, forecast limits analogous to (6.33) can be obtained using the diagonal elements of (10.30).

10.3.2 Spectrum of a VAR(p) Model

In the VAR(p) case, the multivariate spectrum, $\mathbf{P_X}(f)$, defined in (10.7), is the $m \times m$ matrix given, for each $-0.5 \le f \le 0.5$, by

$$\mathbf{P_X}(f) = \mathbf{\Phi}^{-1}(e^{2\pi i f})\mathbf{\Gamma}_{\mathbf{a}}[\mathbf{\Phi}^{-1}(e^{-2\pi i f})]', \tag{10.31}$$

where for real or complex z, the polynomial $\mathbf{\Phi}(z) = \mathbf{I} - \mathbf{\Phi}_1 z - \cdots - \mathbf{\Phi}_p z^p$ is the algebraic equivalent of the operator $\mathbf{\Phi}(B)$ defined in (10.16). Note that (10.31) is the multivariate counterpart to the AR(p) spectrum given in (3.53). We mention again that the use of the term spectrum and spectral density is typically used synonymously in the multivariate case to indicate the Fourier transform of the autocovariance.

10.3.3 Estimating the Coefficients of a VAR(p) Model

10.3.3.1 Yule–Walker Estimation

The parameters of a VAR(p) model can be estimated using methods analogous to the univariate estimation methods discussed in Chapter 7. As a first method, we discuss multivariate Yule–Walker estimation based on the multivariate Yule–Walker equations given in (10.18). Replacing $\mathbf{\Gamma}(k)$ in the Yule–Walker equations with $\hat{\mathbf{\Gamma}}(k)$ defined in (10.11) yields the sample multivariate Yule–Walker equations

$$\hat{\mathbf{\Gamma}}(k) = \hat{\mathbf{\Gamma}}(k-1)\hat{\mathbf{\Phi}}_1' + \cdots + \hat{\mathbf{\Gamma}}(k-p)\hat{\mathbf{\Phi}}_p', \quad k = 1, 2, \ldots, p. \tag{10.32}$$

Letting $\hat{\mathbf{\Phi}}_{(p)}$ and $\hat{\mathbf{\Gamma}}_{(p)}$ be the $mp \times m$ matrices given by $\hat{\mathbf{\Phi}}_{(p)} = [\hat{\mathbf{\Phi}}_1, \ldots, \hat{\mathbf{\Phi}}_p]'$, $\hat{\mathbf{\Gamma}}_{(p)} = [\hat{\mathbf{\Gamma}}(1)', \ldots, \hat{\mathbf{\Gamma}}(p)']'$, and letting $\hat{\mathbf{\Sigma}}_p$ denote the $mp \times mp$ matrix

$$\hat{\mathbf{\Sigma}}_p = \begin{pmatrix} \hat{\mathbf{\Gamma}}(0) & \hat{\mathbf{\Gamma}}(1) & \cdots & \hat{\mathbf{\Gamma}}(p-1) \\ \hat{\mathbf{\Gamma}}(1) & \hat{\mathbf{\Gamma}}(0) & \cdots & \hat{\mathbf{\Gamma}}(p-2) \\ \vdots & \vdots & \ddots & \vdots \\ \hat{\mathbf{\Gamma}}(p-1) & \hat{\mathbf{\Gamma}}(p-2) & \cdots & \hat{\mathbf{\Gamma}}(0) \end{pmatrix}, \tag{10.33}$$

the equations in (10.32) can be written as $\hat{\boldsymbol{\Sigma}}_p\hat{\boldsymbol{\Phi}}_{(p)} = \hat{\boldsymbol{\Gamma}}_{(p)}$, which have the solution $\hat{\boldsymbol{\Phi}}_{(p)} = \hat{\boldsymbol{\Sigma}}_p^{-1}\hat{\boldsymbol{\Gamma}}_{(p)}$. From (10.19), the Yule–Walker estimate of $\boldsymbol{\Gamma}_{\mathbf{a}}$ is given by $\hat{\boldsymbol{\Gamma}}_{\mathbf{a}} = \hat{\boldsymbol{\Gamma}}(0) - \hat{\boldsymbol{\Gamma}}(-1)\hat{\boldsymbol{\Phi}}_1' - \cdots - \hat{\boldsymbol{\Gamma}}(-p)\hat{\boldsymbol{\Phi}}_p'$. As with the univariate case, the multivariate Yule–Walker estimates always produce a stationary model.

10.3.3.2 Least Squares and Conditional Maximum Likelihood Estimation

Least squares estimates of the parameters of a VAR(p) process can be obtained by expressing the VAR(p) model as

$$\begin{aligned}\mathbf{X}_t &= \boldsymbol{\alpha} + \sum_{j=1}^{p}\boldsymbol{\Phi}_j\mathbf{X}_{t-j} + \mathbf{a}_t \\ &= \boldsymbol{\alpha} + \boldsymbol{\Phi}_{(p)}'\mathbf{X}_t(p) + \mathbf{a}_t, \quad t = p+1,\ldots,n,\end{aligned} \tag{10.34}$$

where $\mathbf{X}_t(p)$ is the $mp \times 1$ matrix given by

$$\mathbf{X}_t(p) = (\mathbf{X}_{t-1}',\ldots,\mathbf{X}_{t-p}')'.$$

The equations in (10.34) are analogous to those in (7.12) in the univariate case. We estimate $\boldsymbol{\mu}$ with $\overline{\mathbf{X}}$ as given in (10.10) and analyze the zero-mean process $\mathbf{X}_t - \overline{\mathbf{X}}$. In this section, we will for convenience refer to this mean adjusted process as $\mathbf{X}_t$. Thus, the model in (10.34) does not have a constant term, i.e., $\boldsymbol{\alpha} = \mathbf{0}$, and the least squares estimates of the autoregressive parameters are given in the $mp \times m$ matrix

$$\begin{aligned}\hat{\boldsymbol{\Phi}}_{(p)} &= \left[\hat{\boldsymbol{\Phi}}_1,\ldots,\hat{\boldsymbol{\Phi}}_p\right]' \\ &= (\mathbf{X}'\mathbf{X})^{-1}\mathbf{X}'\mathbf{X}_{(p)},\end{aligned}$$

where $\mathbf{X}_{(p)}$ is an $(n-p) \times m$ matrix with rows $(X_{t1},\ldots,X_{tm})$, $t = p+1,\ldots,n$, and $\mathbf{X}$ is the $(n-p) \times mp$ matrix with rows

$$\left(\mathbf{X}_{t-1}',\ldots,\mathbf{X}_{t-p}'\right), \quad t = p+1,\ldots,n,$$

where $\mathbf{X}_{t-j}' = (X_{t-j,1},\ldots,X_{t-j,m})$. As in the univariate case, under a normality assumption, the conditional ML estimates are equivalent to the least squares estimates.

10.3.3.3 Burg-Type Estimation

Nuttal (1976a,b) and Strand (1977) obtained Burg-type estimates of the autoregressive parameters by considering the multivariate backward model analogous to (7.16) as well as the forward model. See also Jones (1978). As with the univariate Burg estimates, the Burg-type autoregressive estimates always produce a stationary solution. Marple and Nuttall (1983) have shown through simulation results that the Burg-type (Nuttal–Strand) estimator has better properties than the Yule–Walker.

10.3.4 Calculating the Residuals and Estimating $\boldsymbol{\Gamma}_{\mathbf{a}}$

Analogous to (9.2), the residuals associated with an estimated VAR(p) model are given by

$$\hat{\mathbf{a}}_t = \mathbf{X}_t - \sum_{j=1}^{p} \hat{\boldsymbol{\Phi}}_j \mathbf{X}_{t-j} - \hat{\boldsymbol{\alpha}}, \quad t = p+1, \ldots, n. \tag{10.35}$$

The least squares estimator of $\boldsymbol{\Gamma}_{\mathbf{a}}$ is given by

$$\tilde{\boldsymbol{\Gamma}}_{\mathbf{a}} = \frac{1}{n-2p-1} \sum_{t=p+1}^{n} \hat{\mathbf{a}}_t \hat{\mathbf{a}}_t', \tag{10.36}$$

and the conditional maximum likelihood estimator is

$$\hat{\boldsymbol{\Gamma}}_{\mathbf{a}} = \frac{1}{n} \sum_{t=p+1}^{n} \hat{\mathbf{a}}_t \hat{\mathbf{a}}_t'. \tag{10.37}$$

Extension of the backcasting procedure discussed in Section 7.4 can be used to obtain residuals for $t = 1, \ldots, n$ and consequently improve the ability to estimate $\boldsymbol{\Gamma}_{\mathbf{a}}$.

Recall that under normality, the least squares and conditional maximum likelihood estimates of $\boldsymbol{\Phi}_1, \ldots, \boldsymbol{\Phi}_p$ are the same. If backcasting is used to calculate $\hat{\mathbf{a}}_1, \ldots, \hat{\mathbf{a}}_p$, then an unconditional ML estimator is given by

$$\hat{\boldsymbol{\Gamma}}_{\mathbf{a}} = \frac{1}{n} \sum_{t=1}^{n} \hat{\mathbf{a}}_t \hat{\mathbf{a}}_t'. \tag{10.38}$$

An estimator of $\boldsymbol{\Gamma}_{\mathbf{a}}$ can be based on residuals obtained from any estimation procedure (e.g., the Burg-type estimators, Yule–Walker, etc.) using (10.38) whenever backcasting is used to calculate $\hat{\mathbf{a}}_1, \ldots, \hat{\mathbf{a}}_p$.

10.3.5 VAR(p) Spectral Density Estimation

The multivariate spectrum (spectral density) of a VAR(p) process is given in (10.31). The VAR spectral estimate is simply the spectral density associated with the VAR(p) model fit to a set of multivariate time series data. That is, the VAR spectral estimator is given by

$$\hat{\mathbf{P}}_X(f) = \hat{\boldsymbol{\Phi}}^{-1}(e^{2\pi i f}) \hat{\boldsymbol{\Gamma}}_{\mathbf{a}} [\hat{\boldsymbol{\Phi}}^{-1}(e^{-2\pi i f})]', \tag{10.39}$$

where $\hat{\boldsymbol{\Phi}}$ and $\hat{\boldsymbol{\Gamma}}_{\mathbf{a}}$ are estimated from the data as discussed earlier.

10.3.6 Fitting a VAR(p) Model to Data

In the following, we briefly summarize the steps involved in fitting a VAR(p) model to a vector time series realization.

10.3.6.1 Model Selection

As in the univariate case, the first step in fitting a VAR(p) model to a set of data is to identify the order p. The multivariate extension of AIC (based on maximum likelihood estimation of the parameters) is given by

$$AIC = \ln\left|\hat{\mathbf{\Gamma}}_{\mathbf{a}}\right| + \frac{2m^2 p}{n}. \tag{10.40}$$

As in the univariate case, versions of AIC can be found using (10.40) based on other estimation procedures (e.g., Burg-type estimators). The model order is selected as the minimum value of AIC calculated using (10.40) for the range of model orders $0 \le p \le P$ for some integer P. Multivariate versions of alternatives to AIC such as BIC and AICC are also available (see Box et al. 2008).

10.3.6.2 Estimating the Parameters

Once the order p is identified, the estimates of the parameters $\mathbf{\Phi}_1, \ldots, \mathbf{\Phi}_p$, $\mathbf{\Gamma}_{\mathbf{a}}$, and $\boldsymbol{\mu}$ associated with the selected model are obtained. The ML estimates of these parameters have already been obtained in the AIC step. Recall that it is typical to estimate $\boldsymbol{\mu}$ with $\overline{\mathbf{X}}$ and then estimate the remaining parameters using the series $\mathbf{X}_t - \overline{\mathbf{X}}$. We will use this procedure in the following.

10.3.6.3 Testing the Residuals for White Noise

The Ljung–Box test discussed in Section 9.1 has been extended to the multivariate setting (see Hosking, 1981). Suppose we want to test whether the multivariate time series $\mathbf{Z}_t$ is multivariate white noise. Here, $\mathbf{Z}_t$ may simply be an observed time series or it may be the residual series based on a VAR(p) fit to a set of data. A portmanteau test of the hypotheses $H_0: \boldsymbol{\rho}(k) = \mathbf{0}, k = 1, \ldots, K$ against $H_a: \boldsymbol{\rho}(k) \neq \mathbf{0}$ for some k between 1 and K is based on the test statistic

$$L = n^2 \sum_{k=1}^{K} \frac{1}{n-k} \operatorname{tr}[\hat{\mathbf{\Gamma}}'(k)\hat{\mathbf{\Gamma}}^{-1}(0)\hat{\mathbf{\Gamma}}(k)\hat{\mathbf{\Gamma}}^{-1}(0)], \tag{10.41}$$

where $\hat{\mathbf{\Gamma}}(k)$ is defined in (10.11), and tr[$\mathbf{M}$] denotes the trace of the matrix $\mathbf{M}$ (i.e., the sum of its diagonal elements). The statistic L in (10.41) is approximately distributed as χ^2 with m^2K degrees of freedom when the data are multivariate white noise, and we reject $H_0: \boldsymbol{\rho}(k) = \mathbf{0}$, if the calculated L is sufficiently large. When the data to be tested are residuals from a fitted VAR(p) process, then the test statistic is approximately χ^2 with $m^2(K-p)$ degrees of freedom under H_0.

Example 10.4 AIC and Parameter Estimation for the Data in Example 10.2

In this example, we use PROC VARMAX in SAS/ETS to fit a VAR(p) model to the bivariate VAR(1) data shown in Figure 10.3. First, using AIC to select the order of the model (allowing $1 \le p \le 8$), AIC correctly selected $p = 1$. The least-squares estimates of the parameters of the VAR(1) model are $\hat{\mathbf{\Phi}}_1 = \begin{pmatrix} 0.90 & 0.57 \\ -0.27 & 0.63 \end{pmatrix}$ with $\hat{\mathbf{\Gamma}}_{\mathbf{a}} = \begin{pmatrix} 0.81 & 1.53 \\ 1.53 & 3.8 \end{pmatrix}$. Comparing these to (10.20) and (10.21), respectively, we see that these parameter estimates perform reasonably well in this case. Further evidence regarding the appropriateness of the model is provided by testing the residuals. We used SAS to report portmanteau tests using (10.41) for $K \le 10$, and, in all cases, $H_0: \boldsymbol{\rho}(k) = \mathbf{0}, k = 1, \ldots, K$ is not rejected.

Computational Notes: As of the writing of this book, `GW-WINKS` does not have multivariate time series capabilities. In the examples we have used SAS/ETS but several other major packages support multivariate time series analysis.

10.4 Nonstationary VARMA Processes

The ARUMA(p,d,q) processes discussed in Section 5.2 can be generalized to the multivariate case. Specifically, we consider the model

$$\mathbf{\Phi}(B)\boldsymbol{\lambda}(B)(\mathbf{X}_t - \boldsymbol{\mu}) = \mathbf{\Theta}(B)\mathbf{a}_t, \tag{10.42}$$

where

(a) $\boldsymbol{\lambda}(B) = \begin{pmatrix} \lambda_1(B) & 0 & \cdots & 0 \\ 0 & \lambda_2(B) & & \vdots \\ \vdots & & \ddots & 0 \\ 0 & \cdots & 0 & \lambda_m(B) \end{pmatrix}$ and $\lambda_j(z) = 0$ has all its roots on the unit circle, $j = 1, \ldots, m$.

(b) The roots of the determinantal equations $|\mathbf{\Phi}(z)| = 0$ and $|\mathbf{\Theta}(z)| = 0$ lie outside the unit circle.

For example, it might be that $\lambda_1(B) = (1 - B)$, $\lambda_2(B) = 1 - \sqrt{3}B + B^2$, $\lambda_3(B) = (1 - B)^2(1 + B^2)$, etc. The model in (10.42) is called a VARUMA model and (10.42) implies that the series $Y_{tj} = \lambda_j(B)(X_t - \mu_j)$ is a univariate stationary process and additionally that the process $\mathbf{Y}_t = (Y_{t1}, \ldots, Y_{tm})'$ satisfies the stationary VARMA(p,q) model $\mathbf{\Phi}(B)(\mathbf{Y} - \boldsymbol{\mu}_\mathbf{Y}) = \mathbf{\Theta}(B)\mathbf{a}_t$, where $\boldsymbol{\mu}_\mathbf{Y} = (\mu_{Y_{t1}}, \ldots, \mu_{Y_{tm}})'$.

The VARUMA model that has received the most attention is the particular case in which

$$\boldsymbol{\lambda}(B) = \begin{pmatrix} (1-B)^{d_1} & 0 & \cdots & 0 \\ 0 & (1-B)^{d_2} & & \vdots \\ \vdots & & \ddots & 0 \\ 0 & \cdots & 0 & (1-B)^{d_m} \end{pmatrix}, \tag{10.43}$$

which is the multivariate counterpart to the ARIMA(p,d,q) model. Notice that if X_{tj} is a stationary process, then $d_j = 0$, i.e., no differencing is needed to stationarize X_{tj}.

10.5 Testing for Association between Time Series

It is often of interest to the time series analyst to ascertain whether there is a relationship between two time series. In Figures 10.7 and 10.8, we show two pairs of time series. For each pair we investigate whether there is a relationship between the two time series in each pair. In Figure 10.7a, we show the sunspot data for the years 1850–2009 while, in Figure 10.7b, we show the global temperature data (from Figure 1.24a) for the same years. Some scientists have speculated about the existence of a relationship between sunspot activity and the earth's temperature. In fact, it can be seen in Figure 10.7 that the peak amplitude of the sunspot cycles has been increasing during the same time span in which the temperatures show an increasing trend. Can it be concluded that there an association between recent sunspot activity and temperature data?

In Figure 10.8, we show the annual incidence rates (1970–2005) in the United States for influenza (dashed line), and, with a solid line, we show the incidence rates for nonperforating appendicitis (i.e., appendicitis cases in which the appendix has not yet ruptured) as reported by Alder et al. (2010). The incidence rates are given as annual number of hospitalizations per 10,000 population. In the figure, it can be seen that the incidence rates for both diseases declined until about 1995 after which time there has been a gradual increase. A question of interest is whether there is a relationship between these two seemingly unrelated diseases?

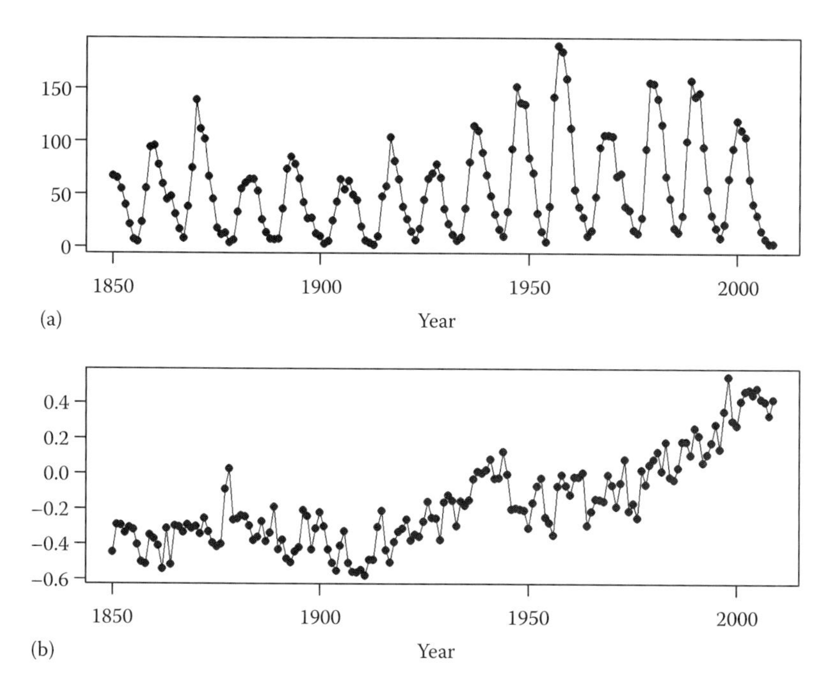

FIGURE 10.7
(a) Annual sunspot numbers and (b) Hadley global temperature data from 1850 to 2009.

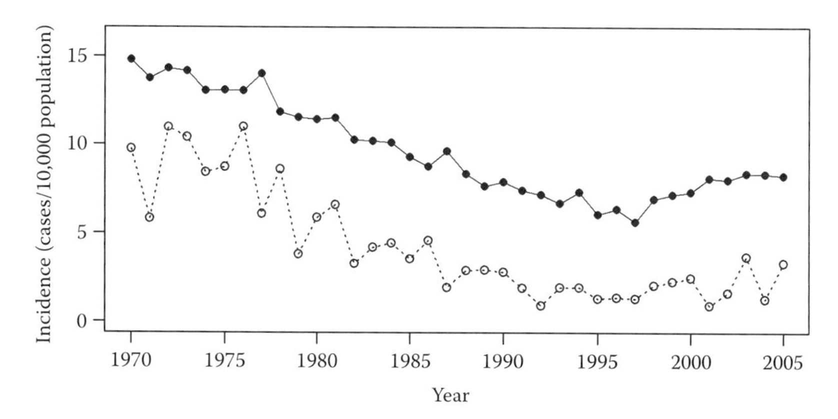

FIGURE 10.8
Annual incidence of nonperforated appendicitis (solid line) and influenza (dotted line) for the years 1970–2005.

To provide answers to these types of questions, we need to be able to test for a relationship between two time series. In this section, we consider testing for a relationship between

(a) Stationary time series
(b) Time series that have the same number of unit roots (testing for co-integration)

10.5.1 Testing for Independence of Two Stationary Time Series

We consider the setting in which X_{t1} and X_{t2} are stationary univariate time series that each have a general linear process representation for which the ψ-weights are absolutely convergent. Let $\hat{\rho}_{12}(k)$ be the estimated cross-correlation at lag k as defined in (10.12). It can be shown that $\hat{\rho}_{12}(k)$ is asymptotically normal with mean 0 and variance (AV_{12}) given by

$$AV_{12} = \frac{1}{n} \sum_{j=-\infty}^{\infty} \rho_{11}(j)\rho_{22}(j). \qquad (10.44)$$

See Brockwell and Davis (1991).

If it is desired to test $H_0 : \rho_{12}(k) = 0$ versus $H_a : \rho_{12}(k) \neq 0$, it can be seen that, the asymptotic variance in (10.44) will not in general be known. Note however that if either or both of X_{t1} and X_{t2} is white noise, then the asymptotic variance becomes $AV_{12} = \frac{1}{n}$. Haugh (1976) recommended a procedure for testing $H_0 : \rho_{12}(k) = 0$ that involves the following steps:

(a) Fit an ARMA(p,q) model to each of X_{t1} and X_{t2} (these two models may be of different orders p_1, q_1, p_2, and q_2.
(b) Calculate the residual series $\hat{a}_{t1}$ and $\hat{a}_{t2}$, $t = 1, \ldots, n$ based on each model fit.
(c) Calculate the estimated cross-correlations, $\hat{\rho}_{12}^{(\hat{a})}(k)$, $k = 1, \ldots, K$ between the two residual series.
(d) For each k, conclude that $\hat{\rho}_{12}^{(\hat{a})}(k)$ is significantly different from zero at the $\alpha = 0.05$ level if $|\hat{\rho}_{12}^{(\hat{a})}(k)| > 2(1/\sqrt{n})$.

This procedure is very similar to the one described in Section 8.1 for testing $H_0 : \rho(k) = 0$ versus $H_a : \rho(k) \neq 0$. As in that case, it should be noted that the significance level $\alpha = 0.05$ applies separately for each k, and it would not be unusual for about 5% of the $\hat{\rho}_{12}^{(\hat{a})}(k)'s$ to exceed $2(1/\sqrt{n})$ when the null hypothesis is true.

Example 10.5 Testing Whether Sunspot Activity and Global Temperatures Are Independent

We let X_{1t} denote the annual sunspot numbers and X_{2t} denote the global temperature data from 1850 to 2009 as displayed in Figure 10.7. In this example, we will consider these two time series to each be stationary (see Chapter 9 for alternative models). In Figure 10.9a, we show the estimated cross correlations, $\hat{\rho}_{12}(k)$ for $k = -20, -19, \ldots, -1, 0, 1, \ldots, 20$ between annual sunspot numbers and temperature data from 1850 to 2009. There it can be seen that the largest cross correlations occur at positive lags, with the largest cross correlation being $\hat{\rho}_{12}(6) = 0.271$. This makes intuitive sense, because if there is a relationship, we would expect current and/or previous sunspot activity to influence the earth's temperature and would not expect that the earth's temperature would have any influence on current or future solar activity. However, the question still remains whether an estimated sample correlation of this size is significantly different from zero, since, in this case, we do not know the asymptotic variance in (10.44). Thus, we use the procedure suggested by Haugh (1976) and find cross correlations of the residuals based on fitting ARMA(p,q) models to each data set.

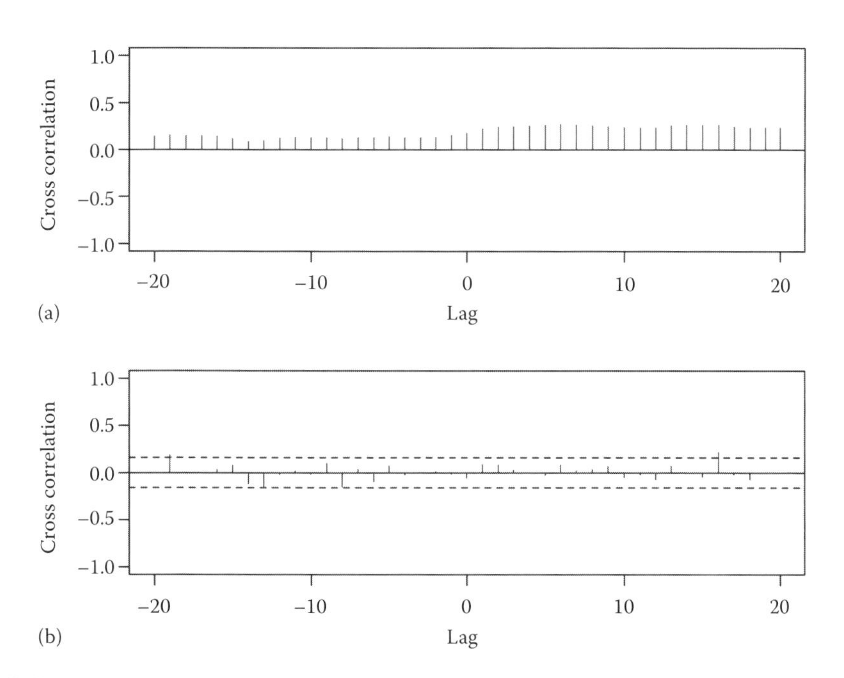

FIGURE 10.9
(a) Estimated cross correlations, $\hat{\rho}_{12}(k)$, where X_{t1} is the sunspot data and X_{t2} is global temperature data from 1850 to 2009 and (b) estimated cross correlations between the residuals from autoregressive models fit to each series.

AIC selects an AR(9) for the sunspot data between 1850 and 2009, and the fitted model based on Burg estimates is

$$(1 - 1.073B + 0.218B^2 + 0.333B^3 - 0.188B^4 + 0.071B^5 - 0.035B^6 \\ - 0.027B^7 + 0.181B^8 - 0.360B^9)(X_t - 55.04) = a_t$$

where $\hat{\sigma}_a^2 = 223.7$. In Example 9.2, we discussed models that might be appropriate for the temperature data. In this example, we consider the temperature data to be stationary and use the AR(4) model, $(1 - 0.646B + 0.026B^2 - 0.097B^3 - 0.24B^4)(X_t + 0.17) = a_t$ with $\hat{\sigma}_a^2 = 0.0108$, previously given in factored form in (9.8). The residuals for this fit were previously shown in Figure 9.4a. Figure 10.9b shows the estimated cross correlations, $\hat{\rho}_{12}^{(a)}$, between the residuals $\hat{a}_{t1}$ from the AR(9) fit to the sunspot data and $\hat{a}_{t2}$ from the AR(4) fit to the temperature data. Also, in Figure 10.9b, we show the limit lines $\pm 2/\sqrt{160} = \pm 0.158$, and it can be seen that there is no real indication of a relationship. So, at least based on this method of analysis, it cannot be concluded that there is an association between sunspot activity and global temperatures.

Clearly, a VAR(p) model allows for a correlation structure among the variables. In order to connect the discussion in this section with the VAR(p) model, in Example 10.6 we consider again the bivariate VAR(1) model in Example 10.2.

Example 10.6 Testing Components of the VAR(1) Model in Example 10.2 for Independence

Suppose the two realizations in Figure 10.3 were observed without the knowledge that they represented a bivariate VAR(1) realization pair. Suppose also that the only question of interest is whether there is a relationship between the two stationary time series. While one approach for assessing relationships between the two series would be to fit a VAR(p) model to the data as was done in Example 10.4, in this example we simply consider the data to represent realizations from two stationary time series. We will model each realization as a univariate AR(p) model, find the residuals from the fitted models, and test whether the residual series are independent using the procedure recommended by Haugh (1976). Using AIC with $0 \le p \le 8$, we select an AR(3) for X_{t1} and an AR(4) fit to X_{t2}. The sample cross correlations of the residuals along with limit lines $\pm 2/\sqrt{75} = \pm 0.231$ are shown in Figure 10.10. There it can be seen that the residual series are positively correlated at lag $k = 0$ while the cross correlations at other lags are within the 95% limit lines except for a couple of isolated borderline cases. This is consistent with the model used to generate the VAR(1) data where from (10.5), it is seen that the contemporaneous cross correlation between the residuals (of the VAR(1) model) is 0.89 while the cross correlations at nonzero lags are zero by construction.

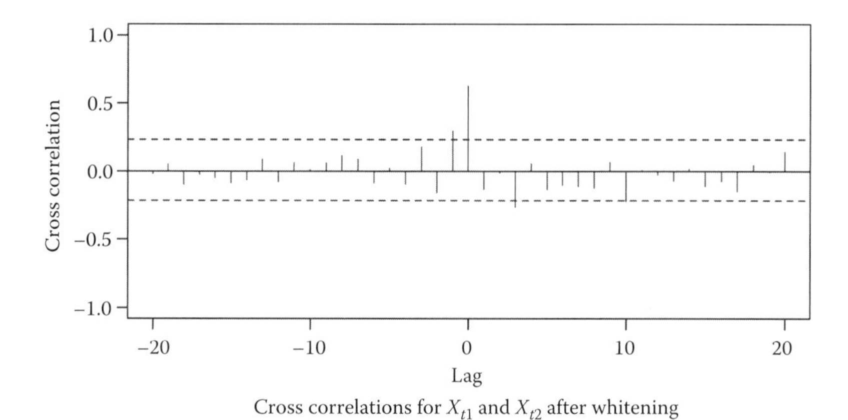

FIGURE 10.10
Estimated cross correlations between the residuals from autoregressive models fit to realizations X_{t1} and X_{t2} in Figure 10.3.

10.5.2 Testing for Cointegration between Nonstationary Time Series

We let X_{t1} and X_{t2} denote the annual incidence of nonperforated appendicitis and influenza shown in Figure 10.8. The disease incidence data show a wandering behavior that is indicative of models with unit roots. The question of interest here is whether the two series tend to be "tethered together," i.e., tend to wander together. Engle and Granger (1987) used the term *cointegration* to describe this "tethering together" idea, which, in our setting, implies that each series has a single unit root and there exists some linear combination of the differences ΔX_{t1} and ΔX_{t2} that is stationary. The cointegration concept has been used extensively in the econometrics literature for analyzing relationships. For an extensive bibliography in this area, see Engle and Granger (1991). Engle and Granger (1987) show that it is useful to use an error correction form similar to that in (8.23) (see also Problem 8.10) for describing the relationship between X_{t1} and X_{t2}. Letting $\mathbf{X}_t = (X_{t1}, X_{t2})'$, we consider the model

$$\mathbf{X}_t = \boldsymbol{\alpha} + \boldsymbol{\Phi}_1\mathbf{X}_{t-1} + \cdots + \boldsymbol{\Phi}_p\mathbf{X}_{t-p} + \mathbf{a}_t,$$

where it is assumed, in this case, that $\boldsymbol{\Phi}(B) = \mathbf{I} - \boldsymbol{\Phi}_1 B - \cdots - \boldsymbol{\Phi}_p B^p$ has a factor

$$\lambda(B) = \begin{pmatrix} 1-B & 0 \\ 0 & 1-B \end{pmatrix}. \tag{10.45}$$

The multivariate error correction model has the form

$$\nabla\mathbf{X}_t = \boldsymbol{\beta}_0 + \mathbf{C}\mathbf{X}_{t-1} + \boldsymbol{\beta}_1\nabla\mathbf{X}_{t-1} + \cdots + \boldsymbol{\beta}_{p-1}\nabla\mathbf{X}_{t-p+1} + \mathbf{a}_t, \tag{10.46}$$

where $\nabla\mathbf{X}_{t-j}=\mathbf{X}_{t-j}-\mathbf{X}_{t-j-1}, \boldsymbol{\beta}_0=\boldsymbol{\mu}(\mathbf{I}-\boldsymbol{\Phi}_1-\cdots-\boldsymbol{\Phi}_p), \mathbf{C}=-(\mathbf{I}-\boldsymbol{\Phi}_1-\cdots-\boldsymbol{\Phi}_p)$, and $\boldsymbol{\beta}_j=-(\boldsymbol{\Phi}_{j+1}+\cdots+\boldsymbol{\Phi}_p)$, for $j=1,\ldots,p-1$. Since it is assumed that $\boldsymbol{\Phi}(\mathbf{B})=\mathbf{I}-\boldsymbol{\Phi}_1-\cdots-\boldsymbol{\Phi}_p$ has a factor of $\lambda(B)$ defined in (10.45), then $|\mathbf{I}-\boldsymbol{\Phi}_1 z-\cdots-\boldsymbol{\Phi}_p z^p|=0$ for $z=1$, i.e., the matrix $\mathbf{C}$ is singular and, in this bivariate case, has rank 0 or 1. Johansen (1988, 1991) and Johansen and Juselius (1990) developed tests for cointegration based on the rank of the matrix $\mathbf{C}$. Specifically, in the bivariate case, if rank($\mathbf{C}$)$=1$, then there is a cointegrating relationship between X_{t1} and X_{t2}; if rank($\mathbf{C}$)$=0$, the variables are not cointegrated, i.e., are not "tethered together," and rank($\mathbf{C}$)$=2$ suggests that the model is stationary (and does not have unit roots). Letting $\lambda_1 \leq \lambda_2$ denote the 2 eigenvalues of the matrix $\mathbf{C}$, then the trace test statistic (see Johansen, 1988, 1991) for testing H_0:rank($\mathbf{C}$)$=0$ (i.e., $\lambda_1=\lambda_2=0$) vs. H_a:rank($\mathbf{C}$)>0 is $\lambda_{trace}=-n\sum_{j=1}^{2}\log(1-\lambda_j)$ while the test statistic for testing H_0:rank($\mathbf{C}$)$=1$ vs. H_a:rank($\mathbf{C}$)>1 is $\lambda_{trace}=-n\log(1-\lambda_2)$. In both cases. H_0 is rejected for λ_{trace} sufficiently large. The conclusion of cointegration follows from rejecting the H_0:rank($\mathbf{C}$)$=0$ vs. H_a:rank($\mathbf{C}$)>0 but failing to reject H_0:rank($\mathbf{C}$)$=1$ vs. H_a:rank($\mathbf{C}$)>1. See Johansen (1988, 1991) for discussion of the distribution of the test statistic and critical values.

It should be noted that these tests can be applied to find more complicated cointegrating relationships. For example, generalizations of (10.46) have been developed to test whether or not a constant term is included and whether a trend term or exogenous variables are included. We only discussed the case $m=2$, and all unit roots are of multiplicity one. The results here can be extended to the case in which unit roots have multiplicities greater than one and in which the multiplicities differ. For further details, see the Johansen references cited earlier.

Example 10.7 Testing for a Cointegrating Relationship between Incidences of Nonperforated Appendicitis and Influenza

We now consider the disease incidence data in Figure 10.8. Let X_{t1} denote incidence of nonperforated appendicitis and X_{t2} denote influenza incidence. AIC selects an AR(1) model for the appendicitis data (X_{t1}) and an AR(3) for the influenza (X_{t2}) data. The augmented Dickey–Fuller test gives $t_{DF}=-1.79$ with p-value 0.38 for the appendicitis data and $t_{DF}=-2.52$ with p-value 0.12 for the influenza incidence data. Thus, the hypothesis of a unit root is not rejected in either case. As noted in Section 8.3.5.2, failure to reject the null hypothesis of a unit root does not provide strong evidence that a unit root exists. However, since, in each case, a unit root cannot be rejected, we proceed to test for cointegration. Based on a VAR(1) model for $\mathbf{X}_t=(X_{t1},X_{t2})'$, and using SAS PROC VARMAX, the hypothesis H_0: rank($\mathbf{C}$) $=0$ vs. H_a: rank($\mathbf{C}$) >0 is rejected but H_0: rank($\mathbf{C}$) $=1$ vs. H_a: rank($\mathbf{C}$) >1 is not rejected, leading to a conclusion of cointegration. See Alder et al. (2010).

10.6 State-Space Models

The state-space model (sometimes called the dynamic linear model) was originally developed by Kalman (1960) and Kalman and Bucy (1961) for control of linear systems and has more recently become widely used for the analysis of time series data. The ARUMA models (univariate and multivariate) can be seen to be special cases of the broad class of time series models that can be obtained via the state-space approach. In this section, we give a brief and introductory treatment of these models and illustrate their usefulness.

State-space formulations depend on a *state equation* and an *observation equation*.

10.6.1 State Equation

The state equation is a VAR(1) type expression giving the rule by which the $v \times 1$ random vector $\mathbf{X}_t$ is generated from $\mathbf{X}_{t-1}$ for time points $t = 1, 2, \ldots, n$. Specifically, the *state equation* is given by

$$\mathbf{X}_t = \mathbf{F}\mathbf{X}_{t-1} + \mathbf{V}_t, \tag{10.47}$$

where

(a) $\mathbf{F}$ is an $v \times v$ matrix

(b) $\mathbf{V}_t \sim \text{MVN}(\mathbf{0}, \boldsymbol{\Gamma}_\mathbf{V})$ is multivariate normal white noise (MVN stands for multivariate normal)

(c) The process is initiated with a vector $\mathbf{X}_0 \sim \text{MVN}(\boldsymbol{\mu}_0, \boldsymbol{\Gamma}_0)$.

It should be noted that a more general form of the state equation replaces the constant matrix $\mathbf{F}$ with the matrices $\mathbf{F}_t$.

10.6.2 Observation Equation

The assumption in state-space analysis is that observations are made on $\mathbf{Y}_t$ which is related to $\mathbf{X}_t$ via the *observation equation*.

$$\mathbf{Y}_t = \mathbf{G}_t\mathbf{X}_t + \mathbf{W}_t, \tag{10.48}$$

where

(a) $\mathbf{Y}_t$ is $w \times 1$ (w can be equal to, larger than, or smaller than v)

(b) $\mathbf{G}_t$ is a $w \times v$ measurement or observation matrix

(c) $\mathbf{W}_t \sim \text{MVN}(\mathbf{0}, \boldsymbol{\Gamma}_\mathbf{W})$, where $\boldsymbol{\Gamma}_\mathbf{W}$ is $w \times w$ is multivariate white noise

(d) Noise terms $\mathbf{W}_t$ and $\mathbf{V}_t$ are uncorrelated

(e) $\mathbf{X}_0$ is uncorrelated with $\mathbf{W}_t$ and $\mathbf{V}_t$

More general assumptions, among other things allowing for correlated noise terms, can be used, but the models and assumptions listed here are sufficient for our purposes.

Example 10.8 State-Space Formulation of a Stationary AR(1) Model

Let Y_t be a univariate stationary AR(1) process given by $Y_t = \phi_1 Y_{t-1} + a_t$. A univariate state-space representation is straightforward using the following:

$$X_t = \phi_1 X_{t-1} + V_t \quad (\textit{State Equation})$$
$$Y_t = X_t \quad (\textit{Observation Equation})$$

That is, in this case, $\mathbf{F} = \phi_1$, $\mathbf{G}_t = 1$, $V_t = a_t$, and $W_t = 0$.

Example 10.9 State-Space Formulation of a Stationary AR(1) plus Noise Model

Consider the state and observation equations

$$X_t = \phi_1 X_{t-1} + V_t \quad (\textit{State Equation})$$
$$Y_t = X_t + W_t \quad (\textit{Observation Equation})$$

which are the same as those in Example 10.8 except that in this example the noise W_t is not identically zero. Consider, for example, the case in which $\phi_1 = 0.9, \mathbf{\Gamma_V} = \sigma_V^2 = 1$, and $\mathbf{\Gamma_W} = \sigma_W^2 = 0.75$. In this case, X_t is AR(1), but the observed Y_t is not AR(1). In Figure 10.11, we show plots of realizations from X_t and Y_t.

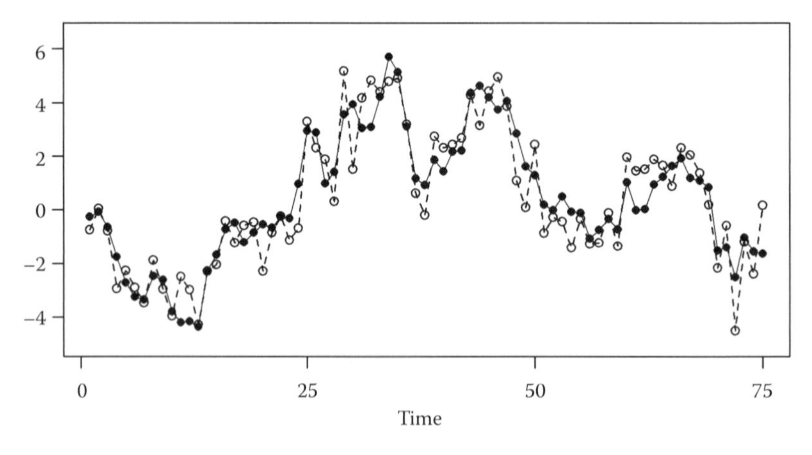

FIGURE 10.11
Realization of length $n = 75$ from the AR(1) model for X_t in Example 10.9 (solid line) along with the corresponding realization (dashed line) from the AR(1) plus noise model Y_t.

Example 10.10 State-Space Formulations of Stationary AR(p) and VAR(p) Models

Let Y_t be a stationary univariate, zero-mean AR(p) process given by $Y_t = \phi_1 Y_{t-1} + \cdots \phi_p Y_{t-p} + a_t$. A state-space representation is obtained using the (vector) equations

$$\mathbf{X}_t = \mathbf{\Phi}\mathbf{X}_{t-1} + \mathbf{V}_t \quad (\textit{State Equation})$$
$$\mathbf{Y}_t = (0\ 0 \ldots 1)\mathbf{X}_t \quad (\textit{Observation Equation})$$

where

$$\mathbf{X}_t = \begin{pmatrix} Y_{t-p+1} \\ Y_{t-p+2} \\ \vdots \\ Y_t \end{pmatrix}, \quad \mathbf{\Phi} = \begin{pmatrix} 0 & 1 & 0 & \cdots & 0 \\ 0 & 0 & 1 & \cdots & 0 \\ \vdots & \vdots & \vdots & & \vdots \\ 0 & 0 & 0 & \cdots & 1 \\ \phi_p & \phi_{p-1} & \phi_{p-2} & \cdots & \phi_1 \end{pmatrix}, \quad \mathbf{V}_t = \begin{pmatrix} 0 \\ 0 \\ \vdots \\ 0 \\ 1 \end{pmatrix} a_t.$$

Note that in this case, $v = p$, $w = 1$, and $\mathbf{W}_t = \mathbf{0}$. This example illustrates the fact that the AR(1)-type form of the state equation does not place the limitations on the applicable settings that might be expected.

If $\mathbf{Y}_t$ satisfies the zero-mean VAR(p) model $\mathbf{Y}_t = \mathbf{\Phi}_1 \mathbf{Y}_{t-1} + \cdots \mathbf{\Phi}_p \mathbf{Y}_{t-p} + \mathbf{a}_t$, then a state-space formulation is obtained using the following equations

$$\mathbf{X}_t = \mathbf{\Phi}\mathbf{X}_{t-1} + \mathbf{V}_t \quad (\textit{State Equation})$$
$$\mathbf{Y}_t = (\mathbf{0}\ \mathbf{0} \ldots \mathbf{I})\mathbf{X}_t \quad (\textit{Observation Equation})$$

where $\mathbf{X}_t$, $\mathbf{\Phi}$, and $\mathbf{V}_t$ are the $pm \times 1, pm \times pm,$ and $pm \times m$ matrices given respectively by where

$$\mathbf{X}_t = \begin{pmatrix} \mathbf{Y}_{t-p+1} \\ \mathbf{Y}_{t-p+2} \\ \vdots \\ \mathbf{Y}_t \end{pmatrix}, \quad \mathbf{\Phi} = \begin{pmatrix} \mathbf{0} & \mathbf{I} & \mathbf{0} & \cdots & \mathbf{0} \\ \mathbf{0} & \mathbf{0} & \mathbf{I} & \cdots & \mathbf{0} \\ \vdots & \vdots & \vdots & & \vdots \\ \mathbf{0} & \mathbf{0} & \mathbf{0} & \cdots & \mathbf{1} \\ \mathbf{\Phi}_p & \mathbf{\Phi}_{p-1} & \mathbf{\Phi}_{p-2} & \cdots & \mathbf{\Phi}_1 \end{pmatrix}, \quad \mathbf{V}_t = \begin{pmatrix} \mathbf{0} \\ \mathbf{0} \\ \vdots \\ \mathbf{0} \\ \mathbf{I} \end{pmatrix} \mathbf{a}_t.$$

Example 10.11 State-Space form of an ARMA(p,q) Process

A state-space formulation of the univariate, zero-mean ARMA(p,q) process

$$Y_t - \phi_1 Y_{t-1} - \cdots \phi_p Y_{t-p} = a_t - \theta_1 a_{t-1} - \cdots - \theta_q a_{t-q},$$

is given by

$$\mathbf{X}_t = \mathbf{\Phi}\mathbf{X}_{t-1} + \mathbf{V}_t \qquad \textit{(State Equation)}$$

$$\mathbf{Y}_t = (-\theta_{r-1}, -\theta_{r-2}, \ldots, -\theta_1, 1)\mathbf{X}_t \qquad \textit{(Observation Equation)}$$

where

$$\mathbf{X}_t = \begin{pmatrix} U_{t-r+1} \\ U_{t-r+2} \\ \vdots \\ U_t \end{pmatrix}, \quad \mathbf{\Phi} = \begin{pmatrix} 0 & 1 & 0 & \cdots & 0 \\ 0 & 0 & 1 & \cdots & 0 \\ \vdots & \vdots & \vdots & & \vdots \\ 0 & 0 & 0 & \cdots & 1 \\ \phi_r & \phi_{r-1} & \phi_{r-2} & \cdots & \phi_1 \end{pmatrix}, \quad \mathbf{V}_t = \begin{pmatrix} 0 \\ 0 \\ \vdots \\ 0 \\ 1 \end{pmatrix} a_t,$$

where U_t is the AR(p) process given by $\phi(B)U_t = a_t$, $r = \max(p, q+1)$, $\phi_j = 0$ for $j > p$, and $\theta_j = 0$ for $j > q$. (See Problem 10.9.)

10.6.3 Goals of State-Space Modeling

Recall that in the case of ARMA and ARUMA modeling, we typically analyze the data in a realization in order to estimate the model parameters and then use the estimated model for purposes of forecasting/predicting future values, spectral estimation, etc. Given a state-space model

$$\mathbf{X}_t = \mathbf{F}\mathbf{X}_{t-1} + \mathbf{V}_t$$

$$\mathbf{Y}_t = \mathbf{G}_t\mathbf{X}_t + \mathbf{W}_t,$$

the fundamental goal is to estimate the underlying unobserved signal $\mathbf{X}_t$ given the observations $\mathbf{Y}_1, \ldots, \mathbf{Y}_s$ for some time s. The typical procedure used to accomplish this goal is to

(a) Estimate the parameters ($\boldsymbol{\mu}_0, \mathbf{\Gamma}_0, \mathbf{F}, \mathbf{\Gamma}_\mathbf{V}, \mathbf{G}_t,$ and $\mathbf{\Gamma}_\mathbf{W}$) of the state-space model

(b) Use the estimated model to find the best linear estimate of $\mathbf{X}_t$ (for some time t) in the mean square sense. Specifically, we want to find $E(\mathbf{X}_t | \mathbf{Y}_1, \ldots, \mathbf{Y}_s)$. For simplicity, we will use $\hat{\mathbf{X}}_t^s$ to denote $E(\mathbf{X}_t | \mathbf{Y}_1, \ldots, \mathbf{Y}_s)$, and we denote the mean square error by $\mathbf{\Gamma}_t^s = E[(\mathbf{X}_t - \hat{\mathbf{X}}_t^s)(\mathbf{X}_t - \hat{\mathbf{X}}_t^s)']$. This step has three subcategories based on the relationship between time t and the time of the last observed value, s.

   (i) *Prediction*: This is the case in which $s < t$. An example would be the one step ahead forecasts $\hat{\mathbf{X}}_2^1, \hat{\mathbf{X}}_3^2, \ldots$.

   (ii) *Filtering*: The case in which $s = t$.

(iii) *Smoothing*: The case in which $s > t$. This case is called smoothing, since a time plot of $\hat{\mathbf{X}}_1^s, \hat{\mathbf{X}}_2^s, \ldots, \hat{\mathbf{X}}_s^s$ tends to be smoother than plots of the one-step ahead forecasts $\hat{\mathbf{X}}_1^0, \hat{\mathbf{X}}_2^1, \ldots, \hat{\mathbf{X}}_s^{s-1}$ or plots of the "filters" $\hat{\mathbf{X}}_1^1, \hat{\mathbf{X}}_2^2, \ldots, \hat{\mathbf{X}}_s^s$.

10.6.4 Kalman Filter

The Kalman filter is a very useful iterative procedure for obtaining the estimates $\hat{\mathbf{X}}_t^s$ for the state-space model in (10.47) and (10.48) with initial conditions $\mathbf{X}_0^0 = \boldsymbol{\mu}$ and $\boldsymbol{\Gamma}_0^0 = \boldsymbol{\Gamma}_0$. In the following, we give the Kalman filter recursions for computing predictions, filters, and smoothers of $\mathbf{X}_t$. In Appendix 10.A, we give a proof of the prediction and filtering recursion.

10.6.4.1 Prediction (Forecasting)

The following equations provide one-step ahead predictions, $\hat{\mathbf{X}}_t^{t-1}$ (or equivalently $\hat{\mathbf{X}}_{t+1}^t$). Specifically, we have observed $\mathbf{Y}_1, \ldots, \mathbf{Y}_{t-1}$ and want to use this information to predict $\mathbf{X}_t$. The Kalman recursion is as follows:

$$\hat{\mathbf{X}}_t^{t-1} = \mathbf{F}\hat{\mathbf{X}}_{t-1}^{t-1}, \tag{10.49}$$

$$\boldsymbol{\Gamma}_t^{t-1} = \mathbf{F}\boldsymbol{\Gamma}_{t-1}^{t-1}\mathbf{F}' + \boldsymbol{\Gamma}_{\mathbf{V}}. \tag{10.50}$$

10.6.4.2 Filtering

The filtering and prediction recursions are run simultaneously. For example, note that the prediction in (10.49) involves the filter term $\hat{\mathbf{X}}_{t-1}^{t-1}$. The equations in the Kalman recursions for filtering, i.e., for estimating $\mathbf{X}_t$ based on $\mathbf{Y}_1, \ldots, \mathbf{Y}_t$. are given by

$$\hat{\mathbf{X}}_t^t = \hat{\mathbf{X}}_t^{t-1} + \mathbf{K}_t(\mathbf{Y_t} - \mathbf{G}_t\hat{\mathbf{X}}_t^{t-1}), \tag{10.51}$$

$$\boldsymbol{\Gamma}_t^t = [\mathbf{I} - \mathbf{K}_t\mathbf{G}_t]\boldsymbol{\Gamma}_t^{t-1}, \tag{10.52}$$

where

$$\mathbf{K}_t = \boldsymbol{\Gamma}_t^{t-1}\mathbf{G}_t'\left[\mathbf{G}_t\boldsymbol{\Gamma}_t^{t-1}\mathbf{G}_t' + \boldsymbol{\Gamma}_{\mathbf{W}}\right]^{-1}. \tag{10.53}$$

K_t is called the *Kalman gain*.

Proof:

A proof of the validity of the prediction and filtering recursions is given in Appendix 10.A.

10.6.4.3 Smoothing Using the Kalman Filter

In this setting, values of $\mathbf{Y}_1, \ldots, \mathbf{Y}_n$ are observed and the goal is to estimate $\mathbf{X}_t$ where $t < n$, i.e., we want to calculate $\hat{\mathbf{X}}_t^n$. The recursion proceeds by calculating $\hat{\mathbf{X}}_t^t, \hat{\mathbf{X}}_t^{t-1}, \Gamma_t^t$, and $\mathbf{\Gamma}_t^{t-1}, t = 1, \ldots, n$ using the (10.49) through (10.53). Then, for $t = n, n-1, \ldots$ we calculate

$$\hat{\mathbf{X}}_{t-1}^n = \hat{\mathbf{X}}_{t-1}^{t-1} + \mathbf{J}_{t-1}\left(\hat{\mathbf{X}}_t^n - \hat{\mathbf{X}}_t^{t-1}\right) \tag{10.54}$$

$$\mathbf{\Gamma}_{t-1}^n = \mathbf{\Gamma}_{t-1}^{t-1} + \mathbf{J}_{t-1}\left(\mathbf{\Gamma}_t^n - \mathbf{\Gamma}_t^{t-1}\right)\mathbf{J}_{t-1}', \tag{10.55}$$

where

$$\mathbf{J}_{t-1} = \mathbf{\Gamma}_{t-1}^{t-1}\mathbf{F}'\left[\mathbf{\Gamma}_t^{t-1}\right]^{-1}. \tag{10.56}$$

10.6.4.4 h-Step Ahead Predictions

It should be noted that the predictions in (10.49) are 1-step ahead predictions (or forecasts), i.e., predictors of X_t given observations of the random variables $Y_1, \ldots, Y_{t-1}$. However, the ARMA-based forecasting discussed in Chapter 6 provided not only 1-step ahead forecasts but also h-step ahead forecasts. State-space methodology can also be used to find h-step ahead predictors, i.e., the predictor of X_{t+h-1} given observations on the random variables $Y_1, \ldots, Y_{t-1}$. The h-step ahead predictor, $\hat{\mathbf{X}}_{t+h-1}^{t-1}$, is defined to be

$$\hat{\mathbf{X}}_{t+h-1}^{t-1} = E(\mathbf{X}_{t+h-1} | \mathbf{Y}_1, \ldots, \mathbf{Y}_{t-1}). \tag{10.57}$$

From (10.47), it follows that $\mathbf{X}_{t+h-1} = \mathbf{F}\mathbf{X}_{t+h-2} + \mathbf{V}_{t+h-1}$, so, from (10.57), we have

$$\begin{aligned}
\hat{\mathbf{X}}_{t+h-1}^{t-1} &= E(\mathbf{F}\mathbf{X}_{t+h-2} + \mathbf{V}_{t+h-1} | \mathbf{Y}_1, \ldots, \mathbf{Y}_{t-1}) \\
&= \mathbf{F}E(\mathbf{X}_{t+h-2} | \mathbf{Y}_1, \ldots, \mathbf{Y}_{t-1}) \\
&= \mathbf{F}\hat{\mathbf{X}}_{t+h-2}^{t-1}.
\end{aligned}$$

Continuing recursively, it follows that

$$\hat{\mathbf{X}}_{t+h-1}^{t-1} = \mathbf{F}^{h-1}\hat{\mathbf{X}}_t^{t-1}. \tag{10.58}$$

Example 10.12 Predicting, Filtering, and Smoothing Using Data in Figure 10.11

In this example, we illustrate the concepts of predicting, filtering, and smoothing using the data in Figure 10.11. In Example 10.9, it was seen that the series $\mathbf{X}_t$ is from an AR(1) model while the observations $\mathbf{Y}_t$ follow an AR(1) plus noise model. We will use (10.49) through (10.53) to find the predictions, $\hat{\mathbf{X}}_t^{t-1}$,

TABLE 10.1

Predictors, Filters, and Smoothers for Data in Figure 10.11

T	X_t	Y_t	$\hat{X}_t^{t-1}$	Γ_t^{t-1}	$\hat{X}_t^t$	Γ_t^t	$\hat{X}_t^{75}$	Γ_t^{75}
0					0.00	5.26	0.90	1.40
1	−0.25	−0.74	0.00	5.26	−0.65	0.66	−0.60	0.49
2	−0.08	0.05	−0.58	1.53	−0.16	0.50	−0.47	0.40
3	−0.65	−0.78	−0.15	1.41	−0.56	0.49	−1.11	0.39
4	−1.77	−2.94	−0.50	1.40	−2.09	0.49	−2.25	0.39
5	−2.72	−2.28	−1.88	1.40	−2.14	0.49	−2.40	0.39

and filters, $\hat{\mathbf{X}}_t^t$, (along with the corresponding mean square errors) for the underlying signal, $\mathbf{X}_t$. These predictions and filters are based on the observed signal, $\mathbf{Y}_t$, and using the model parameters $\mathbf{F} = \phi_1 = 0.9$, $\mathbf{\Gamma_V} = \sigma_V^2 = 1$, $\mathbf{G}_t = 1$, and $\mathbf{\Gamma_W} = \sigma_W^2 = 0.75$. After the predictions and filters are computed for $t = 1, \ldots, n$, we use the smoothing equations (10.54) through (10.56) to calculate the smoothers, $\hat{\mathbf{X}}_t^n$, along with corresponding mean squares error.

In Table 10.1, we show the underlying (unobservable) values of X_t along with the observed values of Y_t for $t = 1, \ldots, 5$. (Recall that the full realization in Figure 10.11 has $n = 75$ observations.) The recursion is initialized with x_0, which is considered to be an observation from the AR(1) model $X_t - 0.9X_{t-1} = v_t$, where $\sigma_V^2 = 1$. Since $E(X_t) = 0$, we assign $\hat{X}_0^0 = x_0 = 0$. Also, since $\text{var}(X_t) = \sigma_w^2/(1 - \phi_1^2) = 1/(1 - 0.81) = 5.26$, we initialize with $\Gamma_0^0 = 5.26$. In this simple example, Equations 10.49 through 10.53 become $\hat{X}_t^{t-1} = 0.9\hat{X}_{t-1}^{t-1}$, $\Gamma_t^{t-1} = 0.81\Gamma_{t-1}^{t-1} + 1$, $\hat{X}_t^t = \hat{X}_t^{t-1} + K_t(y_t - \hat{X}_t^{t-1})$, $\Gamma_t^t = (1 - K_t)\Gamma_t^{t-1}$, and $K_t = \Gamma_t^{t-1}/(\Gamma_t^{t-1} + 0.75)$, respectively. So, it follows that

$$\hat{X}_1^0 = 0.9\hat{X}_0^0 = 0$$

$$\Gamma_1^0 = 0.81\Gamma_0^0 + 1 = 5.263$$

$$K_1 = \frac{\Gamma_1^0}{\Gamma_1^0 + \sigma_w^2} = \frac{5.623}{5.623 + 0.75} = 0.875$$

$$\hat{X}_1^1 = \hat{X}_1^0 + K_1(y_1 - \hat{X}_1^0) = 0 + 0.875(-0.74 - 0) = -0.648$$

$$\Gamma_1^1 = (1 - K_1)\Gamma_1^0 = (1 - 0.875)5.263 = 0.658$$

$$\hat{X}_2^1 = 0.9\hat{X}_1^1 = -0.583$$

$$\Gamma_2^1 = 0.81\Gamma_1^1 + 1 = 1.533$$

$$K_2 = \frac{\Gamma_2^1}{\Gamma_2^1 + 0.75} = \frac{1.533}{1.533 + 0.75} = 0.671$$

$$\hat{X}_2^2 = \hat{X}_2^1 + K_2(y_2 - \hat{X}_2^1) = -0.583 + 0.671(0.05 + 0.583) = -0.158$$

and so forth.

In Table 10.1, we show the predictors ($\hat{X}_t^{t-1}$), filters ($\hat{X}_t^t$), and smoothers ($\hat{X}_t^n$) along with associated mean squared errors (Γ_t^s) for the data in Figure 10.11 for $t = 1, \ldots, 5$. There it can be seen, as would be expected, that the mean-squared errors for prediction are greater than those for filtering (which are based on the additional data value for Y_t), and these are in turn greater than those for the smoothers (which are based on all $n = 75$ data values in the realization). In Figure 10.12a–c, we show the (assumed

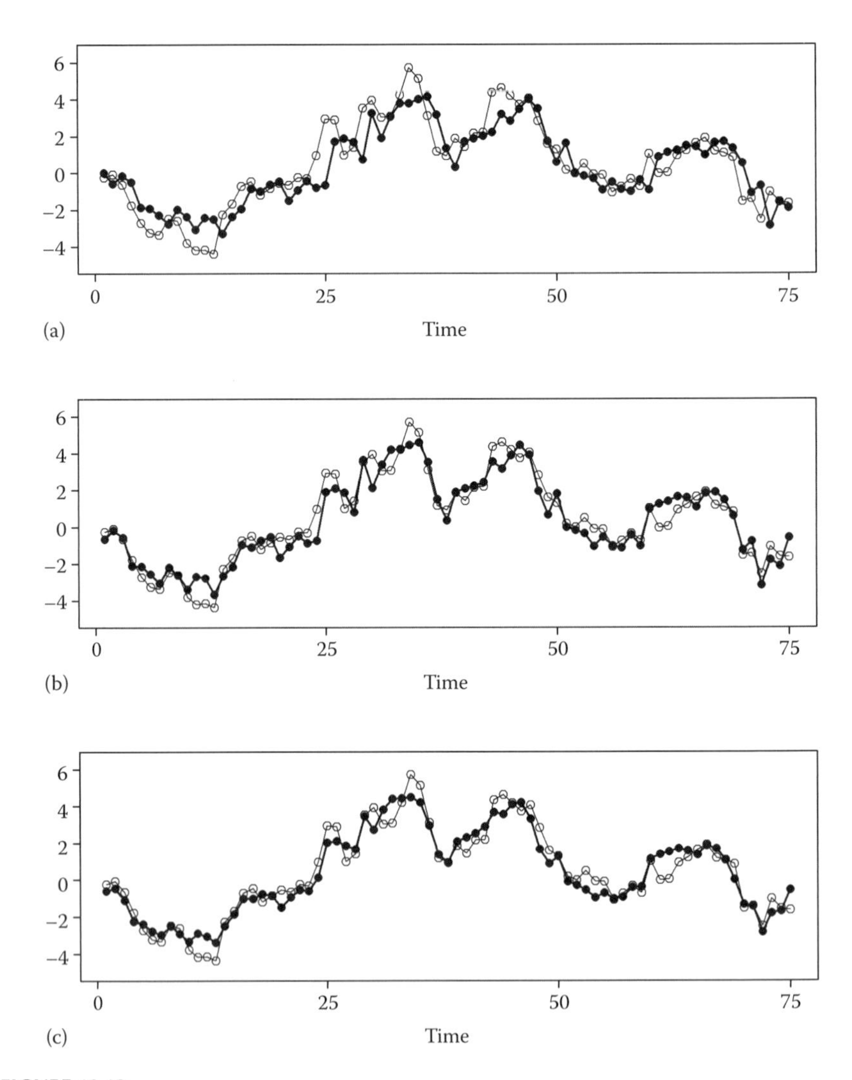

FIGURE 10.12
X_t (light line) overlaid with (a) predictors $\hat{X}_t^{t-1}$, (b) filters $\hat{X}_t^t$, and (c) smoothers $\hat{X}_t^n$.

unobserved) realization from X_t (light line) shown in Figure 10.11 along with the predicted, filtered, and smoothed estimates, respectively. There it can visually be seen that the filtered values (dark line) are somewhat closer to the true values (dark line) than are the predictors and that the smoothers in Figure 10.12c are smoother than the predictors and filters. The observed mean-squared errors, $MSE = \frac{1}{75}\sum_{j=1}^{75}\left(\hat{X}_t^s - x_t\right)^2$, are 1.25 for predictors ($s = t - 1$), 0.51 for filters ($s = t$), and 0.43 for smoothers ($s = n = 75$). Thus, the smoothers tend to be marginally closer to the true values than do the filters while both filters and smoothers offer a substantial improvement over the predictors.

10.6.5 Kalman Filter and Missing Data

Suppose, we observe $\{y_{t_1}, y_{t_2}, \ldots, y_{t_r}\}$, where $1 \le t_1 < t_2 < \cdots < t_r \le n$ and the t_i's are integers. That is, there are $n - r$ missing observations. Such a situation (which is all too common in practice) would cause serious problems with attempts to analyze this type of data using an ARMA(p,q) model based on techniques discussed in previous chapters of this book. One of the advantages of the state-space model is that it provides the ability to analyze such data. Specifically, state-space modeling provides at least two methods for dealing with this situation: (a) estimation of model parameters (using ML) even with some missing data and (b) estimation of missing values themselves.

In order to construct the state and observation equations appropriate for this missing data setting, we first define

$$y_t^* = \begin{cases} y_t & \text{if } t \in \{t_1, t_2, \ldots, t_r\} \\ 0 & \text{otherwise.} \end{cases}$$

Then, the state-space model is

$$\mathbf{X}_t = \mathbf{F}\mathbf{X}_{t-1} + \mathbf{V}_t, \quad \textit{State Equation (Unchanged)}$$

$$\mathbf{Y}_t = \mathbf{G}_t^*\mathbf{X}_t + \mathbf{W}_t^* \quad \textit{New Observation Equation}$$

where

$$\mathbf{G}_t^* = \begin{cases} \mathbf{G}_t, & t \in \{t_1, t_2, \ldots, t_r\} \\ \mathbf{0}, & \text{otherwise} \end{cases}$$

$$\mathbf{W}_t^* = \begin{cases} \mathbf{W}_t, & t \in \{t_1, t_2, \ldots, t_r\} \\ N_t & \text{otherwise,} \end{cases}$$

and $\{N_t\}$ are iid $N(0,1)$ and uncorrelated with $\mathbf{V}_t$ and $\mathbf{W}_t$.

Example 10.13 Example 10.12 Revisited with Two Missing Values

Suppose the observations, y_t, for $t = 2$ and $t = 5$ are missing. That is, we define

$$y_t^* = \begin{cases} y_t & \text{if } t \in \{1,3,4,6 \text{ through } 75\} \\ 0 & \text{otherwise,} \end{cases}$$

since $E(Y_t) = 0$. The revised state-space equations are

$$X_t = \phi_1 X_{t-1} + V_t$$
$$Y_t^* = G_t^* X_t + W_t^*,$$

where

$$Y_t^* = \begin{cases} y_t & \text{if } t \neq 2,5 \\ 0 & \text{if } t = 2,5 \end{cases}, \quad G_t^* = \begin{cases} 1, & t \neq 2,5 \\ 0, & t = 2,5 \end{cases}, \quad W_t^* = \begin{cases} W_t, & \text{if } t \neq 2,5 \\ W_t + N_t, & \text{if } t = 2,5. \end{cases}$$

We let $N_t \sim N(0,1)$ denote noise that is uncorrelated with both V_t and W_t. Consequently,

$$\sigma_{W^*}^2 = \begin{cases} 0.75, & \text{if } t \neq 2,5 \\ 1.75, & \text{if } t = 2,5. \end{cases}$$

The missing values are estimated using the Kalman iteration to calculate $\hat{X}_t^t, \hat{X}_t^{t-1}, \Gamma_t^t$, and $\Gamma_t^{t-1}, t = 1, \ldots, 75$, and then for $t = 75, 74, \ldots, 2, 1$, the smoothing iterations are used to calculate the values for $\hat{X}_2^{75}$ and $\hat{X}_5^{75}$, which are the desired estimates. In Table 10.2, we show the revised version of Table 10.1 when y_2 and y_5 are missing. As expected, it can be seen that $\hat{X}_2^2$ and $\hat{X}_2^{75}$ are associated with higher mean square errors than for the cases with available data on Y_t.

TABLE 10.2

Predictors, Filters, and Smoothers for Data in Figure 10.11

T	x_t	y_t	$\hat{X}_t^{t-1}$	Γ_t^{t-1}	$\hat{X}_t^t$	Γ_t^t	$\hat{X}_t^{75}$	Γ_t^{75}
0					0.00	5.26	0.90	1.40
1	−0.25	−0.74	0.00	5.26	−0.65	0.66	−.83	0.55
2	−0.08	—	−0.58	1.53	−0.58	1.53	−1.07	0.85
3	−0.65	−0.78	−0.52	2.24	−0.71	0.56	−1.31	0.44
4	−1.77	−2.94	−0.64	1.46	−2.16	0.49	−2.36	0.43
5	−2.72	—	−1.94	1.40	−1.94	1.40	−2.58	0.81

10.6.6 Parameter Estimation

Let $\boldsymbol{\Theta}$ denote the vector of parameters to be estimated in the state-space model specified in (10.47) and (10.48). It will be convenient for us to write the likelihood function $L(\boldsymbol{\Theta}; \mathbf{Y}_1, \ldots, \mathbf{Y}_n)$ (conditional on $\mathbf{Y}_0$) as $L(\boldsymbol{\Theta}; \mathbf{Y}_1, \ldots, \mathbf{Y}_n) = \prod_{t=1}^{n} f_t(\mathbf{Y}_t|\mathbf{Y}_{t-1}, \ldots, \mathbf{Y}_1)$, where $f_t(\mathbf{Y}_t|\mathbf{y}_{t-1}, \ldots, \mathbf{y}_1)$ is the conditional density of $\mathbf{Y}_t$, given that $\mathbf{Y}_j = \mathbf{y}_j, j = 1, \ldots, t-1$ (and conditional that $\mathbf{Y}_0 = \mathbf{y}_0$). Calculation of the likelihood is complicated but can be made more tractable if the $\mathbf{V}_t$ and $\mathbf{W}_t$ are jointly normally distributed and uncorrelated and if the initial state $\mathbf{Y} \sim MVN(\boldsymbol{\mu}_0, \boldsymbol{\Gamma}_0)$. The *innovations*, $\mathbf{I}_t$, are calculated where

$$\mathbf{I}_t = \mathbf{Y}_t - \mathbf{G}_t\hat{\mathbf{X}}_t^{t-1}, \tag{10.59}$$

and

$$\begin{aligned}\boldsymbol{\Gamma}_{\mathbf{I}_t} \equiv \text{Var}[\mathbf{I}_t] &= \text{Var}[\mathbf{G}_t(\mathbf{X}_t - \hat{\mathbf{X}}_t^{t-1}) + \mathbf{W}_t] \\ &= \mathbf{G}_t\boldsymbol{\Gamma}_t^{t-1}\mathbf{G}_t' + \boldsymbol{\Gamma}_{\mathbf{W}}.\end{aligned} \tag{10.60}$$

(see also proof of Kalman prediction and filtering in Appendix 10.A). Based on the aforementioned normality assumptions, the likelihood can be written as

$$L(\boldsymbol{\Theta}; \mathbf{Y}_1, \ldots, \mathbf{Y}_n) = \frac{1}{(2\pi)^{nw/2}}\left(\prod_{t=1}^{n}|\boldsymbol{\Gamma}_{\mathbf{I}_t}|\right)^{-1/2} \exp\left[-\frac{1}{2}\sum_{t=1}^{n}\mathbf{I}_t'\boldsymbol{\Gamma}_{\mathbf{I}_t}^{-1}\mathbf{I}_t\right]. \tag{10.61}$$

Numerical maximization procedures are used to find the maximum likelihood estimates from (10.61). Shumway and Stoffer (1982, 2006) discuss the use of Newton–Raphson and Expectation Maximization (EM) algorithms for performing this maximization.

10.6.7 Using State-Space Methods to Find Additive Components of a Univariate Autoregressive Realization

In Section 3.4, we discussed the decomposition of an autoregressive realization into additive components, $x_t = x_t^{(1)} + \cdots + x_t^{(m)}$, where m is the number of rows in the factor table. That is, m is equal to the number of real roots and pairs of complex conjugate roots of the characteristic equation. In that section, we gave a partial fractions derivation to show that the additive components are realizations from an AR(1) model for real roots and an ARMA (2,1) model for complex conjugate roots. We showed that in each case the system frequency associated with row j of the factor table is the system frequency inherent in the AR(1) or ARMA(2,1) model that corresponds to the jth component. However, the derivation in Section 3.4 did not lead to a method of calculation. In this example, we consider a state-space formulation of this problem, and in addition to showing that the components are AR(1) and ARMA(2,1) as before, the state-space approach also provides a simple method for calculating these components.

In Example 10.10, we gave a set of state-space equations for a univariate, zero-mean AR(p) process $Y_t = \phi_1 Y_{t-1} + \cdots \phi_p Y_{t-p} + a_t$. The state and observation equations were seen to be

$$\mathbf{X}_t = \mathbf{\Phi}\mathbf{X}_{t-1} + \mathbf{V}_t \quad (\textit{State Equation})$$

$$\mathbf{Y}_t = (0\ 0 \ldots 1)\mathbf{X}_t \quad (\textit{Observation Equation})$$

where

$$\mathbf{X}_t = \begin{pmatrix} Y_{t-p+1} \\ Y_{t-p+2} \\ \vdots \\ Y_t \end{pmatrix}, \quad \mathbf{\Phi} = \begin{pmatrix} 0 & 1 & 0 & \cdots & 0 \\ 0 & 0 & 1 & \cdots & 0 \\ \vdots & \vdots & \vdots & & \vdots \\ 0 & 0 & 0 & \cdots & 1 \\ \phi_p & \phi_{p-1} & \phi_{p-2} & \cdots & \phi_1 \end{pmatrix}, \quad \mathbf{V}_t = \begin{pmatrix} 0 \\ 0 \\ \vdots \\ 0 \\ 1 \end{pmatrix} a_t.$$

In this case, $v = p$, $w = 1$, and $\mathbf{W}_t = \mathbf{0}$, and we assume $a_t \sim N(0, \sigma_a^2)$. In the following, we will assume that the eigenvalues of $\mathbf{\Phi}$ are distinct. These eigenvalues are $\frac{1}{r_1}, \frac{1}{r_2}, \ldots, \frac{1}{r_p}$ where $r_1, \ldots, r_p$ are the roots of the characteristic equation $\phi(z) = 1 - \phi_1 z - \phi_2 z^2 - \cdots - \phi_p z^p = 0$. The eigenvector corresponding to $\frac{1}{r_j}$ can be chosen to be $\mathbf{u}_j' = \left(r_j^{p-1}\ r_j^{p-2} \ldots r_j\ 1\right)$, and we have $\mathbf{\Phi} = \mathbf{U}\mathbf{D}\mathbf{U}^{-1}$ where $\mathbf{U} = [\mathbf{u}_1 \quad \mathbf{u}_2 \quad \ldots \quad \mathbf{u}_p]$ and $\mathbf{D} = \text{diag}\left(\frac{1}{r_1} \quad \frac{1}{r_2} \quad \cdots \quad \frac{1}{r_p}\right)$ i.e., a diagonal matrix with the list elements along the diagonal. Note that due to the choice of eigenvectors, the last row of $\mathbf{U}$ is a vector of 1s. For our purposes, we arrange $\mathbf{D}$ (and $\mathbf{U}$) so that $|r_1| \geq |r_2| \geq \cdots$.

Defining $\boldsymbol{\psi}_t = \mathbf{U}^{-1}\mathbf{X}_t$ (i.e., $\mathbf{X}_t = \mathbf{U}\boldsymbol{\psi}_t$), the state-space model can be rewritten so that the rows of $\mathbf{U}^{-1}$ provide the filter coefficients needed to calculate the components.

10.6.7.1 Revised State-Space Model

State Equation: $\mathbf{X}_t = \mathbf{\Phi}\mathbf{X}_{t-1} + \mathbf{V}_t$ becomes $\mathbf{X}_t = (\mathbf{U}\mathbf{D}\mathbf{U}^{-1})(\mathbf{U}\boldsymbol{\psi}_{t-1}) + \mathbf{V}_t$, i.e., $U\boldsymbol{\psi}_t = UD\boldsymbol{\psi}_{t-1} + \mathbf{V}_t$, so that

$$\boldsymbol{\psi}_t = \mathbf{D}\boldsymbol{\psi}_{t-1} + \mathbf{U}^{-1}\mathbf{V}_t \quad (\textit{New State Equation}) \tag{10.62}$$

Observation Equation: In this case $\mathbf{Y}_t = (0 \;\; 0 \;\; \ldots \;\; 0 \;\; 1)\mathbf{X}_t$ becomes $\mathbf{Y}_t = (0 \;\; 0 \;\; \ldots \;\; 0 \;\; 1)\mathbf{U}\boldsymbol{\psi}_t$, i.e.,

$$\mathbf{Y}_t = (1\;1\ldots1)\boldsymbol{\psi}_t = \sum_{j=1}^{p} \psi_{tj} \quad (\textit{New Observation Equation}) \tag{10.63}$$

Each additive component of $\boldsymbol{\psi}_t$ can be thought of as a "state variable" that is unobserved. The ψ_{tj} are the desired additive components except for the fact that some of the ψ_{tj} may be complex. We consider two cases.

10.6.7.2 ψ_j Real

If a root of the characteristic equation, r_j, is real, then the corresponding $\psi_{tj}, t = 1, \ldots, n$ is a sequence of real numbers. From (10.62), it follows in this case that

$$\psi_{tj} = \frac{1}{r_j}\psi_{t-1,j} + c_j a_t,$$

where c_j is the last value in the jth row of $\mathbf{U}^{-1}$. Thus, ψ_{tj} follows an AR(1) model, and the corresponding system frequency is 0 or 0.5 depending on whether r_j is positive or negative, respectively. Since $\boldsymbol{\psi}_t = \mathbf{U}^{-1}\mathbf{X}_t$, the additive component associated with current realization can be obtained numerically be multiplying the jth row of $\mathbf{U}^{-1}$ by $\mathbf{x}_t$ where $\mathbf{x}_t$ is the realization written in vector form. In the notation of Section 3.4, this multiplication produces the additive component $x_t^{(j)}$.

10.6.7.3 ψ_j Complex

Complex roots of the characteristic equation, $\phi(z) = 0$, occur in complex conjugate pairs, and if r_j and r_k are complex conjugates, i.e., $r_k = r_j^*$, then it follows that $\psi_{tk} = \psi_{tj}^*$ for every t. We assume that complex conjugate pairs occur in adjacent rows of $\boldsymbol{\psi}_t$. The following assumes that ψ_{tj} and $\psi_{t,j+1}$ are complex conjugate pairs, and ψ_{tj}^* is used to denote $\psi_{t,j+1}$. Thus, the real sequence

$$x_t^{(j)} = \psi_{tj} + \psi_{tj}^*, \tag{10.64}$$

is the additive component that corresponds to the quadratic factor, $1 - \alpha_{1j}z - \alpha_{2j}z^2$, of the characteristic polynomial (associated with the pair of complex conjugate roots, r_j and $r_{j+1} = r_j^*$). Using the fact that $\boldsymbol{\psi}_t = \mathbf{U}^{-1}\mathbf{X}_t$, the component, $x_t^{(j)}$, can be obtained by adding the jth and $(j+1)$st rows of $\mathbf{U}^{-1}$ and multiplying the resulting p real coefficients by $\mathbf{x}_t$. Appendix 10.A shows that $x_t^{(j)}$ in (10.64) is a realization from an ARMA(2,1) model.

Example 10.14 Finding the Additive Components of the AR(3) Discussed in Example 3.17

Figure 3.24 shows the additive components associated with the AR(3) model $X_t - 2.55X_{t-1} + 2.42X_{t-2} - 0.855X_{t-3} = a_t$, for which the factored form is $(1 - 0.95B)(1 - 1.6B + 0.9B^2)X_t = a_t$. For this AR(3) model, it follows that $\mathbf{\Phi} = \begin{pmatrix} 0 & 1 & 0 \\ 0 & 0 & 1 \\ 0.855 & -2.42 & 2.55 \end{pmatrix}$, and from the factor table in Table 3.1, it is seen that

$$\mathbf{U} = \begin{pmatrix} \dfrac{1}{(0.89+0.57i)^2} & \dfrac{1}{(0.89-0.57i)^2} & 0.95^2 \\ \dfrac{1}{0.89+0.57i} & \dfrac{1}{0.89-0.57i} & 0.95 \\ 1 & 1 & 1 \end{pmatrix}, \quad (10.65)$$

and it follows that

$$\mathbf{U}^{-1} = \begin{pmatrix} -1.76-0.53i & 2.81+1.82i & -1.08-1.25i \\ -1.76+0.53i & 2.81-1.82i & -1.08+1.25i \\ 3.52 & -5.61 & 3.15 \end{pmatrix}. \quad (10.66)$$

We let $x_t^{(1)}$ and $x_t^{(2)}$ denote the components associated with the complex conjugate roots and the real root, respectively. Thus, adding the first two rows of $\mathbf{U}^{-1}$ in (10.66) gives the row vector $(-3.52 \quad 5.62 \quad -2.16)$, and the corresponding additive component is

$$x_t^{(1)} = (-3.52 \quad 5.62 \quad -2.16)\begin{pmatrix} x_{t-2} \\ x_{t-1} \\ x_t \end{pmatrix},$$

for $t = 3, \ldots, 200$. From the third row of $\mathbf{U}^{-1}$, we obtain

$$x_t^{(2)} = (3.52 \quad -5.61 \quad 3.15)\begin{pmatrix} x_{t-2} \\ x_{t-1} \\ x_t \end{pmatrix}.$$

The components $x_t^{(1)}$ and $x_t^{(2)}$ were shown in Figure 3.24 where it can be seen that $x_t^{(1)}$ captures the pseudo-cyclic behavior with periods of about 11 (1/0.09) units while $x_t^{(2)}$ depicts the wandering behavior associated with the positive real root.

Using `GW-WINKS`: Decompositions of autoregressive realizations can be obtained using the `Factoring Routine` option in the `Time Series Analysis` menu. For a data set showing in the current GW data editor, you can request Burg estimates of order p that you specify and then request that a factor table be calculated. After the factor table is displayed, select the option `Create Additive Components`. The `GW-WINKS` program will produce columns in the data matrix associated with components1 through m where m is the number of lines in the factor table. These components will have variable names `COMP1, COMP2`, etc.

10.A Appendix: Derivation of State-Space Results

10.A.1 Proof of Kalman Recursion for Prediction and Filtering

From probability theory, it is known that if $Y_1, \ldots, Y_s, X_t$ are univariate random variables that are jointly normal, then the best linear predictor of X_t in the mean square sense based on $Y_1, Y_2, \ldots, Y_s$ is the conditional expectation $E[X_t|Y_1, \ldots, Y_s]$. That is, $E[X_t|Y_1, \ldots, Y_s]$ is linear in the Y_j's, call it $E[X_t|Y_1, \ldots, Y_s] = \sum_{j=1}^{s} a_j Y_j$, and among all linear combination of the Y_j's, the mean square error $E[(X_t - \sum_{j=1}^{s} a_j Y_j)^2]$ is a minimum. A similar result follows when the univariate random variables are replaced by their vector counterparts. That is, if $\mathbf{Y}_1, \ldots, \mathbf{Y}_s, \mathbf{X}_t$ are jointly multivariate normal, then the conditional expectation $E[\mathbf{X}_t|\mathbf{Y}_1, \ldots, \mathbf{Y}_s]$ is the linear combination $\sum_{j=1}^{s} \mathbf{A}_j \mathbf{Y}_j$, such that $E[(\mathbf{X}_t - \sum_{j=1}^{s} \mathbf{A}_j \mathbf{Y}_j)(\mathbf{X}_t - \sum_{j=1}^{s} \mathbf{A}_j \mathbf{Y}_j)']$ is a minimum. For prediction, we want to find

$$\hat{\mathbf{X}}_t^{t-1} = E[\mathbf{X}_t|\mathbf{Y}_1, \ldots, \mathbf{Y}_{t-1}],$$

which has the associated mean square error

$$\Gamma_t^{t-1} = E\left[\left(\mathbf{X}_t - \hat{\mathbf{X}}_t^{t-1}\right)\left(\mathbf{X}_t - \hat{\mathbf{X}}_t^{t-1}\right)'\right].$$

From (10.47), it follows that

$$\begin{aligned}
\hat{\mathbf{X}}_t^{t-1} &= E[\mathbf{X}_t|\mathbf{Y}_1, \ldots, \mathbf{Y}_{t-1}] \\
&= E[\mathbf{F}\mathbf{X}_{t-1} + \mathbf{V}_t|\mathbf{Y}_1, \ldots, \mathbf{Y}_{t-1}] \\
&= \mathbf{F}\hat{\mathbf{X}}_{t-1}^{t-1},
\end{aligned}$$

since $E[\mathbf{V}_t|\mathbf{Y}_1,\ldots,\mathbf{Y}_{t-1}] = \mathbf{0}$, giving the relationship in (10.49). Now,

$$\begin{aligned}
\Gamma_t^{t-1} &= E\left[\left(\mathbf{X}_t - \hat{\mathbf{X}}_t^{t-1}\right)\left(\mathbf{X}_t - \hat{\mathbf{X}}_t^{t-1}\right)'\right] \\
&= E\left[\left(\mathbf{F}\mathbf{X}_{t-1} + \mathbf{V}_t - \mathbf{F}\hat{\mathbf{X}}_{t-1}^{t-1}\right)\left(\mathbf{F}\mathbf{X}_{t-1} + \mathbf{V}_t - \mathbf{F}\hat{\mathbf{X}}_{t-1}^{t-1}\right)'\right] \\
&= E\left[\left(\mathbf{F}(\mathbf{X}_{t-1} - \hat{\mathbf{X}}_{t-1}^{t-1}) + \mathbf{V}_t\right)\left(\mathbf{F}(\mathbf{X}_{t-1} - \hat{\mathbf{X}}_{t-1}^{t-1}) + \mathbf{V}_t\right)'\right] \\
&= E\left[\mathbf{F}\left(\mathbf{X}_{t-1} - \hat{\mathbf{X}}_{t-1}^{t-1}\right)\left(\mathbf{X}_{t-1} - \hat{\mathbf{X}}_{t-1}^{t-1}\right)'\mathbf{F}'\right] + E\left[\mathbf{V}_t\mathbf{V}_t'\right] \\
&= \mathbf{F}\mathbf{\Gamma}_{t-1}^{t-1}\mathbf{F}' + \mathbf{\Gamma}_{\mathrm{V}},
\end{aligned}$$

giving (10.50). Note also that

$$\begin{aligned}
\hat{\mathbf{Y}}_t^{t-1} &= E[\mathbf{Y}_t|\mathbf{Y}_1,\ldots,\mathbf{Y}_{t-1}] \\
&= E[\mathbf{G}\mathbf{X}_t + \mathbf{W}_t|\mathbf{Y}_1,\ldots,\mathbf{Y}_{t-1}] \\
&= \mathbf{G}\hat{\mathbf{X}}_t^{t-1},
\end{aligned}$$

since $E[\mathbf{W}_t|\mathbf{Y}_1,\ldots,\mathbf{Y}_{t-1}] = \mathbf{0}$.

For filtering, we define the innovations

$$\begin{aligned}
\mathbf{I}_t &= \mathbf{Y}_t - \hat{\mathbf{Y}}_t^{t-1} \\
&= \mathbf{Y}_t - \mathbf{G}_t\hat{\mathbf{X}}_t^{t-1} \\
&= \mathbf{G}_t\mathbf{X}_t + \mathbf{W}_t - \mathbf{G}_t\hat{\mathbf{X}}_t^{t-1} \\
&= \mathbf{G}_t(\mathbf{X}_t - \hat{\mathbf{X}}_t^{t-1}) + \mathbf{W}_t,
\end{aligned} \tag{10.A.1}$$

$t = 1,\ldots,n$. Note that $E(\mathbf{I}_t) = \mathbf{0}$ and

$$\begin{aligned}
\mathrm{Var}[\mathbf{I}_t] &= \mathrm{Var}[\mathbf{G}_t\left(\mathbf{X}_t - \hat{\mathbf{X}}_t^{t-1}\right) + \mathbf{W}_t] \\
&= \mathbf{G}_t\mathbf{\Gamma}_t^{t-1}\mathbf{G}_t' + \mathbf{\Gamma}_{\mathbf{W}}.
\end{aligned}$$

Using the fact that the $\mathbf{I}_t$'s are uncorrelated with past observations, i.e., $E[\mathbf{I}_t\mathbf{Y}_s] = 0$ for $s \le t-1$, it follows from (10.A.1) that $\mathbf{X}_t - \hat{\mathbf{X}}_t^{t-1}$ is also uncorrelated with past observations. Because of normality it follows that $\mathbf{X}_t - \hat{\mathbf{X}}_t^{t-1}$ is independent of past observations. Thus,

$$\begin{aligned}
\mathrm{Cov}[\mathbf{X}_t, \mathbf{I}_t|\mathbf{Y}_1,\ldots,\mathbf{Y}_{t-1}] &= \mathrm{Cov}[\mathbf{X}_t - \hat{\mathbf{X}}_t^{t-1}, \mathbf{I}_t|\mathbf{Y}_1,\ldots,\mathbf{Y}_{t-1}] \\
&\quad \text{(since } \hat{\mathbf{X}}_t^{t-1} \text{ is a function of } \mathbf{Y}_1,\ldots,\mathbf{Y}_{t-1}) \\
&= \mathrm{Cov}[\mathbf{X}_t - \hat{\mathbf{X}}_t^{t-1}, \mathbf{G}_t(\mathbf{X}_t - \hat{\mathbf{X}}_t^{t-1}) + \mathbf{W}_t|\mathbf{Y}_1,\ldots,\mathbf{Y}_{t-1}] \\
&= \mathrm{Cov}[\mathbf{X}_t - \hat{\mathbf{X}}_t^{t-1}, \mathbf{G}_t(\mathbf{X}_t - \hat{\mathbf{X}}_t^{t-1}) + \mathbf{W}_t] \\
&= \mathbf{\Gamma}_t^{t-1}\mathbf{G}_t',
\end{aligned}$$

with the third equality following from the independence between $\mathbf{X}_t - \hat{\mathbf{X}}_t^{t-1}$ and past observations.

From the aforementioned development, the conditional distribution of $\begin{pmatrix} \mathbf{X}_t \\ \mathbf{I}_t \end{pmatrix}$ is multivariate normal with mean $\begin{pmatrix} \hat{\mathbf{X}}_t^{t-1} \\ \mathbf{0} \end{pmatrix}$ and covariance matrix $\begin{pmatrix} \mathbf{\Gamma}_t^{t-1} & \mathbf{\Gamma}_t^{t-1}\mathbf{G}_t' \\ \mathbf{G}_t\mathbf{\Gamma}_t^{t-1} & \mathbf{G}_t\mathbf{\Gamma}_t^{t-1}\mathbf{G}_t' + \mathbf{\Gamma}_\mathbf{W} \end{pmatrix}$. We will use the result from multivariate statistics that if $\mathbf{Z}_1$ and $\mathbf{Z}_2$ are m_1 and m_2 dimensional random variables, where $\text{var}[\mathbf{Z}_1] = \mathbf{\Gamma}_{\mathbf{Z}_1}$, $\text{var}[\mathbf{Z}_2] = \mathbf{\Gamma}_{\mathbf{Z}_2}$, and $\text{Cov}[\mathbf{Z}_1, \mathbf{Z}_2] = \mathbf{\Gamma}_{\mathbf{Z}_1\mathbf{Z}_2}$, and if the vector $\begin{pmatrix} \mathbf{Z}_1 \\ \mathbf{Z}_2 \end{pmatrix}$ is multivariate normal, then

$$\text{E}[\mathbf{Z}_1|\mathbf{Z}_2] = \boldsymbol{\mu}_{\mathbf{Z}_1} + \mathbf{\Gamma}_{\mathbf{Z}_1\mathbf{Z}_2}\mathbf{\Gamma}_{\mathbf{Z}_2\mathbf{Z}_2}^{-1}(\mathbf{Z}_2 - \boldsymbol{\mu}_{\mathbf{Z}_2}], \tag{10.A.2}$$

and

$$\text{var}[\mathbf{Z}_1|\mathbf{Z}_2] = \Gamma_{\mathbf{Z}_1\mathbf{Z}_1} - \mathbf{\Gamma}_{\mathbf{Z}_1\mathbf{Z}_2}\mathbf{\Gamma}_{\mathbf{Z}_2\mathbf{Z}_2}^{-1}\mathbf{\Gamma}_{\mathbf{Z}_2\mathbf{Z}_1}. \tag{10.A.3}$$

Using (10.A.2), it follows that

$$\begin{aligned} \hat{\mathbf{X}}_t^t &= E[\mathbf{X}_t|\mathbf{Y}_1, \ldots, \mathbf{Y}_{t-1}, \mathbf{Y}_t] \\ &= E[\mathbf{X}_t|(\mathbf{Y}_1, \ldots, \mathbf{Y}_{t-1}), \mathbf{I}_t] \\ &= \hat{\mathbf{X}}_t^{t-1} + \mathbf{\Gamma}_t^{t-1}\mathbf{G}_t'(\mathbf{G}_t\Gamma_t^{t-1}\mathbf{G}_t' + \mathbf{\Gamma}_\mathbf{W})^{-1}\mathbf{I}_t. \end{aligned}$$

Letting $\mathbf{K}_t = \mathbf{\Gamma}_t^{t-1}\mathbf{G}_t'[\mathbf{G}_t\mathbf{\Gamma}_t^{t-1}\mathbf{G}_t' + \mathbf{\Gamma}_\mathbf{W}]^{-1}$ and noting that the second equality in (10.A.1) shows that $\mathbf{I}_t = \mathbf{Y}_t - \mathbf{G}_t\hat{\mathbf{X}}_t^{t-1}$, then the filtering result (10.51) follows. Also, from (10.A.3) we obtain

$$\begin{aligned} \mathbf{\Gamma}_t^t &= \text{Var}[\mathbf{X}_t|\mathbf{Y}_1, \ldots, \mathbf{Y}_{t-1}, \mathbf{Y}_t] \\ &= \text{Var}[\mathbf{X}_t|(\mathbf{Y}_1, \ldots, \mathbf{Y}_{t-1}), \mathbf{I}_t] \\ &= \mathbf{\Gamma}_t^{t-1} - \mathbf{\Gamma}_t^{t-1}\mathbf{G}_t'\left(\mathbf{G}_t\Gamma_t^{t-1}\mathbf{G}_t' + \mathbf{\Gamma}_\mathbf{W}\right)^{-1}\mathbf{G}_t\mathbf{\Gamma}_t^{t-1}, \end{aligned}$$

and the result in (10.52) follows.

10.A.2 Derivation of the Fact That $x_t^{(j)}$ in (10.64) Is a Realization from an ARMA(2,1) Model

We want to show that the real sequence $x_t^{(j)} = \psi_{tj} + \psi_{tj}^*$ as defined in (10.64) is a realization from an ARMA(2,1) model and is in fact the additive component corresponding to the quadratic factor $1 - \alpha_{1j}z - \alpha_{2j}z^2$ of the characteristic polynomial associated with the pair of complex conjugate roots, r_j and $r_{j+1} = r_j^*$. In the following, it is assumed that the component $x_t^{(j)}$ is obtained by adding the jth and $(j+1)$st rows of $\mathbf{U}^{-1}$ and multiplying the resulting p real coefficients by $\mathbf{X}_t$.

First, we define

$$\mathbf{x}_t^{(j)} = \begin{pmatrix} x_{t-1}^{(j)} \\ x_t^{(j)} \end{pmatrix} = \begin{pmatrix} \psi_{t-1,j} + \psi_{t-1,j}^* \\ \psi_{tj} + \psi_{tj}^* \end{pmatrix}, \tag{10.A.4}$$

and note from (10.62) that ψ_{tj} and ψ_{tj}^* are both complex-valued AR(1) realizations, i.e.,

$$\begin{aligned} \psi_{tj} &= \frac{1}{r_j}\psi_{t-1,j} + c_j a_t \\ \psi_{tj}^* &= \frac{1}{r_j^*}\psi_{t-1,j}^* + c_j^* a_t, \end{aligned} \tag{10.A.5}$$

where c_j and c_j^* are the last values in the jth row and $(j+1)$st row, respectively, of $\mathbf{U}^{-1}$. From (10.A.4) and (10.A.5),

$$\begin{aligned} \mathbf{x}_t^{(j)} &= \begin{pmatrix} \psi_{t-1,j} + \psi_{t-1,j}^* \\ \psi_{tj} + \psi_{tj}^* \end{pmatrix} \\ &= \begin{pmatrix} \psi_{t-1,j} + \psi_{t-1,j}^* \\ \frac{1}{r_j}\psi_{t-1,j} + \frac{1}{r_j^*}\psi_{t-1,j}^* + (c_j + c_j^*)a_t \end{pmatrix} \\ &= \begin{pmatrix} 1 & 1 \\ \frac{1}{r_j} & \frac{1}{r_j^*} \end{pmatrix} \boldsymbol{\psi}_{t-1,j} + \begin{pmatrix} 0 \\ 1 \end{pmatrix} c a_t \\ &= \mathbf{C}\boldsymbol{\psi}_{t-1,j} + \begin{pmatrix} 0 \\ 1 \end{pmatrix} c a_t, \end{aligned}$$

where $\mathbf{C} = \begin{pmatrix} 1 & 1 \\ \frac{1}{r_j} & \frac{1}{r_j^*} \end{pmatrix}$, $\boldsymbol{\psi}_{t-1,j} = \begin{pmatrix} \psi_{t-1,i} \\ \psi_{t-1,j}^* \end{pmatrix}$, and $c = c_j + c_j^*$. Thus, if $\mathbf{D}_j = \begin{pmatrix} \frac{1}{r_j} & 0 \\ 0 & \frac{1}{r_j^*} \end{pmatrix}$, then $\mathbf{C}\mathbf{D}_j\mathbf{C}^{-1} = \boldsymbol{\Phi}_j$ where $\boldsymbol{\Phi}_j = \begin{pmatrix} 0 & 1 \\ \alpha_{2j} & \alpha_{1j} \end{pmatrix}$, and $1 - \alpha_{1j}z - \alpha_{2j}z^2$ is the irreducible second order factor of $\phi(z) = 0$ associated with the roots r_j and r_j^*. Specifically, using (3.5.7) and (3.5.8), it can be seen that $\alpha_{1j} = \frac{1}{r_j} + \frac{1}{r_j^*}$ and $\alpha_{2j} = -\frac{1}{r_j r_j^*}$. That is, $\boldsymbol{\Phi}_j$ is a 2×2 matrix of the same form as $\boldsymbol{\Phi}$. Also, note that $\boldsymbol{\psi}_{t-1,j}$ can be written in terms of $\mathbf{x}_t^{(j)}$ and a_t as

$$\boldsymbol{\psi}_{t-1,j} = \mathbf{C}^{-1}\left(\mathbf{x}_t^{(j)} - \begin{pmatrix} 0 \\ 1 \end{pmatrix} ca_t\right)$$
$$\boldsymbol{\psi}_{t-2,j} = \mathbf{C}^{-1}\left(\mathbf{x}_{t-1}^{(j)} - \begin{pmatrix} 0 \\ 1 \end{pmatrix} ca_{t-1}\right). \tag{10.A.6}$$

Taking the relevant portion of the new state equation in (10.62), it follows that

$$\boldsymbol{\psi}_{tj} = \mathbf{D}_j \boldsymbol{\psi}_{t-1,j} + \begin{pmatrix} c_j \\ c_j^* \end{pmatrix} a_t$$
$$\boldsymbol{\psi}_{t-1,j} = \mathbf{D}_j \boldsymbol{\psi}_{t-2,j} + \begin{pmatrix} c_j \\ c_j^* \end{pmatrix} a_{t-1}. \tag{10.A.7}$$

Substituting (10.A.6) into (10.A.7), we obtain

$$\mathbf{C}^{-1}\left(\mathbf{x}_t^{(j)} - \begin{pmatrix} 0 \\ 1 \end{pmatrix} ca_t\right) = \mathbf{D}_j \mathbf{C}^{-1}\left(\mathbf{x}_{t-1}^{(j)} - \begin{pmatrix} 0 \\ 1 \end{pmatrix} ca_{t-1}\right) + \begin{pmatrix} c_j \\ c_j^* \end{pmatrix} a_{t-1},$$

$$\mathbf{x}_t^{(j)} - \begin{pmatrix} 0 \\ 1 \end{pmatrix} ca_t = \mathbf{C}\mathbf{D}_j \mathbf{C}^{-1}\left(\mathbf{x}_{t-1}^{(j)} - \begin{pmatrix} 0 \\ 1 \end{pmatrix} ca_{t-1}\right) + \mathbf{C}\begin{pmatrix} c_j \\ c_j^* \end{pmatrix} a_{t-1},$$

$$\mathbf{x}_t^{(j)} - \begin{pmatrix} 0 \\ 1 \end{pmatrix} ca_t = \boldsymbol{\Phi}_j \mathbf{x}_{t-1}^{(j)} - c\boldsymbol{\Phi}_j \begin{pmatrix} 0 \\ 1 \end{pmatrix} a_{t-1} + \begin{pmatrix} c_j + c_j^* \\ \dfrac{c_j}{r_j} + \dfrac{c_j^*}{r_j^*} \end{pmatrix} a_{t-1}.$$

That is,

$$\mathbf{x}_t^{(j)} = \boldsymbol{\Phi}_j \mathbf{x}_{t-1}^{(j)} + \begin{pmatrix} 0 \\ 1 \end{pmatrix} ca_t + \boldsymbol{\Theta} a_{t-1},$$

where

$$\begin{aligned}
\boldsymbol{\Theta} &= -c\boldsymbol{\Phi}_j \begin{pmatrix} 0 \\ 1 \end{pmatrix} + \begin{pmatrix} c_j + c_j^* \\ \dfrac{c_j}{r_j} + \dfrac{c_j^*}{r_j^*} \end{pmatrix} \\
&= \begin{pmatrix} -c \\ -c\alpha_{1j} \end{pmatrix} + \begin{pmatrix} c \\ \dfrac{c_j}{r_j} + \dfrac{c_j^*}{r_j^*} \end{pmatrix} \\
&= \begin{pmatrix} -c \\ -c\left(\dfrac{1}{r_j} + \dfrac{1}{r_j^*}\right) \end{pmatrix} + \begin{pmatrix} c \\ \dfrac{c_j}{r_j} + \dfrac{c_j^*}{r_j^*} \end{pmatrix} \\
&= \begin{pmatrix} 0 \\ \dfrac{-c_j}{r_j^*} - \dfrac{c_j^*}{r_j} \end{pmatrix}.
\end{aligned}$$

Thus, $x_t^{(j)}$ is a realization from an ARMA(2,1) process.

Exercises

GW-WINKS does not have multivariate capabilities. To do the computer-based exercises in the following, you may need to use a package such as SAS or write your own code.

10.1 Generate bivariate Gaussian white noise realizations, $\mathbf{a}_t$, of length $n = 100$ with the following values of $\mathbf{\Gamma_a}$:

(A) $\mathbf{\Gamma_a} = \begin{pmatrix} 1 & 0 \\ 0 & 4 \end{pmatrix}$

(B) $\mathbf{\Gamma_a} = \begin{pmatrix} 1 & 1.2 \\ 1.2 & 3 \end{pmatrix}$

(C) $\mathbf{\Gamma_a} = \begin{pmatrix} 1 & -2 \\ -2 & 5 \end{pmatrix}$

(a) Explain the technique you used to generate your realizations.
(b) In each case, use the realizations you generated to illustrate the contemporaneous correlation.
(c) For the model in (C) what is $\text{Cov}[a_{t1}, a_{t+2,2}]$?

10.2 Determine whether the following VAR(1) processes, $(\mathbf{I} - \mathbf{\Phi}_1 B)\mathbf{X}_t = \mathbf{a}_t$, are stationary. In each case, let $\mathbf{\Gamma_a} = \mathbf{I}_p$, i.e., a $p \times p$ identity matrix..

(A) $\mathbf{\Phi}_1 = \begin{pmatrix} 0.5 & 0.7 \\ 0.4 & 0.9 \end{pmatrix}$

(B) $\mathbf{\Phi}_1 = \begin{pmatrix} 0.5 & 0.4 \\ 0.1 & 0.9 \end{pmatrix}$

(C) $\mathbf{\Phi}_1 = \begin{pmatrix} 0.5 & -0.4 \\ 0.1 & 0.9 \end{pmatrix}$

10.3 Consider the zero-mean bivariate VAR(1) process $(\mathbf{I} - \mathbf{\Phi}_1 B)\mathbf{X}_t = \mathbf{a}_t$ where $\mathbf{\Phi}_1 = \begin{pmatrix} 0.5 & -0.7 \\ 0.4 & 0.9 \end{pmatrix}$ and $\mathbf{\Gamma_a} = \begin{pmatrix} 1 & 1.2 \\ 1.2 & 3 \end{pmatrix}$.

(a) Show that $\mathbf{X}_t$ is stationary.
(b) Find $\mathbf{\Gamma}(0), \mathbf{\Gamma}(1), \text{and} \mathbf{\Gamma}(2)$.
(c) Find $\boldsymbol{\rho}(0), \boldsymbol{\rho}(1)$, and $\boldsymbol{\rho}(2)$.
(d) Plot a realization of length $n = 100$ from this bivariate process, using the realization for bivariate white noise series (C) in Problem 10.1.

10.4 File PROB10.4.XLS contains a realization of length $n = 100$ from the bivariate VAR(1) model in Problem 10.3. For this realization:

(a) Using the true model, perform calculations such as those in (10.27) to find forecasts $\hat{X}_{100,1}(1), \hat{X}_{100,2}(1), \hat{X}_{100,1}(2), \hat{X}_{100,2}(2)$
(b) Find the Yule–Walker estimates of $\mathbf{\Phi}_1$ assuming a bivariate VAR(1) model.

(c) Find the Yule–Walker estimates of $\mathbf{\Phi}_1$ and $\mathbf{\Phi}_2$ assuming a bivariate VAR(2) model.
(d) Verify that in both (b) and (c), the resulting model is stationary.
(e) For each model found in (b) and (c) find $\hat{\mathbf{\Gamma}}_p$.
(f) Find the residuals using (10.35) based on the fitted VAR(1) model.

10.5 Use the procedure discussed in Section 10.5.1, to test whether the following two pairs of time series are independent (i.e., assume that all series involved come from stationary models).
(a) The annual incidence of nonperforated appendicitis and influenza for the years 1970–2005 (file `APPYFLU.XLS`).
(b) 6-month CD and Moody's Seasonal Aaa corporate bond rates (file `ECODATA.XLS`)
(c) 6-month CD-rate and 30-year mortgage rates (file `ECODATA.XLS`).

10.6 File `PROB10.6.XLS` contains observed data, Y_t, generated from a state space model with $\phi_1 = 0.95$, $\mathbf{\Gamma}_{\mathbf{v}} = \sigma_v^2 = 0.5$, and $\mathbf{\Gamma}_{\mathbf{w}} = \sigma_w^2 = 0.5$. Also in file `PROB10.6.XLS` are the (unobservable) X_t. Complete a table such as that in Table 10.1 for these data. Show the entire table, not simply the first five lines.

10.7 File `PROB10.7.XLS` contains the same data in `PROB10.6.XLS` except that the values for Y_2 and Y_4 are missing. Complete a table such as that in Table 10.1 for these data.

10.8 Generate realizations of length $n = 100$ from the following stationary AR(p) models
(A) $(1 - 0.83B + 0.98B^2 - 0.79B^3)(X_t - 25) = a_t$
(B) $(1 - 1.47B + 0.12B^2 + 1.12B^3 - 0.75B^4)(X_t - 200) = a_t$
For each of these realizations
(a) Use `GW-WINKS` to plot the realization along with the additive autoregressive components (see Example 10.14 and associated discussion along with Section 3.4).
(b) Discuss your results relating each additive component to one of the lines in the factor table.

10.9 Show that a state-space formulation of the univariate, zero-mean ARMA(p,q) process

$$Y_t - \phi_1 Y_{t-1} - \cdots \phi_p Y_{t-p} = a_t - \theta_1 a_{t-1} - \cdots - \theta_q a_{t-q},$$

is given by

$$\mathbf{X}_t = \mathbf{\Phi}\mathbf{X}_{t-1} + \mathbf{V}_t \quad (\textit{State Equation})$$

$$\mathbf{Y}_t = (-\theta_{r-1}, -\theta_{r-2}, \ldots, -\theta_1, 1)\mathbf{X}_t \quad (\textit{Observation Equation})$$

where

$$\mathbf{X}_t = \begin{pmatrix} U_{t-r+1} \\ U_{t-r+2} \\ \vdots \\ U_t \end{pmatrix}, \quad \mathbf{\Phi} = \begin{pmatrix} 0 & 1 & 0 & \cdots & 0 \\ 0 & 0 & 1 & \cdots & 0 \\ \vdots & \vdots & \vdots & & \vdots \\ 0 & 0 & 0 & \cdots & 1 \\ \phi_r & \phi_{r-1} & \phi_{r-2} & \cdots & \phi_1 \end{pmatrix}, \quad \mathbf{V}_t = \begin{pmatrix} 0 \\ 0 \\ \vdots \\ 0 \\ 1 \end{pmatrix} a_t,$$

U_t is the AR(p) process given by $\phi(B)U_t = a_t$, $r = \max(p, q+1)$, $\phi_j = 0$ for $j > p$, and $\theta_j = 0$ for $j > q$.

11

Long-Memory Processes

In the previous chapters, we have considered ARMA, ARUMA, seasonal, signal-plus-noise, and ARCH/GARCH processes. For the current discussion, we restrict ourselves to nonseasonal ARUMA(p,d,q) models and signal-plus-noise models. Stationary ARMA(p,q) processes have autocorrelations that decay (go to zero) exponentially as k gets large. That is, $\rho_k \simeq \alpha^{|k|}, |\alpha| < 1$ as $k \to \infty$ where k is lag. We use the notation $f_k \simeq g_k$ as $k \to \infty$ to indicate that $\lim_{k\to\infty} (f_k/g_k) = c$ where c is a finite nonzero constant. For example, for an AR(1) model with $\phi_1 = 0.9$, $\rho_k = 0.9^{|k|}$. In Example 3.8, it was shown that the autocorrelation function for the AR(3) model $(1 - 0.95B)(1 - 1.6B + 0.9B^2)X_t = a_t$ is given by

$$\rho_k = 0.66(0.95)^k + 0.3518(0.95)^k \sin[2\pi(0.09)k + 1.31],$$

so that in both cases $\rho_k \simeq \alpha^{|k|}$ as $k \to \infty$. In Chapter 5 we viewed ARUMA(p,d,q) processes as limits of ARMA processes as one or more roots of the characteristic equation approach the unit circle uniformly. Since the autocorrelation of an ARMA process decays exponentially, such processes are referred to as "short-memory" processes. It was noted in Section 5.2 that the correlation function decays more slowly as one or more roots of the characteristic equation approach the unit circle, and in fact the extended autocorrelations of the limiting ARUMA(p,d,q) process with integer $d > 0$ do not decay as k goes to infinity. Signal-plus-noise processes also have autocorrelations that do not decay, but these models involve functional restrictions that often prove to be too limiting in practice. In Chapter 6, we showed that the correlation-driven (ARMA and ARUMA) models are more adaptive than the signal-plus-noise models in that the ARMA/ARUMA models "react" strongly to the most current data and very little to the data of the distant past.

In many physical problems, in such diverse areas as hydrology, economics, seismic monitoring, atmospheric analysis, brain imaging, and others, the correlations for data that otherwise appear to be stationary in nature are seen to decay much slower than those of short-memory (e.g., ARMA) models. See, for example, Beran (1992), Mandelbrot (1983), Keshner (1982), Taqqu (1986), and van der Ziel (1988). Hurst (1951, 1955) observed long-memory behavior (which became known as the "Hurst effect") in hydrological data. This discovery prompted researchers to develop models that would adequately account for this long-memory behavior.

11.1 Long Memory

In this section, we give definitions and characterizations of long-memory models. We begin with the following definition.

Definition 11.1 A stationary time series defined over $t = 0, \pm 1, \pm 2, \ldots$ is said to be *long memory* if $\sum_{k=0}^{\infty} |\gamma_k|$ diverges, where γ_k is the autocovariance of the process. Otherwise, the time series is said to be *short memory*.

For example, if the autocovariance decays hyperbolically, that is, $\gamma_k \simeq k^{-\alpha}$ as $k \to \infty$ for some $0 \le \alpha \le 1$, then the process is long memory. Of course, other correlation structures are possible that would satisfy the conditions of long memory in Definition 11.1. A frequency-domain definition of long memory that is a slight extension of Definition 11.1 is given in Definition 11.2.

Definition 11.2 A stationary time series defined over $t = 0, \pm 1, \pm 2, \ldots$ is said to be *long memory* if for some frequency, $f \in [0, 0.5]$, the power spectrum, $P(f)$, becomes unbounded.

This definition characterizes memory in the frequency domain as the existence of a persistent cyclical component within the process. If the spectrum becomes unbounded, then the autocovariances are not absolutely summable since by (1.29) it follows that

$$|P(f)| < 2\left|\sum_{k=0}^{\infty} \gamma_k \cos(2\pi k f)\right| \le 2\sum_{k=0}^{\infty} |\gamma_k|,$$

for all $f \in [0, 0.5]$. Miller (1994) gives a counter example to show that the converse is not true. From Definition 11.2 it follows that a short-memory process (e.g., stationary ARMA) has a bounded spectrum.

An example of long-memory processes (based on Definition 11.2) is the family of the so-called $1/f$ processes ($1/f$ noise, flicker noise, pink noise) whose spectrum behaves like $1/|f|^{\alpha}$ where the primary interest is $0 < \alpha < 1$. See Bullmore et. al. (2001) for a discussion of $1/f$ noise in fMRI brain imaging data, and Wornell (1993, 1996) for a discussion of the use of the wavelet transform (to be discussed in Chapter 12) for decorrelating data with $1/f$ noise. $1/f$-type data have been modeled using fractional Gaussian noise (*fGn*), which is a discrete version of fractional Brownian motion. See Mandelbrot and van Ness (1968) and Mandelbrot and Wallis (1969). The fractional difference and FARMA models, discussed in Section 11.2, are useful tools for modeling $1/f$-type data. Notice that Definition 11.2 defines long-memory behavior at frequencies different from zero, that is, "cyclic long memory." In Section 11.3, we discuss the Gegenbauer and GARMA models for modeling cyclic long-memory behavior, and in Section 11.4, we discuss the k-factor Gegenbauer and k-factor GARMA models for data with a variety of types of long-memory behavior. Finally, in Sections 11.5 and 11.6, we discuss estimation and forecasting for k-factor GARMA models, and in Section 11.7, we apply the results to atmospheric CO_2 data.

11.2 Fractional Difference and FARMA Processes

A widely used tool for analyzing long-memory time series is the fractional difference, $(1 - B)^\alpha$, which for real (noninteger) α is defined by

$$\begin{aligned}(1 - B)^\alpha f(t) &= \sum_{k=0}^{\infty} (-1)^k \binom{\alpha}{k} B^k f(t) \\ &= \sum_{k=0}^{\infty} (-1)^k \binom{\alpha}{k} f(t-k),\end{aligned} \tag{11.1}$$

where

$$\binom{\alpha}{k} = \frac{\Gamma(\alpha+1)}{\Gamma(\alpha-k+1)k!}, \tag{11.2}$$

and $\Gamma(z) = \int_0^\infty t^{z-1}e^{-z}dt$ is the gamma function. Using the fact that $\Gamma(x+1) = x\Gamma(x)$ for real x, it follows that

$$\begin{aligned}(1 - B)^\alpha f(t) &= \sum_{k=0}^{\infty} (-1)^k \frac{\Gamma(\alpha+1)}{\Gamma(\alpha-k+1)k!} f(t-k) \\ &= \sum_{k=0}^{\infty} \frac{(k-\alpha-1)(k-\alpha-2)\cdots(-\alpha)}{k!} f(t-k) \\ &= \sum_{k=0}^{\infty} \frac{\Gamma(k-\alpha)}{\Gamma(-\alpha)k!} f(t-k).\end{aligned} \tag{11.3}$$

Remark: The fractional difference can be extended to the case where α is any real or complex number. For our purposes here it suffices for α to be real. For a more general definition of the fractional difference and its applications, see Gray and Zhang (1988).

Granger and Joyeux (1980), Hosking (1981, 1984), and Geweke and Porter-Hudak (1983) proposed the fractional difference model defined in Definition 11.3.

Definition 11.3 If $\sum_{k=0}^{\infty} \frac{|\Gamma(k+d)|}{|\Gamma(d)|k!} < \infty$, then the general linear process defined by

$$\begin{aligned} X_t - \mu &= (1-B)^{-d} a_t \\ &= \sum_{k=0}^{\infty} \frac{\Gamma(k+d)}{\Gamma(d)k!} a_{t-k}, \quad t = 0, \pm 1, \pm 2, \ldots \end{aligned} \tag{11.4}$$

where a_t is white noise, is called a *fractional process*. If X_t is invertible, then the fractional model is written

$$(1-B)^d (X_t - \mu) = a_t. \tag{11.5}$$

We typically express the fractional model in the form given in (11.5). Theorem 11.1 gives basic properties of the fractional model including conditions for stationarity and invertibility.

Theorem 11.1

Let X_t be a fractional process of the form (11.4).

(a) If $d < 0.5$, then X_t is causal and stationary.
(b) If $d > -0.5$, then X_t is invertible.
(c) The spectral density of X_t is given by $S(f) = [4(\sin \pi f)^2]^{-d}$, for $0 < f \le 0.5$, and thus $\lim_{f \to 0} f^{2d} S(f)$ exists and is finite.
(d) If $d \in (-0.5, 0.5)$, then the autocorrelation function is given by $\rho_k = \frac{\Gamma(-d+1)\Gamma(k+d)}{\Gamma(d)\Gamma(k-d+1)}$, $k = 0, \pm 1, \ldots$, and $\rho_k \simeq k^{2d-1}$ as $k \to \infty$. Also, $\sigma_X^2 = \Gamma(-2d+1)/[\Gamma(-d+1)]^2$.
(e) If $0 < d < 0.5$ then X_t is long memory.

Proof:

See Hosking (1981). The long-memory result in (e) follows from (d) using Definition 11.1 and from (c) using Definition 11.2.

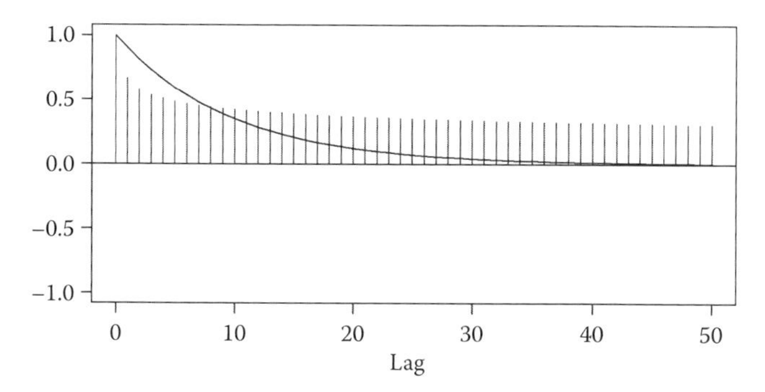

FIGURE 11.1
True autocorrelations for a fractional model with $d = 0.4$ (vertical bars) and an AR(1) model with $\phi_1 = 0.9$ (solid curve).

Example 11.1 Fractional Models

(a) Figure 11.1 shows autocorrelations for an AR(1) model with $\phi_1 = 0.9$ (solid line) along with those for a fractional model with $d = 0.4$ (vertical bars). For the AR(1) model, $\rho_1 = 0.9$ while $\rho_1 = 0.67$ for the fractional model, and for small lags ($k \leq 7$) the autocorrelations for the AR(1) model are larger. However, the AR(1) autocorrelations damp much more quickly, and ρ_{25} is 0.07 for the AR(1) model and 0.35 for the fractional model.

(b) Figure 11.2 shows autocorrelations for fractional models with $d = 0.2, 0.35, 0.45$, and 0.48. For $d = 0.2$ the autocorrelations are quite small (i.e., $\rho_1 = 0.25$ and $\rho_2 = 0.17$) but in all cases the autocorrelations damp slowly. As d approaches 0.5 the process approaches nonstationarity and autocorrelations move closer to 1. As an example, $\rho_{50} = 0.79$ when $d = 0.48$. The AR(1) model with this same autocorrelation at lag 50 has $\phi_1 = 0.9953$ which is very near the nonstationary region.

In Definition 11.4, we extend the idea of a fractional process to a FARMA process that can have both short and long-memory behavior.

Definition 11.4 If X_t is a causal, stationary, and invertible process satisfying

$$\phi(B)(1 - B)^d (X_t - \mu) = \theta(B)a_t, \tag{11.6}$$

where a_t is white noise, and where $\phi(z)$ and $\theta(z)$ are pth and qth-order polynomials, respectively, that share no common factors, then X_t is called a *fractional* ARMA (FARMA(p,d,q)) process.

To be more specific we state the following theorem.

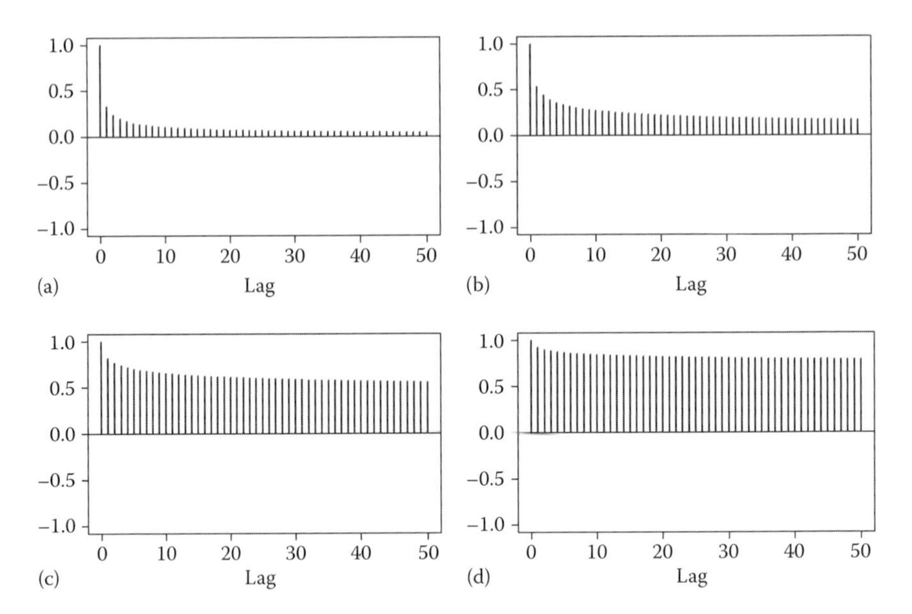

FIGURE 11.2
True autocorrelations for fractional processes with various values of d. (a) $d = 0.2$. (b) $d = 0.35$. (c) $d = 0.45$. (d) $d = 0.48$.

Theorem 11.2

Let X_t be a FARMA(p,d,q) process as defined in (11.6).

(a) If $d < 0.5$ and all the roots of $\phi(z) = 0$ are outside the unit circle, then X_t is causal and stationary.

(b) If $d > -0.5$ and all the roots of $\theta(z) = 0$ are outside the unit circle, then X_t is invertible.

(c) If $d \in (-0.5, 0.5)$ with $d \neq 0$, then the autocorrelation function ρ_k satisfies

$$\rho(k) \simeq k^{2d-1} \quad \text{as } k \to \infty. \tag{11.7}$$

(d) If $d \in (-0.5, 0.5)$ and $d \neq 0$, then $\lim_{f \to 0} f^{2d} S(f)$ exists and is finite.

Proof:

See Hosking (1981, Theorem 2) or Brockwell and Davis (1991).

NOTE: Hosking (1981) and others call X_t in (11.6) an ARIMA(p,d,q) process, but we reserve the expression ARIMA(p,d,q) to refer to processes for which d is a positive integer. The model is also referred to as an ARFIMA(p,d,q) model.

Example 11.2 Fractional and FARMA Models

(a) A realization of length $n = 500$ was generated from the fractional model with $d = 0.2$ whose theoretical autocorrelations are shown in Figure 11.2a. Figure 11.3b shows the sample autocorrelations, when $d = 0.2$, computed from the realization of length $n = 500$, and Figure 11.3a shows a sub-realization of length $n = 100$ from the original realization in order to show the behavior more clearly. The data have a similar appearance to white noise except that there is a tendency for wandering/random trending behavior. The sample autocorrelations generally stay within the 95% lines for testing white noise. Note however that the first 14 sample autocorrelations are positive, and 41 of the first 50 sample autocorrelations are positive. This is consistent with the theoretical autocorrelations in Figure 11.2a but inconsistent with white noise sample autocorrelations which, as discussed in Section 8.1, should be approximately uncorrelated. This behavior is consistent with $1/f$ noise encountered, for example, in fMRI data (Bullmore et al., 2001).

Figure 11.3c and d relate to a realization of length $n = 500$ from the fractional model with $d = 0.45$ (theoretical autocorrelations are

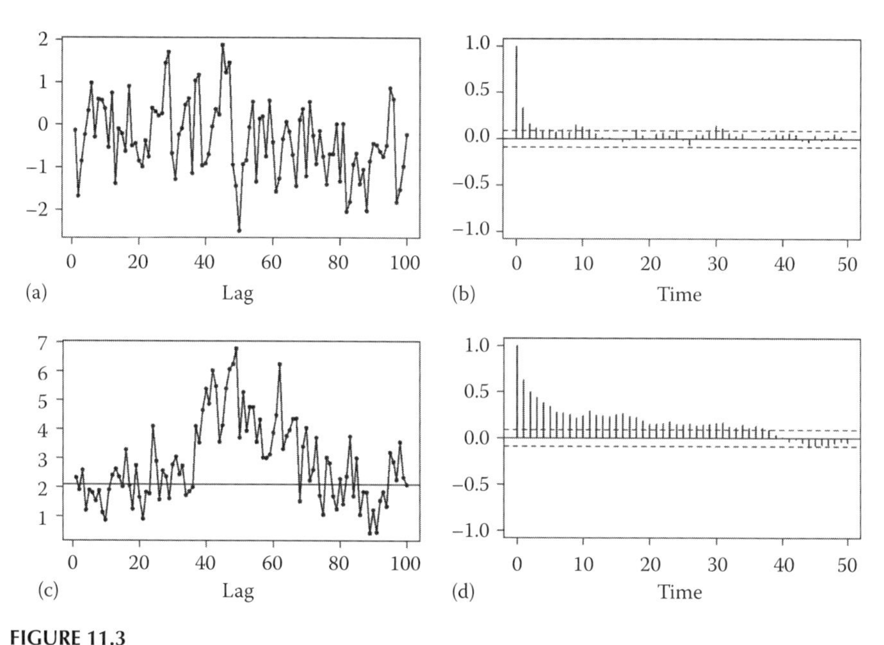

FIGURE 11.3
(a) Subrealization of length 100 from a realization of length $n = 500$ from a fractional model with $d = 0.2$. (b) Sample autocorrelations based on full realization of length 500. (c) Subrealization of length 100 from a realization of length $n = 500$ from a fractional model with $d = 0.45$. (d) Sample autocorrelations based on full realization of length 500.

shown in Figure 11.2c). Figure 11.3d shows the sample autocorrelations computed from the realization of length $n = 500$, and Figure 11.3c shows a sub-realization of length $n = 100$ from the original realization. The data show wandering trend behavior similar to that seen in AR(1) models with positive ϕ_1. Note however, even though the autocorrelations (and sample autocorrelations) damp slowly, the lag 1 sample autocorrelation ($\hat{\rho}_1 = 0.62$) is not sufficiently high to cause the type of highly correlated behavior seen for the realization from an AR(1) with $\phi_1 = 0.95$ shown in Figure 3.5a. It is interesting to note that for lag 10 and beyond, the true autocorrelations for the fractional model with $d = 0.45$ are larger than those for the AR(1) with $\phi_1 = 0.95$. In some ways, the realization in Figure 11.3c is similar to the AR(1) realization in Figure 3.5c for $\phi_1 = 0.7$. However, to illustrate the long-memory behavior, a horizontal line is shown at $\bar{x} = 2.1$, the mean of the 500-point realization. It is seen that, although the lag 1 autocorrelation is similar to that seen in Figure 3.5c, closer examination of the realization in Figure 11.3c shows that it contains longer stages of wandering above and below the mean. This reflects the fairly sizable autocorrelations at longer lags for the fractional model. It is worth noting that for the fractional model with $\lambda = 0.45$ the true autocorrelations at lags 10 and 20 are 0.64 and 0.61, while those for an AR(1) model with $\phi_1 = 0.7$ are 0.04 and 0.0008, respectively.

(b) Figure 11.4a shows a realization of length $n = 100$ from the FARMA (2,0) model

$$(1 + 0.7B^2)(1 - B)^{0.45}X_t = a_t, \tag{11.8}$$

and Figure 11.4b shows the true autocorrelations. It is informative to note the sense in which the autocorrelations of the FARMA(2,0.45,0) model shown in Figure 11.4b reflect the behaviors of the fractional and AR(2) factors. Figure 11.4c shows the slowly damping true autocorrelations for the fractional model $(1 - B)^{0.45}X_t = a_t$ previously shown in Figure 11.2c. Figure 11.4d shows the true autocorrelations associated with the AR(2) model $(1 + 0.7B^2)X_t = a_t$, and based on factor tables, we know that the second-order autoregressive factor $1 + 0.7B^2$ is associated with a system frequency of $f = 0.25$ (i.e., a period of 4). These autocorrelations display a quickly damping period of 4. The true FARMA(2,0.45,0) autocorrelations in Figure 11.4b show that for small lags both behaviors are present, but after about lag 20 there is very little evidence of the (short memory) behavior associated with the AR(2) factor. This is consistent with the fact that the AR(2) autocorrelations are essentially damped to zero by lag 20. Beyond lag 20 the autocorrelations of the

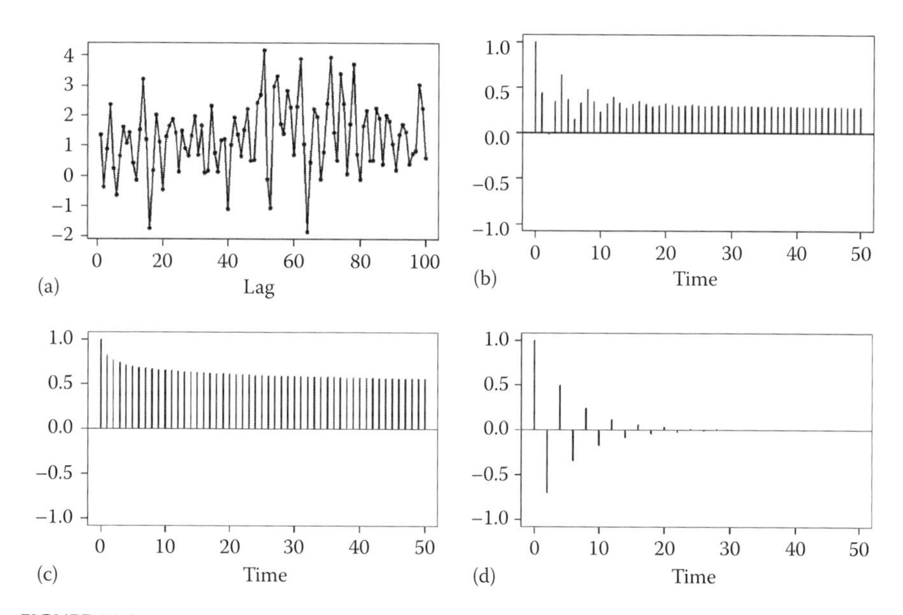

FIGURE 11.4
(a) Realization from the FARMA(2,0) model in (11.8), (b) true autocorrelations for the FARMA model in (11.8), (c) true autocorrelations for the fractional model with $d = 0.45$, and (d) true autocorrelations for AR(2) model $(1 + 0.7B^2)X_t = a_t$.

FARMA(2,0) model have only a slowly damping behavior although the effect of the AR(2) component has been to weaken the actual autocorrelations. The data in Figure 11.4a show both the wandering behavior associated with the fractional component and the period 4 behavior associated with the AR(2) factor.

11.3 Gegenbauer and GARMA Processes

Even though the fractional and FARMA models have been widely studied and have shown to be useful long-memory models, it can be seen that the long-memory fractional difference $(1 - B)^d$ can only model "first-order" long-memory behavior. For example, according to Theorem 11.1(c) the peak in the spectrum associated with $(1 - B)^d$ will always be at $f = 0$. Also the autocorrelations damp like $\rho(k) \simeq k^{2d-1}$ as $k \to \infty$ and are noncyclic in nature. The important point is that the fractional and FARMA models, as defined in Section 11.2, cannot model cyclic long-memory behavior associated with a spectrum that is unbounded for some $f \in (0, 0.5)$ or for which the autocorrelation damps like $\rho_k \simeq k^{\alpha} \cos(2\pi f)$ as $k \to \infty$ where $0 < \alpha < 1$.

The restriction to long-memory behavior associated with system frequency $f = 0$ is analogous, in the setting of AR(p) models, to being restricted to first-order autoregressive models with $\phi_1 > 0$. Recall that in Section 3.2 it was shown that cyclic behavior associated with $f \in (0, 0.5)$ in AR(p) models is associated with irreducible second-order factors of $\phi(z)$. In this section, we discuss a second-order extension of the fractional process in (11.5) which is based on the second-order factor $(1 - 2uB + B^2)^\lambda$ for $|u| \le 1$. In order to understand properties associated with such factors, we first consider the Gegenbauer polynomials.

11.3.1 Gegenbauer Polynomials

Gegenbauer polynomials are among the classical orthogonal polynomials of applied mathematics and physics. The Gegenbauer polynomial is defined by

$$C_n^{(\lambda)}(u) = \sum_{k=0}^{[n/2]} \frac{(-1)^k (\lambda)_{n-k} (2u)^{n-2k}}{k!(n-2k)!}, \tag{11.9}$$

where

$$(\alpha)_n = \frac{\Gamma(\alpha + n)}{\Gamma(\alpha)},$$

and where $[x]$ denotes the greatest integer less than or equal to x. Although these polynomials can be defined as in (11.9), it is also useful to define them through their generating function (see Rainville, 1960; Magnus et al., 1966).

Definition 11.5 Let $\lambda \ne 0$ and $|z| < 1$. Then for $|u| \le 1$ the Gegenbauer polynomials, $C_u^{(\lambda)}(u)$, satisfy

$$(1 - 2uz + z^2)^{-\lambda} = \sum_{j=0}^{\infty} C_j^{(\lambda)}(u) z^j. \tag{11.10}$$

11.3.2 Gegenbauer Process

In Definition 11.6 that follows we define the Gegenbauer process, which is an extension of the fractional process of Definition 11.3. See Gray et al. (1989, 1994).

Definition 11.6 If $\sum_{k=0}^{\infty} \left|C_k^{(\lambda)}(u)\right| < \infty$ where $C_k^{(\lambda)}(u)$ is defined in (11.9), then the general linear process defined by

$$\begin{aligned} X_t - \mu &= (1 - 2uB + B^2)^{-\lambda} a_t \\ &= \sum_{j=0}^{\infty} C_j^{(\lambda)}(u) a_{t-j}, \end{aligned} \tag{11.11}$$

where a_t is white noise, is called a *Gegenbauer process*. If X_t is invertible, then the Gegenbauer model is written

$$(1 - 2uB + B^2)^{\lambda}(X_t - \mu) = a_t. \tag{11.12}$$

NOTES: In the remainder of this section, we will assume without loss of generality that $\mu = 0$. It should be noted that if $u = 1$

$$(1 - B)^{2\lambda} X_t = a_t. \tag{11.13}$$

That is, X_t is a fractional process of order 2λ. If $u = -1$ then (11.12) becomes

$$(1 + B)^{2\lambda} X_t = a_t. \tag{11.14}$$

Theorem 11.3 gives the stationarity, long memory, and invertibility conditions for the Gegenbauer process.

Theorem 11.3

Let X_t be a Gegenbauer process of the form (11.11).

(a) If $|u| < 1$ and $\lambda < 1/2$ or if $u = \pm 1$ and $\lambda < 1/4$, then X_t is causal and stationary.

(b) If $|u| < 1$ and $\lambda > -1/2$ or if $u = \pm 1$ and $\lambda > -1/4$, then X_t is invertible.

(c) The spectrum of a stationary and invertible Gegenbauer process is given by

$$\begin{aligned} P(f) &= \sigma_a^2 \left|1 - 2u \exp(2\pi i f) + \exp(4\pi i f)\right|^{-2\lambda} \\ &= \sigma_a^2 \left[4(\cos 2\pi f - u)^2\right]^{-\lambda}. \end{aligned} \tag{11.15}$$

The frequency $f_G = \dfrac{1}{2\pi} \cos^{-1} u$ is referred to as the *Gegenbauer frequency (G-frequency)*.

(d) If $0 < \lambda < \dfrac{1}{2}$ and $|u| < 1$ or if $|u| = 1$ and $0 < \lambda < \dfrac{1}{4}$, then X_t is long memory.

(e) The autocorrelations of a stationary and invertible Gegenbauer process, X_t, are given in the following:

(i) If $u = 1$ and $0 < \lambda < \dfrac{1}{4}$, the autocorrelation function of X_t is

$$\rho_k = \frac{\Gamma(1 - 2\lambda)(k + 2\lambda)}{\Gamma(2\lambda)\Gamma(k - 2\lambda + 1)}.$$

Thus $\rho_k \simeq k^{4\lambda-1}$ as $k \to \infty$. Also,

$$\sigma_X^2 = \Gamma(-4\lambda + 1)/(\Gamma(-2\lambda + 1))^2.$$

(ii) If $u = -1$ and $0 < \lambda < \frac{1}{4}$, the autocorrelation function of X_t is

$$\rho_k = (-1)^k \frac{\Gamma(1 - 2\lambda)\Gamma(k + 2\lambda)}{\Gamma(2\lambda)\Gamma(k - 2\lambda + 1)}.$$

Thus $\rho_k \simeq (-1)^k k^{4\lambda-1}$ as $k \to \infty$. Also,

$$\sigma_X^2 = \frac{\Gamma(-4\lambda + 1)}{(\Gamma(-2\lambda + 1))^2}.$$

(iii) If $|u| < 1$ and $0 < \lambda < \frac{1}{2}$, then $\rho_k \simeq k^{2\lambda-1} \cos(2\pi k f_G)$, as $k \to \infty$ where f_G is the G-frequency defined in (c). Also, $\sigma_X^2 = \sum_{j=0}^{\infty} [C_j^{(\lambda)}(u)]^2 \sigma_a^2$.

Proof:

See Gray et al. (1989, 1994) and Andel (1986). Note that (d) follows from the fact that the spectrum in (11.15) becomes unbounded at the G-frequency, f_G.

As a result of Theorem 11.3, for a stationary, long-memory Gegenbauer process, it follows that $\rho_k \simeq k^{2\lambda-1} \cos(2\pi f_G k)$ as $k \to \infty$, and the spectrum is unbounded at f_G where $f_G \in [0, 0.5]$. Note that when $u = 1$, it can be shown using half-angle formulas that the spectrum given by (11.15) simplifies to the spectrum of the fractional process given in Theorem 11.1(c). That is, when $u = 1$, the spectrum in (11.15) has an infinite peak at $f = 0$.

Figure 11.5 (originally given by Gray et al., 1989) shows the stationary long-memory region (range of u and λ) for the Gegenbauer processes. The stationary long-memory region for u and λ is the shaded region ($0 < \lambda < 0.5$ and $|u| < 1$) along with the values of u and λ along the solid bold horizontal borders ($u = \pm 1$ and $0 < \lambda < 0.25$. The fractional process is represented by the solid bold horizontal line segment at the top of the rectangle. This figure illustrates the fact that the Gegenbauer process represents a considerable generalization of the fractional process. As u and λ move along a path that approaches a dashed boundary of the rectangle, the corresponding Gegenbauer processes approach nonstationarity. For example, if λ is held constant at $\lambda = 0.4$, then as u approaches 1 from below, the processes associated with these u and λ values will increasingly display "near-nonstationary" behavior. Thus while u determines the Gegenbauer frequency, it also effects the degree of "near nonstationarity."

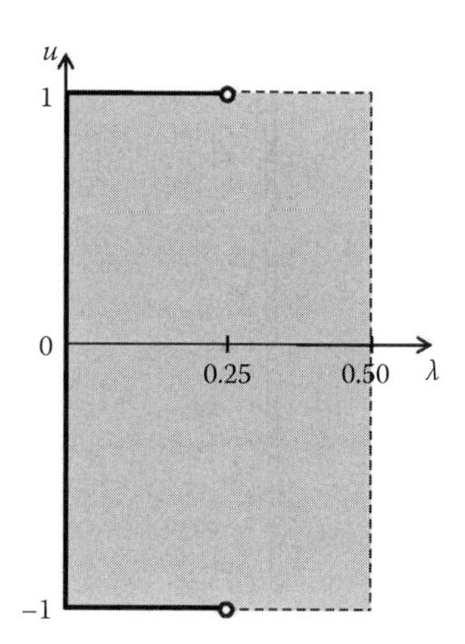

FIGURE 11.5
Stationary long-memory region of the Gegenbauer process.

Example 11.3 Gegenbauer Processes

(a) Figure 11.6 shows true autocorrelations for Gegenbauer processes with $u = 0.9$ and a range of values for λ. The Gegenbauer frequency is given by $f_G = \dfrac{1}{2\pi}\cos^{-1} 0.9 = 0.07$, and thus it is associated with periods a little over 14. As $\lambda \to 0.5$ the cyclic behavior becomes more dominant. However, the slowly damping behavior can be seen even when $\lambda = 0.25$.

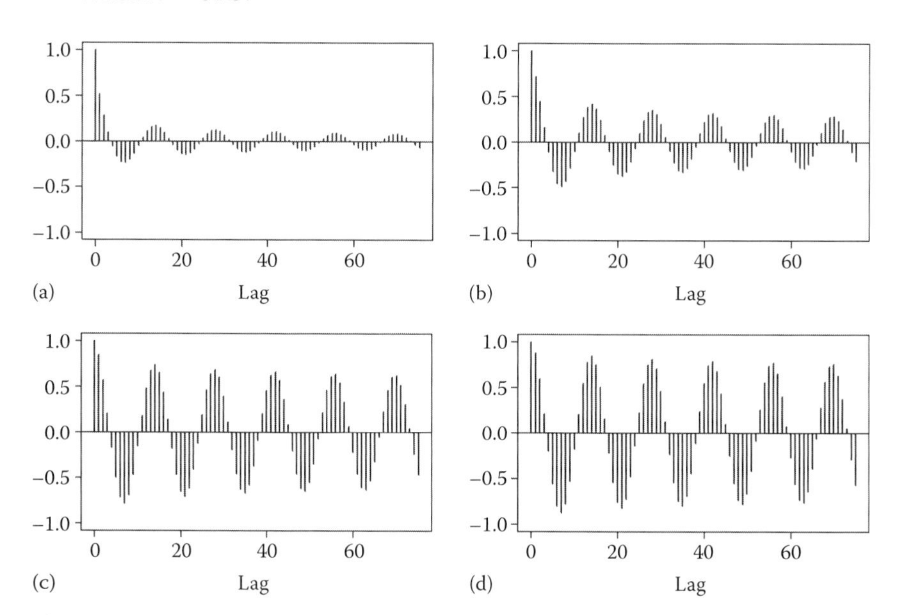

FIGURE 11.6
True autocorrelations for Gegenbauer processes with $u = 0.9$ and various values of λ (a) $\lambda = 0.25$. (b) $\lambda = 0.35$. (c) $\lambda = 0.45$. (d) $\lambda = 0.495$.

11.3.3 GARMA Process

In the same manner that the fractional process has been extended to FARMA process, Definition 11.7 which follows extends the Gegenbauer process by combining it with the ARMA process.

Definition 11.7 If X_t is a causal, stationary, and invertible process satisfying

$$\phi(B)(1 - 2uB + B^2)^{\lambda}(X(t) - \mu) = \theta(B)a(t), \tag{11.16}$$

where $\phi(z)$ and $\theta(z)$ share no common factors, then X_t is a Gegenbauer ARMA (GARMA) process which will be denoted GARMA(p,λ,u,q).

Theorem 11.4 summarizes the properties of the GARMA process.

Theorem 11.4

Let X_t be a GARMA process ($\lambda \neq 0$) and let all roots of $\phi(z) = 0$ and $\theta(z) = 0$ lie outside the unit circle. Then we have the following:

(a) X_t is stationary if $\lambda < \frac{1}{4}$ when $u = \pm 1$ or $\lambda < \frac{1}{2}$ when $|u| < 1$.

(b) X_t is invertible if $\lambda > -\frac{1}{4}$ when $u = \pm 1$ or $\lambda > -\frac{1}{2}$ when $|u| < 1$.

(c) The spectrum is given by

$$P(f) = \sigma_a^2 \frac{\left|\theta(e^{-2\pi i f})\right|^2}{\left|\phi(e^{-2\pi i f})\right|^2} \left|1 - 2u\exp(-2\pi i f) + \exp(-4\pi i f)\right|^{-2\lambda}$$

$$= \sigma_a^2 \frac{\left|\theta(e^{-2\pi i f})\right|^2}{\left|\phi(e^{-2\pi i f})\right|^2} [4(\cos 2\pi f - \cos 2\pi f_G)^2]^{-\lambda}.$$

(d) X_t is long memory if $0 < \lambda < \frac{1}{4}$ when $u = \pm 1$ or $0 < \lambda < \frac{1}{2}$ when $|u| < 1$.

(e) As $k \to \infty$,

(i) $\rho_k \simeq k^{4\lambda - 1}$ when $u = 1$ and $0 < \lambda < \frac{1}{4}$

(ii) $\rho_k \simeq (-1)^k k^{4\lambda - 1}$ when $u = -1$ and $0 < \lambda < \frac{1}{4}$

(iii) $\rho_k \simeq k^{2\lambda - 1}\cos(2\pi f_G k)$ when $|u| < 1$ and $0 < \lambda < \frac{1}{4}$.

Proof:

See Gray et al. (1989, 1994).

Example 11.4 GARMA Processes

In this example we show true autocorrelations for Gegenbauer and GARMA models. The procedure for calculating these autocorrelations will be discussed in Section 11.4.1.

(a) Figure 11.7 shows true autocorrelations for GARMA processes with $u = 0.9$ and $\lambda = 0.45$. Figures 11.7a and b are for GARMA (1,0.45,0.9,0) processes with $\phi_1 = 0.9$ and $\phi_1 = 0.95$, respectively. These autocorrelations have the general appearance of the autocorrelations in Figure 3.14b for the AR(3) process $(1 - 0.95B)(1 - 1.6B + 0.9B^2)X_t = a_t$. That is, the AR(1) factor with positive ϕ_1 induces a damped exponential behavior, and the appearance in both settings is that of a damped sinusoidal behavior along a damped exponential path. It is instructive to compare Figures 3.14b and 11.7b. In Figure 3.14b, the damped exponential and damped sinusoidal behaviors have almost disappeared by lag 40. However, in Figure 11.7b, while the damped exponential behavior is almost nonexistent by lag 40, the sinusoidal behavior associated with the Gegenbauer frequency, $f_G = 0.07$, is still visible (and "going strong") at lag 70. Figure 11.7c and d show very unusual autocorrelation behavior. The GARMA(1,0.45,0.9,0) model associated with the autocorrelations in Figure 11.7c has an AR(1) factor with $\phi_1 = -0.9$

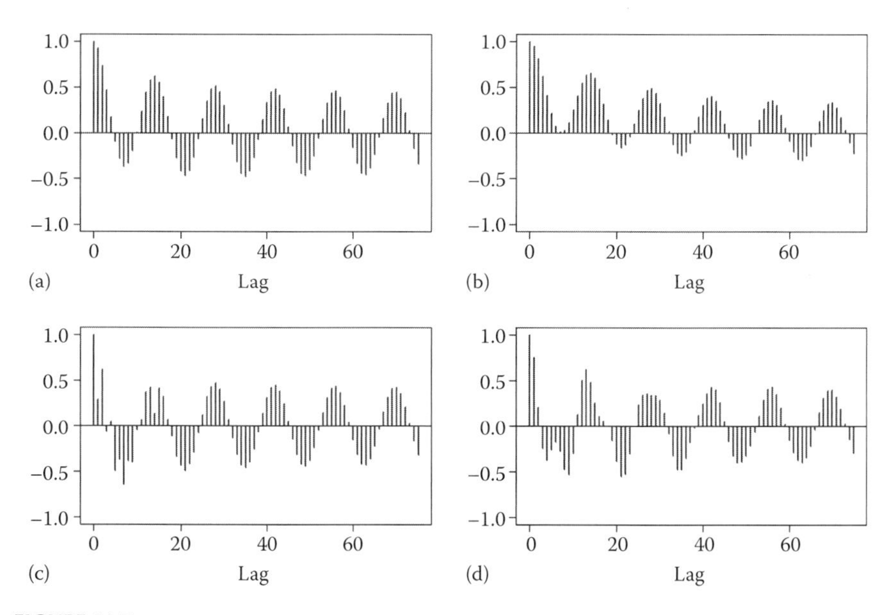

FIGURE 11.7
Autocorrelations for GARMA processes with $u = 0.9$ and $\lambda = 0.45$: (a) $\phi_1 = 0.9$, (b) $\phi_1 = 0.95$, (c) $\phi_1 = -0.85$, and (d) $\phi_1 = 1$, $\phi_2 = -0.9$.

which introduces oscillating damped exponential behavior. In Figure 11.7c it is seen that this high-frequency/oscillating behavior is noticeable for low lags, but after $k = 20$ the only behavior visible is that due to the long-memory damped sinusoidal. Figure 11.7d shows autocorrelations for a GARMA(2,0.45,0.9,0) with AR(2) factor $1 - B + 0.9B^2$ which, from (3.63) is associated with a system frequency of $f_0 = 0.16$ (or period of 6.25). The shorter period of about 6 can be seen within the first 15 lags, but for larger lags, the only observable behavior is due to the Gegenbauer frequency, $f_G = 0.07$ (i.e., periods of about 14).

(b) A realization of length $n = 500$ was generated from the GARMA (1,0.45,0.9,0) model

$$(1 + 0.85B)(1 - 1.8B + B^2)^{0.45} X_t = a_t, \qquad (11.17)$$

that is, with $u = 0.9$, $\lambda = 0.45$, and $\phi_1 = -0.85$. The theoretical autocorrelations for this model are shown in Figure 11.7c. Figure 11.8a shows a realization of length $n = 500$, but since this realization is too compact to be easily interpreted we show the first 200 points in Figure 11.8b. The high-frequency (short memory) behavior associated with $(1 + 0.85B)$ is visible in Figure 11.8b along with

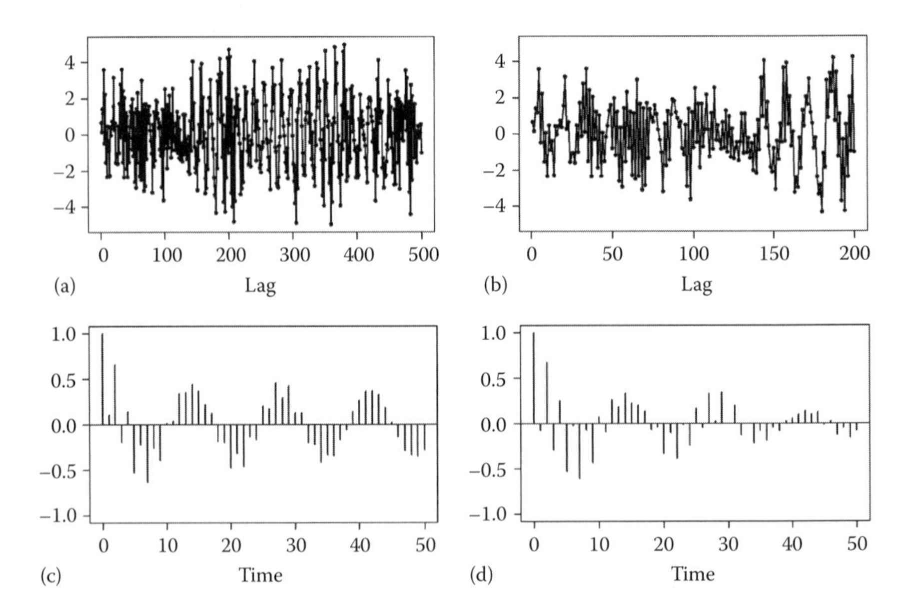

FIGURE 11.8
(a) Realization of length $n = 500$ from GARMA(1,0.45,0) model (11.17), (b) first 200 points in (a), (c), and (d) sample autocorrelations for realizations in (a) and (b), respectively.

some indication, especially for $t \geq 150$, of the Gegenbauer periodic behavior with $f_G = 0.07$. The persistence of the Gegenbauer period of 14 is not apparent in the 200-point realization in Figure 11.8b, and it is not surprising that the sample autocorrelations in Figure 11.8d (based on the 200-point realization) primarily pick up the oscillatory behavior. However, the sample autocorrelations for the full realization of length $n = 500$ are a fairly close approximation to the theoretical autocorrelations in Figure 11.7c, which show a mixture of oscillatory and 14-period cyclic behavior for early lags with only the persistent autocorrelation from the Gegenbauer factor showing up in the autocorrelations for $k \geq 20$. In general, the natural damping associated with $\hat{\rho}_k$ (based on division by n instead of $n - k$ as discussed in Section 1.4) causes it to be difficult for persistent autocorrelations to be well estimated for moderate realization lengths.

11.4 *k*-Factor Gegenbauer and GARMA Processes

As we introduced the Gegenbauer process in Section 11.3, we noted that the long-memory fractional difference $(1 - B)^d$ can only model "first-order" long-memory behavior associated with $f = 0$. It was noted that the restriction to this type of first-order long-memory behavior is analogous in the setting of AR(p) models to being restricted to first-order autoregressive models with $\phi_1 > 0$. The Gegenbauer factor $(1 - 2uB + B^2)^\lambda$ introduced in Section 11.3 provides a model for "second order" or cyclical long memory when $|u| < 1$. Continuing with the AR(p) analogy, this corresponds to an irreducible second-order factor. In Section 3.2, we noted that a pth-order autoregressive operator, $\phi(B) = 1 - \phi_1 B - \cdots - \phi_p B^p$, can be written as $\phi(B) = \sum_{j=1}^{k} (1 - \alpha_{1j}B - \alpha_{2j}B^2)$ where $\alpha_{2j} = 0$ for first-order factors. Thus, the first- and second-order factors are viewed as building blocks of the AR(p) model, and the properties of an AR(p) process represent a combination of the behaviors induced by the first- and second-order factors of $\phi(B)$. In this section, we discuss the k-factor Gegenbauer process that allows for long-memory behavior to be associated with each of k frequencies in [0, 0.5]. This model includes the fractional and Gegenbauer as special cases. We also consider the k-factor GARMA model that similarly generalizes the GARMA and FARMA models.

In Definition 11.8, we define the k-factor Gegenbauer process which was considered in the $k = 2$ case by Cheng (1993) and in general by Woodward et al. (1998) and Giraitis and Leipius (1995).

Definition 11.8 If X_t is a causal, stationary, and invertible process satisfying

$$\prod_{j=1}^{k}(1-2u_jB+B^2)^{\lambda_j}(X_t-\mu)=a_t. \tag{11.18}$$

where a_t is white noise, then it is called a *k-factor Gegenbauer process.*

Theorem 11.5 gives the stationarity, long memory, and invertibility conditions for a k-factor Gegenbauer process.

Theorem 11.5

Let X_t be a k-factor Gegenbauer process of the form (11.18).

(a) If the u_j are distinct, $|\lambda_j|<1/2$ when $|u_j|\neq 1$ and $|\lambda_j|<1/4$ when $|u_j|=1$, then X_t is causal, stationary, and invertible.

(b) The spectrum of a causal, stationary, and invertible k-factor Gegenbauer process is given by

$$\begin{aligned} P(f) &= \sigma_a^2\left|\prod_{j=1}^{k}\{1-2u_je^{-2\pi if}+e^{-4\pi if})\}^{-\lambda_j}\right|^2. \\ &= \sigma_a^2\prod_{j=1}^{k}[4(\cos 2\pi f-u_j)^2]^{-\lambda_j}. \end{aligned} \tag{11.19}$$

The frequencies $f_{G_j}=\dfrac{1}{2\pi}\cos^{-1}u_j$ are called Gegenbauer frequencies or G-frequencies.

(c) If the conditions of (a) are satisfied along with $\lambda_j>0$ for any $j=1,\ldots,k$, then X_t is long memory.

Proof:

See Giraitis and Leipus (1995) and Woodward et al. (1998). Note that (c) follows from the fact that the spectrum in (11.19) becomes unbounded at the G-frequencies f_{G_j}.

In the same manner that the fractional and Gegenbauer processes have been extended to include ARMA factors, the k-factor Gegenbauer process can be extended to the k-factor GARMA process which is defined in Definition 11.9.

Definition 11.9 If X_t is a causal, stationary, and invertible process satisfying

$$\phi(B)\prod_{j=1}^{k}(1-2u_jB+B^2)^{\lambda_j}(X_t-\mu)=\theta(B)a_t, \tag{11.20}$$

where $\phi(z)$ and $\theta(z)$ share no common factors, then X_t is a k-factor GARMA process.

In Theorem 11.6 we give the stationarity, long memory, and invertibility conditions for a k-factor GARMA process.

Theorem 11.6

(a) X_t in (11.20) is causal, stationary, and invertible if u_j and λ_j $(j=1,\ldots,k)$ satisfy the conditions in Theorem 11.5(a) and all the roots of the equation $\phi(z)=0$ and $\theta(z)=0$ lie outside the unit circle.

(b) The spectrum of a causal, stationary, and invertible k-factor GARMA process is given by

$$P(f)=\sigma_a^2\frac{|\theta(e^{2\pi if})|^2}{|\phi(e^{2\pi if})|^2}\prod_{j=1}^{k}[4(\cos 2\pi f-u_j)^2]^{-\lambda_j}. \tag{11.21}$$

(c) A process X_t satisfying the conditions in (a) is also long memory if $\lambda_j>0$ for any $j=1,\ldots,k$.

Proof:

See Woodward et al. (1998) and Giraitis and Leipus (1995).

Example 11.5 2-Factor Gegenbauer and 2-Factor GARMA Processes

(a) Figure 11.9a shows a realization of length $n=500$ from the 2-factor Gegenbauer model

$$(1-1.8B+B^2)^{0.45}(1+1.6B+B^2)^{0.3}X_t=a_t, \tag{11.22}$$

with $\sigma_a^2=1$, and Figure 11.9b shows the first 100 observations in order to show more detail. The Gegenbauer frequencies associated with u_1 and u_2 are $f_{G_1}=0.07$ and $f_{G_2}=0.4$, respectively, and in Figure 11.9b it can be seen that there is both low-frequency and high-frequency periodic behavior in the data. Figure 11.9c shows the true

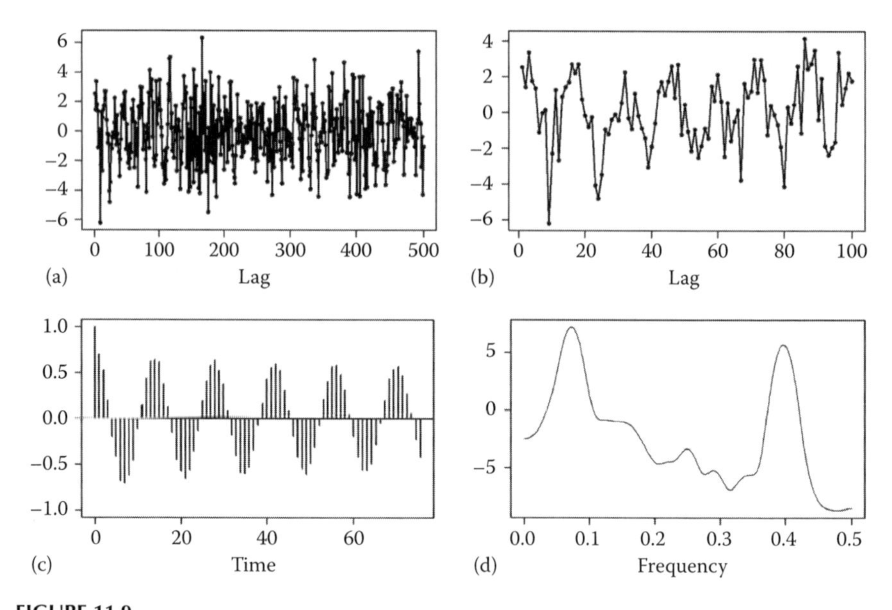

FIGURE 11.9
(a) Realization of length $n = 500$ from 2-factor Gegenbauer model in (11.22), (b) first 100 points in (a), (c) true autocorrelations for the model in (11.22), and (d) Parzen spectral density estimator for data in (a).

autocorrelations which have a slowly damping sinusoidal behavior associated with the low-frequency Gegenbauer frequency ($f_G = 0.07$). Little evidence of the higher Gegenbauer frequency behavior is seen in the autocorrelations. Figure 11.9d shows the estimated spectral density function (using a Parzen window on the data in Figure 11.9a) which shows two peaks at about 0.07 and 0.4.

(b) Figure 11.10a shows a realization of length $n = 500$ from the 2-factor GARMA model

$$(1 - 0.9B)(1 - 1.8B + B^2)^{0.45}(1 + 1.6B + B^2)^{0.3}X_t = a_t, \tag{11.23}$$

where $\sigma_a^2 = 1$. Figure 11.10b shows the first 100 observations. In both realizations, the wandering behavior induced by the AR(1) factor $1 - 0.9B$ can be seen, and Figure 11.10b shows low-frequency and high-frequency periodic behaviors. Figure 11.10c shows the true autocorrelations which display a slowly damping sinusoidal behavior associated with the low-frequency Gegenbauer frequency ($f_G = 0.07$) along a damped exponential path induced by $1 - 0.9B$. Again, little evidence of the higher Gegenbauer frequency behavior is seen in the autocorrelations. Figure 11.10d shows the estimated spectral density function (using a Parzen window on the data in Figure 11.10a).

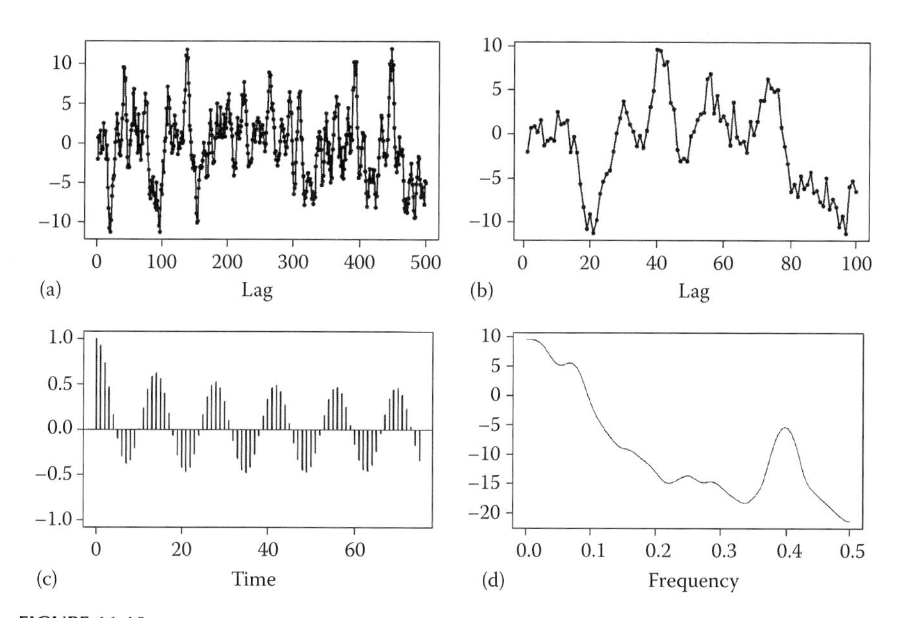

FIGURE 11.10
(a) Realization of length $n = 500$ from 2-factor GARMA model in (11.23), (b) first 100 points in (a), (c) true autocorrelations for the model in (11.22), and (d) Parzen spectral density estimator for data in (a).

The estimated spectrum shows the peaks associated with Gegenbauer frequencies, 0.07 and 0.4, which were seen in Figure 11.9d. The peaks at 0.07 and 0 (associated with $1 - 0.9B$) are somewhat smeared together, but clearly present.

11.4.1 Calculating Autocovariances

The autocorrelation function of a Gegenbauer process with $u = 1$ (i.e., a fractional model) is given in Theorem 11.3(e)(i), and the autocorrelation for a Gegenbauer process with $u = -1$ is given in Theorem 11.3(e)(ii). No simple closed form expression for the autocovariance or autocorrelation exists for a Gegenbauer process with $|u| < 1$, but from (2.7) it follows that

$$\begin{aligned} \gamma_k &= \sigma_a^2 \sum_{j=0}^{\infty} \psi_j \psi_{j+k} \\ &= \sigma_a^2 \sum_{j=0}^{\infty} C_j^{(\lambda)}(u) C_{j+k}^{(\lambda)}(u). \end{aligned} \tag{11.24}$$

One technique for calculating the autocovariances is to truncate the series in (11.24) after a sufficient number of terms. Table 11.1 (given by Ford, 1991)

TABLE 11.1

Approximating γ_0 by Truncating (11.24) When $u = 1$ and $\sigma_a^2 = 1$

$\lambda(=d/2)$	Exact	Sum to 200,000	Sum to 10 Million
0.05	1.0195	1.0195	1.0195
0.125	1.1803	1.1800	1.1803
0.24	8.4061	3.9004	4.5527

illustrates that, depending on the parameters, the terms in this series may not damp quickly enough to make truncation feasible.

The table considers the problem of using (11.24) to calculate γ_0 for the case $u = 1$. In this case using Theorem 11.1(d) and the fact that $d = 2\lambda$, the sum of the infinite series in (11.24) is known in closed form to be $\gamma_0 = \Gamma(-4\lambda + 1)/[\Gamma(-2\lambda + 1)]^2$. Table 11.1 compares values of the infinite sum in (11.24) to the results of truncating the sum after 200,000 terms and after 10 million terms. In the table it is seen that for $\lambda = 0.05$ and $\lambda = 0.125$ the two truncated sums approximate the infinite sum quite well. However, in the near nonstationary case $\lambda = 0.24$, the convergence is extremely slow and a sum to 10 million terms is a very poor approximation! Chung (1996a) expressed γ_k for a (1-factor) Gegenbauer process as a recursion involving Legendre functions. Using (1.28) and (11.19), an expression for the autocovariance of a Gegenbauer process is given as the inverse Fourier transform of the spectrum,

$$\gamma_k = \int_{-0.5}^{0.5} e^{2\pi i f k} P(f) df. \qquad (11.25)$$

In the 1-factor Gegenbauer setting, Ford (1991) examined the use of Levin and Sidi's (1981) d-transform for increasing the rate of convergence of the sum in (11.24) and compared the resulting approximations with those obtained by numerically integrating (11.25). As noted, $P_X(f)$ is unbounded at $f_{G_j} = \frac{1}{2\pi}\cos^{-1} u_j$. Ford (1991) concluded that careful numerical integration using the use of (11.25) is the preferred method. Based on these findings, we approximate the autocovariances of k-factor Gegenbauer and k-factor GARMA models by numerically integrating the integral (11.25). It should be noted that both methods produce approximate autocorrelations, and for some parameter configurations can be relatively poor.

Using `GW-WINKS`: R routines are included in the ATSA website for calculating the true autocorrelations of a given 1-factor or 2-factor GARMA model using (11.25). These R routines can be run from R or from within `GW-WINKS`. To calculate the theoretical autocorrelations for a 1-factor or 2-factor GARMA(p,q) model using `GW-WINKS`, go to the `Time Series Analysis` menu of

GW-WINKS and choose Select R Routine. On the pull-down menu choose Calculate GARMA Autocorrelations. Specify the model parameters along with the number of autocorrelations to be computed. A new column in the Data Editor will be created that contains these autocorrelations using the convention that the value in row one is $\rho_0(=1)$ and in general the value in row k is ρ_{k-1}. The process variance is printed on the Output Viewer. For additional information and for instructions on calling R programs from within GW-WINKS, see the ATSA website.

11.4.2 Generating Realizations

Gray et al. (1989, 1994) generate realizations from Gegenbauer and GARMA models using the strategy of truncating the general linear process given in (11.11) to obtain

$$X_t \approx \mu + \sum_{h=0}^{M} C_h^{(\lambda)}(u)a_{t-h}, \tag{11.26}$$

using $M = 290{,}000$. Woodward et al. (1998) suggest a more efficient algorithm due to Ramsey (1974) based on conditional expectations and variances of X_t given $X_0, X_1, \ldots, X_{t-1}$. However, the more efficient method involves the true autocovariances, and our experience is that due to the computational difficulties described earlier, some of the conditional quantities become corrupted. Consequently, the realizations shown here will be based on the truncated general linear process.

In order to generate realizations from a k-factor GARMA, we let Z_t be the corresponding k-factor Gegenbauer process, which we express as

$$Z_t = \prod_{j=1}^{k} (1 - 2u_jB + B^2)^{-\lambda_j}a_t,$$

and we write (11.20) as

$$\phi(B)X_t = \mu(1 - \phi_1 - \cdots - \phi_p) + \theta(B)Z_t.$$

A realization from the k-factor Gegenbauer process, Z_t, is generated using the truncation method described earlier, and the k-factor GARMA processes is then simulated using the recursion

$$X_t = \mu(1 - \phi_1 - \cdots - \phi_p) + \sum_{j=1}^{p} \phi_jX_{t-j} + Z_t - \sum_{j=1}^{q} \theta_jZ_{t-j}.$$

See Hosking (1984), Gray et al. (1989, 1994), and Woodward et al. (1998).

Using `GW-WINKS`: R routines are included in the ATSA website for generating realizations from a given 1-factor or 2-factor GARMA model. These R routines can be run from R or from within `GW-WINKS`. To generate a realization from a 1-factor or 2-factor GARMA(p,q) model using `GW-WINKS`, go to the `Time Series Analysis` menu of `GW-WINKS` and choose `Select R Routine`. On the pull-down menu choose `Generate GARMA Realization`. Specify the model parameters along with the realization length. A new column in the `Data Editor` will be created that contains the generated realization. See the ATSA website for instructions on calling R programs from within `GW-WINKS`.

11.5 Parameter Estimation and Model Identification

As in Chapter 7, we estimate μ by $\overline{X}$ and treat the mean corrected data (denoted by X_t) as the realization to be analyzed. It should be noted that the properties of $\overline{X}$ as an estimator of μ_X depend on the values of $|u_j|$. For example, in the case of a (1-factor) GARMA model, the sample mean converges at the usual $(1/\sqrt{n})$ rate whenever $|u| < 1$, while it has a slower convergence rate when $|u| = 1$. See Chung (1996b).

If X_t satisfies the model in (11.20) with $\mu = 0$, then $Y_t = \prod_{j=1}^{k}(1 - 2u_jB + B^2)^{\lambda_j}X_t$ satisfies the ARMA(p,q) model $\phi(B)Y_t = \theta(B)a_t$. Woodward et al. (1998) recommend the use of a grid search procedure for finding ML estimates of the parameters of the k-factor GARMA model. We assume that k is known. In practice, k will be estimated by identifying peaks in a nonparametric spectral estimator, or by examining factor tables. Then, based on positions and magnitudes of the peaks in the preliminary spectral estimate, a grid of possible values for $\lambda_1, u_1, \ldots, \lambda_k, u_k$ is selected to be used in the search. For each set of values $\lambda_1, u_1, \ldots, \lambda_k, u_k$ in the grid search, the data are transformed to obtain

$$\begin{aligned} Y_t &= \prod_{j=1}^{k}(1 - 2u_jB + B^2)^{\lambda_j}(X_t - \overline{X}) \\ &\approx \prod_{j=1}^{k}\left\{\sum_{h=0}^{t+K-1} C_h^{(-\lambda_j)}(u_j)(X_{t-h} - \overline{X})\right\}. \end{aligned} \tag{11.27}$$

In order to make the transformation in (11.27), K values $X_{-1}, X_{-2}, \ldots, X_{-K+1}$ are obtained by backcasting as discussed in Section 7.4 using a high-order AR model. The resulting process, Y_t, is modeled as an ARMA(p,q), an associated likelihood function is obtained, and the combination of $\lambda_1, u_1, \ldots, \lambda_k, u_k$ associated with the largest likelihood value is the approximate maximum likelihood estimator (for the current values of p and q). To identify p and q this procedure is performed for each $p = 1, \ldots, P$ and $q = 1, \ldots, Q$, and the

ARMA-based AIC associated with the ML estimator is obtained in each case. The model (i.e., $p, q, \lambda_1, u_1, \ldots, \lambda_k, u_k$) associated with minimum AIC value is selected as the model. See Hosking (1984), Gray et al. (1989, 1994), and Woodward et al. (1998).

Giraitis and Leipus (1995), Ferrara and Guégan (2001), and Diongue and Guegan (2008) use the Whittle likelihood to estimate the parameters of a k-factor Gegenbauer model. The Whittle likelihood is a frequency-domain based likelihood based on a class of models parameterized by $\boldsymbol{\xi}$ (possibly vector valued). Letting $P_X(f;\boldsymbol{\xi})$ denote the spectrum at frequency f under the assumed model with parameter value $\boldsymbol{\xi}$, a useful approximation to -2log-likelihood based on Whittle (1953) is

$$-2\ln L_W(X;\boldsymbol{\xi}) = \sum_{j=1}^{\left[\frac{n}{2}\right]} \frac{I(f_j)}{P_X(f_j;\boldsymbol{\xi})}, \tag{11.28}$$

where $[z]$ is the greatest integer less than or equal to z, $f_j = \frac{j}{n}, j = 1, 2, \ldots, \left[\frac{n}{2}\right]$, and $I(f_j)$ is the periodogram defined in (1.38). Model identification can be accomplished using AIC. In the current setting, $P_X(f)$ is defined in (11.21). Kouamé and Hili (2008) proposed minimum distance estimators for the parameters of the k-factor GARMA model which are easy to compute and are not based on an assumption that the a_t's are normally distributed.

Using `GW-WINKS`: R routines are included in the ATSA website for estimating the parameters of a 1-factor or 2-factor GARMA model using the grid-search procedure recommended by Hosking (1984), Gray et al. (1989, 1994), and Woodward et al. (1998). These R routines can be run from R or from within `GW-WINKS`. To use `GW-WINKS` to estimate the parameters of a 1-factor or 2-factor GARMA(p,q) model fit to a realization occupying a column in the current Data Editor, go to the `Time Series Analysis` menu of `GW-WINKS` and choose `Select R Routine`. On the pull-down menu choose `Estimate GARMA Parameters`. Specify the grid-search parameters to be used. The parameter estimates will be shown in the `Output Viewer`. See the ATSA website for instructions on calling R programs from within `GW-WINKS`.

Example 11.6 Estimating the Parameters of GARMA Processes

In this example, we estimate the parameters of two GARMA processes based on simulated realizations.

(a) *A 1-factor GARMA(1,0) model*
For this example, we use a realization of length 500 from the 1-factor GARMA model

$$(1 - \phi_1 B)(1 - 2uB + B^2)^{\lambda} X_t = a_t, \tag{11.29}$$

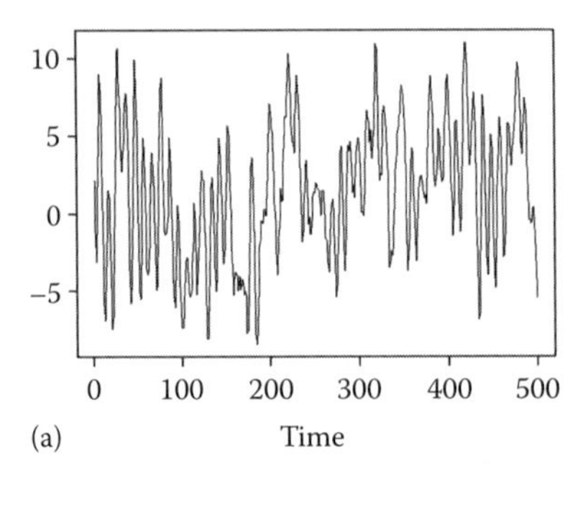

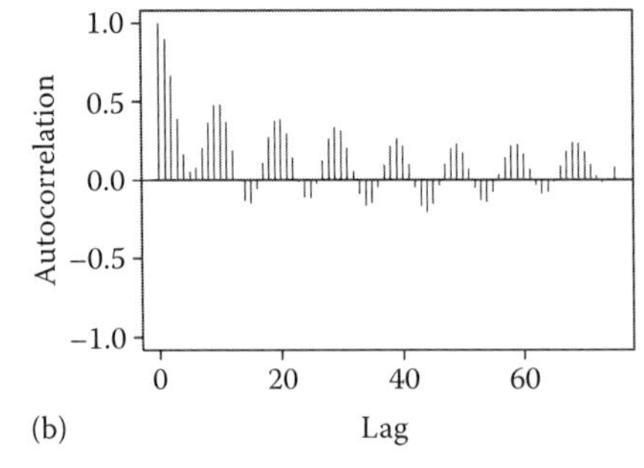

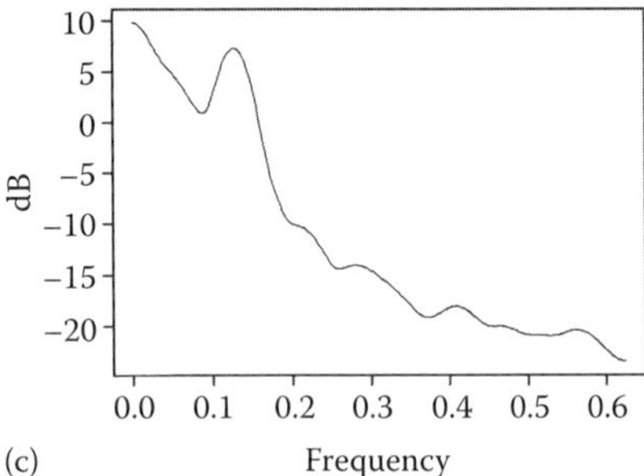

FIGURE 11.11
(a) Realization of length $n = 500$ from 1-factor GARMA(1,0) model in (11.29) along with (b) sample autocorrelations, and (c) Parzen spectral density estimate for data in (a).

where $\phi_1 = 0.9, u = 0.8, \lambda = 0.45$, and $\mu = 0$. The data used in this simulation are shown in Figure 11.11a. The first four rows of the factor table associated with an AR(20) fit to the data are shown in Table 11.2, and it is seen that the factor associated with the frequency $f = 0.1$ is clearly the top contender for the Gegenbauer frequency, suggesting that $u = 0.8$. The Parzen spectral density estimator in Figure 11.11c has a relatively sharp peak at $f = 0.1$, and the sample autocorrelations shown in Figure 11.11b have a slowly damping behavior associated with a frequency of 0.1.

TABLE 11.2
First 4 Rows of Factor Table for AR(20) Fit to Realization from (11.29)

AR Factors	Roots (r_j)	$\lvert r_j^{-1}\rvert$	f_{0j}
$1 - 1.57B + 0.97B^2$	$0.81 \pm 0.61i$	0.98	0.10
$1 - 0.94B$	1.06	0.94	0.0
$1 - 1.74B + 0.81B^2$	$1.07 \pm 0.28i$	0.90	0.04
$1 + 0.84B + 0.79B^2$	$-0.53 \pm 0.99i$	0.89	0.33
⋮			

Since the sample autocorrelations and spectral estimate suggest that λ is not small, we limit λ to the range 0.2 to 0.85 with an increment of 0.02. We use the range $0.6 \le u \le 0.9$ with a 0.05 increment. We also use the restrictions $p \le 3$ and $q = 0$, and AIC picks a GARMA(1,0). The ML estimates of the final GARMA(1,0) model are $\hat{u} = 0.8$, $\hat{\lambda} = 0.4$, and $\hat{\phi}_1 = 0.91$, yielding the estimated model

$$(1 - 0.91B)(1 - 1.6B + B^2)^{0.4}\{X(t) - \overline{X}\} = a(t). \tag{11.30}$$

(b) *A 2-factor GARMA(1,0) model*
A realization of length $n = 500$ was generated from the 2-factor GARMA model

$$(1 - 0.5B)(1 - 1.6B + B^2)^{0.45}(1 + 1.6B + B^2)^{0.3}X_t = a_t. \tag{11.31}$$

Thus, $\phi_1 = 0.5, u_1 = 0.8, \lambda_1 = 0.45, u_2 = -0.8, \lambda_2 = 0.3$, and $\mu = 0$. The realization, sample autocorrelations, and Parzen spectral density estimate are shown in Figure 11.12 where it can be seen that the Parzen spectral density has two distinct peaks at approximately 0.1 and 0.4, suggesting $\hat{u}_1 = 0.8$ and $\hat{u}_2 = -0.8$. Based on a

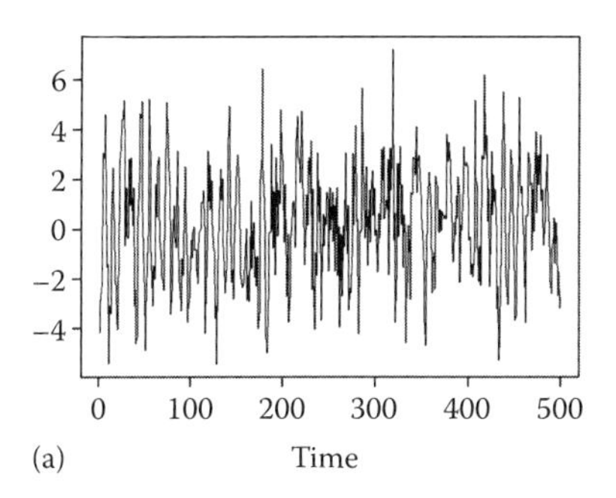

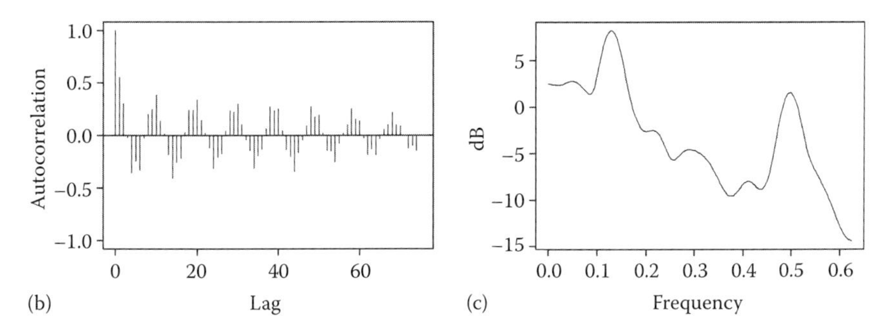

FIGURE 11.12
(a) Realization of length $n = 500$ from 2-factor GARMA(1,0) model in (11.31) along with (b) sample autocorrelations, and (c) Parzen spectral density estimate for data in (a).

grid search restricting possible values of $\hat{u}_1$ and $\hat{u}_2$ to neighborhoods of 0.8 and -0.8, respectively, letting $\hat{\lambda}_1$ and $\hat{\lambda}_2$ each range between 0.2 and 0.45, and $p \leq 2$ with $q = 0$ the model associated with the minimum AIC is

$$(1 - 0.52B)(1 - 1.6B + B^2)^{0.35}(1 + 1.6B + B^2)^{0.3}\{X(t) - \overline{X}\} = a(t),$$

which is close to the true model.

11.6 Forecasting Based on the k-Factor GARMA Model

In this section we consider the problem of forecasting using the k-factor GARMA model. While the use of the difference equation provides easy-to-use forecasts in the case of the ARMA and ARUMA models using (6.24) and (6.36), respectively, this method is not suitable for GARMA processes because there is no finite-order difference equation form for $\prod_{j=1}^{k}(1 - 2u_jB + B^2)^{\lambda_j}$.

Brockwell and Davis (1991) use the π-weights version of the classical Box-Jenkins forecasts (see Section 6.4) to forecast the FARMA process $\phi(B)(1 - B)^d(X_t - \mu) = \theta(B)a_t$, and Woodward et al. (1998) use this method for forecasting based on a k-factor GARMA model. The π-weights of the k-factor GARMA model can be calculated using the identity

$$\begin{aligned}\sum_{m=0}^{\infty} \pi_m B^m &= \theta^{-1}(B)\phi(B)\prod_{j=1}^{k}(1 - 2u_jB + B^2)^{\lambda_j} \\ &= \theta^{-1}(B)\phi(B)\prod_{j=1}^{k}\left\{\sum_{h=0}^{\infty} C_h^{(-\lambda_j)}(u_j)B^h\right\}.\end{aligned}$$

The truncated form (assuming only the data $X_1, \ldots, X_{t_0}$ are available) is given by

$$\hat{X}_{t_0}^{(T)}(\ell) = \overline{X} - \sum_{j=1}^{t_0-1+\ell} \pi_j\left(\hat{X}_{t_0}^{(T)}(\ell - j) - \overline{X}\right). \tag{11.32}$$

In Section 6.4 we noted that forecasts based on (11.32) are based on the assumption that the π_j are sufficiently close to zero for $j > t_0 - 1 + \ell$. This will be the case for most ARMA models and for t_0 of moderate size. However, for long-memory models, the π-weights damp very slowly which can cause the truncation in (11.32) to have a substantial effect. For information concerning the relationship between the best linear prediction (based on the available data) and forecasts using the truncated π-weights in the context of the fractional and FARMA models and GARMA models. See Miller (1994) and Fu (1995). (See also Ferrara and Guégan, 2001.)

Using GW-WINKS: R routines are included in the ATSA website for forecasting based on a 1-factor or 2-factor GARMA model. The R routines can be run from R or from within GW-WINKS. To use GW-WINKS to find forecasts based on a 1-factor or 2-factor GARMA(p,q) model fit to a realization occupying a column in the current Data Editor, go to the Time Series Analysis menu of GW-WINKS and choose Select R Routine. On the pull-down menu choose GARMA Forecasts and specify the requested information. After running the forecasting routine, a new column in the Data Editor will contain the original realization up to the time at which forecasts begin, followed by the GARMA forecasts. The Output Viewer will contain additional information about the forecasts. See the ATSA website for instructions on calling R programs from within GW-WINKS.

11.7 Modeling Atmospheric CO_2 Data Using Long-Memory Models

Figure 11.13a shows 382 monthly atmospheric CO_2 readings collected from March 1958 until December 1989 at the summit of Mauna Loa in Hawaii (Keeling et al., 1989). The CO_2 data show a clear increase during this time frame of 31 years and 10 months. Close examination of Figure 11.13a shows that there is an annual pattern in CO_2 levels which tend to be high in the summer and low in the winter. Figure 11.13b and c show the sample autocorrelations and a spectral density estimate based on the Parzen window with truncation point $M = 50$. Clearly the sample autocorrelations damp very slowly, and there are three distinct peaks in the spectral density. As a first step in modeling the data we use the overfit procedure discussed in Chapter 3 and fit an AR(16) model for which the associated factor table is shown in Table 11.3. Other overfit models were examined and the results were consistent with those observed here. The three sets of complex conjugate roots with system frequencies 0.084, 0.167, and 0.001 are very close to the unit circle. The first two correspond to 12 and 6 month cycles since $1/0.084 \approx 12$ and $1/0.167 \approx 6$, while the third pair corresponds to either a very long cycle or, more likely, to two unit real roots. We will examine the use of a k-factor GARMA to fit these data which clearly have long-memory or nonstationary characteristics. The fact that these are monthly data and there is some annual seasonal behavior in the data suggests a model with a

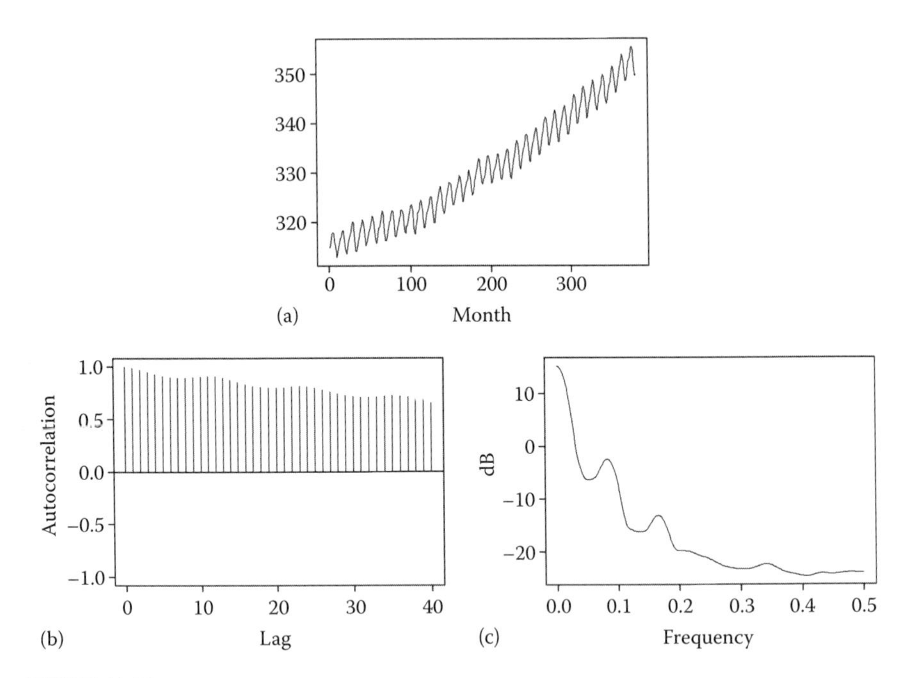

FIGURE 11.13
(a) Mauna CO_2 data, (b) sample autocorrelations of data in (a), and (c) Parzen spectral density estimate of data in (a) using $M = 50$.

TABLE 11.3
Factor Table for AR(16) Fit to CO_2 Data

AR Factors	Roots (r_j)	$\lvert r_j^{-1} \rvert$	f_{0j}
$1 - 1.729B + 0.999B^2$	$0.87 \pm 0.50i$	0.999	0.084
$1 - 0.999B + 0.992B^2$	$0.50 \pm 0.87i$	0.996	0.167
$1 - 1.982B + 0.983B^2$	$1.01 \pm 0.01i$	0.991	0.001
$1 - 0.024B + 0.859B^2$	$0.01 \pm 1.08i$	0.927	0.248
$1 + 0.941B + 0.836B^2$	$-0.56 \pm 0.94i$	0.915	0.336
$1 + 1.466B + 0.734B^2$	$-1.00 \pm 0.60i$	0.857	0.413
$1 + 1.474B + 0.640B^2$	$-1.23 \pm 0.22i$	0.800	0.472
$1 - 0.092B + 0.267B^2$	$0.17 \pm 1.93i$	0.517	0.236

factor of the form $1 - B^{12}$ or $(1 - B)(1 - B^{12})$ because of the upward trending behavior. However, comparison of the factor table in Table 11.3 with the factor table for $1 - B^{12}$ in Table 5.5 shows that, although the factors associated with those expected for $1 - B^{12}$ are visible in the table, the first three factors in Table 11.3 are much closer to the unit circle than are other factors. This differs from the factor table for the airline data in Table 8.5 which

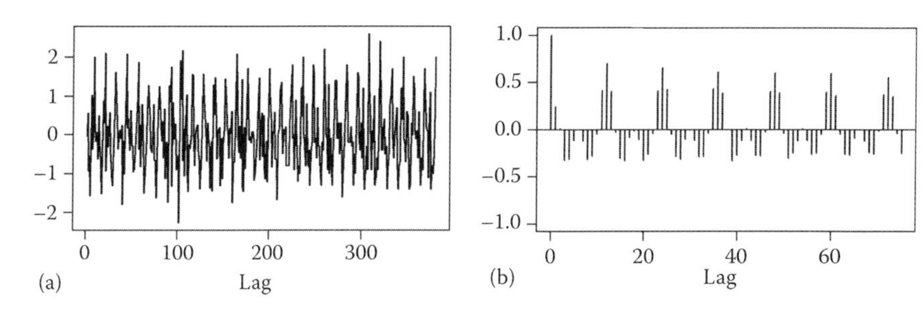

FIGURE 11.14
(a) Twice differenced CO_2 data and (b) sample autocorrelations of the data in (a).

showed a much stronger tendency for the roots of $(1 - z)(1 - z^{12}) = 0$ to be about the same distance from the unit circle. We conclude that an "airline model" (i.e., a model containing the factors $(1 - B)(1 - B^{12})$) is not appropriate for these data.

Since the factor table in Table 11.3 strongly suggests two positive unit roots, we begin modeling by twice differencing the data. The twice differenced data and the associated sample autocorrelations are shown in Figure 11.14a and b. The sample autocorrelations of the twice differenced data, shown in Figure 11.14b, have the form of a slowly damping mixture of two sinusoidal components. This is suggestive of a 2-factor GARMA model, and we will proceed to model the data using this model. The two frequencies associated with roots closest to the unit circle in Table 11.3 are 0.084 and 0.166. Because of the annual cycle, we choose to use $f_{G_1} = 1/12$ and $f_{G_2} = 1/6$. This leads to $u_1 = \cos(2\pi f_{G_1}) = \cos(\pi/6) = 0.866$ and $u_2 = 0.5$. Thus, we will consider models containing either the Gegenbauer factors $(1 - 1.732B + B^2)^{\lambda_1}$ and $(1 - B + B^2)^{\lambda_2}$ or the nonstationary ARMA factors $1 - 1.732B + B^2$ and $1 - B + B^2$. Note that the two differences initially applied to the data will enter the model as a factor $(1 - B)^2$. Therefore, the 2-factor GARMA model suggested for this data set is

$$\phi(B)(1 - B)^2(1 - 1.732B + B^2)^{\lambda_1}(1 - B + B^2)^{\lambda_2}(X_t - \mu) = \theta(B)a_t, \qquad (11.33)$$

where $\phi(B)$ and $\theta(B)$ are of orders p and q, respectively.

Woodward et al. (1998) examined maximum likelihood estimation and argued for the use of the estimates $\hat{\lambda}_1 = \hat{\lambda}_2 = 0.49$ since these did the best job of removing a strong long-memory type 12 month cycle. The data were then transformed by $(1 - B)^2(1 - 1.732B + B^2)^{0.49}(1 - B + B^2)^{0.49}$, and usual model identification techniques selected an AR(17) for the transformed data. The final 2-factor GARMA model for this data set is

$$(1 - 1.732B + B^2)^{0.49}(1 - B + B^2)^{0.49}(1 - B)^2\hat{\phi}(B)(X_t - \overline{X}) = a_t,$$

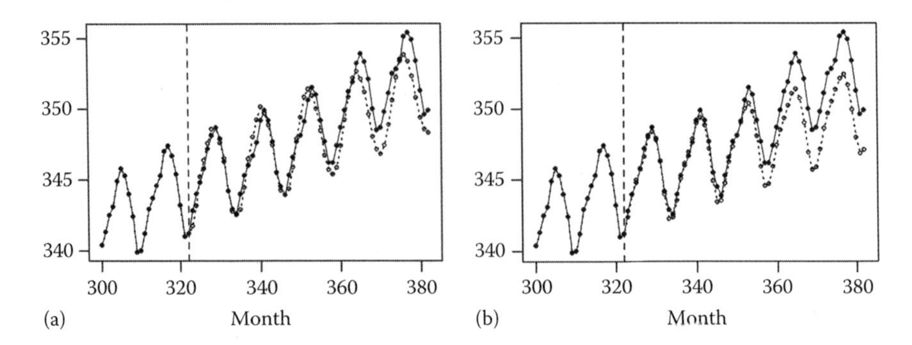

FIGURE 11.15
(a) Forecasts of last 60 months of CO_2 data (a) using the 2-factor GARMA model in (11.33) and (b) using the ARUMA(11,6,0) model in (11.34).

where $\hat{\phi}(B)$ represents the Burg estimates

$$\begin{aligned}\hat{\phi}(B) &= 1 + 2.32B + 3.23B^2 + 3.52B^3 + 3.42B^4 + 3.27B^5 + 3.25B^6 \\ &\quad + 3.31B^7 + 3.29B^8 + 3.09B^9 + 2.86B^{10} + 2.54B^{11} + 1.83B^{12} \\ &\quad + 0.91B^{13} + 0.17B^{14} - 0.21B^{15} - 0.20B^{16} - 0.08B^{17},\end{aligned}$$

with $\overline{X} = 327.57$ and $\hat{\sigma}_a^2 = 1.5$.

Forecasts based on this model for the last 60 months are shown in Figure 11.15a along with the actual data values. It can be seen that the forecasts are excellent, maintaining the cycles and upward trending present in the data.

As an alternative model for the CO_2 data, we consider the ARUMA(11,6,0) process

$$\phi_1(B)(1 - B)^2(1 - 1.732B + B^2)(1 - B + B^2)(X_t - \mu) = \theta_1(B)a_t \qquad (11.34)$$

suggested by the factor table in Table 11.3. The data transformed by the sixth-order nonstationary factor $(1 - B)^2(1 - 1.732B + B^2)(1 - B + B^2)$ was modeled as an AR(11), and $\hat{\phi}_1(B)$ based on the Burg estimates is given by

$$\begin{aligned}\hat{\phi}_1(B) &= 1 + 3.64B + 6.99B^2 + 8.92B^3 + 7.84B^4 + 4.07B^5 - 0.28B^6 \\ &\quad -2.99B^7 - 3.29B^8 - 2.14B^9 - 0.90B^{10} - 0.20B^{11}.\end{aligned}$$

Figure 11.15b shows the forecasts for the same 60 observations considered in Figure 11.15a. The forecasts based on (11.34) have the same cyclic behavior as the data, but they fail to follow the same upward trend after one or two cycles. We considered the use of an AR(17) to model the transformed data in this case, and the associated forecasts were poorer than those obtained with the AR(11). Woodward et al. (1998) show that for this data set the ARUMA model tends to have smaller MSE in short-term forecasts while the forecasts based on the 2-factor GARMA model perform better in the long term.

NOTE: The model in (11.33) is not a 2-factor GARMA model as we have defined in this chapter since it also contains the nonstationary factor $(1-B)^2$. That is, by an extension of our previous notation, this could be called a 2-factor GARIMA model. However, $Y_t = (1-B)^2(X_t - \overline{X})$ is a 2-factor GARMA model and our previous conditions for stationarity and estimation then apply.

Exercises

Applied Problems

In the ATSA website we include R programs which can be run from R or from with GW-WINKS for long-memory data. The following problems can be worked using either these programs, or other long-memory code (e.g., the mostly wavelet-based package `Waveslim` that is available on CRAN). Also, in some cases, it may be best to create your own code.

11.1 Consider the three models

(A) $(1-B)^{0.3}X_t = a_t$

(B) $(1-B)^{0.4}X_t = a_t$

(C) $(1-0.95B)X_t = a_t$.

(a) Calculate and plot the autocorrelations, $\rho_k, k = 0, \ldots, 100$ from each of the models.

(b) Generate and plot realizations of length $n = 500$ from each of the three models in Problem 11.1.

(c) For each realization in (a) find and plot sample autocorrelations, $\hat{\rho}_k$, $k = 0, \ldots, 100$.

(d) Discuss the differences among the three models based on

(i) The autocorrelations in (a)

(ii) The realizations in (b)

(iii) The sample autocorrelations in (c).

11.2 Repeat Problem 11.1 for the models

(A) $(1-0.5B+0.9B^2)(1-B)^{0.3}X_t = a_t$

(B) $(1-0.5B+0.9B^2)(1-B)^{0.4}X_t = a_t$

(C) $(1-0.5B+0.9B^2)(1+B+0.9B^2)(1-B)^{0.4}X_t = a_t$

11.3 Repeat Problem 11.1 for the models

(A) $(1-1.6B+B^2)^{0.3}X_t = a_t$

(B) $(1-1.6B+B^2)^{0.45}X_t = a_t$

(C) $(1-1.54B+0.9B^2)X_t = a_t$

11.4 Repeat Problem 11.1 for the models
(A) $(1 - 0.95B)(1 - 1.6B + B^2)^{0.3}X_t = a_t$
(B) $(1 - 0.95B)(1 - 1.6B + B^2)^{0.45}X_t = a_t$
(C) $(1 - 0.95B)(1 - 1.54B + 0.9B^2)X_t = a_t$

11.5 File `PROB11.5.XLS` contains a file of length $n = 10$ from the fractional model $(1 - B)^{0.4}X_t = a_t$.
(a) Find π_k, $k = 1, \ldots, 10$
(b) Use (11.32) to find forecasts $\hat{X}_n^{(T)}(\ell), \ell = 1, \ldots, 5$
(c) Discuss the form of the forecast function.

Theoretical Problems

11.6 Show that a process with autocovariance function
(a) $\gamma_k = k^{-\alpha}, 0 \leq \alpha \leq 1$ is long memory
(b) $\gamma_k = k^{-\alpha}, \alpha > 1$ is short memory.

11.7 Use (1.28) to show that $P(f) \leq 2\sum_{k=0}^{\infty} |\gamma_k|$.

11.8 Verify (11.3).

11.9 Verify Theorem 11.1(c).

11.10 Using the notation $f_k \simeq g_k$ to indicate $\lim_{k\to\infty} (f_k/g_k) = c$ where c is a finite constant, show that for a stationary fractional model $(1 - B)^d X_t = a_t$ it follows that for $d \in (-0.5, 0.5)$
(a) $\rho_k \simeq k^{2d-1}$
(b) $\psi_k \simeq k^{d-1}$.

11.11 Show that if a fractional process is invertible, then the π-weights in its infinite autoregressive representation, $\pi(B)X_t = \sum_{k=0}^{\infty} \pi_k X_{t-k}$, are given by $\pi_k = \dfrac{(k - d - 1)!}{k!(-d - 1)!}$.

11.12 Verify (11.15) in Theorem 11.3(c).

11.13 Verify (11.21) in Theorem 11.6(b).

12

Wavelets

As we know, a (covariance) stationary process, $X(t)$, is defined to be a process for which the following conditions hold:

$$\begin{aligned}
&1.\ E[X(t)] = \mu \quad \text{(constant for all } t\text{)} \\
&2.\ \text{Var}[X(t)] = \sigma^2 < \infty \quad \text{(i.e., a finite constant for all } t\text{)} \\
&3.\ \text{Cov}[X(t), X(t+h)] = \gamma(h) \quad \text{(i.e., the covariance} \\
&\quad \text{between random variables) } X(t) \text{ and } X(t+h) \\
&\quad \text{depends only on } h \text{ (and not on } t\text{)}
\end{aligned} \tag{12.1}$$

To this point in this book, we have discussed processes that are stationary, processes that are not stationary because the process mean changes with time (signal + noise, etc.), and the ARUMA models. Time series for which Condition 3 in (12.1) is not satisfied are characterized by realizations whose correlation structure (or frequency behavior) changes with time. In Chapters 12 and 13, we will discuss techniques for analyzing data that are nonstationary because they have frequency (correlation) behavior that changes with time. Such data are called time varying frequency (TVF) data and include a wide number of applications such as seismic signals, animal and insect noises (e.g., whale clicks, bat echolocation, insect chirps), and astronomical signals. In Section 12.1, we will discuss shortcomings of classical spectral analysis when applied to data with time-varying frequencies. In Section 12.2, we discuss modifications of Fourier-based spectral analysis techniques designed to capture TVF behavior in data, and in the remainder of the chapter we discuss wavelet analysis as a much more general tool that can be used for analyzing such data. In Chapter 13, we discuss recent developments in the analysis of TVF data that change regularly across time.

12.1 Shortcomings of Traditional Spectral Analysis for TVF Data

Consider the realizations in Figure 12.1a and b. The nature of Realization 1 in Figure 12.1a changes near $t = 100$. Actually, there are two frequencies in the data: (a) a frequency of about $f = 0.08$ (period of about 12) for $t = 1$ to 100 and (b) a frequency of about $f = 0.2$ (period about 5) for $t = 101$ to 200.

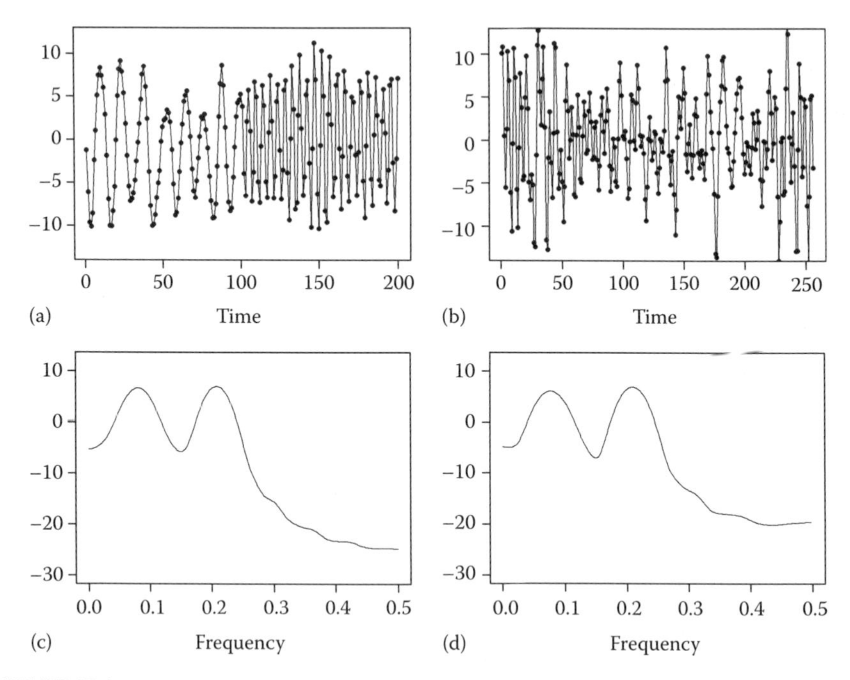

FIGURE 12.1
Two distinctly different realizations resulting in very similar Parzen spectral density estimates. (a) Realization 1. (b) Realization 2. (c) Spectral estimate for realization 1. (d) Spectral estimate for realization 2.

The Parzen window-based spectral density estimate using $M = 28$ in Figure 12.1c shows distinct peaks at these two frequencies. Clearly, however, the realization in Figure 12.1a is not a realization from a stationary time series because $\text{Cov}[X_t, X_{t+h}]$ for $t \in [1, 100]$ is different than it is for $t \in [101, 200]$. That is, Condition 3 in (12.1) is not met. Figure 12.1b shows a realization from the AR(4) model:

$$(1 - 1.7B + 0.95B^2)(1 - 0.5B + 0.985B^2)X_t = a_t. \tag{12.2}$$

Visually there seem to be two frequency behaviors, a longer periodic component superimposed with a higher-frequency component. The factor table in Table 12.1 shows that (12.2) is a stationary model and that the two pairs of

TABLE 12.1
Factor Table for AR(4) Model in (12.2)

AR Factors	Roots (r_j)	$\lvert(r_j^{-1})\rvert$	f_{oj}
$1 - 0.5B + 0.985B^2$	$0.25 \pm 0.98i$	0.99	0.21
$1 - 1.7B + 0.95B^2$	$0.89 \pm 0.50i$	0.97	0.08

complex conjugate roots of the characteristic equation are associated with system frequencies of about $f = 0.08$ and $f = 0.21$. The associated Parzen spectral density estimate with $M = 28$ is shown in Figure 12.1d. The spectral density estimates in Figure 12.1c and d are extremely similar even though the two realizations have distinctly different structures.

NOTE: It is important to note that although the mathematical formulas involved in calculating the smoothed spectral density estimates can be applied to any realization, the spectrum and spectral density (as functions of the autocovariance and autocorrelation functions) are only defined for stationary processes. That is, since the time series generating the realization in Figure 12.1a is not stationary, the spectral estimation resulting in Figure 12.1c was inappropriately applied. Also, since the spectrum and spectral density are defined for stationary processes, they do not show time-frequency behavior. That is, the spectral density estimates shown in Figure 12.1 do not give any indication that Realization 1 has two distinct frequencies, each over a separate time period. In Realization 2 the two frequencies are present simultaneously throughout the entire realization.

To further illustrate this point consider the linear chirp signal in Figure 12.2a. The frequency (autocorrelation) structure is clearly changing, and it is seen that the realization becomes increasingly higher frequency in time. For example, the first cycle in the data is roughly of length 50 while the last cycle is of about length 6. That is, the frequency behavior goes from approximately $f = 0.02$ at the beginning of the realization to around $f = 0.14$ at the end. Although this is clearly a realization from a nonstationary process, we again apply the Parzen window spectral density estimator, and the resulting spectral density estimate is shown in Figure 12.2b, where we can see how the spectrum has been spread in an attempt to adjust for the changing frequencies. In Figure 12.2b the "peak" in the spectrum extends from about $f = 0$ to a little below $f = 0.2$ reflecting the fact that the frequency behavior varies within this range across time. The concept of a "spread spectrum" was introduced in Example 1.16.

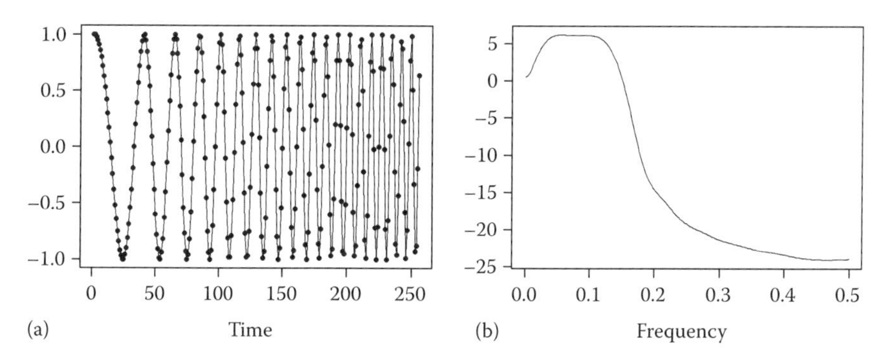

FIGURE 12.2
(a) Linear chirp. (b) Spectral estimate for realization in (a).

In summary, we note that since the spectrum "assumes" that the time series is stationary, i.e., that its frequency behavior does not change over time, it does not provide insight regarding any time-frequency behavior in the data. Note, for example, that if the realization in Figure 12.2a were simply reversed in direction so that the frequencies decreased in time, the Parzen spectral density estimate would not reflect the change in direction. In essence using classical spectral estimation methods, the frequency content at a given frequency is found by averaging over time. If characteristics of the signal depend on time, then these classical methods are not appropriate and can lead to misleading results.

12.2 Window-Based Methods That Localize the "Spectrum" in Time

A method of obtaining frequency information that is localized in time is needed in order to address the concerns raised in Section 12.1. In this section we discuss two such methods: the Gabor spectrogram and the Wigner–Ville spectrum. These are both window-based modifications of usual Fourier transform–based methods. In Section 12.3 we discuss wavelets which provide an entirely different method for dealing with this situation.

The objective of window-based methods is to deal with the problem of time-varying frequency behavior by breaking the time axis into time segments (windows) that are sufficiently narrow that stationarity is a reasonable assumption in the window. For example, the data in Figure 12.2a show a dramatic change in frequency behavior from the beginning of the series to the end. However, when restricted to a smaller time window (say of width 25) there is not much frequency change. As a consequence, Fourier-based methods are reasonably appropriate within this time window.

12.2.1 Gabor Spectrogram

The *Gabor spectrum* (or *spectrogram)* is a widely used window-based tool for spectral analysis. The continuous Fourier transform of the square integrable function, $g(x)$, is given by

$$G(f) = \int_{-\infty}^{\infty} e^{-2\pi i f x} g(x) dx. \tag{12.3}$$

Window-based methods, such as the Gabor spectrogram, are based on a window function, $h(t)$, for which $|h(t)| \to 0$ as $|t| \to \infty$. The short time Fourier transform (STFT) is defined by

$$G(t,f) = \int_{-\infty}^{\infty} e^{-2\pi i f x} g(x) h(x - t) dx. \quad (12.4)$$

The function $h(t)$ is defined in such a way that $h(x - t)g(x)$ localizes the signal around t so that for a given t, $G(t,f)$ given in (12.4) represents a "Fourier transform" localized around t. We represent this by expressing the Fourier transform in (12.4) as a function of two variables: t and f. The *Gabor transform* is a special case of the STFT that uses a Gaussian window

$$h(t) = \frac{1}{2\sqrt{\pi\alpha}} e^{-t^2/4\alpha}. \quad (12.5)$$

The parameter, α, controls the window width, and the *Gabor spectrogram* is a two-dimensional plot of

$$S(t,f) = |G(t,f)|^2. \quad (12.6)$$

Figure 12.3a and b show Gabor spectrograms for the data in Figure 12.1a and b, respectively. Note that time is plotted along the horizontal axis,

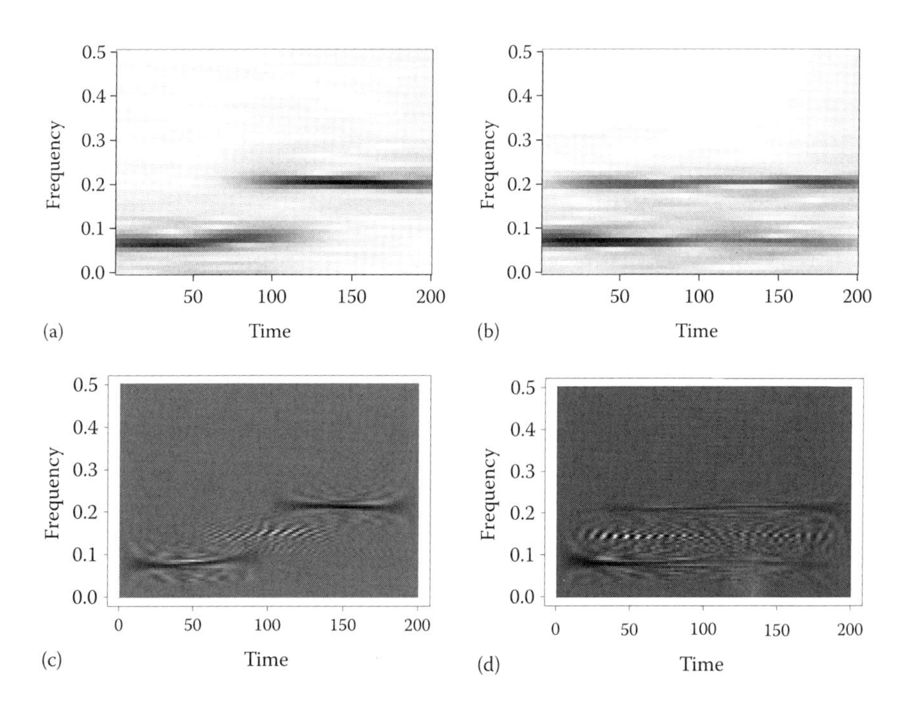

FIGURE 12.3
Time frequency representations for the data in Figure 12.1. (a) Gabor spectrum for Figure 12.1a, (b) Gabor spectrum for Figure 12.1b, (c) Wigner–Ville spectrum for Figure 12.1a, and (d) Wigner–Ville spectrum for Figure 12.1b.

frequency along the vertical axis, and $S(t,f)$ is plotted using a gray scale where darker shades indicate larger spectral values. For example, in Figure 12.3a we see that for $t = 1$ through 100, the largest values of $S(t,f)$ are around $f = 0.08$ while for $t = 101$ through 200 the largest values seem to be at about $f = 0.2$. This is consistent with the observations regarding the spectral behavior of the data in Figure 12.1a. In Figure 12.3b it is seen that two strong frequency behaviors are observed for $t \in [1, 200]$ at about $f = 0.08$ and $f = 0.2$ which is in agreement with the fact that the data in Figure 12.1b were from an AR(4) model with system frequencies at about $f = 0$ and $f = 0.2$. Since the AR(4) model is stationary, the frequency behavior does not change with time and these two frequencies consistently appear for all t.

Figure 12.4a shows the Gabor transform for the chirp data in Figure 12.2a. The largest values of $S(t,f)$ change linearly from about $f = 0$ at time $t = 1$ to about $f = 0.17$ at $t = 256$. The frequency at $t = 256$ is consistent with the previous observation that the last cycle in the data is of length 6, i.e., $f = 1/6 = 0.167$. A "snapshot" of the spectrogram at $t = 256$ could be graphed by plotting frequency on the horizontal axis and $S(256,f)$ (ind B) on the vertical axis. The resulting "spectral-type" plot would have a peak at about $f = 0.17$. The corresponding snapshot at $t = 100$ would have a peak at about $f = 0.08$. See Figure 13.9 for examples of snapshot plots.

It is clear that the time-frequency plot in Figure 12.4a does a much better job of describing the frequency behavior in Figure 12.2a than does the usual spectral plot in Figure 12.2b. For more discussion on the Gabor transform see Chui (1992), Kumar and Foufoula-Georgiou (1994), Ogden (1997), and Carmona et al. (1998).

Computational notes: The Gabor transform can be obtained using function `cgt` in the package `R-wave` which is available on `CRAN` to R users.

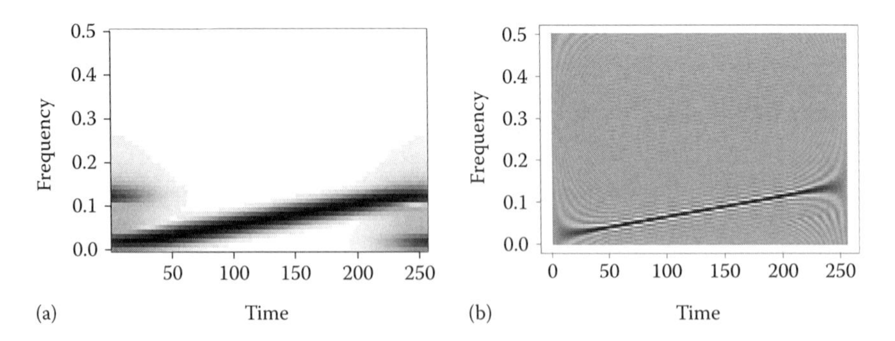

FIGURE 12.4
Time frequency representations for the data in Figure 12.2a: (a) Gabor spectrum and (b) Wigner–Ville spectrum.

12.2.2 Wigner–Ville Spectrum

Recall that the spectrum for a stationary discrete time series defined on $t = 0, \pm 1, \pm 2, \ldots$ is defined in (1.25) to be

$$P(f) = \sum_{k=-\infty}^{\infty} \gamma(k) e^{-2\pi i f k}, \tag{12.7}$$

for $-0.5 < f < 0.5$. However, for time series whose autocorrelation changes with time, Condition 3 in (12.1) does not hold. Letting $\gamma(t_1, t_2)$ denote the autocovariance between $X(t_1)$ and $X(t_2)$, the discrete Wigner–Ville spectrum can be defined by

$$W(t,f) = \sum_{k=-\infty}^{\infty} \gamma(t + k/2, t - k/2) e^{-2\pi i f k}. \tag{12.8}$$

If X_t is approximately stationary in a neighborhood of t, then $\gamma(t + k/2, t - k/2)$ can be approximated by use of a (time domain) window centered at t. See Martin and Flandrin (1985) and Boashash (2003). Figure 12.3c and (d) show the Wigner–Ville time frequency representations associated with the data in Figure 12.2a and b, respectively. It is seen that the information concerning the time frequency relationships are consistent with those shown using the Gabor spectrum shown in Figure 12.3a and b. Similarly, the Wigner–Ville plot in Figure 12.4b for the chirp data shows the same linearly increasing time-frequency relationship shown in Figure 12.4a.

Computational notes: The Wigner–Ville plot can be obtained using the function `WV` that is provided on the ATSA website.

12.3 Wavelet Analysis

In this section we give an overview of wavelet analysis and its application to statistical time series analysis of nonstationary data. Wavelets provide an alternative to the window methods for investigating frequency behavior that changes across time. Wavelet analysis is a relatively recent methodology developed primarily in the mathematical sciences and engineering, and it has proved to be a powerful tool for the analysis of localized behavior of data whose characteristics change with time. These methods can be viewed as variations of classical Fourier analysis in which short waves (or "wavelets") are used in place of the classical sines and cosines.

12.3.1 Fourier Series Background

In order to fully appreciate wavelet methods, it is necessary to have some familiarity with the classical Fourier methods of which wavelet analysis is an extension. For a brief review of Fourier series methods, see Appendix 12.A.

As discussed in Section 1.5 and Appendix 12.A, Fourier analysis involves expressing functions as linear combination of sines and cosines. Theorem 12.A.1 states that a well-behaved function, g, defined on $[-\pi, \pi]$ can be expressed as a linear combination of the *basis functions* in the orthogonal systems $S_1 = \{1, \cos kx, \sin kx; k = 1, 2, \ldots,\}$ or $S_2 = \{e^{ikx}, k = 0, \pm 1, \pm 2, \ldots\}$. The advantage of using S_2 is that the basis functions actually consist of one "fundamental" function, namely, $\varphi(x) = e^{ix}$, and the individual basis functions are "integral dilations" of this function, i.e., $e^{ikx} = \varphi(kx)$. Fourier analysis is based on the fact that when certain conditions hold, functions can be written as a linear combination of dilations of the single function e^{ix}. See Appendix 12.A.

12.3.2 Wavelet Analysis Introduction

Suppose that g is square integrable over the real line, i.e., that $\int_{-\infty}^{\infty} |g(x)|^2 dx < \infty$. The class of such functions over the real line, $\mathbb{R}$, is denoted $L^2(\mathbb{R})$. We desire a collection of basis functions so that functions $g \in L^2(\mathbb{R})$ can be approximated arbitrarily closely using a finite linear combination of these basis elements. It should be noted that $\sin x$, $\cos x$, and e^{ix} are not candidates for the desired basis functions since they are not in $L^2(\mathbb{R})$ themselves. In fact, a necessary condition for $\int_{-\infty}^{\infty} |g(x)|^2 dx < \infty$ is $\lim_{|t| \to \infty} |g(t)| = 0$. That is, the basis functions must decay to zero as $|t| \to \infty$, so intuitively short waves or "wavelets" are needed. As in the Fourier series setting, it is preferable to have a single function, say ψ, from which to work. In the wavelet setting, it is common to use $\psi(2^{-j}x)$ where j is an integer. Approximating functions g, that are defined over the entire real line would intuitively be difficult to accomplish simply using dilations of a single, quickly damping function. Thus, in wavelet analysis, we consider dilations *and translations* of the basis function for purposes of estimating localized frequency behavior. That is, given a function $\psi(x)$, we use as our basis collection the functions

$$\psi_{j,k}(x) = 2^{-j/2}\psi(2^{-j}x - k), \tag{12.9}$$

where j and k are integers that are referred to as the *diliation index* and *translation index*, respectively. The function $\psi(x)$ is called the mother wavelet since it "gives birth" to the entire collection of basis elements.

Example 12.1 Haar Wavelets

As a first example of a wavelet we consider the Haar wavelet defined by

$$\psi(x) = \begin{cases} 1, & 0 \le x < 1/2 \\ -1, & 1/2 \le x < 1 \\ 0, & \text{elsewhere.} \end{cases} \tag{12.10}$$

Figure 12.5a shows the Haar mother wavelet along with three dilations and translations. If, for example, $j = 2$, $k = 0$ then $\psi_{2,0}(x)$ is given by $\psi_{2,0}(x) = 2^{-1}\psi(2^{-2}x - 0)$, i.e.,

$$\psi_{2,0}(x) = \begin{cases} 0.5, & 0 \le x < 2 \\ -0.5, & 2 \le x < 4 \\ 0, & \text{elsewhere,} \end{cases} \tag{12.11}$$

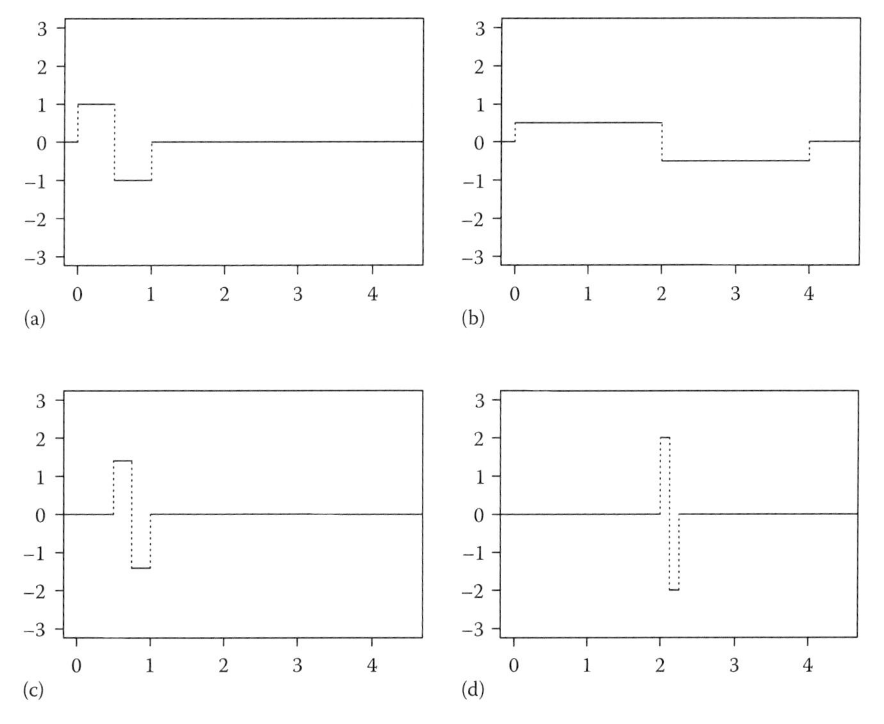

FIGURE 12.5
(a) Haar mother wavelet and various dilations and translations, (b) $j = 2$, $k = 0$, (c) $j = -1$, $k = 1$, (d) $j = -2$, $k = 8$.

which is shown in Figure 12.5b. Figure 12.5c and d show $\psi_{-1,1}$ and $\psi_{-2,8}$, respectively. Note that

$$\int_{-\infty}^{\infty} \psi(x)dx = 0, \tag{12.12}$$

and consequently for each j and k

$$\int_{-\infty}^{\infty} \psi_{j,k}(x)dx = 0. \tag{12.13}$$

The Haar wavelets have compact support, i.e., they are equal to zero outside the closed bounded interval $[k2^j, (k+1)2^j]$. Specifically,

$$\begin{aligned} \psi_{j,k}(x) &= 2^{-j/2}, \quad && 2^j k \le x < 2^j(k+0.5) \\ &= -2^{-j/2}, && 2^j(k+0.5) \le x < 2^j(k+1) \\ &= 0, && \text{otherwise.} \end{aligned} \tag{12.14}$$

From (12.14) it follows that the support of $\psi_{j,k}$ is of length $2^j(k+1) - 2^j k = 2^j$. Dilates and translates of a wavelet are designed to describe localized frequency behavior. Thus the wavelet $\psi_{j,k}$ describes localized behavior related to pseudo-cyclic or detail behavior of duration (or "period") about 2^j for x in the neighborhood of $2^j(k+0.5)$. The support of the Haar wavelet $\psi_{j,k}$ is also called the *scale* of the wavelet. In general the scale of a wavelet is a measure of its width.

A note about frequencies: It is important to note that since we are not using sines and cosines as the basis elements, the strict interpretation of frequency is no longer valid. Since scale is a measure of the "period" of a wavelet we use "1/scale" as a measure of "frequency."

There are many types of wavelets, and the Haar is shown here for simplicity. Properties (12.12) and (12.13) are true for all wavelets. While the Haar wavelet has compact support, this is not a requirement. However, all wavelets have the property that $|\psi_{j,k}(x)| \to 0$ as $|x| \to \infty$. Figure 12.6 shows several wavelets. The Daublet(4), Daublet(8), and Daublet(12) mother wavelets in Figure 12.6a, c, and e are from the Daublet family of wavelets suggested by Daubechies (1988). The Daublet(2) wavelet, i.e., the second-order Daublet, is the Haar wavelet. The higher the order, the smoother the wavelet, and the broader the support. The Daublet(4) wavelet is sometimes called the shark-fin wavelet for obvious reasons. Figure 12.6g shows the Symlet(8) wavelet. This wavelet is widely used and is a member of the Symlet (sometimes called least asymmetric) family of wavelets also proposed by Daubechies (1988). The Symlet and Daublet families are both orthogonal families. That is, for

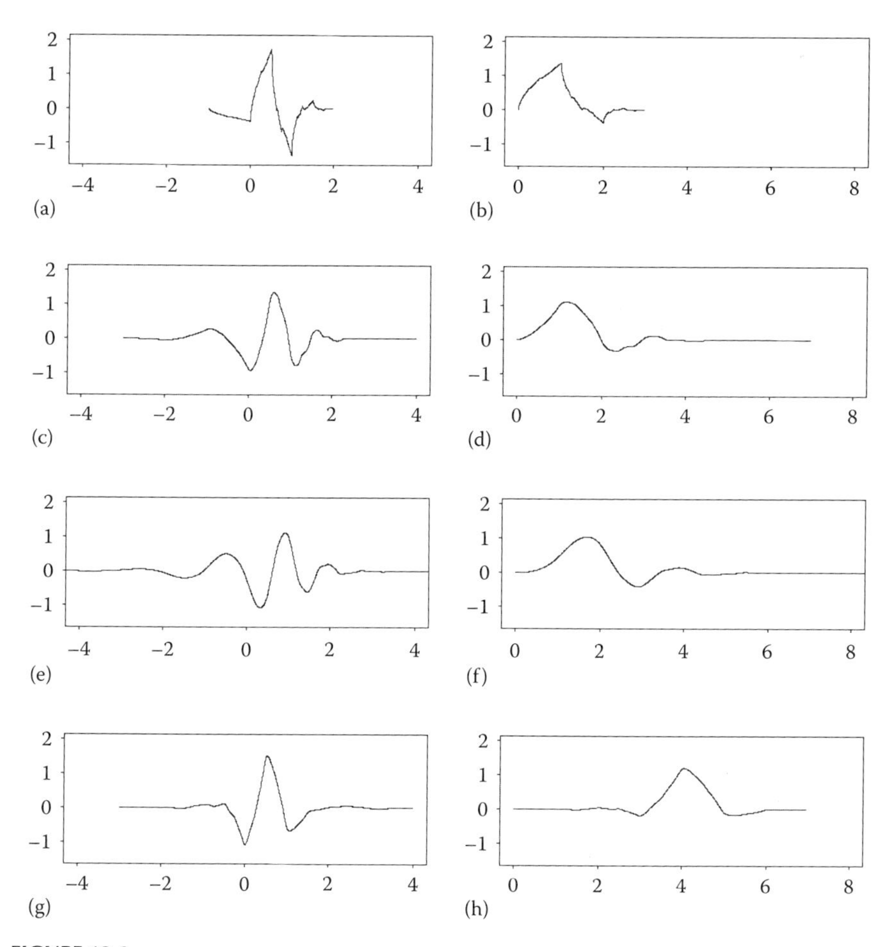

FIGURE 12.6
Four mother wavelets and their corresponding father wavelets.(a) D4 mother wavelet. (b) D4 father wavelet. (c) D8 mother wavelet. (d) D8 father wavelet. (e) D12 mother wavelet. (f) D12 father wavelet. (g) S8 mother wavelet. (h) S8 father wavelet.

example, the Symlet(8) wavelet and its collection of dilates and translates given in (12.9) is an orthogonal collection of functions that form a basis for functions $g \in L^2(\mathbb{R})$ in the sense of Theorem 12.1 that follows. Comparing Figure 12.6a, c, and e it can be seen that as k increases, the support of the Daublet($2k$) wavelet (interval on which the wavelet takes on nonzero values) gets longer and the wavelet becomes smoother.

NOTE: The discretized versions of Daublet($2k$) and Symlet($2k$) wavelets take on $2k$ values and some authors refer to these as Daublet($2k$) (and Symlet($2k$)) wavelets, as we have here. However, others refer to them as Daublet(k) and Symlet(k) wavelets. That is, the Daublet(4) (shark-fin) wavelet in Figure 12.6a is sometimes referred to as a Daublet(2) wavelet. Be careful!

12.3.3 Fundamental Wavelet Approximation Result

The fundamental wavelet-based approximation result analogous to the Fourier series approximation is given in Theorem 12.1 that follows.

Theorem 12.1

Let $\{\psi_{j,k}\}$ be an orthogonal wavelet basis, and let $g \in L^2(\mathbb{R})$. Then,

$$g(x) = \sum_{j=-\infty}^{\infty} \sum_{k=-\infty}^{\infty} d_{j,k}\psi_{j,k}(x), \qquad (12.15)$$

where the equality is in the mean square sense and the wavelet transform coefficients, $d_{j,k}$, are given by

$$d_{j,k} = \frac{1}{\|\psi\|^2} \int_{-\infty}^{\infty} g(t)\psi_{j,k}(t)dt, \qquad (12.16)$$

where $\|\psi\|^2$ denotes $\int_{-\infty}^{\infty} |\psi(x)|^2 dx$.

Proof:

See Daubechies (1992) and Ogden (1997). Appendix 12.A gives a formal derivation to show that if (12.15) holds, then $d_{j,k}$ is given by (12.16).

Equation 12.15 is analogous to the Fourier series representation in (12.A.3). Also, the coefficients, $d_{j,k}$, in (12.16) correspond to the Fourier series coefficients, α_k in (12.A.4). In Fourier analysis large values of α_k^2 are associated with time series that have periodic content associated with the frequency $f = k/n$. Similarly, large values of $|d_{j,k}|$ in (12.15) are associated with detail behavior of duration about 2^j, i.e., the "frequency" or "1/scale" is about 2^{-j}. In Fourier analysis, α_k's as defined in (12.4) are Fourier transforms of $g(x)$, and similarly the coefficients, $d_{j,k}$, defined in (12.16) are referred to as (continuous) *wavelet transforms*.

The expression in (12.15) is sometimes expressed as

$$g(x) = \sum_{j=J+1}^{\infty} \sum_{k=-\infty}^{\infty} d_{j,k}\psi_{j,k}(x) + \sum_{j=-\infty}^{J} \sum_{k=-\infty}^{\infty} d_{j,k}\psi_{j,k}(x),$$

which is often written as

$$g(x) = \sum_{k=-\infty}^{\infty} s_{J,k}\varphi_{J,k}(x) + \sum_{j=-\infty}^{J} \sum_{k=-\infty}^{\infty} d_{j,k}\psi_{j,k}(x), \qquad (12.17)$$

where the $\varphi_{j,k}$ are dilates and translates of a *father wavelet* or *scaling function*, $\varphi(x)$, corresponding to the mother wavelet, $\psi(x)$. Scaling functions (father wavelets) are functions that have the property

$$\int_{-\infty}^{\infty} \varphi(x)dx = 1, \tag{12.18}$$

and their dilates and translates are defined analogously to (12.9), i.e.,

$$\varphi_{j,k}(x) = 2^{-j/2}\varphi(2^{-j}x - k), \tag{12.19}$$

for integer j and k. Consequently, it follows that for all integer j and k,

$$\int_{-\infty}^{\infty} \varphi_{j,k}(x)dx = 1. \tag{12.20}$$

As an example, the Haar father wavelet is

$$\begin{aligned} \varphi(x) &= 1, \quad x \in [0,1) \\ &= 0, \quad \text{otherwise.} \end{aligned}$$

The right-hand side of Figure 12.6 shows the father wavelets corresponding to each of the mother wavelets shown there. Their behavior mimics that of the mother wavelets in that as k gets larger, the father wavelet becomes smoother and the support widens.

It should be noted that wavelets (mother and father) usually have no analytic form. They are typically calculated using the dilation equation. See Daubechies (1992), Bruce and Gao (1996), and Strang and Nguyen (1997).

Equation 12.17 is sometimes written as

$$g(x) = S_J(x) + \sum_{j=-\infty}^{J} D_j(x), \tag{12.21}$$

where $S_J(x) = \sum_{k=-\infty}^{\infty} s_{J,k}\varphi_{J,k}(x)$ and $D_j(x) = \sum_{k=-\infty}^{\infty} d_{j,k}\psi_{j,k}(x)$. The first term in (12.21) gives a "smooth" approximation to $g(x)$ to level J (i.e., involves the higher values of j). The $s_{j,k}$ are called *smooth* coefficients. As we have previously discussed, the $d_{j,k}$ are called the detail coefficients, and smaller values of j correspond to finer detail.

12.3.4 Discrete Wavelet Transform for Data Sets of Finite Length

For clarity of presentation, the discussion up to this point regarding wavelets has involved data on a continuum. However, for analyzing discrete data, $\mathbf{x} = (x_1, x_2, \ldots, x_n)$, of finite length (such as a realization from a discrete time series), we will use the *discrete wavelet transform* (DWT). Given an orthogonal wavelet basis set, the DWT maps the vector $\mathbf{x} = (x_1, x_2, \ldots, x_n)$ to a vector of n wavelet coefficients. Letting $\mathbf{w}$ denote the vector of wavelet coefficients, then it can be shown that

$$\mathbf{w} = \mathbf{M}\mathbf{x}, \tag{12.22}$$

where $\mathbf{M}$ is an orthogonal matrix that can be written as

$$\mathbf{M} = \begin{pmatrix} \mathbf{M}_1 \\ \mathbf{M}_2 \\ \vdots \\ \mathbf{M}_{J_0} \\ \mathbf{V}_{J_0} \end{pmatrix}, \tag{12.23}$$

where

$$\mathbf{M}_1 = \begin{pmatrix} \tilde{\psi}_{1,1}(1) & \tilde{\psi}_{1,1}(2) & \cdots & \tilde{\psi}_{1,1}(n) \\ \tilde{\psi}_{1,2}(1) & \tilde{\psi}_{1,2}(2) & \cdots & \tilde{\psi}_{1,2}(n) \\ \vdots & \vdots & & \vdots \\ \tilde{\psi}_{1,n/2}(1) & \tilde{\psi}_{1,n/2}(2) & \cdots & \tilde{\psi}_{1,n/2}(n) \end{pmatrix},$$

$$\mathbf{M}_2 = \begin{pmatrix} \tilde{\psi}_{2,1}(1) & \tilde{\psi}_{2,1}(2) & \cdots & \tilde{\psi}_{2,1}(n) \\ \tilde{\psi}_{2,2}(1) & \tilde{\psi}_{2,2}(2) & \cdots & \tilde{\psi}_{2,2}(n) \\ \vdots & & & \\ \tilde{\psi}_{2,n/2^2}(1) & \tilde{\psi}_{2,n/2^2}(2) & \cdots & \tilde{\psi}_{2,n/2^2}(n) \end{pmatrix},$$

$$\mathbf{M}_{J_0} = \left(\tilde{\psi}_{J_0,1}(1) \quad \tilde{\psi}_{J_0,1}(2) \cdots \tilde{\psi}_{J_0,1}(n) \right),$$

$$\mathbf{V}_{J_0} = \left(\tilde{\varphi}_{J_0,1}(1) \quad \tilde{\varphi}_{J_0,1}(2) \cdots \tilde{\varphi}_{J_0,1}(n) \right).$$

Since $\mathbf{M}$ is orthogonal then it also follows that

$$\mathbf{x} = \mathbf{M}'\mathbf{w}, \tag{12.24}$$

where $\mathbf{M}'$ is the transpose of $\mathbf{M}$. That is, knowing $\mathbf{M}$ allows us to transform from the data to the wavelet coefficients (the DWT) and also from the wavelet coefficients to the data, i.e., the *inverse discrete wavelet transform* (IDWT).

NOTE: While the DWT can be applied to data sets whose length is not a power of 2, in the illustrative material here, we will assume that $n = 2^{J_0}$ for some J_0.

For a discussion of discrete wavelets and scaling functions see Percival and Walden (2000), Strang and Nguyen (1997), and Bruce and Gao (1996). It should be noted that the discrete version of a wavelet takes the general form of the continuous wavelet, but is not simply a discretized version of it. Most authors refer to the elements of $\mathbf{M}$ as wavelet and scaling filter coefficients. Since the wavelet and scaling filter coefficients are close approximations to the corresponding discretized versions of the continuous dilated and translated wavelets, we use the notation $\tilde{\psi}_{j,k}$ and $\tilde{\varphi}_{j,k}$ (see Percival and Walden, 2000, equations (469) and (477a)).

Using the partitioning in (12.23) we can write (12.22) as

$$\begin{pmatrix} \mathbf{w}_1 \\ \mathbf{w}_2 \\ \vdots \\ \mathbf{w}_{J_0} \\ \mathbf{v}_{J_0} \end{pmatrix} = \begin{pmatrix} \mathbf{M}_1 \\ \mathbf{M}_2 \\ \vdots \\ \mathbf{M}_{J_0} \\ \mathbf{V}_{J_0} \end{pmatrix} \mathbf{x}, \tag{12.25}$$

where

$$\mathbf{w}_1 = \begin{pmatrix} d_{1,1} & d_{1,2} & d_{1,3} & d_{1,n/2} \end{pmatrix}',$$
$$\mathbf{w}_2 = \begin{pmatrix} d_{2,1} & d_{2,2} & \cdots & d_{2,n/2^2} \end{pmatrix}',$$
$$\mathbf{w}_{J_0} = d_{J_0,1} \quad \mathbf{v}_{J_0} = s_{J_0,1}.$$

See Percival and Walden (2000) for a discussion of the issues involved in constructing the matrix $\mathbf{M}$.

Example 12.2 Discrete Haar Wavelets

Consider a data set of size $n = 8$ for which the Haar wavelet is used. In this case $J_0 = 3$, the eight wavelet coefficients are $\mathbf{w} = (d_{1,1}, d_{1,2}, d_{1,3}, d_{1,4}, d_{2,1}, d_{2,2}, d_{3,1}, s_{3,1})'$, and (12.22) takes the form

$$\begin{pmatrix} d_{11} \\ d_{12} \\ d_{13} \\ d_{14} \\ d_{21} \\ d_{22} \\ d_{31} \\ s_{31} \end{pmatrix} = \begin{pmatrix} -1/\sqrt{2} & 1/\sqrt{2} & 0 & 0 & 0 & 0 & 0 & 0 \\ 0 & 0 & -1/\sqrt{2} & 1/\sqrt{2} & 0 & 0 & 0 & 0 \\ 0 & 0 & 0 & 0 & -1/\sqrt{2} & 1/\sqrt{2} & 0 & 0 \\ 0 & 0 & 0 & 0 & 0 & 0 & -1/\sqrt{2} & 1/\sqrt{2} \\ -1/2 & -1/2 & 1/2 & 1/2 & 0 & 0 & 0 & 0 \\ 0 & 0 & 0 & 0 & -1/2 & -1/2 & 1/2 & 1/2 \\ -1/\sqrt{8} & -1/\sqrt{8} & -1/\sqrt{8} & -1/\sqrt{8} & 1/\sqrt{8} & 1/\sqrt{8} & 1/\sqrt{8} & 1/\sqrt{8} \\ 1/\sqrt{8} & 1/\sqrt{8} & 1/\sqrt{8} & 1/\sqrt{8} & 1/\sqrt{8} & 1/\sqrt{8} & 1/\sqrt{8} & 1/\sqrt{8} \end{pmatrix} \begin{pmatrix} x_1 \\ x_2 \\ x_3 \\ x_4 \\ x_5 \\ x_6 \\ x_7 \\ x_8 \end{pmatrix}. \tag{12.26}$$

(see Percival and Walden, 2000). Thus we see that $\tilde{\psi}_{1,1}(1) = -1/\sqrt{2}$, $\tilde{\psi}_{1,1}(2) = 1/\sqrt{2}$, and $\tilde{\psi}_{1,1}(t) = 0$, $t = 3, \ldots, 8$. Similarly, $\tilde{\psi}_{31}(t) = -1/\sqrt{8}$, $t = 1, \ldots, 4$ and $\tilde{\psi}_{31}(t) = 1/\sqrt{8}$, $t = 5, \ldots, 8$ while $\tilde{\varphi}_{3,1} = 1/\sqrt{8}$, $t = 1, \ldots, 8$. Using (12.23) we write (12.26) as

$$\begin{pmatrix} \mathbf{w}_1 \\ \mathbf{w}_2 \\ \mathbf{w}_3 \\ \mathbf{v}_3 \end{pmatrix} = \begin{pmatrix} \mathbf{M}_1 \\ \mathbf{M}_2 \\ \mathbf{M}_3 \\ \mathbf{V}_3 \end{pmatrix} \mathbf{x},$$

where

$$\mathbf{M}_1 = \begin{pmatrix} -1/\sqrt{2} & 1/\sqrt{2} & 0 & 0 & 0 & 0 & 0 & 0 \\ 0 & 0 & -1/\sqrt{2} & 1/\sqrt{2} & 0 & 0 & 0 & 0 \\ 0 & 0 & 0 & 0 & -1/\sqrt{2} & 1/\sqrt{2} & 0 & 0 \\ 0 & 0 & 0 & 0 & 0 & 0 & -1/\sqrt{2} & 1/\sqrt{2} \end{pmatrix},$$

$$\mathbf{M}_2 = \begin{pmatrix} -1/2 & -1/2 & 1/2 & 1/2 & 0 & 0 & 0 & 0 \\ 0 & 0 & 0 & 0 & -1/2 & -1/2 & 1/2 & 1/2 \end{pmatrix},$$

$$\mathbf{M}_3 = \begin{pmatrix} -1/\sqrt{8} & -1/\sqrt{8} & -1/\sqrt{8} & -1/\sqrt{8} & 1/\sqrt{8} & 1/\sqrt{8} & 1/\sqrt{8} & 1/\sqrt{8} \end{pmatrix},$$

and

$$\mathbf{V}_3 = \begin{pmatrix} 1/\sqrt{8} & 1/\sqrt{8} & 1/\sqrt{8} & 1/\sqrt{8} & 1/\sqrt{8} & 1/\sqrt{8} & 1/\sqrt{8} & 1/\sqrt{8} \end{pmatrix}.$$

A few of the elements of $\mathbf{w}$ are given in (12.27) that follows:

$$\begin{aligned} d_{1,1} &= -\frac{1}{\sqrt{2}}x_1 + \frac{1}{\sqrt{2}}x_2, \\ d_{2,1} &= -\frac{1}{2}(x_1 + x_2) + \frac{1}{2}(x_3 + x_4), \\ d_{3,1} &= -\frac{1}{\sqrt{8}}(x_1 + x_2 + x_3 + x_4) + \frac{1}{\sqrt{8}}(x_5 + x_6 + x_7 + x_8), \\ s_{3,1} &= \frac{1}{\sqrt{8}}(x_1 + x_2 + x_3 + x_4 + x_5 + x_6 + x_7 + x_8). \end{aligned} \tag{12.27}$$

To illustrate the IDWT, we note that using (12.24) and (12.26), the value, x_1, is given by

$$x_1 = -\frac{1}{\sqrt{2}}d_{1,1} - \frac{1}{2}d_{2,1} - \frac{1}{\sqrt{8}}d_{3,1} + \frac{1}{\sqrt{8}}s_{3,1}, \tag{12.28}$$

which can be verified using (12.27) (see Problem 12.5).

12.3.5 Pyramid Algorithm

While the wavelet coefficients can be computed from (12.22), the pyramid algorithm introduced in the wavelet setting by Mallat (1989) provides an efficient method for calculating them. The pyramid algorithm is outlined in the following for a data set of length $n = 2^{J_0}$:

Step 1: Beginning with the data values $x_1, x_1, \ldots, x_n$ (which it is convenient to denote as $s_{0,1}, s_{0,1}, \ldots, s_{0,n}$), "down-sample" these into two sets, each containing $n/2$ values, to obtain

$n/2$ coefficients: $s_{1,k}, k = 1, \ldots, n/2$
$n/2$ coefficients: $d_{1,k}, k = 1, \ldots, n/2$

Step 2: Retain $d_{1,k}, k = 1, \ldots, n/2$, and down-sample $s_{1,k}, k = 1, \ldots, n/2$ to obtain

$n/4$ coefficients: $s_{2,k}, k = 1, \ldots, n/4$
$n/4$ coefficients: $d_{2k}, k = 1, \ldots, n/4$

Step 3: Retain $d_{2,k}, k = 1, \ldots, n/4$, and down-sample $s_{2,k}, k = 1, \ldots, n/4$ to obtain

$n/8$ coefficients: $s_{3,k}, k = 1, \ldots, n/8$
$n/8$ coefficients: $d_{3,k}, k = 1, \ldots, n/8$
$\vdots$

Step J_0: Retain $d_{J_0-1,k}, k = 1, 2$ and down-sample $s_{J_0-1,k}, k = 1, 2$ to obtain

1 coefficient $s_{J_0,1}$
1 coefficient $d_{J_0,1}$

Consider the case $J_0 = 3$. Then the eight retained coefficients are

$$d_{1,k}, \quad k = 1, \ldots, 4$$
$$d_{2,k}, \quad k = 1, 2$$
$$d_{3,1} \quad \text{and} \quad s_{3,1},$$

which are the components of the vector $\mathbf{w}$ in (12.22).

A useful way to view the pyramid algorithm is that the coefficients $s_{m,k}$ and $d_{m,k}, k = 1, \ldots, n/2^m$ obtained at the mth step of the pyramid algorithm are the outputs from a low-pass and high-pass filter, respectively, applied to $s_{m-1,k}, k = 1, \ldots, n/2^{m-1}$. For a thorough discussion of the pyramid algorithm see Percival and Walden (2000).

12.3.6 Multiresolution Analysis

The pyramid algorithm not only is an efficient algorithm for performing the matrix multiplication in (12.22), but it also provides insight into a useful wavelet decomposition of the data, referred to as *multiresolution analysis* (MRA). Using the partitioning in (12.23) we can write (12.24) as

$$\mathbf{x} = \begin{pmatrix} \mathbf{M}_1' & \mathbf{M}_2' & \cdots & \mathbf{M}_{J_0}' & \mathbf{V}_{J_0}' \end{pmatrix} \begin{pmatrix} \mathbf{w}_1 \\ \mathbf{w}_2 \\ \vdots \\ \mathbf{w}_{J_0} \\ \mathbf{v}_{J_0} \end{pmatrix}$$

$$= \mathbf{M}_1'\mathbf{w}_1 + \mathbf{M}_2'\mathbf{w}_2 + \cdots + \mathbf{M}_{J_0}'\mathbf{w}_{J_0} + \mathbf{V}_{J_0}'\mathbf{v}_{J_0}. \tag{12.29}$$

So, for example, if $n = 2^3$, the first element, x_1, of $\mathbf{x}$ is given by

$$x_1 = \sum_{k=1}^{4} d_{1,k}\tilde{\psi}_{1,k}(1) + \sum_{k=1}^{2} d_{2,k}\tilde{\psi}_{2,k}(1) + d_{3,1}\tilde{\psi}_{3,1}(1) + s_{3,1}\tilde{\varphi}_{3,1}(1), \tag{12.30}$$

and in general for data of length $n = 2^{J_0}$, the data value x_t is given by

$$x_t = \sum_{k=1}^{n/2} d_{1,k}\tilde{\psi}_{1,k}(t) + \sum_{k=1}^{n/2^2} d_{2,k}\tilde{\psi}_{2,k}(t) + \cdots + d_{J_0,1}\tilde{\psi}_{J_0,1}(t) + s_{J_0,1}\tilde{\varphi}_{J_0,1}(t)$$

$$= \sum_{j=1}^{J_0} D_j(t) + S_{J_0}(t), \tag{12.31}$$

for $t = 1, \ldots, n$ where $D_j(t) = \sum_{k=1}^{2/n^j} d_{j,k}\tilde{\psi}_{j,k}(t)$ and in this case $S_{J_0}(t) = s_{J_0,1}\tilde{\varphi}_{J_0,1}(t)$.

Note that (12.31) is the DWT version of (12.21). Each of the discrete wavelets, $\tilde{\psi}_{j,k}$ in (12.31) is associated with a "short wave" behavior. For example, in the discussion following (12.14) we noted that the Haar wavelet, $\psi_{j,k}$, is associated with behavior in the neighborhood of $t = 2^j(k + 0.5)$ with a duration of about 2^j. Thus, larger values of $|d_{j,k}|$ are associated with time periods in which the data display some sort of "cyclic-type" behavior, called *detail* behavior, with a duration of about 2^j. The $d_{j,k}$'s are referred to as *detail coefficients*. Small values of j are associated with finer detail, i.e., they describe more localized behavior, while the larger values of j are associated with coarser detail behavior (i.e., longer-span behavior).

The partitioning in (12.31) is referred to as a *multiresolution decomposition* (MRD). In the case $J_0 = 4$ (i.e., $n = 16$) considered in Example 12.3 which follows, this decomposition becomes

$$x_t = D_1(t) + D_2(t) + D_3(t) + D_4(t) + S_4(t),$$

where the coefficients needed for $D_j(t), j = 1, \ldots, 4$ are obtained in the jth step of the pyramid algorithm. If the pyramid algorithm is terminated at Step 3, then 16 coefficients $d_{1,k}$, $k = 1, \ldots, 8$; $d_{2,k}$, $k = 1, \ldots, 4$; $d_{3,k}, k = 1, 2$; and $s_{3,k}$, $k = 1, 2$ have been calculated. Letting $S_2(t) = D_3(t) + S_3(t)$, it follows that x_t can be written as $x_t = D_1(t) + D_2(t) + S_2(t)$, and the coefficients needed for the term $S_2(t)$ are $s_{2,k}$, $k = 1, 2$ which were obtained in the second step of the pyramid algorithm. In general, for $n = 2^{J_0}$ and where J is an integer such that $J \leq J_0$, the decomposition in (12.31) can be expressed as

$$x_t = D_1(t) + D_2(t) + \cdots + D_J(t) + S_J(t),$$

which can in turn be written as

$$x_t = D_1(t) + D_2(t) + \cdots + D_{J-1}(t) + S_{J-1}(t), \tag{12.32}$$

where $S_{J-1}(t) = D_J(t) + S_J(t)$. Letting $\mathbf{x}$, $\mathbf{D}_j$, and $\mathbf{S}_j$ denote the n-component vectors that have $x_t, D_j(t)$, and $S_j(t)$ as the tth components, respectively, then from (12.32) it follows that

$$\begin{aligned} \mathbf{x} &= \mathbf{D}_1 + \mathbf{D}_2 + \cdots + \mathbf{D}_J + \mathbf{S}_J \\ &= \mathbf{D}_1 + \mathbf{D}_2 + \cdots + \mathbf{D}_{J-1} + \mathbf{S}_{J-1}, \end{aligned} \tag{12.33}$$

where $\mathbf{S}_{J-1} = \mathbf{D}_J + \mathbf{S}_J$. Successively applying (12.33) we obtain

$$\begin{aligned} \mathbf{x} &= \mathbf{D}_1 + \mathbf{D}_2 + \mathbf{S}_2 \\ &= \mathbf{D}_1 + \mathbf{S}_1 \\ &= \mathbf{S}_0, \end{aligned} \tag{12.34}$$

where $\mathbf{S}_0$ is simply the data. The vectors $\mathbf{S}_{J_0}, \mathbf{S}_{J_0-1}, \ldots, \mathbf{S}_1, \mathbf{S}_0 = \mathbf{x}$ represent increasingly accurate (and less smooth) approximations to $\mathbf{x}$.

Example 12.3 DWT of a 16 Point Time Series Realization

The top row of Figure 12.7 shows a time series realization of length $n = 16$. The $16 = 2^4$ (i.e., $J_0 = 4$) coefficients obtained via the pyramid algorithm and the Haar wavelet are plotted on the remaining rows. The row labeled `d1` plots $d_{1,k}$, $k = 1, \ldots, 8$ (equally spaced). That is, the eight values of $d_{1,k}$ are plotted at 1.5, 3.5, 5.5, ..., 13.5, 15.5. The row `d2` plots $d_{2,k}, k = 1, \ldots, 4$, row `d3` plots $d_{3,k}$, $k = 1, 2$, while rows `d4` and `s4` plot only the single values $d_{4,1}$ and $s_{4,1}$, respectively. Note that although the time series contains 16 data values, only about 10 of the wavelet and scaling coefficients are sufficiently different from zero to show clearly on the plot. One of the features of the wavelet decomposition is that it often produces a *sparse* representation of the original data. Using (12.24) the original data can be reconstructed from the wavelet coefficients,

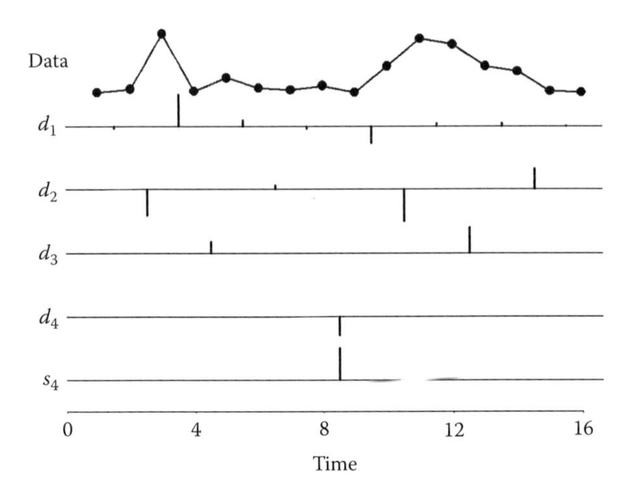

FIGURE 12.7
DWT for 16-point time series realization plotted on the top row.

implying that the information in the data (16 data values) is conveyed using about 10 nonzero wavelet and scaling coefficients. The extent of sparseness is often much more striking than in this simple example. Recalling that the coefficient, $d_{j,k}$, is a measure of the detail of duration about 2^j that occurs in the vicinity of its location on the time axis, it follows that larger values of $|d_{j,k}|$ indicate locations of "activity" in the data. Detail coefficients for large values of j (i.e., $d_{3,1}$, $d_{3,2}$, $d_{4,1}$, and $s_{4,1}$) measure the strength of smoother (longer duration) behavior. In Figure 12.7 it can be seen that the jump at $t = 3$ is described by detail coefficients $d_{1,2}$ and $d_{2,1}$ while the wider bump around $t = 12$ produces sizable detail coefficients at levels 1, 2, and 3. The DWT in Figure 12.7 was obtained using the `S-Plus Wavelets` module.

Example 12.4 DWT, MRD, and MRA of Bumps Data

The top graph of the left panel of Figure 12.8 shows the bumps signal generated using `S-Plus Wavelets` module with $n = 256$. This is a smooth signal with a sequence of "bumps" at various locations and of various sizes (see Bruce and Gao, 1996). The DWT, based on the Symlet(8) wavelet, shows that the role of the detail and smooth coefficients is primarily to describe the 11 bumps of various sizes throughout the signal. In the right panel we show a noisy bumps signal, also of length $n = 256$, generated using `S-Plus` where the signal-to-noise ratio is 2. The signal on the top right is equal to the signal on the top left along with additive noise. The smoother (coarser) behavior of the noisy bumps signal associated with rows `d3`, `d4`, and `s4` is very similar to that on the left-hand side for the original `Bumps` signal. That is, the high-frequency noise is primarily described using the `d1` and `d2` rows. Note that the bumps signal at the top left in Figure 12.8 is relatively smooth and that this signal, which has 256 points, can be reproduced with only about

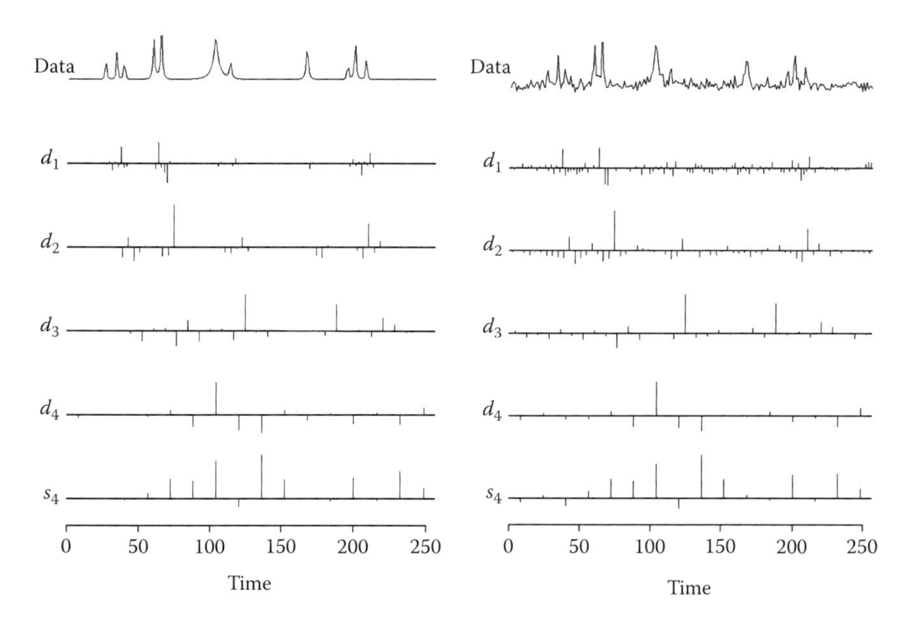

FIGURE 12.8
Left panel: Bumps signal (top left) and DWT for levels 1, ..., 4. Right panel: noisy-bumps signal (top left) and DWT for levels 1, ..., 4.

80 wavelet and scaling coefficients. This again points out the "sparseness" of the wavelet transform, and it seems reasonable that eliminating the zero and near zero wavelet coefficients before reconstruction of the `Bumps` signal would have negligible effect. It is seen that more wavelet coefficients are needed to reproduce the `Noisy Bumps` signal. Inspection of the wavelet coefficients in the right panel of Figure 12.8 shows that the noise in the time series is primarily described by relatively small wavelet coefficients in levels 1 and 2.

In the left panel of Figure 12.9 we show, from bottom to top: $\mathbf{S}_4, \mathbf{D}_4, \mathbf{D}_3, \mathbf{D}_2, \mathbf{D}_1$ (defined in (12.31) and (12.33) associated with the DWT for the bumps data shown in the top left of Figure 12.8. From (12.33) it follows that the bumps signal at the top is the sum of the very crude "smooth approximation" $\mathbf{S}_4$ and the increasingly finer details $\mathbf{D}_4, \mathbf{D}_3, \mathbf{D}_2$, and $\mathbf{D}_1$. The plot in the right panel of Figure 12.9 shows the smooth approximations $\mathbf{S}_4, \mathbf{S}_3, \mathbf{S}_2, \mathbf{S}_1$, and $\mathbf{S}_0 = \mathbf{x}$ using the relationship $\mathbf{S}_{J-1} = \mathbf{D}_J + \mathbf{S}_J$. That is, $\mathbf{D}_4 + \mathbf{S}_4 = \mathbf{S}_3$, $\mathbf{S}_3 + \mathbf{D}_3 = \mathbf{S}_4 + \mathbf{D}_4 + \mathbf{D}_3 = \mathbf{S}_2$, etc. In the figure it can be seen that the smooth approximations, $\mathbf{S}_j$ are quite poor for $j > 1$. The plots on the left panel of Figure 12.9 illustrate an MRD while the right panel shows an MRA. See Ogden (1997).

The MRA in the right panel of Figure 12.9 provides a visualization of the fact that the wavelet approximations, $\mathbf{S}_j$, can be viewed as output of low-pass filters applied to the original data. The quality of the associated filtering can depend on the wavelet selected. The MRD and MRA in Figure 12.9 are based on the relatively smooth Symlet(8) wavelet. In Figure 12.10 (left panel) we

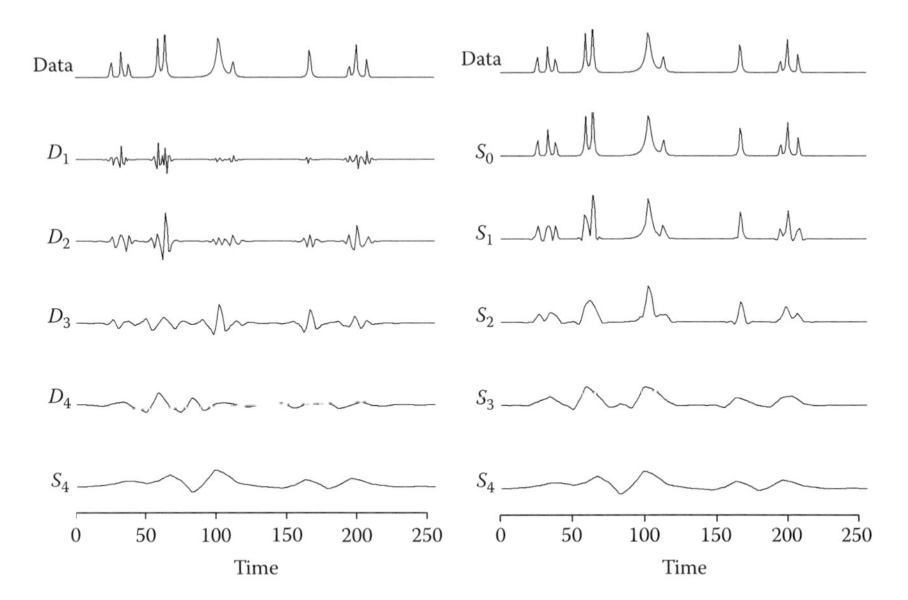

FIGURE 12.9
Left panel: MRD for bumps signal, for levels $1, \ldots, 4$. Right panel: MRA for bumps signal for levels $1, \ldots, 4$.

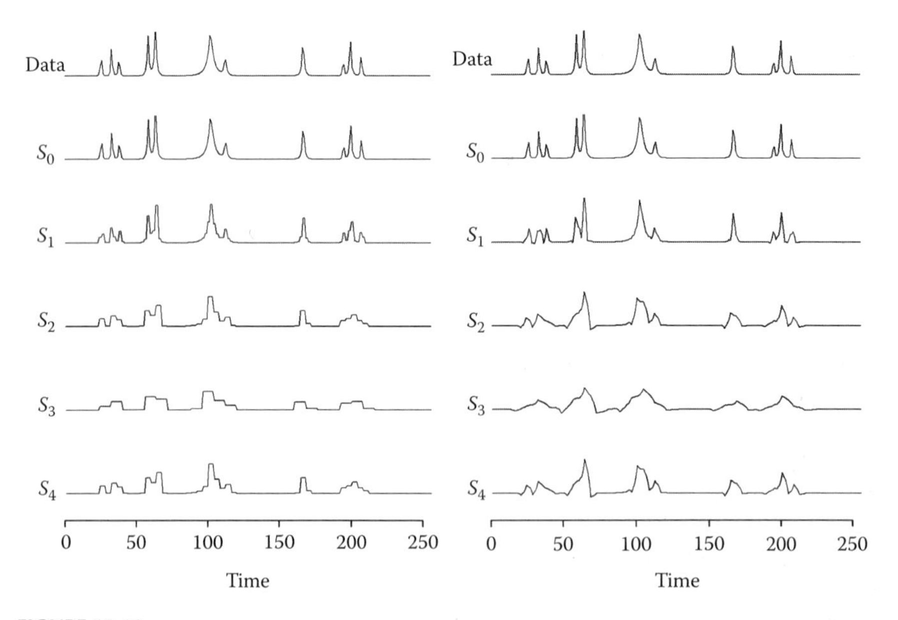

FIGURE 12.10
Left panel: MRA for bumps signal using Haar wavelet. Right panel: MRA for bumps signal using the D(4) wavelet.

show the MRA of the bumps signal using the Haar wavelet, and Figure 12.10 (right panel) shows the MRA based on the Daublet(4) wavelet. While each of these wavelets eventually reproduce the data (in $\mathbf{S}_0$) the preceding "filtered versions" show the behavior of the underlying wavelets. For example, the Haar version is "blocky" and the Daublet(4) version shows vestiges of the shark fin. Consequently, for the purpose of approximation, a relatively smooth wavelet (such as the Symlet(8)) is best applied to a signal that is relatively smooth (as is the bumps signal). However, Bruce and Gao (1996) show a "blocks" signal for which the Haar wavelet is preferable.

Computational notes: The computations leading to Figures 12.6 through 12.9 were performed using the `Waveslim` package in `R` that is available on CRAN. They can also be obtained using the `Wavelets` module of `S-Plus`.

12.3.7 Wavelet Shrinkage

Smoothing techniques have been widely used for extracting the underlying signal from a data set containing a signal buried in noise. The low-pass filters discussed in Section 2.4 are often used for this purpose. Nonparametric statistical methods such as regression and smoothing splines, kernel smoothers, and generalized additive models have become quite popular for signal extraction. See, for example, Wahba (1990), Hastie and Tibshirani (1990), Eubank (1999), Green and Silverman (1993), and Hart (1997).

To illustrate wavelet-based smoothing, refer back to the discussion in Example 12.4 concerning the bumps DWT representations of the bumps and noisy bumps signal. Starting with the noisy bumps signal (bumps signal plus additive noise) shown in the upper right panel of Figure 12.8, it seems reasonable that the true bumps signal would be well approximated by eliminating the nearly zero wavelet coefficients in the DWT in the right panel of Figure 12.8 and then reconstructing the signal (i.e., calculating $\mathbf{S}_0$). This basic idea is the foundation of smoothing techniques based on the DWT. See, for example, Donoho (1993) and Donoho and Johnstone (1994, 1995). The basic model in the wavelet domain is

$$d_{j,k} = \theta_{j,k} + z_{j,k}, \tag{12.35}$$

where

$d_{j,k}$ is the detail coefficient calculated from the data
$\theta_{j,k}$ is the unknown "true" detail coefficient associated with the underlying signal
$z_{j,k}$ is the zero-mean white noise with variance σ_z^2

The key idea for extracting "smooth" signals from noisy data such as the upper right frame of Figure 12.8 is that most of the $\theta_{j,k}$ will be zero or approximately zero. Although many of the wavelet coefficients, d_{jk}, of the noisy signal are nonzero, only a relatively few of these nonzero $d_{j,k}$'s are actually attributable to the true bumps signal. The goal is to reconstruct the "signal" based on estimates $\hat{\theta}_{j,k}$. This is typically accomplished by selecting a threshold and "zeroing-out" coefficients less than the threshold. Two thresholding techniques suggested by Donoho and Johnstone (1994) based on a positive threshold λ are as follows:

Hard threshold ("Keep or Kill")

$$\begin{aligned} \hat{\theta}_{j,k} &= 0, && \text{if } |d_{j,k}| \le \lambda \\ &= d_{j,k}, && \text{otherwise.} \end{aligned}$$

Soft threshold

$$\begin{aligned} \hat{\theta}_{j,k} &= 0, && \text{if } |d_{j,k}| \le \lambda \\ &= d_{j,k} - \lambda, && \text{if } d_{j,k} > \lambda \\ &= d_{j,k} + \lambda, && \text{if } d_{j,k} < -\lambda. \end{aligned}$$

The underlying signal is then reconstructed using the $\hat{\theta}_{j,k}$ as the detail coefficients.

A variety of techniques have been used for obtaining the threshold λ. One popular method is to use a single (universal) threshold, λ_u, across all levels of j given by

$$\lambda_u = \hat{\sigma}_z \sqrt{2 \log n}, \tag{12.36}$$

where $\hat{\sigma}_z^2$ is an estimate of the noise variance σ_z^2. Note that level $\mathbf{d}_1$ coefficients of the DWT of the noisy signal in Figure 12.8 are almost all due to noise. This is typical for smooth signals, and consequently σ_z^2 is often estimated using only the $\mathbf{d}_1$ level coefficients. Donoho and Johnstone (1994) set $\hat{\sigma}_z$ to be the median absolute deviation (MAD) of the $\mathbf{d}_1$ level coefficients. (The MAD of a list of numbers is the median of the absolute differences between the values in the list and the median of the list.) Alternatively, separate thresholds can be calculated for each row of detail coefficients, and this may be a preferable approach in the case of correlated noise.

Example 12.5 Smoothing the Noisy Bumps Signal

The left panel of Figure 12.11 shows the noisy bumps data and the associated DWT that was previously shown in the right panel of Figure 12.8. A universal hard threshold is applied to the first three levels of wavelet coefficients using $\mathbf{d}_1$ level coefficients to estimate σ_z. The DWT after thresholding is shown in the right panel of Figure 12.11 where it can be seen that only 11 of the $128 + 64 + 32 = 224$ coefficients in these three rows are retained. Note that

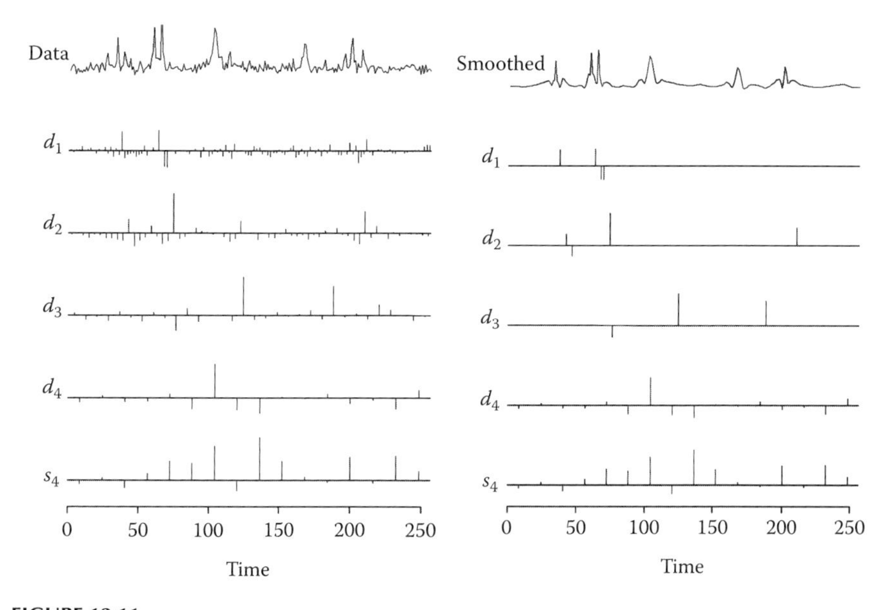

FIGURE 12.11
Left panel: Noisy bumps signal and DWT from right panel of Figure 12.8. Right panel: Wavelet coefficients after universal hard threshold along with reconstructed smoothed noisy-bumps signal (top right).

the DWT after thresholding is quite similar to the DWT of the underlying smooth bumps signal shown in the right panel of Figure 12.8. The smoothed signal is reconstructed from the thresholded DWT using (12.24) and is shown at the top right of Figure 12.11.

Computational notes: The wavelet smoothing shown in Figure 12.11 was done using the `Wavethresh` `R` package which is available on CRAN.

12.3.8 Scalogram: Time-Scale Plot

Considering the data vector $\mathbf{x}$ as a time series realization, the wavelet representation at time t assigns "large" values of $d_{j,k}$ when the corresponding $\psi_{j,k}$ has "frequency" behavior similar to that of the realization in the vicinity of time t. The discrete *scalogram* or *time-scale plot* is a time-frequency plot similar to those shown in Figures 12.3 and 12.4 in which the horizontal axis represents time, t, and the vertical axis represents the "frequency" measure (1/scale). In the discussion following Example 12.1 it was noted that we use 1/scale as a measure of "frequency." In the DWT the $d_{j,k}$'s are coefficients that correspond to $\psi_{j,k}$ for $j = 1, \ldots, J$ where $J \leq J_0$ and $n = 2^{J_0}$. Since the support of the Haar mother wavelet is of length 1 and the support of $\psi_{j,k}$ is of width 2^j, then it follows that in the DWT setting, 1/scale reaches an upper limit of 1 instead of 0.5 as is the case for standard frequency measure. In general it is useful to associate a 1/scale value of u with a frequency of $u/2$.

The third dimension in the scalogram is a gray-scale where the value of $d_{j,k}^2$ is plotted on the rectangle

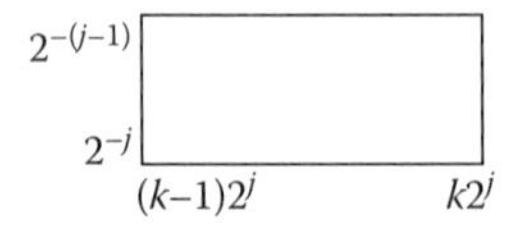

and $s_{J,k}^2$ is plotted on the rectangle.

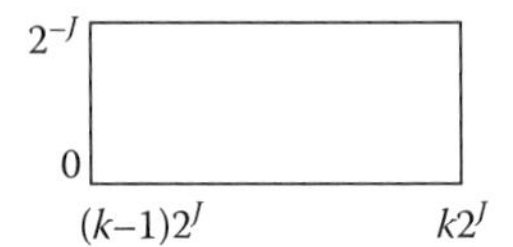

Figure 12.12a shows a template for the time-scale plot for the case $n = 16$ and $J = J_0 = 4$. The value $s_{4,1}^2$ is plotted on the rectangle of height

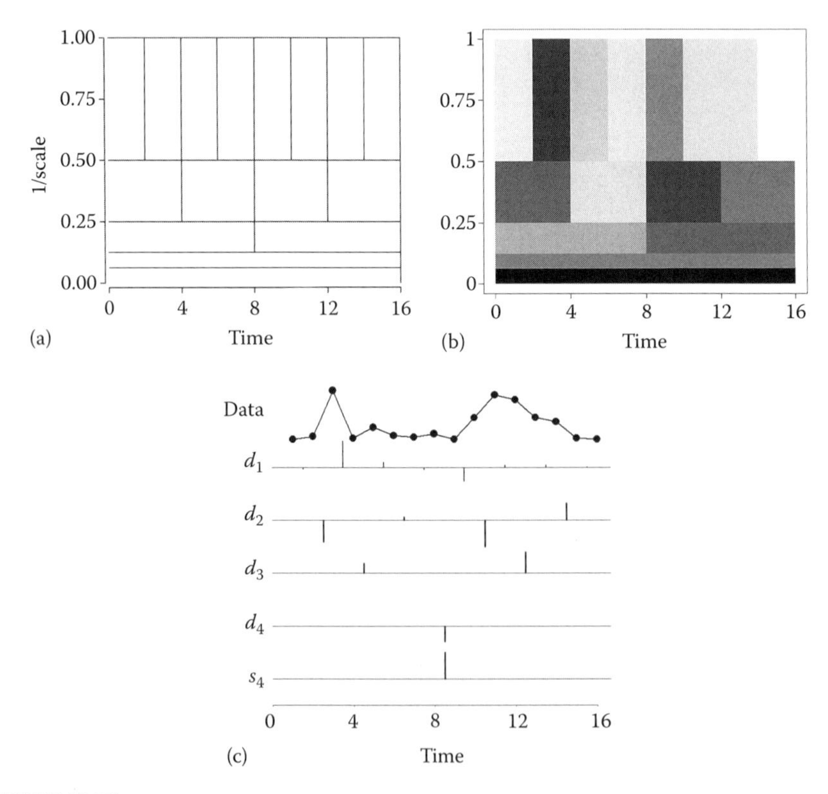

FIGURE 12.12
(a) Template for time-scale plot for $n = 16$, (b) time scale plot for the series of length $n = 16$ whose DWT is shown in (c) and previously in Figure 12.7.

$2^{-4} = 0.0625$ and width 16 at the bottom of the time-scale plot. This rectangle is shown at the bottom of Figure 12.12a. Sitting on top of this rectangle is a rectangle of the same dimensions in which the value $d_{4,1}^2$ is plotted. The grayscale of this rectangle (darker indicates larger $d_{4,1}^2$) is a measure of the detail in the data of duration about $2^4 = 16$ time points. The value of $d_{3,1}^2$ is plotted on a rectangle whose base extends from $(k-1)2^3 = 0$ to $k2^3 = 8$ and whose height extends from $2^{-3} = 1/8 = 0.125$ to $2^{-2} = 1/4 = 0.25$. The shading of the rectangle associated with $d_{3,1}^2$ can be usefully interpreted as a measure of strength of detail behavior of extent between 4 (1/scale = 0.25) and 8 (1/scale = 0.125) that occurs in the time range $t \le 8$. The rectangle associated with $d_{3,1}^2$ is paired with a similar rectangle that extends from 8 to 16 of the same height over which $d_{3,2}^2$ is plotted. The rectangle associated with $d_{3,2}^2$ measures the detail behavior of duration between 4 and 8 time units that occurs in the time range $t \ge 8$. The remaining rectangles are constructed analogously.

Note that the rectangles associated with $d_{1,k}^2, k = 1, \ldots, 8$ are shaped much differently from those for $d_{3,1}^2$ and $d_{3,2}^2$. For example, $d_{1,1}^2$ measures the detail behavior of duration between 2 and 1 time units (i.e., 1/scale between 0.5 and 1) that occurs for $t \le 2$. Thus, the shading of this bar describes a broad "frequency" range but a narrow time range (i.e., $t \le 2$). We say that the bars for $d_{1,k}^2$ have good time localization but poor frequency localization. Correspondingly, the bars for $d_{3,k}^2$ have good frequency localization (i.e., 1/scale covers a relatively short range, 0.125 to 0.25) while the time localization is relatively poor, i.e., it covers the time frame $t \le 8$, i.e., half of the entire time span.

Figure 12.12b shows the time scale plot associated with the DWT for the data set of length $n = 16$ shown in Figure 12.7 and for convenience again in Figure 12.12c. The sharp peak at $t = 3$ is "high-frequency" behavior measured by the relatively large (negative) value of $d_{1,2}$ and correspondingly by the tall darker gray block toward the top left of Figure 12.12b. Similarly, the second (wider) peak at about $t = 12$ corresponds to the three visible values of $d_{2,3}$, $d_{2,4}$, and $d_{3,2}$ and the corresponding three bars of varying shades of gray toward the middle right of Figure 12.12b. Finally, the signal is generally relatively smooth, and this is reflected by darker bars toward the bottom of Figure 12.12b.

Example 12.6 Time-Scale Plot for Bumps and Noisy Bumps Series

While Figure 12.12 and the related discussion provide information concerning time-scale plots and how they relate to the DWT, the 16 point signal is fairly uninteresting. In the current example, we show the time-scale plots for the 256 point bumps and noisy bumps time series whose DWTs with $J = 4$ using the Symlet(8) wavelet are shown in Figure 12.8. In this example we use Daublet(6) wavelet with $J = 5$ in the calculation of the time-scale plots and the DWTs shown in Figure 12.13. The correspondence between the larger

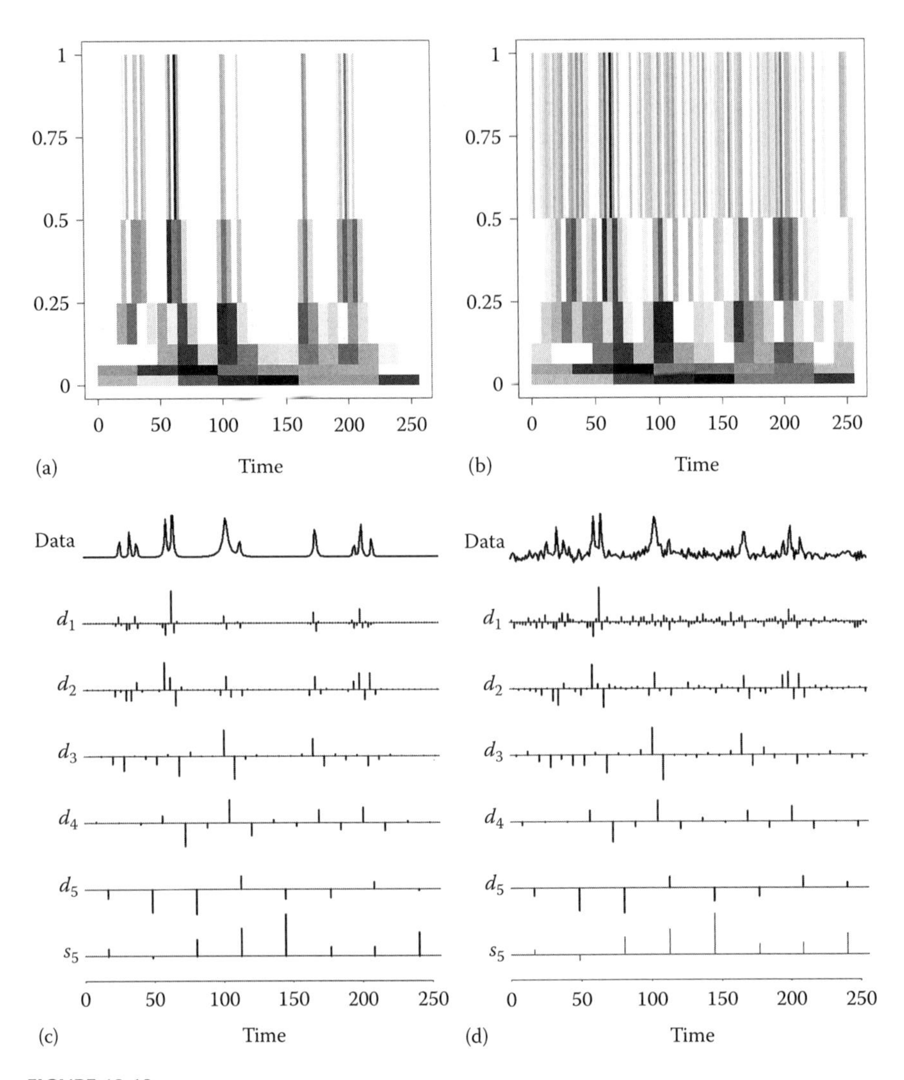

FIGURE 12.13
Time-scale plots for 256 point (a) bumps signal and (b) noisy bumps signal whose DWTs with $J = 5$ are shown in (c) and (d), respectively.

values of $|d_{j,k}|$ associated with time values near which bumps occur is visible in the DWT and the time-scale plots. The major difference between the two time-scale plots is the existence in the noisy bumps time-scale plot of light shading associated with 1/scale 0.5 to 1 that occurs consistently across time. This is related to the high-frequency noise that is present throughout the data set.

Example 12.7 DWT and Time-Scale plot for the Chirp Data Shown in Figure 12.3a

We return to the 256 point linear chirp signal shown in Figure 12.2a. The DWT using a Daublet(6) wavelet with $J = 5$ is shown in Figure 12.14a where it can be seen that the lower frequency portion of the signal, which occurs toward the beginning of the realization, is represented with wavelet coefficients primarily at levels 4 and 5. The higher frequency (finer detail) behavior toward the end of the realization, however, is represented in the wavelet domain by coefficients at levels 2 and 3. The time-scale plot in Figure 12.14b shows that as t increases, the associated wavelet behavior is associated with finer detail. Comparing Figure 12.14b with the Gabor and Wigner–Ville plots in Figure 12.4 illustrates the fact that the Gabor and Wigner–Ville plots do a much better job of displaying the linearly increasing frequency behavior in the data especially for higher frequencies. The resolution in the time-scale plots can be improved using wavelet packets which will be discussed in Section 12.3.9.

Computational notes: We found the `S-Plus Wavelets` module to be a very convenient way to plot the time-scale plots. Some of the DWTs shown here were computed using the `S-Plus Wavelets` module and some using `Waveslim`.

12.3.9 Wavelet Packets

Figure 12.14b shows that the linearly increasing frequency behavior known to exist in the linear chirp realization, which is well exhibited in Figure 12.4, is

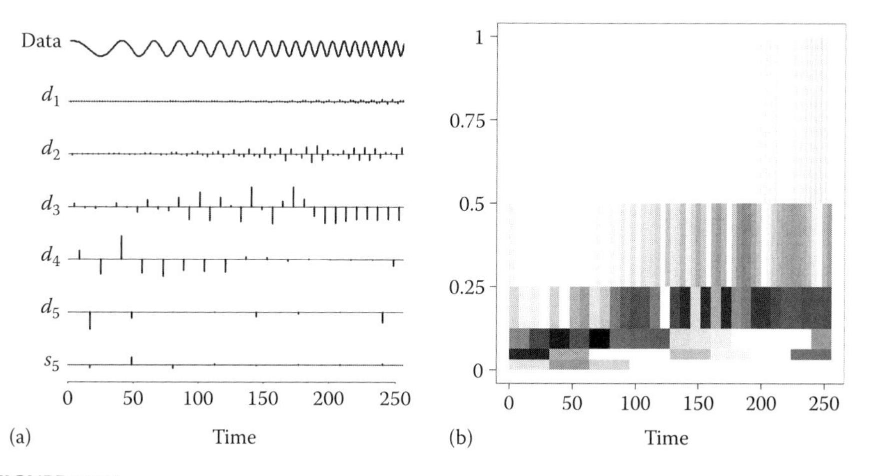

FIGURE 12.14
(a) DWT and (b) time-scale plot for 256 point linear chirps signal.

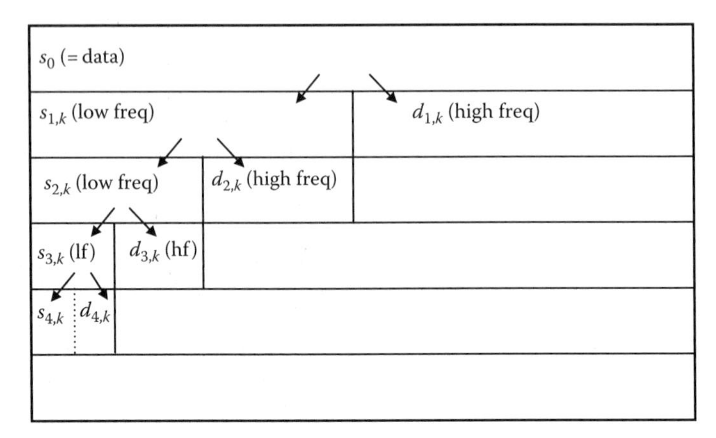

FIGURE 12.15
Diagram of the pyramid algorithm for $J = 4$.

hardly visible in the time-scale plot. The problem relates back to the poor frequency localization associated with higher levels of 1/scale (frequency). A schematic diagram of the DWT calculated using the pyramid algorithm is shown in Figure 12.15. The pyramid algorithm begins by down-sampling the data to obtain two sets containing $n/2$ coefficients each. As mentioned previously, the $s_{1,k}$ and $d_{1,k}$ coefficients are the results of applying a low pass (1/scale = 0 to 0.5) and high pass (1/scale = 0.5 to 1) filter, respectively to the data ($s_{0,k}$). To convey the 1/scale information, the horizontal axis of Figure 12.15 measures 1/scale, so that in this case the $n/2$ coefficients, $s_{j,k}$, are plotted between 0 and 0.5 while the $n/2$ detail coefficients, $d_{1,k}$, are plotted between 0.5 and 1. The next step of the pyramid algorithm is to retain the $d_{1,k}$ coefficients and down-sample the $s_{1,k}$ coefficients, i.e., to treat the $s_{1,k}$ coefficients as "data" of length $n/2$, and then to apply a low pass (1/scale = 0 to 0.25) and a high pass (1/scale = 0.25 to 0.5) filter to compute the $s_{2,k}$ and $d_{2,k}$ coefficients ($n/4$ each). This procedure is continued, at each step retaining the $d_{j,k}$ (those associated with the higher frequency behavior) and then further down-sampling the $s_{j,k}$ coefficients. At step J both the $s_{J,k}$ and $d_{J,k}$ coefficients are retained. Figure 12.15 represents a DWT with $J = 4$. Note that at the kth step the $n/2^k$ $d_{j,k}$ coefficients are positioned within their corresponding frequency band. The arrows from one level to the next indicate coefficients that are further down-sampled.

The important point to note is that the down-sampling involved in the DWT is done in such a way that the lowest frequency behavior is further localized at each step. However, the $d_{1,k}$ coefficients measure the highest frequency (finer detail) behavior over the 1/scale range from 0.5 to 1 (i.e., half of the entire "frequency" band). Similarly, the $d_{2,k}$ coefficients also cover a broad 1/scale range from 0.25 to 0.5, etc. Close examination of the linear chirp realization in Figure 12.2a shows that at time 100 the period lengths are about 14 while at time 256 they are about 6. That is, the frequency (as measured by 1/period) changes from 0.07 to 0.167 over this time range, so that the 1/scale measure changes from about 0.14 to 0.33. Thus, although frequencies (i.e., detail lengths) change dramatically between times 100 and 256, the primary 1/scale information falls in only two ranges: level 3 (1/scale = 0.125 to 0.25) and level 2 (1/scale = 0.25 to 0.5). Examination of Figure 12.14b shows that these two levels of the time scale plot tend to be the darkest for $t \geq 100$ with a little indication that the level 2 shading is darker for larger values of n. However, the detail concerning the linearly increasing frequencies is lost because of the poor high frequency localization.

As previously mentioned, the DWT of a realization of length n produces a total of n wavelet coefficients. The key feature of wavelet packet analysis is that down-sampling *both* the low frequency and the high frequency coefficients occurs at each step. This procedure is illustrated in Figure 12.16 which is in the same format as the DWT diagram in Figure 12.15. Figure 12.16 shows that in the wavelet packet transform both the $s_{j,k}$ and the $d_{j,k}$ coefficients are further down-sampled. Thus, while the down-sampling procedure for the DWT produces a total of n coefficients, each row of the wavelet packet transform contains n elements. Consequently, a wavelet packet transform for the case $J = 4$, as illustrated in Figure 12.16, has $4n$ coefficients.

In Figure 12.17a we show a diagram of the DWT that illustrates the "finished product" of the down-sampling illustrated in Figure 12.15. The dark bars in Figure 12.17 indicate which coefficients are retained and included in the vector **w** in (12.22), and the span of each bar represents the 1/scale range measured by the coefficients. Notice that the union of all the dark bars "covers" the entire horizontal axis and that there is no overlap in

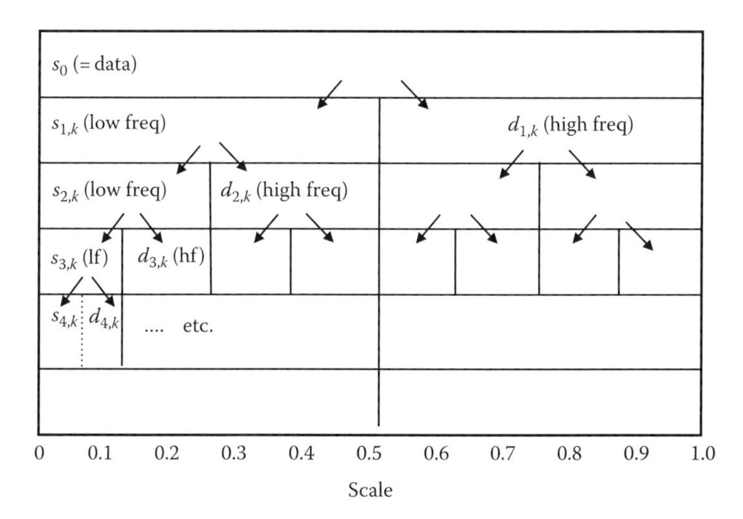

FIGURE 12.16
Diagram of the wavelet packet transform.

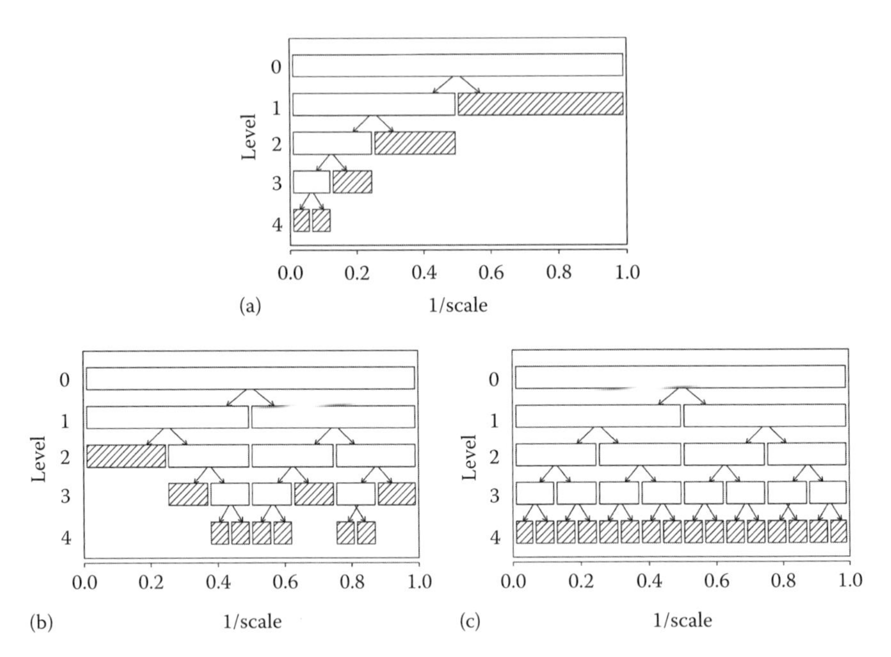

FIGURE 12.17
Schematic diagrams of three wavelet packet transforms. (a) DWT. (b) A wavelet packet transform. (c) Level 4 wavelet packet transform.

the horizontal direction. In the DWT, the $d_{1,k}$ coefficients describe detail behavior in the range 0.5 to 1, the $d_{2,k}$ coefficients describe detail behavior in the range 0.25 to 0.5, and this is repeated until, in the $J = 4$ case illustrated in Figure 12.17, the $d_{4,k}$ coefficients cover 1/16 to 1/8 and the $s_{4,k}$ coefficients describe the smoothest behavior, i.e., 1/scale between 0 and 1/16.

Wavelet packet analysis involves finding other "coverings" such as the one shown in Figure 12.17b. In this covering the shaded bar in level 2 indicates that the $n/4$ coefficients, $s_{2,k}$, are retained (and not further down-sampled) and that these coefficients describe the 1/scale behavior between 0 and 0.25. The entire 1/scale range from 0 to 1 is covered, and there is no overlap, i.e., no 1/scale range is described by more than one set of coefficients. Figure 12.17c shows a "level 4" wavelet packet transform in which all n of the level 4 coefficients are retained. Coifman and Wickerhauser (1992) developed a *best basis algorithm* for selecting the wavelet packet that best matches the characteristics of the signal. Note that the DWT is a special case of a wavelet packet transform.

In describing the coefficients retained in the wavelet packet transform illustrated in Figure 12.17b, we were able to describe the coefficients in the level 2 bar as the collection of $s_{2,k}$ coefficients. However, at this point terminology has not been introduced in this book to describe, for example, the level 3 coefficients obtained by low-pass filtering the $d_{2,k}$ coefficients. Wavelet packet functions are a more flexible family of functions that include as special cases the wavelet and scaling functions. Very briefly, the wavelet packet functions, $w_{j,k}^{m}(x)$, are defined by setting $w^0(x) = \varphi(x)$, $w^1(x) = \psi(x)$, and defining $w_{j,k}^{m}(x) = 2^{-j/2}w^m(2^{-j}x - k)$, where $m = 0, 1, \ldots$ is called the *modulation* parameter. Wavelet packet functions with $m > 1$ are defined recursively. See, for example, Ogden (1997).

Example 12.8 Wavelet Packet Transform for the Linear Chirp Signal

In Example 12.7 we found the DWT and associated time-scale plot for a linear chirp signal, and we noted that the time-scale plot, shown in Figure 12.14b does a poor job of displaying the linearly increasing frequency behavior. Figure 12.18a shows the time-scale plot based on the level 4 wavelet packet transform diagrammed in Figure 12.17c, while in Figure 12.18b we show the corresponding time-scale plot using the level 5 wavelet packet transform. In both cases, the linearly increasing frequency behavior is more clearly shown than it is in Figure 12.14b, especially for $t \geq 100$. Comparison of Figure 12.17a and b illustrates the trade-off between time localization and frequency localization in a time scale plot. The gain in frequency localization (i.e., reduced height of the bars in Figure 12.18b) by going to the level 5 transform resulted in wider bars in the time domain as a result of the Heisenburg uncertainty principle (see Ogden, 1997). In this case it appears that the level 4 transform produces the better time scale plot due to the fact that the loss of time resolution in the level 5 time scale plot causes difficulty in describing the rapidly changing frequency behavior for large values of t.

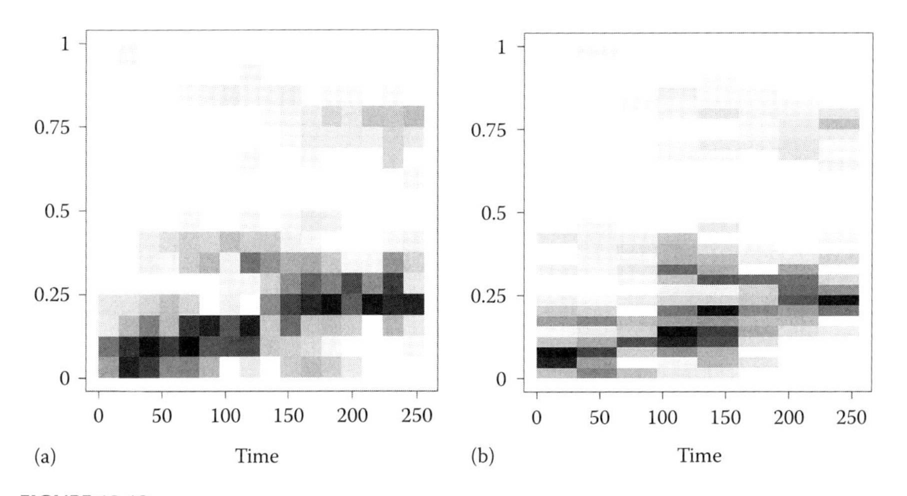

FIGURE 12.18
Time-scale plots for the 256 linear chirp signal (a) based on level 4 wavelet packet transform and (b) based on level 5 wavelet packet transform.

Example 12.9 Wavelet Packet Transform for an AR(2) Realization

As a final example illustrating the use of wavelet packets we consider the realization of length $n = 256$ in Figure 12.19a which is from the AR(2) model

$$(1 + 0.995B^2)X_t = a_t. \tag{12.37}$$

Examination of the factor table shows that this is a stationary model with system frequency $f = 0.25$. The time-scale plot should essentially follow a horizontal line at $1/\text{scale} = 0.5$ (i.e., $f = 0.25$). The time-scale plot based on the DWT is shown in Figure 12.19b where it is seen that the poor frequency

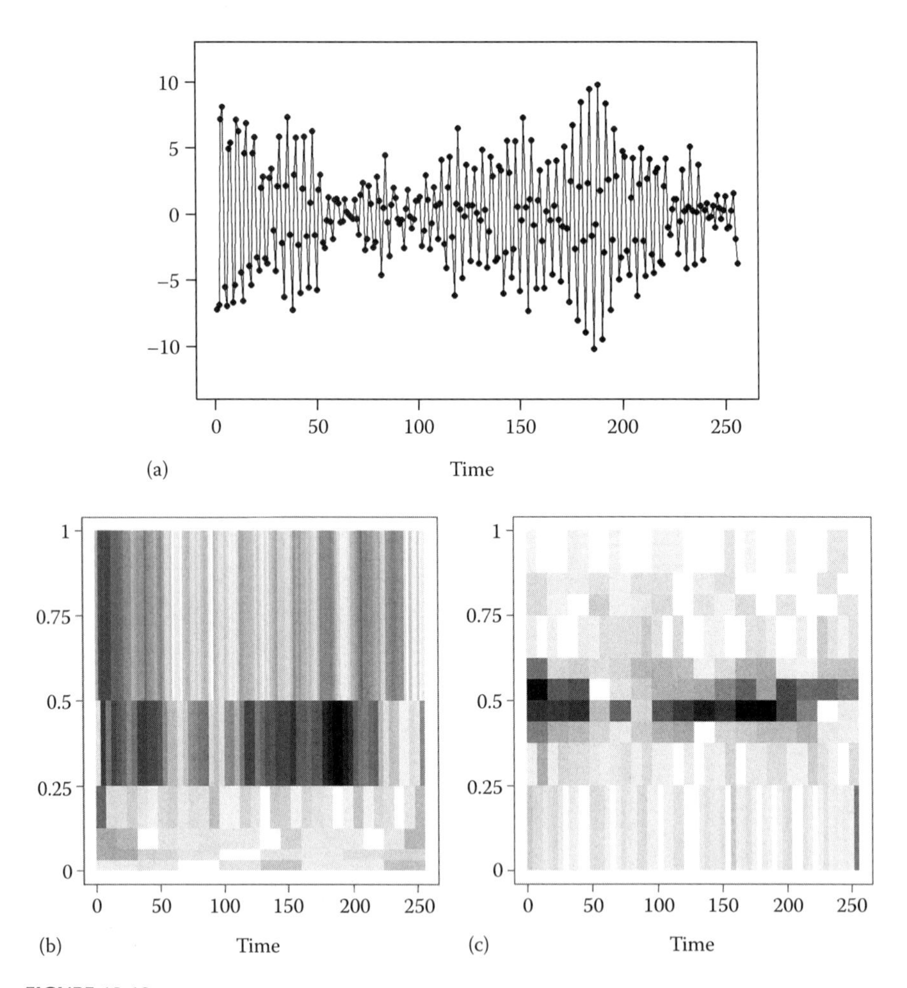

FIGURE 12.19
(a) Realization of length $n = 256$ from the AR(2) model in (12.37), (b) time-scale plot based on the DWT, and (c) time-scale plot based on the best basis algorithm for selecting the wavelet packet basis.

localization for higher frequencies causes the time-scale plot to lose precision. Figure 12.19c shows the time-scale plot based on the best basis algorithm, and the horizontal behavior at about $1/\text{scale} = 0.5$ is shown much more clearly. A schematic diagram of the wavelet packet transform obtained from the best basis algorithm is shown in Figure 12.17b. Note that the wavelet packet coefficients "hone in" on $1/\text{scale} = 0.5$ (i.e., system frequency $f = 0.25 \times 2$) and provide best frequency localization in this region of the 1/scale axis. This is similar to the manner in which DWT coefficients focus in on lower frequencies.

Computational notes: The time-scale plots based on wavelet packets that are shown here were plotted using the `S-Plus Wavelets` module.

12.3.10 Two-Dimensional Wavelets

Although this is a book on time series analysis, we close this section on wavelet analysis with a brief discussion of two-dimensional wavelets. This section briefly covers wavelet analysis of an $m \times n$ image, $F_{m,n}$, where, as in the univariate case, we assume that both dimensions are dyadic (i.e., powers of 2). Examples will be based on a square image, i.e., of size $2^{J_0} \times 2^{J_0}$ for some integer J_0. The two-dimensional DWT maps the $m \times n$ image $F_{m,n}$ to an $m \times n$ image of wavelet coefficients. The univariate DWT is based on two types of wavelets: father wavelets that measure smooth behavior and mother wavelets that measure more detailed behavior. In the case of the two-dimensional DWT, there are four types of wavelets:

$$\Phi(x,y) = \varphi_h(x) \times \varphi_v(y) \quad \text{(captures smooth features of the image)}$$

$$\Psi^v(x,y) = \psi_h(x) \times \varphi_v(y) \quad \text{(captures vertical features of the image)}$$

$$\Psi^h(x,y) = \varphi_h(x) \times \psi_v(y) \quad \text{(captures horizontal features of the image)}$$

$$\Psi^d(x,y) = \psi_h(x) \times \psi_v(y) \quad \text{(captures diagonal features of the image)}$$

The subscripts h and v indicate horizontal and vertical dimensions, respectively, and for example, $\varphi_v(y)$ is the father wavelet obtained by separately considering the one-dimensional Y variable. The two-dimensional counterpart to (12.17) is given by

$$\begin{aligned} g(x,y) = &\sum_{k,m} s_{J,k,m}\Phi_{J,k,m}(x,y) + \sum_{j=1}^{J}\sum_{k,m} d^v_{j,k,m}\Psi^v_{j,k,m}(x,y) \\ &+ \sum_{j=1}^{J}\sum_{k,m} d^h_{j,k,m}\Psi^h_{j,k,m}(x,y) + \sum_{j=1}^{J}\sum_{k,m} d^d_{j,k,m}\Psi^d_{j,k,m}(x,y), \end{aligned} \qquad (12.38)$$

where the first term in (12.38) gives a "smooth" approximation to $g(x,y)$ to level J (i.e., involves the higher values of j). The remaining summations

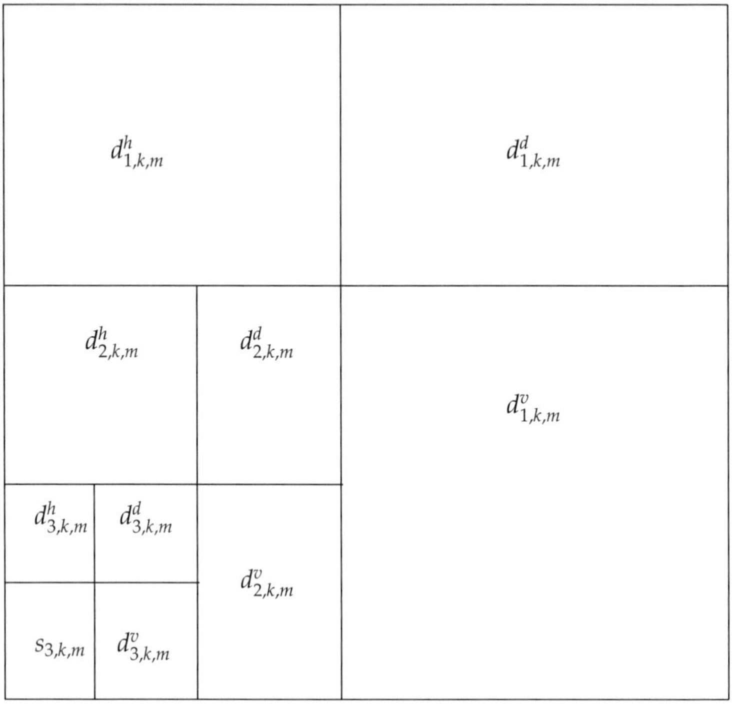

$d^h_{1,k,m}$ $d^d_{1,k,m}$

$d^h_{2,k,m}$ $d^d_{2,k,m}$ $d^v_{1,k,m}$

$d^h_{3,k,m}$ $d^d_{3,k,m}$ $d^v_{2,k,m}$

$s_{3,k,m}$ $d^v_{3,k,m}$

FIGURE 12.20
Schematic drawing of a two-dimensional DWT with $J = 3$.

measure the vertical, horizontal, and diagonal detail, respectively. The $s_{j,k,m}$, $d^v_{j,k,m}$, $d^h_{j,k,m}$, and $d^d_{j,k,m}$ are called the smooth coefficients, vertical detail, horizontal detail, and diagonal detail coefficients, respectively. The two-dimensional DWT consists of the coefficients $s_{J,k,m}$, along with $d^v_{j,k,m}$, $d^h_{j,k,m}$, and $d^d_{j,k,m}$, $j = 1, \ldots, J$. Each of these coefficients is represented as a grayscale (or color) on a pixel in a $2^{J_0} \times 2^{J_0}$ image. In particular, for $J = 3$ the two-dimensional pyramid algorithm produces $n/2 \times n/2$ coefficients of each level 1 detail type $\left(d^v_{1,k,m}, d^h_{1,k,m}, \text{ and } d^d_{1,k,m}\right)$. Also, there are $n/2^2 \times n/2^2$ level 2 detail coefficients and $n/2^3 \times n/2^3$ level 3 coefficients of types h, v, d. Additionally, $n/2^3 \times n/2^3$ level 3 smooth coefficients are retained. Figure 12.20 shows a schematic diagram of a two-dimensional DWT with $J = 3$ where the original image contains $2^{J_0} \times 2^{J_0}$ pixels where $J_0 \geq 3$.

Example 12.10 XBOX and Barbara Images and Associated DWT

(a) *XBOX image*: A classic two-dimensional image used for illustrating two-dimensional wavelets is the XBOX image shown in Figure 12.21a. The XBOX image (not to be confused with the gaming

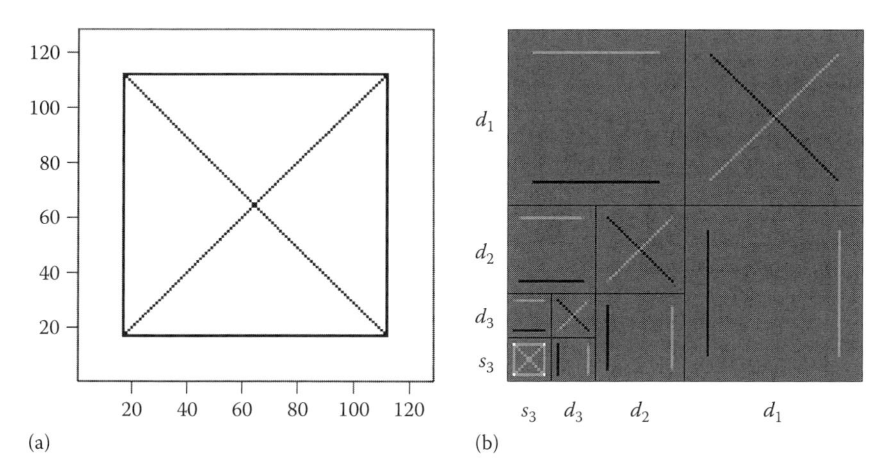

FIGURE 12.21
(a) XBOX Image ($n = 128$) and (b) DWT of XBOX Image.

console) in Figure 12.21a consists of 128×128 pixels, and the image (as the name suggests) is that of an X enclosed in a box. Figure 12.21b shows the DWT using the Haar wavelet and $J = 3$. The small square at the bottom left of Figure 12.21b is analogous to the bottom row (i.e., the smooth coefficients) of the DWT. In this case it is a 16×16 image ($128/2^3 = 16$). The XBOX image is ideally suited for showing the sense in which the $d^v_{j,k,m}$, $d^h_{j,k,m}$, and $d^d_{j,k,m}$ coefficients measure vertical, horizontal, and diagonal behavior, respectively. We reiterate the fact that the image in Figure 12.21b is a DWT, i.e., it is a two-dimensional plot of the coefficients analogous to the DWT plots shown, for example, in Figures 12.7 and 12.8. That is, although the bottom left box has the appearance of an approximating function (such as S4 in Figure 12.9) it is simply a plot of the coefficients, which in this case has the appearance of the actual figure. Note that in Figure 12.21b the gray background represents zero while darker colors represent positive and lighter colors represent negative coefficients.

(b) *Barbara image*: Figure 12.22a shows the ''Barbara'' image (256×256) which is another classic image that is more complex than the XBOX image. The gray background in the DWT again denotes zero, so it can be seen that the DWT is a sparse representation of the Barbara image in that the number of nonzero pixels required to reproduce the image is much smaller than the original 256×256.

Computational notes: The two-dimensional DWT is shown in Figures 12.20 and 12.21 were plotted using the `R` package `Waveslim` available on CRAN.

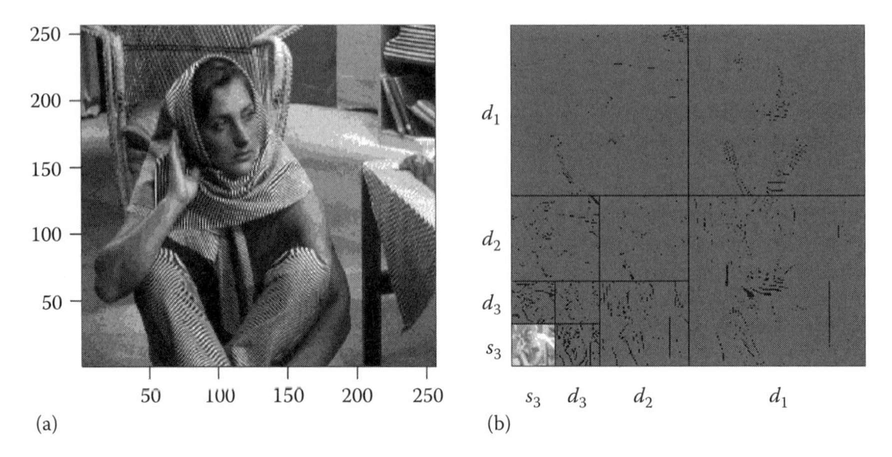

FIGURE 12.22
(a) Barbara Image ($n = 256$) and (b) DWT of Barbara Image.

12.5 Concluding Remarks on Wavelets

This section on wavelets has been a brief introduction of wavelet methods with a focus on time series applications. Wavelets have proven to be very useful in a variety of applications. We have discussed the use of wavelets for time-frequency analysis. The ability of the wavelet transformation to provide a sparse representation, i.e., to characterize a time series or image based on only a small fraction of the number of data values in the original data set, proves to be very useful in applications. For example, these sparse representations lead to useful data compression methods such as the common JPEG image which is a compressed image format based on the discrete cosine transform. See Strang and Nguyen (1997) for further discussion of data compression techniques and applications. Consider also the application of image analysis in which statistical decisions need to be made concerning the detection of certain features of an image. For example, in the analysis of a statistic map from an fMRI experiment, investigators wish to detect regions of brain activity. A typical two-dimensional brain slice from an functional magnetic resonance imaging (FMRI) image has about 3000 volume elements (or voxels), and preprocessing steps produce a statistic map of t-statistics for testing the null hypothesis of no activity at each voxel. The goal is to determine which voxels are activated (show brain activity). Methods such as the false discovery rate (FDR) have been developed to provide a reasonable method of dealing with the multiple testing problem, although in this case the large number of hypotheses to be tested is still a concern. Shen et al. (2002) proposed an expected false discovery rate (EFDR) procedure in which the image is transformed slice-by-slice into the wavelet domain using the two-dimensional DWT, the unnecessary wavelet coefficients are removed, FDR is performed on the remaining (much smaller) set of wavelet coefficients to determine which are nonzero, and then the image is recreated using the reduced set of coefficients. Modifications to the procedure were made by Shen et al. (2002), Sendur and Selesnick (2002), Sendur et al. (2005), and Pavlicova et al. (2008). O'Hair (2010) examined the use of three-dimensional wavelets in an EFDR-type setting by analyzing the entire brain volume rather than taking a slice-by-slice approach.

Another useful feature of the wavelet transformation is that it tends to decorrelate data that are correlated. That is, the wavelet coefficients, especially at the finer levels, tend to be uncorrelated even when the original data are correlated. For example, as mentioned in Chapter 11, Wornell (1993, 1996) has used the wavelet transform to decorrelate data with $1/f$ noise. See also, Percival and Walden (2000), Section 9.1 of Flandrin (1992), and Tewfik and Kim (1992). A key application is in the analysis of two-dimensional or three-dimensional fMRI brain imaging data. The statistic maps discussed in the preceding paragraph often take the form of a signal buried in $1/f$ or other positively correlated noise. Thus, in some situations it is useful to transform to the wavelet domain to avoid the problem of dealing with the correlation structure in the original data. Bullmore et al. (2001) and Breakspear et al. (2004) make use of the decorrelating property to resample uncorrelated units of a given data set by (a) transforming to the wavelet domain, (b) resampling the wavelet coefficients, and (c) creating the bootstrap data set by performing the inverse wavelet transform on the new set of wavelet coefficients. The goal of such a resampling procedure is to estimate the null distribution of the original data while preserving its correlation structure. See also Tang et al. (2008) concerning problems encountered with this "Wavestrap" procedure. Wiechecki (1998) showed that the discrete wavelet packet transform can be used to decorrelate data from $1/(f - f_0)$ processes which are processes whose spectrum behaves like $1/|(f - f_0)|^{\alpha}$, i.e., have infinite peaks at some frequency f_0, not necessarily zero.

Finally, we mention that the wavelet transform has been shown to be useful for estimating long memory parameters. Specifically, McCoy and Walden (1996) and Jensen (1999a,b) have used it to estimate the parameter d in the FARMA model given in (11.6). Additionally, Whitcher (2004) and Wiechecki (1998) use the discrete wavelet packet transform to estimate the parameter λ in the GARMA model in (11.16).

12.A Appendix: Mathematical Preliminaries for This Chapter

The following mathematical review lists some definitions and results that are critical to understanding the material in Section 12.2 on wavelets.

12.A.1 Classical Fourier Series

Consider the basis set $S_1 = \{1, \cos kx, \sin kx, k = 1, 2, \ldots\}$. In 1807 Joseph Fourier made the astounding assertion that, subject to fairly loose restrictions, an arbitrary function $f(x)$ could be written as a linear combination of sines and cosines, i.e., a linear combination of basis elements from S_1 (see James, 2011). Later, it was shown that, a function $f(x)$ satisfying the Dirichlet conditions on a closed interval (i.e., $f(x)$ has only a finite number of discontinuities and only a finite number of maxima and minima on the interval), can be expanded in a Fourier series which converges pointwise to $f(x)$ at points of continuity and to the mean of the left- and right-hand limits at points of discontinuity. Letting the domain of $f(x)$ be $[-\pi, \pi]$, we can write

$$f(x) \sim \frac{a_0}{2} + \sum_{k=1}^{\infty} (a_k \cos kx + b_k \sin kx), \tag{12.A.1}$$

where "$\sim$" denotes the aforementioned type of pointwise convergence under conditions such as the Dirichlet conditions. The coefficients in (12.A.1) are given by

$$\begin{aligned} a_k &= \frac{1}{\pi} \int_{-\pi}^{\pi} f(t) \cos kt\, dt, \quad k = 0, 1, \ldots \\ b_k &= \frac{1}{\pi} \int_{-\pi}^{\pi} f(t) \sin kt\, dt, \quad k = 1, \ldots. \end{aligned} \tag{12.A.2}$$

It is often convenient to express functions, f, defined on $[-\pi, \pi]$ as linear combinations of the basis functions in $S_2 = \{e^{ikx}, k = 0, \pm 1, \pm 2, \ldots,\}$ where in this case the Fourier expansion becomes

$$f(x) \sim \sum_{k=-\infty}^{\infty} \alpha_k e^{ikx}, \tag{12.A.3}$$

where

$$\alpha_k = \frac{1}{2\pi} \int_{-\pi}^{\pi} f(t) e^{-ikt} dt. \tag{12.A.4}$$

The coefficient α_k in (12.A.4) is the Fourier transform of f, and conversely, (12.A.3) expresses the fact that f is the inverse Fourier transform of the sequence α_k.

12.A.2 Vector Spaces

Let $\mathbf{x} = (x_1, x_2, \ldots, x_n)'$ and $\mathbf{y} = (y_1, y_2, \ldots, y_n)'$ be two n-component real vectors.

(a) The *inner product* (or *dot product*) of $\mathbf{x}$ and $\mathbf{y}$ is given by $\sum_{k=1}^{n} x_k y_k$.

(b) The *norm* of $\mathbf{x}$, denoted by $||\mathbf{x}||$, is given by $\left(\sum_{k=1}^{n} x_k^2\right)^{1/2}$.

(c) Vectors $\mathbf{x}$ and $\mathbf{y}$ are *orthogonal* if their inner product is zero.

(d) Vectors $\mathbf{x}$ and $\mathbf{y}$ are *orthonormal* if they are orthogonal and have norm one.

12.A.3 Function Spaces of Real-Valued Functions f:$[-\pi, \pi] \to \mathbb{R}$

(a) The *inner product* of functions f and g is given by $\int_{-\pi}^{\pi} f(x)g(x)dx$.

(b) The *norm* of f, denoted by $|| f ||$, is given by $\int_{-\pi}^{\pi} |f(x)|^2 dx$.

(c) The real-valued function f is said to be in $L^2[-\pi, \pi]$ if $\int_{-\pi}^{\pi} |f(x)|^2 dx < \infty$.

(d) Functions f and g are *orthogonal* if their inner product is zero.

(e) Functions f and g are *orthonormal* if they are orthogonal and have norm one.

12.A.4 Systems of Functions

(a) Let $S = \{f_1, f_2, \ldots\}$ where f_m is orthogonal to f_j when $m \neq j$. Then S is said to be an *orthogonal system*.

(b) An orthogonal system S is said to be *orthonormal* if the norm of each function in S is one.

(c) The functions $S_1 = \{1, \cos kx, \sin kx; k = 1, 2, \ldots,\}$, (or equivalently $S_2 = \{e^{ikx}, k = 0, \pm 1, \pm 2, \ldots\}$) are orthogonal systems of functions.

(d) Fourier series results show that the collections S_1 (and S_2) are basis collections for a wide class of functions in $L^2[-\pi, \pi]$.

12.A.5 Formal Derivation of Wavelet Coefficient

The following is a formal derivation of the fact that

$$g(x) = \sum_{j=-\infty}^{\infty} \sum_{k=-\infty}^{\infty} d_{j,k} \psi_{j,k}(x),$$

then this implies the coefficients d_{jk} are given by

$$d_{j,k} = \frac{1}{|| \psi ||^2} \int_{-\infty}^{\infty} g(t) \psi_{j,k}(t) dt.$$

Formal Derivation: From the equality regarding $g(x)$ given in (12.15), it follows that can

$$g(t)\psi_{\ell,m}(x) = \sum_{j=-\infty}^{\infty} \sum_{k=-\infty}^{\infty} d_{j,k}\psi_{\ell,m}(x)\psi_{j,k}(x),$$

and

$$\int_{-\infty}^{\infty} g(t)\psi_{\ell,m}(t)dt = \sum_{j=-\infty}^{\infty} \sum_{k=-\infty}^{\infty} d_{j,k} \int_{-\infty}^{\infty} \psi_{\ell,m}(t)\psi_{j,k}(t)dt.$$

Note that,

$$\int_{-\infty}^{\infty} \psi_{\ell,m}(t)\psi_{j,k}(t)dt = \begin{cases} 0, & j \neq \ell \text{ or } m \neq k \\ \|\psi\|^2, & j = \ell, m = k \end{cases}.$$

Consequently,

$$d_{j,k} = \frac{1}{\|\psi\|^2} \int_{-\infty}^{\infty} g(t)\psi_{j,k}(t)dt.$$

Exercises

The following problems involve computing capabilities that are not currently available in `GW-WINKS`. The book's website contains a Wigner–Ville `R` package. Other resources that will be useful are

`R-Wave` available on CRAN
`Waveslim` available on CRAN
`S-Plus Wavelets Module`
MATLAB® `Wavelets Toolbox`

12.1 This problem involves analysis of the following data sets:
(A) `LYNX.XLS` (the annual number of lynx trapped in a Canadian district—analyze the logarithm of this series). See Campbell and Walker (1977).
(B) `HADLEY.XLS` (global temperature data)

(C) PROB12.1C.XLS (simulated data)

(a) Plot the data

(b) Plot the periodogram

(c) Plot the Parzen window spectral estimator with the default lag window

(d) Plot the autoregressive spectral estimator

(e) Plot the Gabor (cgt in R-Wave) and Wigner–Ville spectra

In each case, discuss the information in the plots with regard to whether or not there is an indication of stationarity, time-varying frequencies, etc. If the process appears to be stationary, describe the frequency behavior. If the process has time-varying frequencies, discuss the resulting effect on the plots.

12.2 Consider the following mother wavelet:

$$\psi(x) = x - 1, \quad 0 \leq x \leq 1,$$
$$= 0, \qquad \text{elsewhere}$$

Find and plot the following dilations and translations, $\psi_{j,k}(x)$:

(a) $j = 2, k = 0$

(b) $j = 2, k = 3$

(c) $j = -1, k = 3$

(d) $j = -2, k = 3$

12.3 This problem involves the analysis of the following three data sets:

(A) PROB12.2A.XLS

(B) PROB12.2B.XLS

(C) PROB12.1C.XLS

For these data sets:

(a) Plot the following related to the discrete wavelet transform:

i. DWT

ii. MRD

iii. MRA

(b) Plot the following spectral plots:

i. Time scale plot

ii. Level 4 wavelet packet transform

iii. Wavelet packet transform based on the best basis algorithm

For each of these plots, discuss the information that the plot conveys.

12.4 Consider the four data sets in Problem 12.3. For each of these data sets plot the Wigner–Ville spectrum and compare it with the three wavelet-based plots produced in Problem 12.3(b). Compare the plots with regard to the time-frequency spectral information.

12.5 Using (12.24) and the matrix **M** given in (12.26), show that

(a) $x_1 = -\frac{1}{\sqrt{2}}d_{1,1} - \frac{1}{2}d_{2,1} - \frac{1}{\sqrt{8}}d_{3,1} + \frac{1}{\sqrt{8}}s_{3,1},$

(b) Find a similar expression for x_2.

12.6 Analyze the following data sets using wavelet analysis and the Wigner–Ville plot. Describe any time-varying frequency behavior, abrupt changes, etc. that are identified by these analyses.

(A) `BAT.XLS`—echolocation signal of a big brown bat

(B) `DOPPLER.XLS`—a Doppler signal

(C) `PROB12.6C.XLS`—a simulated data set

13

G-Stationary Processes

13.1 Generalized-Stationary Processes

In Chapter 12, we considered window-based methods for analyzing nonstationary processes with time-varying frequencies (TVF). These include short-term Fourier transforms (Gabor), along with Wigner–Ville and wavelet representations. In this chapter, we introduce an altogether different approach for analyzing TVF data in which we extend the definition of stationarity to the class of *generalized stationary* (or *G-stationary*) processes. This approach allows us to transform the time index of many nonstationary processes to an index set upon which they are stationary.

The presentation here will be introductory in nature, and no effort will be made to cover this topic in detail. Our approach is to include sufficient information to give the reader a basic understanding of this methodology. Most of the developments in this area have only recently appeared in the literature, and these references are given here as a source of further details.

Definition 13.1 Let $\{X(t): t \in S\}$ be a stochastic process defined on $S \subset \mathbb{R}$, let $u = g(t)$ be a mapping onto a set $R_g \subset \mathbb{R}$, and let g^{-1} denote a specified inverse. Then $X(t)$ is a *G-stationary process* if

$$\begin{aligned}
&\text{(i) } E[X(t)] = \mu \\
&\text{(ii) } \mathrm{Var}[X(t)] = \sigma^2 < \infty \\
&\text{(iii) } E[(X(t) - \mu)(X(g^{-1}(g(t) + g(\tau))) - \mu)] = R_X(\tau).
\end{aligned} \tag{13.1}$$

The usefulness of this definition may not be immediately apparent. In order to better understand its implications, suppose g is a transformation of the time axis and let $u = g(t)$ with $t = g^{-1}(u)$, so that $X(t) = X(g^{-1}(u)) = Y(u)$. Letting $\xi = g(\tau)$, then

$$X(g^{-1}(g(t) + g(\tau))) = Y(g(t) + g(\tau)) = Y(u + \xi). \tag{13.2}$$

From (13.1) and (13.2), it follows that

$$\begin{aligned} R_X(\tau) &= E[(X(t) - \mu)(X(g^{-1}(g(t) + g(\tau))) - \mu)] \\ &= E[(Y(u) - \mu)(Y(u + \xi) - \mu)]. \end{aligned} \tag{13.3}$$

Consequently, Definition 13.1 gives the conditions on $X(t)$ and $g(t)$ for the *dual process* $Y(u)$, to be stationary in the usual sense (of Definition 1.5). That is, although $X(t)$ may not be stationary on the original index set (on t), it can be mapped onto a new index set on which it is stationary. We will refer to this new process as the stationary dual of $X(t)$. We use the notation $C_Y(\xi) = E[(Y(u) - \mu)(Y(u + \xi) - \mu)]$ to denote the autocovariance of the dual process, $Y(u)$, and $R_X(\tau)$ is called the G-stationary autocovariance. Clearly $C_Y(\xi) = R_X(\tau)$.

Note that stationarity is a special case of G-stationarity since a process is stationary if and only if the corresponding dual process based on the linear time transformation $u = g(t) = a + bt$ is stationary. Again, although in general, a G-stationary process is not stationary, the corresponding dual process is stationary. The basic idea is that in order to analyze a G-stationary process you should use the time transformation, $u = g(t)$, for which the dual process, $Y(u)$, is stationary. The dual process, $Y(u)$, can then be analyzed using the techniques discussed in the first 11 chapters.

13.1.1 General Strategy for Analyzing G-Stationary Processes

The following is an outline of the general procedure we will use for analyzing G-stationary processes.

1. Transform the time axis to obtain a stationary dual realization.
2. Analyze the transformed (dual) realization using methods for stationary time series. For example:
   a. Fit an AR(p) or an ARMA(p,q) model to the dual data using model identification techniques discussed in Chapter 8 and estimation methods covered in Chapter 7.
   b. Compute forecasts, spectral estimates, etc., as desired, for the problem at hand.
3. Transform back to original timescale.

In the next few sections, we will discuss the M-stationary, G(λ)-stationary, and linear chirp processes that are special cases of G-stationary processes. In Section 13.3, we give further discussion of the aforementioned procedure for the case of G(λ)-stationary processes.

13.2 M-Stationary Processes

The first type of G-stationary processes we will discuss is the multiplicative stationary (called M-stationary) process, which is based on a logarithmic time transformation.

13.2.1 Continuous M-Stationary Process

Continuous M-stationary processes were first introduced and studied by Gray and Zhang (1988). The M-stationary process is defined in Definition 13.2.

Definition 13.2 $X(t)$ is a continuous M-stationary process for $t \in (0, \infty)$ if for any $t \in (0, \infty)$ and $t\tau \in (0, \infty)$:

(i) $E[X(t)] = \mu$
(ii) $\text{Var}[X(t)] < \infty$
(iii) $E[(X(t) - \mu)(X(t\tau) - \mu)] = R_X(\tau)$,

where $R_X(\tau)$ is referred to as the M-autocovariance of $X(t)$.

Now let $g(t) = u = \ln t/\ln h$. In this case, $t = g^{-1}(u) = h^u$, and

$$
\begin{aligned}
g^{-1}(g(t) + g(\tau)) &= g^{-1}\left(\frac{\ln t}{\ln h} + \frac{\ln \tau}{\ln h}\right) \\
&= h^{\ln t/\ln h} h^{\ln \tau/\ln h} \\
&= t\tau.
\end{aligned}
$$

Letting $\xi = g(\tau) = \ln \tau/\ln h$, then $R_X(\tau)$ defined in part (iii) of Definition 13.1 becomes

$$
\begin{aligned}
R_X(\tau) &= E[(X(t) - \mu)(X(g^{-1}(g(t) + g(\tau))) - \mu)] \\
&= E[(X(t) - \mu)(X(t\tau) - \mu)] \\
&= E[(Y(u) - \mu)(Y(u + \xi) - \mu)] \\
&= C_Y(u)
\end{aligned}
$$

from (13.3). Thus, a G-stationary process with $u = g(t) = \ln t/\ln h$ is M-stationary. For the development and application of continuous M-stationary processes, see Gray and Zhang (1988).

Remark: We have taken $u = \ln t/\ln h$ for convenience. We could let $u = C \ln t$ where C is any real constant.

Example 13.1 An M-Stationary Cosine Process

A simple example of an M-stationary process is the cosine with argument $\ln(t)$ and random phase shift given by

$$X(t) = A\cos(2\pi\beta\ln(t) + \psi), \tag{13.4}$$

where $t \in (0, \infty)$, and $\psi \sim \text{Uniform}(0, 2\pi)$. In Figure 13.1a, we show a typical realization from (13.4) with the circled values illustrating the evaluation of the curve at evenly spaced time points. Figure 13.1b shows the same curve evaluated at the time points $t_k = ch^k$ for $c > 0$, and then Figure 13.1c displays the dual realization where the circled points represent the circled values in Figure 13.1b indexed on k. Note that the curves shown in Figure 13.1 go through about three cycles. The effect of evaluating the curve in Figure 13.1a at equally spaced points is to over-sample the last (longest) cycle and to take very few samples in the first (shortest) cycle. Evaluating the curve at the points h^k (Figure 13.1b) has the effect of sampling more rapidly for shorter cycles and slower for longer cycles, resulting in an equal number of observations in each of the cycles.

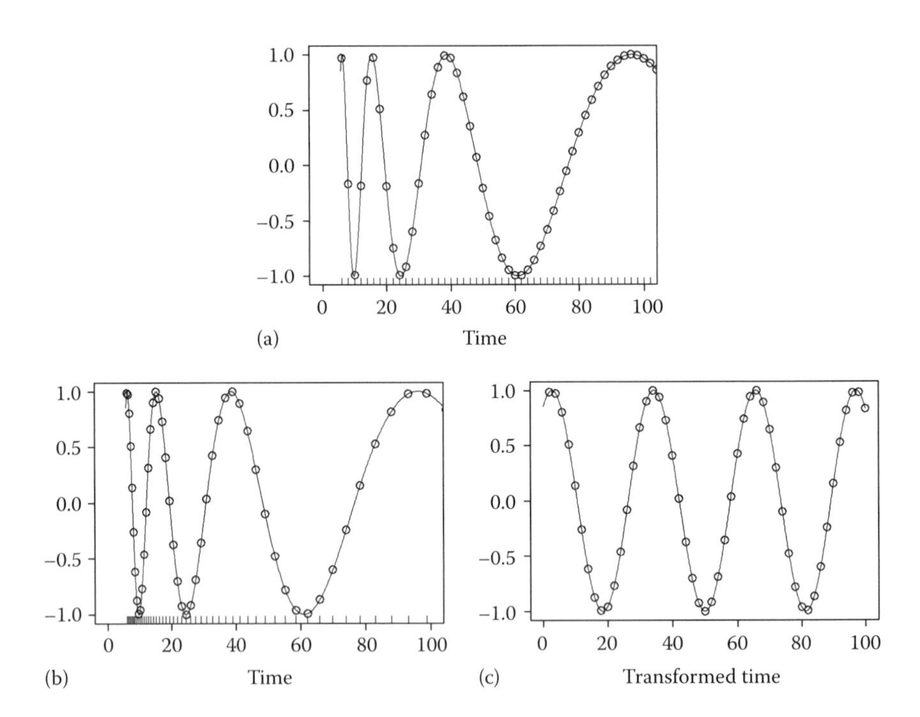

FIGURE 13.1
M-stationary cosine process in (13.4) with $\psi = 0$. (a) Showing 50 equally spaced points, (b) showing the values of the process at ch^k, and in (c) the circled values in (b) are plotted against the index k.

Finally, indexing on k (i.e., taking the logarithm) produces a curve with non-time-varying cycle lengths in Figure 13.1c.

13.2.2 Discrete M-Stationary Process

Example 13.1 motivates the following definition.

Definition 13.3 Let $h > 1$ and $S = \{t_k : t_k = h^k, k = 0, \pm 1, \pm 2, \ldots\}$ and suppose

(i) $E[X(h^k)] = \mu$
(ii) $\text{Var}[X(h^k)] = \sigma^2 < \infty$
(iii) $E[(X(h^k) - \mu)(X(h^k h^j) - \mu)] = R_X(h^j)$,

for all $t_k = h^k \in S$. Then $X(t_k)$ is a *discrete M-stationary process* with the discrete stationary dual process $Y_k = X(h^k)$.

Let $C_Y(j) = E[(Y_k - \mu)(Y_{k+j} - \mu)]$. This can be rewritten $C_Y(j) = E[(X(h^k) - \mu)\ (X(h^{k+j}) - \mu] = R_X(h^j)$. Since $C_Y(j) = C_Y(-j)$, it follows that $R_X(h^j) = R_X(h^{-j})$. If $t_k = h^k$ and $u = g(t) = \ln t/\ln h$, then $u_k = g(t_k) = g(h^k) = \ln h^k/\ln h = k$. Then, $t_k = g^{-1}(u_k) = h^{u_k} = h^k$. So

$$\begin{aligned} g^{-1}(g(t_k) + g(t_j)) &= h^{g(t_k)+g(t_j)} \\ &= h^{k+j} \\ &= h^k h^j. \end{aligned}$$

Thus, a discrete M-stationary process is a discrete G-stationary process with $u_k = g(t_k) = \ln t_k/\ln h$ and $S = \{t_k : t_k = h^k, k = 0, \pm 1, \ldots\}$. That is, $u_k = k$.

13.2.3 Discrete Euler(p) Model

Making use of Definition 13.3, Vijverberg and Gray (2003) and Gray et al. (2005) introduced and studied discrete Euler(p) processes. See also Vijverberg and Gray (2009).

Definition 13.4 Let $t_k \in S = \{t_k = h^k, k = 0, \pm 1, \pm 2, \ldots\}$, and suppose $X(t_k)$ is the M-stationary solution of

$$(X(h^k) - \mu) - \phi_1(X(h^{k-1}) - \mu) - \cdots - \phi_p(X(h^{k-p}) - \mu) = a(h^k), \tag{13.5}$$

where $a(h^k)$ is white noise. Then $X(t_k)$ is called a *discrete Euler(p) process*.

Whenever there is no loss in generality, we will, without comment, generally take $\mu = 0$. So, we write (13.5) as

$$X(h^k) - \phi_1 X(h^{k-1}) - \cdots - \phi_p X(h^{k-p}) = a(h^k). \quad (13.6)$$

If $Y_k = X(h^k)$ and $a(h^k) = Z_k$, then

$$Y_k - \phi_1 Y_{k-1} - \cdots - \phi_p Y_{k-p} = Z_k. \quad (13.7)$$

Consequently, if $X(t)$ is a pth-order discrete Euler process, then its dual process is a discrete AR(p). Choi et al. (2006) extend Definition 13.4 to the discrete Euler(p,q) process.

13.2.4 Time Transformation and Sampling

It should be clear from Definition 13.3 that an M-stationary process requires sample values at unequally spaced points. This is true in general for G-stationary processes. Thus, finding a time transformation, $u = g(t)$, is equivalent to determining a sampling scheme such that $X(t_k) = X(g^{-1}(u_k)) = Y_k$, where Y_k is stationary. That is, $X(t)$ should be sampled at the points $t = t_k = g^{-1}(u_k)$. In the case of M-stationary processes, $t_k = h^{u_k}$ and $u_k = k\Delta$. Without loss of generality, we take $\Delta = 1$, that is, $t_k = h^k$ and $u_k = k$.

In general, data will only be available at equally spaced points and may not necessarily be obtainable at the points t_k. This problem can be overcome by over-sampling or interpolation. By over-sampling, we mean the situation in which data are sampled at sufficiently small increments in time so that the values of the realization are available at the t_k's. If the data are available only at equally spaced points and the data cannot be sufficiently over-sampled, then interpolation and Kalman filter methods can be used to properly process the data. The S-Plus program GWS, which uses linear interpolation, is available through the ATSA website. See Gray et al. (2005), Jiang et al. (2006), and Wang et al. (2009).

Example 13.2 A Discrete Euler(2) Model

Figure 13.2 shows a realization of length $n = 200$ from the discrete Euler(2) process

$$X(h^k) - 1.6X(h^{k-1}) + 0.99X(h^{k-2}) = a(h^k), \quad (13.8)$$

with $h = 1.01$. Unlike stationary processes for which it can always be assumed without loss of generality that the sample origin is at time $t = 0$, for M-stationary processes the *origin offset* or simply the *offset* is a parameter to be estimated. In this example, we use offsets $h^j = 30$ and 60. For a discussion of the origin problem for G-stationary processes in general, see Gray et al. (2005)

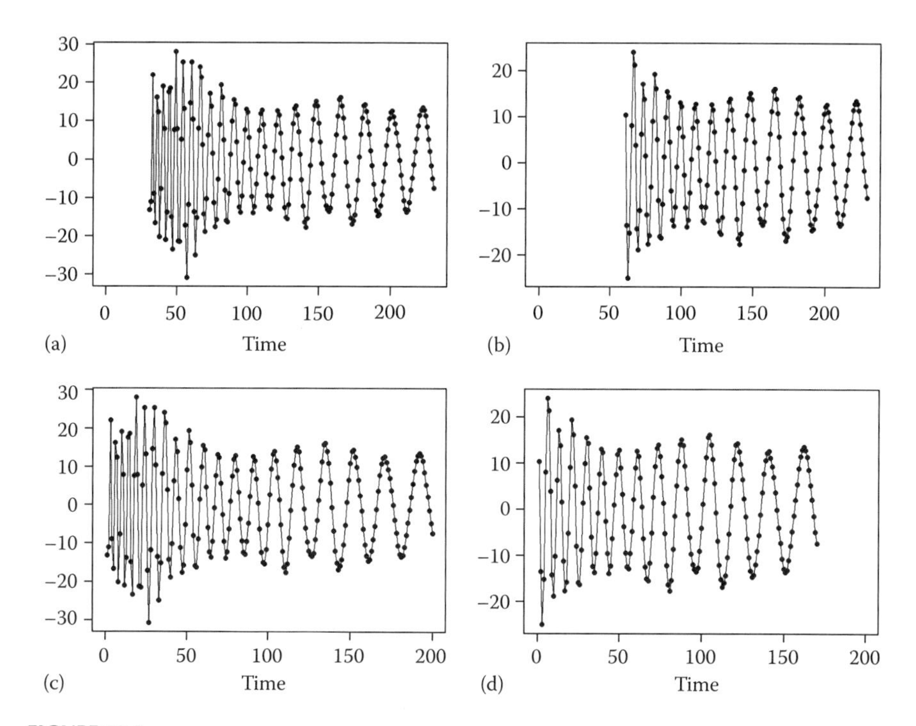

FIGURE 13.2
Realizations from the Euler(2) model in (13.8) (a) with offset = 30 and (b) with offset = 60; (c) and (d) are (a) and (b) shifted to begin at $t = 1$.

and Jiang et al. (2006). The GWS program estimates the offset using the techniques discussed in these references.

Each realization in Figure 13.2 shows decreasing frequency (increasing period length) through time, which is typical of Euler(p) processes. All data plots shown in Figure 13.2 are from the same realization with Figure 13.2a showing the data beginning at offset 30 (i.e., the first observation is at $t = 31$) while Figure 13.2b begins at offset 60. However, when the data are observed, the offset will not be known, and the realizations shown in Figure 13.2a and b would have the appearance of Figure 13.1c and d, respectively. These two realizations, although similar, are different with regard to the degree of frequency change illustrating the point that the offset is a parameter that must be estimated.

Example 13.3 Instantaneous Period and Frequency

Another concept is that of instantaneous period and instantaneous frequency. Loosely speaking, the instantaneous period at time t is the length of the next period. The instantaneous frequency is then defined as the reciprocal of the instantaneous period. In Figure 13.3, we show an enlarged version of the plot shown in Figure 13.2c. From the plot, it is seen that the

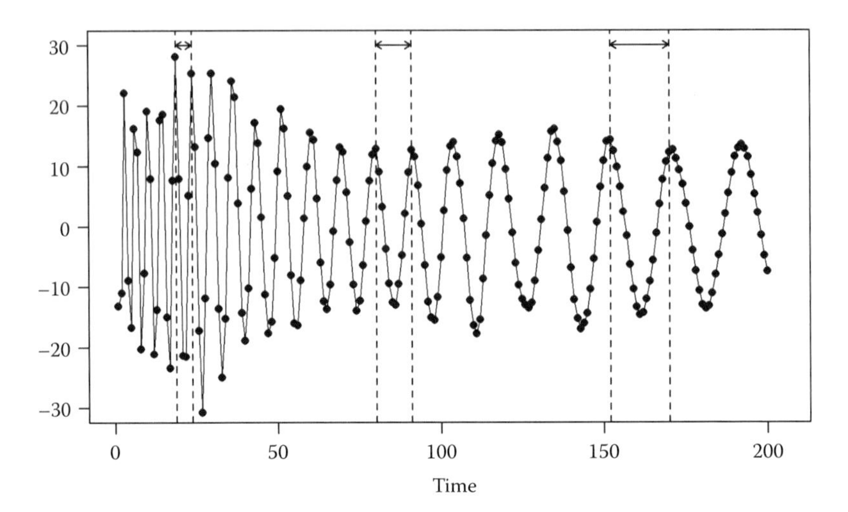

FIGURE 13.3
Realization in Figure 13.2c showing instantaneous period at $t = 19$, 80, and 152.

instantaneous period at $t=19$ is about 5 while the instantaneous periods at $t=80$ and $t=152$ are about 11 and 18, respectively. Consequently, instantaneous frequencies at $t=19$, 80, and 152 are about 0.2, 0.09, and 0.06, respectively. See Appendix 13.A for a mathematical treatment of instantaneous period and frequency.

Example 13.4 Dual Process and Spectral Estimation

In this example, we consider the Euler(2) realization in Figure 13.2c, which is shown again in Figure 13.3. Using the steps outlined earlier (and to be discussed in Section 13.3), we transform the time axis and derive the dual realization that is plotted in Figure 13.4b. Close inspection of Figure 13.4a and b shows that each realization goes through about 20 cycles. However, while the cycle lengths are increasing in Figure 13.4a, each of the cycles in Figure 13.4b is of essentially the same cycle length. There is a 1–1 correspondence between the cycles. For example, cycles 5–8 are characterized by higher peaks in each realization. Basically, the time transformation has transformed the original time axis into one in which the realization has the appearance of stationarity. Figure 13.4c shows the Parzen spectral density estimator, based on truncation point $M = 28$ for the data in Figure 13.4a. As would be expected, the spectral plot does not show strong periodic behavior and has the appearance of a spread spectrum since the cycle length varies throughout the realization. There is clearly a pattern of period (i.e., frequency) change that is not captured in the standard spectral estimate. Figure 13.4d shows the corresponding Parzen spectral density for the dual data in Figure 13.4b. This spectral estimate has a distinct peak at about $f=0.1$ suggesting strong cyclic behavior associated with a period length of around 10.

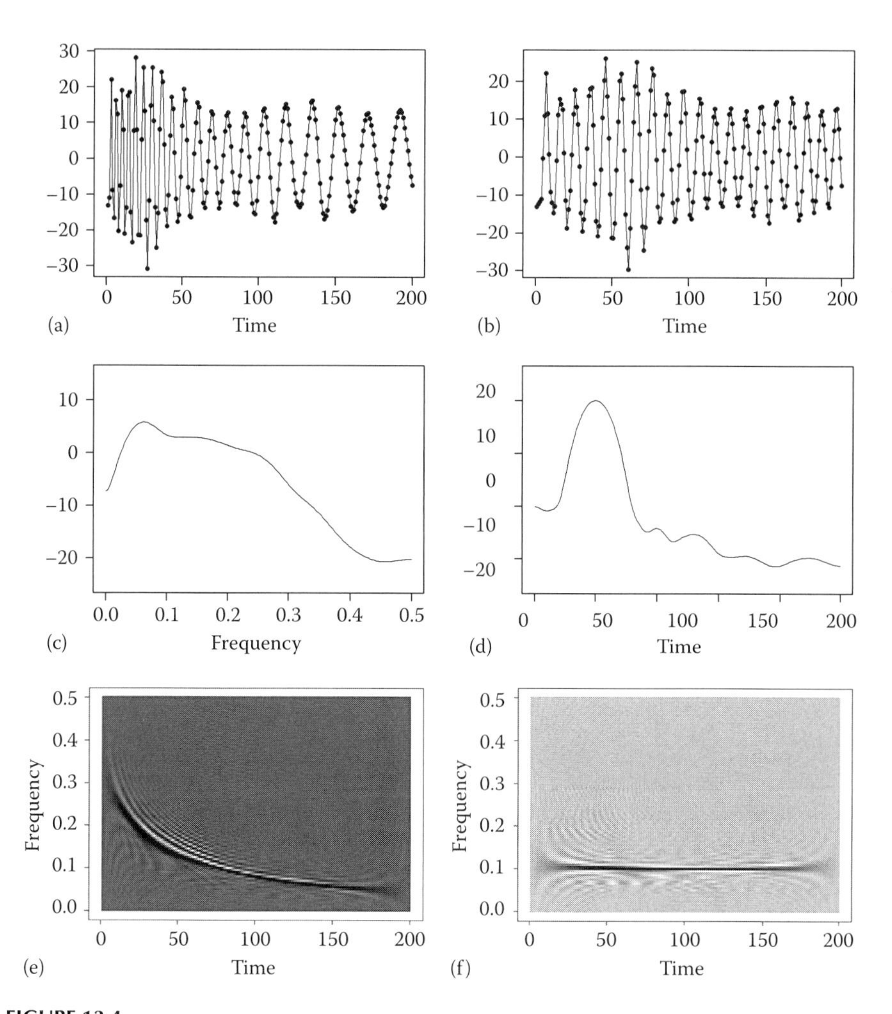

FIGURE 13.4
(a) Euler(2) realization from Figure 13.2a, (b) dual realization, (c) Parzen spectral density estimator for the data in (a), (d) Parzen spectral density estimator for dual data in (b), (e) Wigner–Ville plot for (a), and (f) Wigner–Ville plot for (b).

Figure 13.4c and d illustrates the fact that while on the original timescale, there is a wide range of frequency behavior, the logarithmic time transformation has created a new time axis on which frequency behavior is stable, and there is a single dominant frequency.

To better understand the cyclic behavior in the dual process, note that by (13.6) and (13.7), the dual model associated with (13.8) satisfies the AR(2) model

$$Y_k - 1.6Y_{k-1} + 0.99Y_{k-2} = Z_k \tag{13.9}$$

where $Z_k = a(h^k)$. The factor table (see Section 3.2) for this AR(2) model shows that $1 - 1.6z + 0.99z^2 = 0 + 0.99z^2 = 0$ has a pair of complex conjugate roots $0.81 \pm 0.60i$ that are very close to the unit circle $\left(|r_j|^{-1} = 0.995\right)$, and the associated system frequency is $f = 0.1$, approximately the location of the peak in Figure 13.4d. The "frequencies" of the dual of an M-stationary process are called *dual-frequencies* and are denoted f^*. See Gray et al. (2005) and Choi et al. (2006).

Figure 13.4e shows the Wigner–Ville plot for the data in Figure 13.4a, where it can be seen that the frequency behavior decreases (i.e., the period length increases) with time. Figure 13.4f shows the Wigner–Ville plot for the dual data in Figure 13.4b, and the stationarity is noted by the fact that peak frequency behavior stays constant at about $f = 0.1$.

13.3 G(λ)-Stationary Processes

An important class of G-stationary processes that includes the M-stationary process as a special case is based on the Box–Cox transformation

$$g(t) = \frac{t^\lambda - 1}{\lambda}. \tag{13.10}$$

These processes, referred to as G(λ) processes, are G-stationary processes with $g(t) = \dfrac{t^\lambda - 1}{\lambda}$. For the case $\lambda = 0$, we take $g(t) = \lim\limits_{\lambda \to 0} \dfrac{t^\lambda - 1}{\lambda} = \ln t$, and thus the G(0) process is the M-stationary process. We begin with the definition of a continuous G(λ)-stationary process.

Definition 13.5 Let $X(t)$ be a stochastic process defined for $t, \tau \in [0, \infty)$ such that for any $(t^\lambda + \tau^\lambda - 1) \in [0, \infty)$, and any constant $\lambda \in (-\infty, 0) \cup (0, \infty)$,

$$\begin{aligned} &\text{(i) } E[X(t)] = \mu \\ &\text{(ii) } \operatorname{Var} X(t) = \sigma^2 < \infty \\ &\text{(iii) } E[(X(t^\lambda) - \mu)(X(t^\lambda + \tau^\lambda - 1)^{1/\lambda} - \mu)] = R_X(\tau; \lambda). \end{aligned} \tag{13.11}$$

Then we define $X(t)$ to be a *continuous G(λ)-stationary process* and $R_X(\tau; \lambda)$ to be the G(λ) autocovariance.

NOTES:

1. If $\lambda = 0$, we will refer to the process as an M-stationary process. That is, we will allow $\lambda \in (-\infty, \infty)$, but the case $\lambda = 0$ should be interpreted as the limit as $\lambda \to 0$.

2. A G(λ) process is a G-stationary process with $u = g(t) = \dfrac{t^\lambda - 1}{\lambda}$ and stationary dual $Y(u)$, where $X(t) = X((\lambda u + 1)^{1/\lambda}) = Y(u)$.
3. Property (iii) is based on the fact that

$$X(g^{-1}(g(t) + g(\tau))) = X((t^\lambda + \tau^\lambda - 1)^{1/\lambda}).$$

4. It is easily shown that

$$\lim_{\lambda \to 0} (t^\lambda + \tau^\lambda - 1)^{1/\lambda} = t\tau,$$

so again, the case $\lambda = 0$ yields an M-stationary process.
5. If $\lambda = 1$, a G(λ) process is a standard stationary process.

Clearly, the G(λ) process greatly enlarges the applicability of G-stationary processes beyond the M-stationary process. Before proceeding further, we give the following definition.

Definition 13.6 Let $\varepsilon(u)$ be a zero mean white noise process, and let $\varepsilon\left(\dfrac{t^\lambda - 1}{\lambda}\right) = a(t;\lambda)$ for $t > 0$. We define $a(t;\lambda)$ to be "G(λ)-white noise." For a discussion of G(λ) white noise, see Jiang et al. (2006).

Example 13.5 A G(λ) Cosine-Plus-Noise Process

An example of a G(λ)-stationary processes, $X(t)$, is given by

$$X(t) = A\cos\left(2\pi\beta\left(\frac{t^\lambda - 1}{\lambda}\right) + \psi\right) + a(t), \tag{13.12}$$

where $t \in (0, \infty), \psi \sim \text{Uniform}(0, 2\pi), \lambda \in (-\infty, \infty)$, A and β are constants, $a(t)$ is G(λ)-white noise, and ψ and $a(t)$ are uncorrelated. The dual process that results from the application of the time transformation $u = g(t) = \dfrac{t^\lambda - 1}{\lambda}$ is $Y(u) = A\cos(2\pi\beta u + \psi) + \varepsilon(u), u \in (-\infty, \infty)$, a special case of (4.10) which was shown to be stationary. When $\lambda = 0$, then (13.12) (defined as the limiting process as $\lambda \to 0$) becomes

$$X(t) = A\cos(2\pi\beta \ln t + \psi) + a(t), \tag{13.13}$$

which is a signal-plus-noise version of the continuous M-stationary process considered in Example 13.1.

Figure 13.5 illustrates the TVF behavior in realizations from (13.12) for $\lambda = -1, 0, 1, 2$ where the noise variance is made very small for better

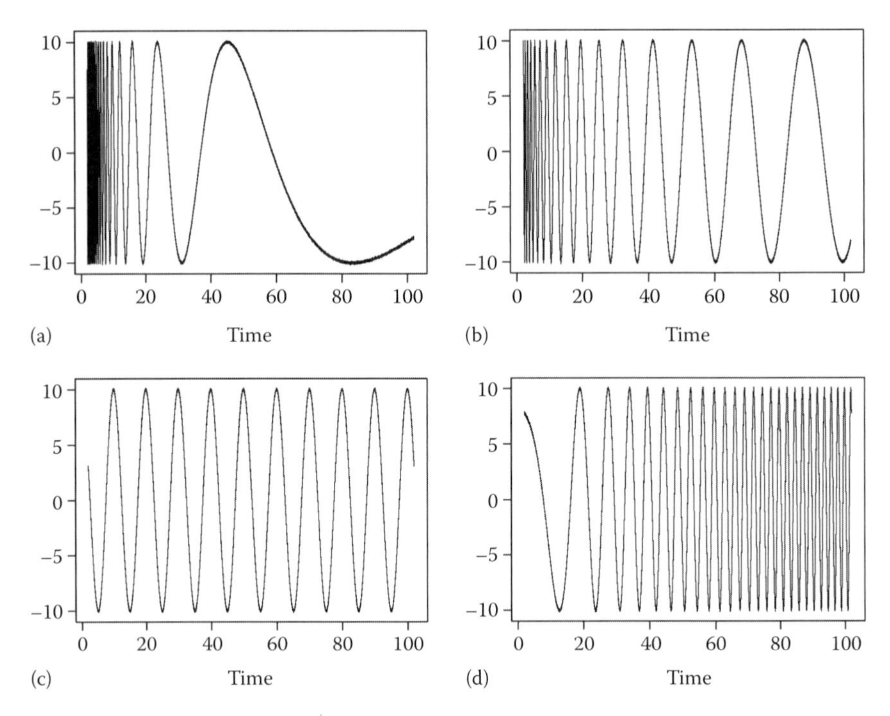

FIGURE 13.5
Realizations from (13.12) in all cases with $A = 10$ and $\psi = 0.2\pi$. (a) $\lambda = -1$, $\beta = 50$; (b) $\lambda = 0$, $\beta = 4$; (c) $\lambda = 1$, $\beta = 0.1$; (d) $\lambda = 2$, $\beta = 0.005$.

visualization of patterns. Note that for $\lambda = -1$ and 0, the frequency is decreasing with time, when $\lambda = 1$, the frequency does not change with time, and when $\lambda = 2$, the frequency increases with time. It is important to note that the change from decreasing frequency behavior to increasing behavior occurs at $\lambda = 1$ (not at $\lambda = 0$).

13.3.1 Continuous G($p;\lambda$) Model

To this point, we have not discussed the derivative process of a stochastic process, $X(t)$, which we will denote by $\frac{dX(t)}{dt}$. Nevertheless, we will proceed formally making use of the fundamental properties as needed since our purpose here is motivational only. Ultimately, we will discretize such processes through proper sampling and will only rely on them in the continuum conceptually. For a rigorous treatment of the derivative of a stochastic process and the continuous ARMA(p,q) processes, see Priestley (1981).

Definition 13.7 A *continuous G(p;λ)-stationary process* (with mean $\mu = 0$) is defined to be the G(λ)-stationary solution of the equation

$$\prod_{j=1}^{p}\left(t^{1-\lambda}D - \alpha_j\right)X(t) = \varepsilon(t), \tag{13.14}$$

where the α_j's are real or complex constants, $\varepsilon(t)$ is G(λ)-white noise (i.e., $\varepsilon((t^\lambda - 1)/\lambda) = a(u)$ is white noise), and its stationary dual process, $Y(u)$, is a stationary continuous AR(p) process where $u = (t^\lambda - 1)/\lambda$. Also, the symbol "$D$" stands for a differential operator, that is, $DX(t) = dX(t)/dt$.

Definition 13.7 says that if $\lambda \neq 0$ and $u = g(t) = \dfrac{t^\lambda - 1}{\lambda}$, or $\lambda = 0$ and $u = \ln t/\ln h, h > 1$, then the G(λ)-process in (13.14) has stationary continuous AR(p) dual $Y(u)$ given by

$$\prod_{j=1}^{p}(D - \alpha_j)Y(u) = a(u). \tag{13.15}$$

As a simple example, take the case $p = 1$ in (13.14), which is

$$\left(t^{1-\lambda}D - \alpha_1\right)X(t) = \varepsilon(t). \tag{13.16}$$

If $u = \dfrac{t^\lambda - 1}{\lambda}$, then $t = (u\lambda + 1)^{1/\lambda}$, and proceeding formally, we obtain

$$\begin{aligned}\frac{dX(t)}{dt} &= \frac{dX(t)}{du}\frac{du}{dt}\\ &= t^{\lambda-1}\frac{dX(t)}{du}.\end{aligned}$$

So, from (13.16), we have $t^{1-\lambda}\dfrac{dX(t)}{dt} - \alpha_1 X(t) = \dfrac{dX(t)}{du} - \alpha_1 X(t) = \varepsilon(t)$. Since $X(t) = Y(u)$, it follows that $\dfrac{dY}{du} - \alpha_1 Y(u) = a(u)$, where $a(u)$ is white noise. Therefore, $Y(u)$ is a continuous AR(1) process as noted in (13.15).

The G(p; λ) process in (13.14) has been extended to the $G(p, q; \lambda)$ process

$$\prod_{j=1}^{p}\left(t^{1-\lambda}D - \alpha_j\right)X(t) = \prod_{j=1}^{q}\left(t^{1-\lambda}D - \delta_j\right)\varepsilon(t). \tag{13.17}$$

We will not pursue this here except to note that for a further discussion of the Euler(p,q) and $G(p, q; \lambda)$ process, see Choi et al. (2006) and Jiang et al. (2006).

13.3.2 Sampling the Continuous G(λ)-Stationary Processes

Suppose $X(t)$ is a continuous G(λ)-stationary process, and suppose

$$t_k = ((k+\xi)\Delta\lambda + 1)^{1/\lambda}, \tag{13.18}$$

where $k = 1, 2, \ldots, n, \Delta > 0$, and $(k+\xi)\Delta\lambda + 1 \geq 0$, where in this case ξ determines the offset and Δ is the sampling interval. Then the dual, $Y_k = X(t_k)$, is a stationary process. Further, if $X(t)$ is a continuous $G(p;\lambda)$ process, the dual process is a discrete ARMA(p,r) with $r \leq p-1$. Also, the Nyquist frequency of the discrete dual, Y_k, is given by $\frac{1}{2\Delta}$. Note that $\lim_{\lambda \to 0} t_k = h^{k+\xi}$. For further discussion and proof of these remarks, see Jiang et al. (2006).

Definition 13.8 If the data are sampled from a continuous G(λ)-stationary process, $X(t)$, at t_k defined in (13.18), then $f_N^*(\Delta;\lambda) = \frac{1}{2\Delta}$ is called the G(λ)-*Nyquist frequency*.

NOTE: When $\lambda = 1$, the G(λ)-Nyquist frequency is the usual Nyquist frequency. When $\lambda = 0$, the G(λ)-Nyquist frequency is $f_N^*(\Delta;0) = \frac{1}{2\Delta} = \frac{1}{2\ln h}$. This follows from the fact that in the transformed time space,

$$\begin{aligned} u_k - u_{k-1} &= \ln h^k - \ln h^{k-1} \\ &= \ln\left(\frac{h^k}{h^{k-1}}\right) \\ &= \ln h. \end{aligned}$$

13.3.2.1 Equally Spaced Sampling from G(p; λ) Processes

As was noted in the case of M-stationary processes, the desirable points at which to sample a G(λ)-stationary process (the t_k's given in (13.18)) are unequally spaced (unless $\lambda = 1$), which can cause difficulties since data are commonly taken at equally spaced points. Just as in the case of Euler processes or M-stationary processes, in general, this difficulty can be overcome by over-sampling or interpolation. When the data are only available at equally spaced points and are not sufficiently over-sampled, interpolation can be used to obtain values at the t_k in (13.18) required for (G(λ) processes. This is the procedure used by the GWS S-Plus program (see the ATSA website). Wang et al. (2009) give a Kalman filtering-based approach to avoid the need for interpolation. Haney (2011) discusses a Fourier interpolation technique that can be preferable to linear interpolation for data having cyclic behavior. However, in the applications considered here, linear interpolation will be satisfactory.

Example 13.6 G(2; λ) Models for a Variety of λ Values

Figure 13.6 shows several realizations from G(2; λ) models associated with the dual process

$$(1 - 1.9B + 0.99B^2)Y_k = a_k, \tag{13.19}$$

with offset 25, $n = 500$, and a different set of λ values than those used in Figure 13.5 to further emphasize the fact that the direction of frequency change occurs at $\lambda = 1$.

13.3.3 Analyzing TVF Data Using the G(p; λ) Model

At the end of Section 13.1, we outlined the techniques for analyzing TVF data using G-stationary methods. In this section, we briefly discuss the items in

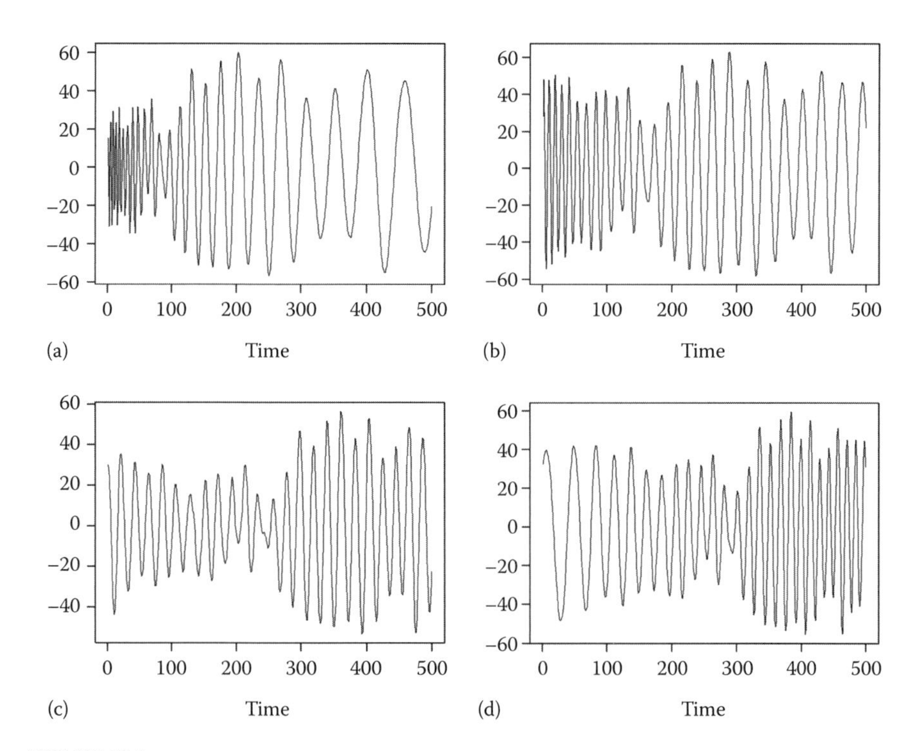

FIGURE 13.6
Realizations from G(2; λ) models associated with the AR(2) dual model in (13.19) and offset = 25. (a) $\lambda = 0$. (b) $\lambda = 0.5$. (c) $\lambda = 1$. (d) $\lambda = 1.5$.

the outline and provide a few examples with a focus on using the G(p;λ) model. These procedures are implemented in the GWS S+ software package and are discussed in much greater detail in Gray et al. (2005), Jiang et al. (2006), and Choi et al. (2006).

Step 1: Transform the Time Axis to Obtain a Stationary Dual Realization

In GWS, a search routine is used to search for the values of λ and offset that produce the "most stationary" dual. For each set of λ, h, and offset values considered in the search, the data, $X(t_k)$, $k = 1, 2, \ldots$, are approximated at the t_k's in (13.18), using interpolation. By then indexing on k, the dual realization associated with the given λ, h, and offset is obtained. For each of these dual realizations, GWS employs techniques for measuring the degree to which the correlation behavior seems to remain constant across the realization. This measure is based on the fact that the correlation structure under stationarity stays constant across the realization while for TVF data, such as that in Figure 13.6a, the correlation behavior is very different at the beginning of the realization than it is toward the end. Based on such a measure of correlation change, the final values of λ, h, and offset are taken to be those that produced the "most stationary behaving" realization. To illustrate these remarks, Figure 13.7a shows the sample autocorrelations for the first third of the data in Figure 13.4a with a solid line, the sample autocorrelations for the middle third of the data with a dotted line, and for the last third of the data with a dashed line. There it can be seen that there is dramatic difference in autocorrelation patterns among the three. For the initial third of the data, there is higher frequency behavior that is changing so rapidly that the sample autocorrelations after the first 8 lags or so are very small. The sample autocorrelations for the other two-thirds of the data each show a cyclic behavior but with a higher frequency behavior for the middle third (dotted line) than the final third (dashed line). The point is that if the data set were from a stationary data set, these three curves would be expected to be similar to

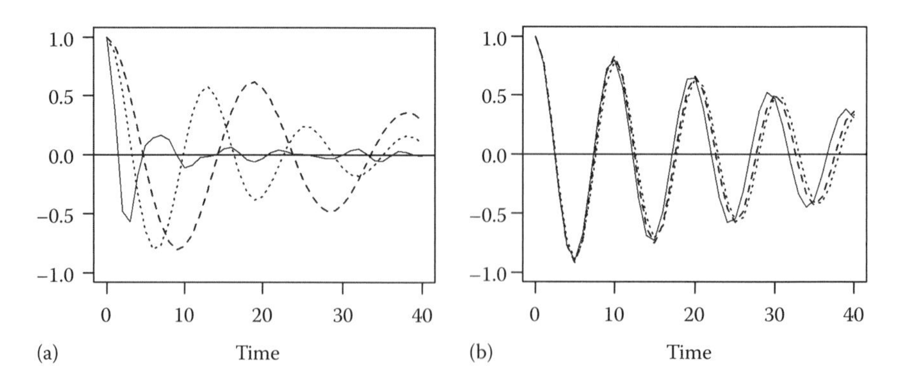

FIGURE 13.7
Sample autocorrelations of the first, second, and last third of (a) the data in Figure 13.4a and (b) the associated dual data in Figure 13.4b.

each other. This is illustrated in Figure 13.7b, where we show the sample autocorrelations for the first, second, and last third of the stationary dual data in Figure 13.4b. See Jiang et al. (2006).

Step 2: Analyze the Transformed (Dual) Realization Using Methods for Stationary Time Series

In the case of fitting a $G(p;\lambda)$ model to a set of TVF data, the dual model obtained in Step 1 is then modeled as an AR(p). Specifically, techniques such as AIC can be used to estimate the model order, and the associated coefficients can be estimated using ML or Burg estimators. Further analyses, as needed, are then performed.

(i) *Forecasting*: If forecasts are desired for the original TVF data, then the next step is to obtain the corresponding forecasts using the dual model. That is, techniques such as those given in Chapter 6 for obtaining forecasts for a stationary AR(p) model can be employed. The forecasts are then to be mapped back to the original timescale as will be discussed in Step 3.

(ii) *Spectral analysis*: In Figure 13.4d, we showed a window-based spectral estimator for a dual realization and noted the interesting and potentially useful fact that the time transformation had changed the wide range of TVF behavior (i.e., the spread spectrum) in the original series to a single dominant frequency in the transformed time. Characterization of the stable frequency behavior in the dual space may provide a useful way of summarizing information in the original series. See Example 13.11 for an example involving seismic data.

13.3.3.1 G(p;λ) Spectral Density

In general, we will refer to the autoregressive spectral density of the dual data as the $G(p;\lambda)$ *spectral density*. The spectral density in Figure 13.8a is an example of an autoregressive spectral estimate based on the dual, i.e., a $G(p;\lambda)$ spectral density. This spectral estimate can be obtained using a variety of methods, but we recommend using the $G(p;\lambda)$ spectral density, which is the AR(p) spectral density associated with the AR(p) model fit to the dual. The special case of a $G(p;\lambda)$ spectral density in the case of $\lambda = 0$ (i.e., the M-stationary case) is sometimes called an *M-spectral density* estimator. See Gray et al. (2005).

Step 3: Transform Back to the Original Timescale

It will often be the case that the forecasts or spectral estimators obtained on the dual space are not the final results that are desired. We consider forecasting and spectral estimation separately.

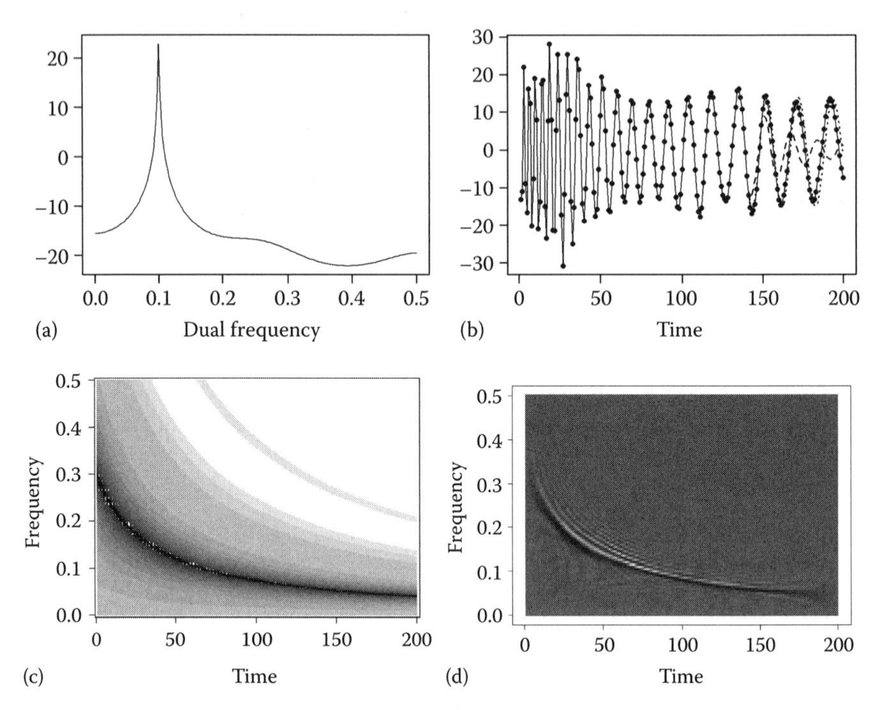

FIGURE 13.8
Plots associated with the G(p;λ) analysis of the data in Figure 13.2c. (a) G(5; −0.1) spectral density, (b) Forecasts based on the G(5; −0.1) fit (doted forecasts) and an AR(11) fit to the original data (dashed lines), (c) instantaneous spectrum associated with the G(5; −0.1) fit, and (d) Wigner–Ville plot.

(i) *Forecasting*: Assume that $X(t)$ is observed at equally spaced points and that forecasts are desired for $X(k+1), X(k+2), \ldots, X(k+h)$ for some forecast origin k. The forecasts for Y_{k+1}, $Y_{k+2}, \ldots,$ which are obtained in Step 2(i), are converted into forecasts for the desired time points using interpolation. The `GWS S+` program performs the necessary steps to obtain $G(p;\lambda)$ forecasts.

(ii) *Spectral analysis*: It has been mentioned that the $G(p;\lambda)$ spectral density can be a very useful tool for summarizing frequency information based on a stationarizing time transformation. However, the frequency scale in the dual space is difficult to relate back to frequencies in the original timescale. For this reason, it is often desirable to transform the frequency information in the $G(p;\lambda)$ spectral density back into the original time and frequency domains using the *Instantaneous Spectrum*. Appendix 13.A provides details concerning the definition of instantaneous frequency and the instantaneous spectrum. By referring back to Figure 13.3, it is seen that the peak in the dual spectrum in Figure 13.4d is associated with an instantaneous frequency of about $f = 0.2$ at $t = 19$, of about 0.09 at $t = 80$, and about 0.06 at $t = 152$. The instantaneous spectrum is a time-frequency plot (similar to the Wigner–Ville plot, Gabor transform, and wavelet spectrograms) that transforms the information in the $G(p;\lambda)$ spectral density back into the original dimensions of time and frequency to display the manner in which frequency changes with time.

Example 13.7 $G(p;\lambda)$ Analysis of Data in Figure 13.2c

We reconsider Example 13.2 and the data in Figure 13.2c, which is also shown in Figures 13.3 and 13.4a. In this example, we fit a $G(p;\lambda)$ model to the data and use this model for forecasting and spectral estimation. The realization was generated from the M-stationary model given in (13.8) (i.e., $\lambda = 0$) with offset 30 and $h = 1.01$. The dual model for the generated realization is the AR(2) model in (13.9), which has a system frequency of $f = 0.1$. We will obtain forecasts for the last 60 values of the time series along with spectral estimates based on the fitted model found using the procedure just described.

Step 1: *Finding a time transformation*. Using the GWS S+ software to search over a range of λ and offset values, the estimates for λ and offset were selected to be -0.1 and 38 respectively. The original data and the dual data are shown in Figure 13.4a and (b) and will not be repeated here.

Step 2: *Analyzing the dual realization*. AIC selects $p = 5$ for the dual realization.

(i) Forecasts (not shown) based on the dual model are obtained.

(ii) The $G(5;-0.1)$ spectral density estimator is obtained as the AR(5) spectral density estimate based on the model fit to the dual. This spectral density estimate is shown in Figure 13.8a displaying a much stronger peak at about $f = 0.1$ than did the corresponding Parzen spectral estimate of the dual data shown in Figure 13.4d.

Step 3: *Transform back to the original timescale* (i) Forecasts based on the $G(5;\ -0.1)$ model for the last 60 data values are shown in Figure 13.8b using dotted lines. For comparison, we fit an AR(p) model to the original data and use it to forecast. Note that this should be a poor fit due to the TVF behavior of the data. AIC selected an AR(11) for the original data, and the AR (11) forecasts for the last 60 data values are shown in Figure 13.8b using dashed lines. Examination of the forecasts shows that the $G(5;-0.1)$ forecasts track the increasing cycle lengths present in the data. The AR(11) model fit to the original data is a stationary model that assumes no frequency change. Consequently, the AR(11) forecasts become poor for larger steps ahead due to the fixed frequency assumption. (ii) The instantaneous spectrum based on the $G(5;\ -0.1)$ fit to the data is shown in Figure 13.8c. This plot illustrates the validity of the previous remark that the peak in the dual spectrum (shown here in Figure 13.8a) is associated with an instantaneous frequency of about $f=0.2$ at $t=19$, a frequency of about 0.09 at $t=80$, and about 0.06 at $t=152$. The associated Wigner–Ville plot was shown in Figure 13.4e, and is shown again in Figure 13.8d for comparison. It can be seen that the Wigner–Ville and instantaneous spectrum show similar time-frequency behavior except that the instantaneous spectrum shows more detail for time values close to 0 and 220.

We remind readers that Figure 13.8c and d are three-dimensional plots visualized using a gray-scale, where black represents larger spectral values (dB). To provide better visualization of the information in the instantaneous spectrum in Figure 13.8c, the plots in Figure 13.9 show "snapshots" of the instantaneous spectrum for $t=19$, 80, and 152. These snapshots should be compared with the discussion in the previous paragraph and that accompanying Figure 13.3. In particular, the plot in Figure 13.9a illustrates the information in the instantaneous spectrum (in Figure 13.8c) in the vertical strip at $t = 19$. Note that the darkest portion of that strip is at $f=0.2$, which is the location of the peak in Figure 13.9a and is a good approximation to the instantaneous frequency at $t=19$, based on Figure 13.3. The instantaneous spectral densities at $t=80$ and $t=152$ are shown in Figure 13.9b and c. These have peaks at about $f=0.09$ and $f=0.06$ respectively, which is consistent with earlier discussions.

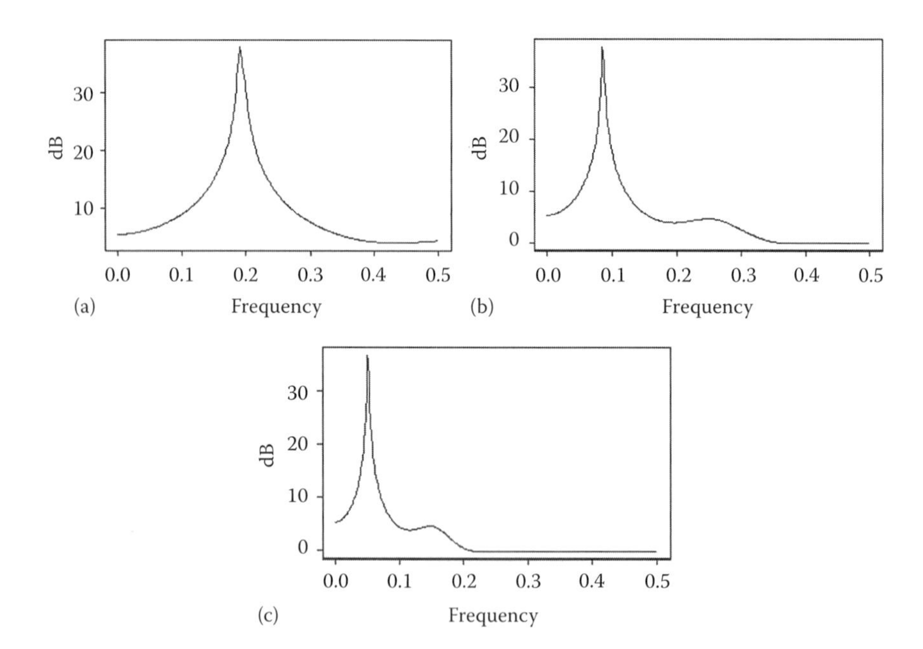

FIGURE 13.9
Snapshots associated with instantaneous spectral density in Figure 12.7c for (a) $t = 19$, (b) 80, and (c) 152.

Example 13.8 G($p;\lambda$) Analysis of a Doppler Signal

In this example, we consider an undamped Doppler signal, $X(t)$, of the type given by Donoho and Johnstone (1994). The signal we consider is given by

$$X(t) = 0.5\sin\left(\frac{840\pi}{t+50}\right) + \sin\left(\frac{2415\pi}{t+50}\right) + 0.05N(t), \qquad (13.20)$$

where $N(t) \sim N(0,1)$. Figure 13.10a shows a realization of length 200 from this model, and Figure 13.10c shows the AR(8) spectral density obtained by

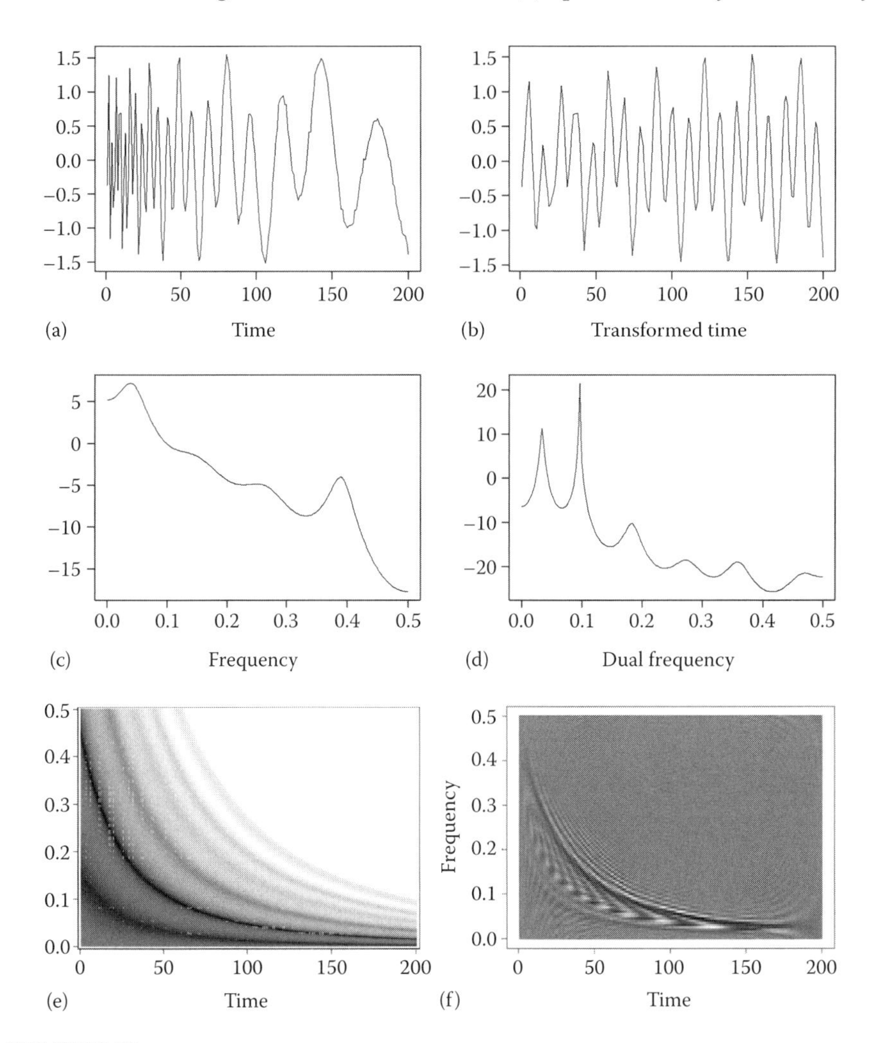

FIGURE 13.10
(a) Realization from Doppler model in (13.20), (b) dual realization based on G(12; −1) fit, (c) AR(8) spectral density estimator of the original data, (d) G(12; −1) spectral density estimate, (e) Instantaneous spectrum, and (f) Wigner–Ville plot for (a).

(unwisely) fitting an AR(8) to the original data. The spectrum is quite spread and is no help in describing the characteristics of the underlying signal. Even though (13.20) cannot be stationarized using a transformation of the form (13.10), we attempt to fit a $G(p;\lambda)$ model because of the flexibility this model allows. Using the GWS software, we fit a $G(12;-1)$ as a "best" fit model. Figure 13.10b shows the dual associated with the $G(12;-1)$ fit, and the decreasing frequency behavior is no longer visible. Figure 13.10d shows the $G(12;-1)$ spectral density (i.e., the AR(12) spectral density of the dual data). Two distinct peaks are visible as would be expected based on the form of the model. Figure 13.10e shows the instantaneous spectrum based on the $G(12;-1)$ model while Figure 13.10e shows the Wigner–Ville distribution. Clearly, the $G(12;-1)$ outperforms the Wigner–Ville, in this case, in that it more clearly shows the two decreasing frequencies.

Example 13.9 $G(p;\lambda)$ Analysis of Bat Echolocation Data

Figure 13.11a shows big brown bat echolocation data furnished courtesy of Al Feng of the Beckman Center at the University of Illinois. This realization consists of 381 data points sampled at 7 microsecond intervals. This realization is very interesting and is quite difficult to analyze. Figure 13.11b shows the first 175 data values, where it can be seen that the basic period length is increasing. Also, at about $t = 100$, a second higher-frequency behavior seems to enter the realization. This realization was previously analyzed by Gray et al. (2005), who fit an Euler(11) model (i.e., a $G(11;0)$) to the data. Using GWS to select a $G(p;\lambda)$ model for the data suggests that the $G(11;0)$ model provides a reasonable fit, and we will use it in the analysis here. Figure 13.11c shows the autoregressive spectral density estimator of the original data based on fitting an (unsatisfactory) AR(20) model to the original bat data. As expected, this is a spread spectrum due to the time-varying behavior of the frequency content in the data. Figure 13.11d shows the $G(11;0)$ spectral density estimate for the bat data. This spectral density estimate (which is the AR(11) spectral density estimate of the dual data) shows a peak at $f = 0$ along with three other distinct peaks. It is clear that a much better understanding of the signal is obtained by making the time transformation.

The instantaneous spectrum associated with the $G(11;0)$ fit is shown in Figure 13.11d in order to better understand the TVF behavior back in the original timescale. It can be seen that the signal is primarily made up of three frequency bands that are decreasing with time along with an additional band, around zero frequency, that is essentially constant. This behavior is much more clearly shown in Figure 13.11e than in the Wigner–Ville plot in Figure 13.11f. Note that the middle frequency band begins to appear at about $t = 10$, while the highest frequency band only begins to appear at around $t = 100$. That is, until that time, this echolocation frequency band was

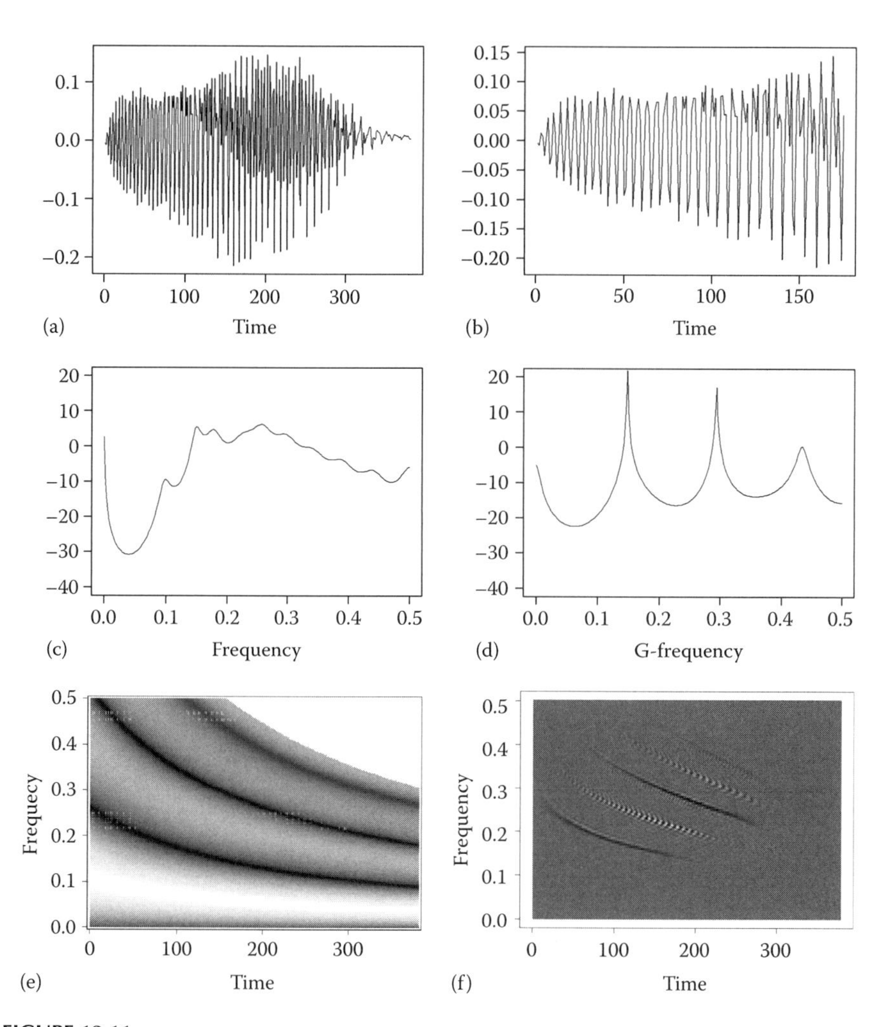

FIGURE 13.11
Plots associated with the G(p;λ) analysis of the bat data. (a) Bat echolocation data, (b) first 175 data points of bat echolocation data showing decreasing frequencies, (c) AR(20) spectral density estimate for bat data, (d) G(11;0) spectral density estimate for bat data, (e) instantaneous spectrum based on G(11;0) fit to bat data, and (f) Wigner–Ville plot for bat data.

above the Nyquist frequency and hence not visible at the sample rate used. However, by about $t = 100$, the highest TVF band had decreased in frequency to the point that it was detectable given the sampling rate. Notice also that the highest frequency band shows up more clearly in the instantaneous spectrum, and for all frequency bands, the frequency behavior along the border of the time-frequency plot is more visible using the instantaneous spectrum in Figure 13.11e than it is in the Wigner–Ville plot in Figure 13.11f.

Example 13.10 Whale Click Data

Figure 13.12a shows 142 points of a whale click signal sampled at 4.8 Hz for a duration of about 0.03 seconds. The increasing frequency behavior can be clearly seen, which suggests that $\lambda > 1$. Using the procedures outlined earlier, a G(11; 3.5) model was fit to the data. The dual data are shown in Figure 13.12b, where the frequency behavior is no longer changing with time and there seems to be one dominant frequency in the data. The G (11; 3.5) spectral density of the dual is shown in Figure 13.12d. It is seen that

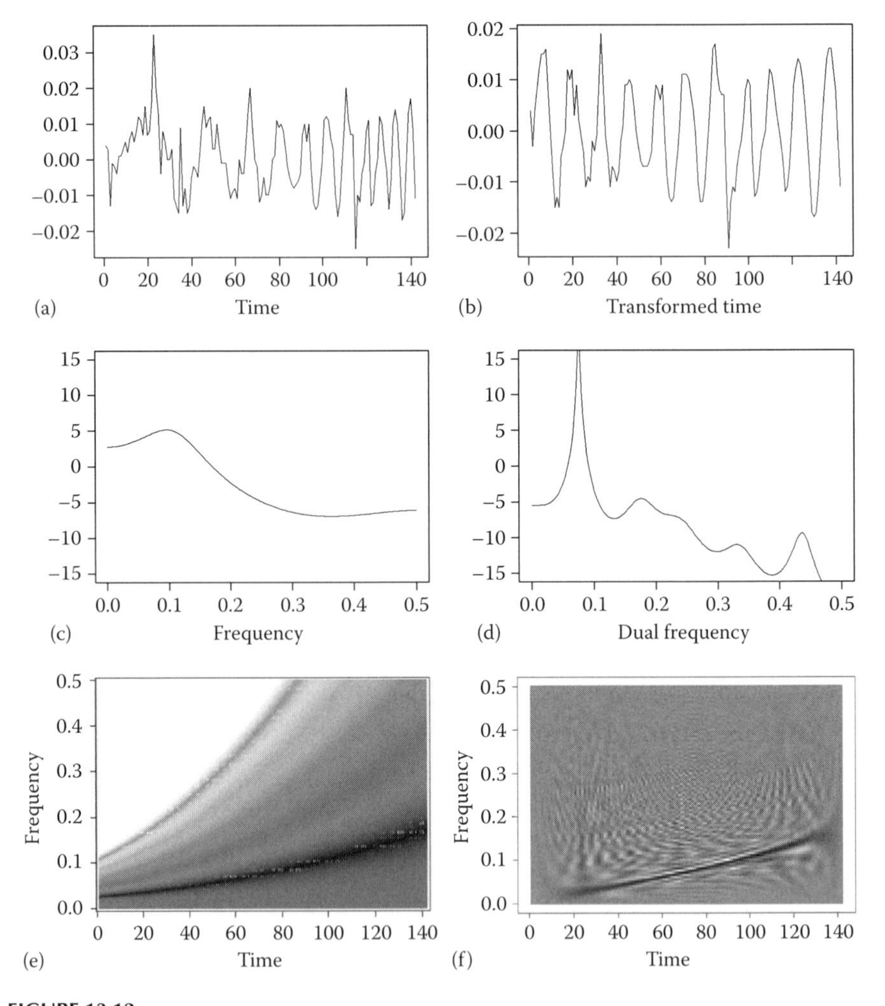

FIGURE 13.12
(a) Whale click data, (b) dual data based on a G(11; 3.5) fit, (c) AR(3) spectral density for data in (a), (d) G(11; 3.5)-spectral density, (e) instantaneous spectrum, and (f) Wigner–Ville Plot for (a).

there is a distinct peak at a dual frequency slightly less than 0.1. If we fit an AR(p) model to the original whale click data, AIC selects an $p = 3$ and the associated AR(3) spectral density estimator is shown in Figure 13.12c which has the form of a spread spectrum. In Figure 13.12e and f, we show a G (11; 3.5) instantaneous spectrum and the Wigner–Ville spectrum, respectively for the whale click data in Figure 13.12a. In both cases, a nonlinear increasing frequency behavior is indicated.

Example 13.11 Seismic Data from Earthquakes and Explosions

In this example, we illustrate the point made earlier that a time transformation produces stable frequency behavior in the dual space that may provide a useful method of summarizing information in the original series. Woodward et al. (2007) examined the problem of distinguishing between earthquakes and explosions based on seismic monitoring. Seismic events such as earthquakes and explosions create a variety of waveforms that can be measured by seismographs. An extremely important area of research involves the use of seismic measurements to distinguish between explosions and earthquakes with the goal of detecting clandestine underground nuclear testing around the world. Seismologists have identified the Lg wave, which is a shear wave that has most of its energy trapped in the Earth's crust, as a useful seismic waveform for this purpose. Figure 13.13a and b shows the Lg waveform for an earthquake (the Massachusetts Mountain (MM) earthquake) and an explosion (the Starwort explosion). See Gupta et al. (2005) for information on the data. It can be seen from the figure that the Lg wave for the earthquake and the explosion have similar time-varying behavior in which the frequency is decreasing with time. For each of the 10 earthquakes and 10 explosions used in the study, Woodward et al. (2007) transformed the corresponding Lg wave into a stationary dual. The duals for the MM earthquake and the Starwort explosion are shown in Figure 13.13c and d, respectively. Each of the 20 dual processes was then modeled as an AR(10) model. The associated G(10; 0.8) spectral densities for MM and Starwort are shown in Figure 13.13e and f. Inspection of the two G (10; 0.8) spectral densities shows that the one associated with the Starwort explosion has a much sharper peak, and thus, indicates a stronger cyclic dual behavior than seen in the earthquake spectral density. That is, after moving to the dual space, the corresponding G(10; λ) spectral densities are quite distinctive. One method for measuring the information shown visually in Figure 13.13e and f is to examine the factor tables associated with the two AR(10) models fit to the duals. In Table 13.1, we show the first three lines of the factor tables associated with the AR(10) models fit to the MM and Starwort dual processes. There it can be seen that the AR(10) fit to the Starwort dual has a root much closer to the unit circle than does the MM. For each of the 20 dual data sets in the study, the strength of the dominant

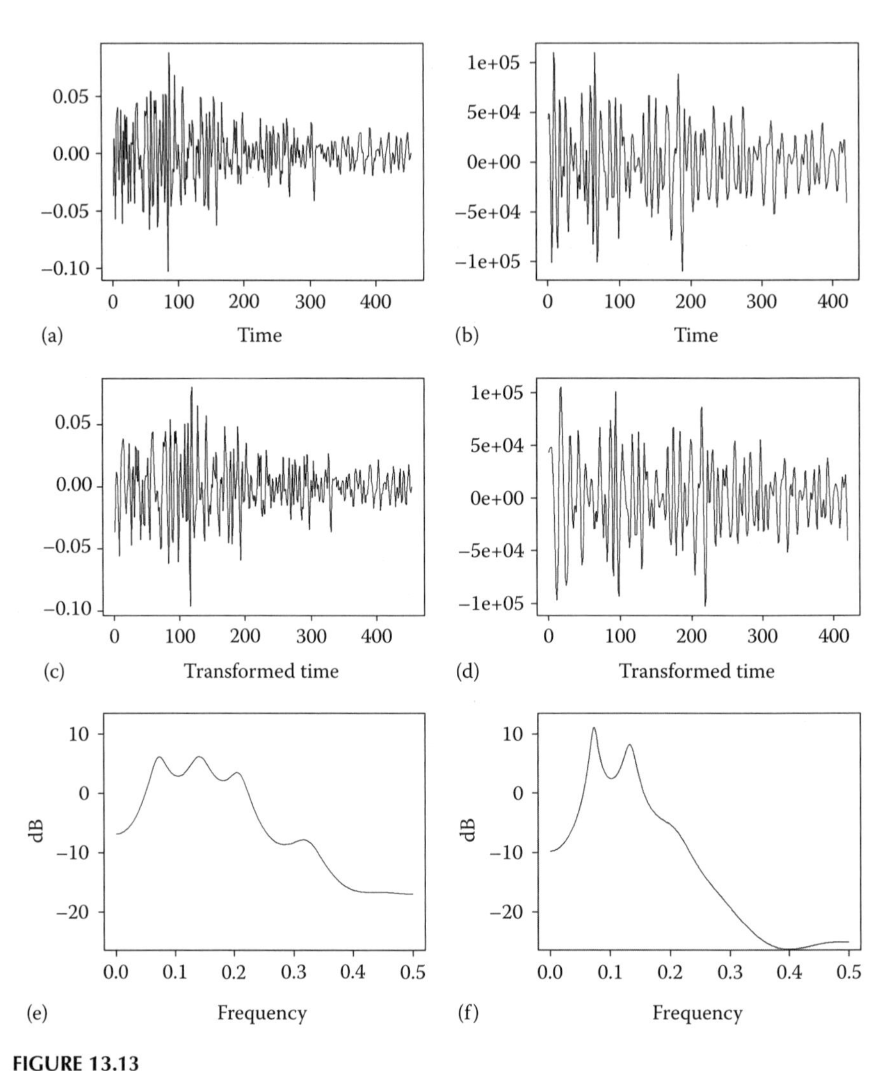

FIGURE 13.13
(a) Lg wave for Massachusetts Mountain earthquake, (b) Lg wave for Starwort explosion, (c) dual data based on G(10; 0.8) fit to (a), (d) dual data based on G(10; 0.8) fit to (b), (e) G(10; 0.8) spectral density estimator for Massachusetts Mountain data, and (f) G(10; 0.8) spectral density estimate for Starwort data.

frequency was quantified using the absolute value of the reciprocal of the associated roots as the measure. Using this measure as a discriminant, Woodward et al. (2007) were able to obtain perfect discrimination between earthquakes and explosions for these 20 events. Thus, moving to the dual space produced a natural measure of the strength of the time-varying cyclic behavior in the original data.

TABLE 13.1

First 3 Lines of the Factor Tables Associated with ML Estimates for an AR(10) Fit to the Dual Data in Figure 13.12c and d

AR Factors	Roots (r_j)	$\lvert r_j^{-1}\rvert$	f_{0j}
(a) MM dual model			
$1 - 1.65B + 0.84B^2$	$0.98 \pm 0.47i$	0.917	0.072
$1 - 0.44B + 0.80B^2$	$0.28 \pm 1.08i$	0.896	0.210
$1 - 1.12B + 0.79B^2$	$0.71 \pm 0.87i$	0.890	0.141
(b) Starwort dual model			
$1 - 1.73B + 0.93B^2$	$0.93 \pm 0.46i$	0.965	0.073
$1 - 1.26B + 0.89B^2$	$0.71 \pm 0.79i$	0.941	0.133
$1 - 0.43B + 0.67B^2$	$0.33 \pm 1.18i$	0.816	0.207

13.4 Linear Chirp Processes

Linear chirp processes have wide applications in fields such as sonar, radar, and optics. Linear chirp processes are so named because they have frequency behavior that changes linearly with time. The data in Figure 12.2a are from a linear chirp process, and the time-frequency plots in Figure 12.4 illustrate the linear change in frequency. In this section, we make use of the fact that linear chirp processes are G-stationary processes with $g(t) = at^2 + bt + c$. Without loss of generality, we take $c = 0$. Then $u = at^2 + bt = g(t)$. So, $at^2 + bt - u = 0$,

$$\begin{aligned} t &= \frac{-b \pm \sqrt{b^2 + 4au}}{2a} \\ &= g^{-1}(u), \end{aligned}$$

and

$$\begin{aligned} g^{-1}(g(t) + g(\tau)) &= \frac{-b \pm \sqrt{b^2 + 4a(g(t) + g(\tau))}}{2a} \\ &= \frac{-b \pm \sqrt{(2at + b)^2 + 4ag(\tau)}}{2a}. \end{aligned}$$

This leads to the following definition given by Robertson et al. (2010).

Definition 13.9 A continuous parameter stochastic process, $X(t)$, is called a generalized linear chirp (GLC) process if

$$\begin{aligned}
&\text{(i) } E[X(t)] = \mu < \infty, \\
&\text{(ii) Var } X(t) = \sigma^2 < \infty, \\
&\text{(iii) } E[(X(t) - \mu)(X(g^{-1}(g(t) + g(\tau))) - \mu)] = R_X(\tau),
\end{aligned}$$

where $g(t) = at^2 + bt$, with $a, b \in \mathbb{R}$ and

$$\begin{aligned}
g^{-1}(g(t) + g(\tau)) &= \frac{-b + \sqrt{(2at + b)^2 + 4ag(\tau)}}{2a}, \quad t \geq \frac{-b}{2a} \\
&= \frac{-b - \sqrt{(2at + b)^2 + 4ag(\tau)}}{2a}, \quad t < \frac{-b}{2a}.
\end{aligned}$$

Example 13.12 A Generalized Linear Chirp Process

Let

$$X(t) = A \cos(2\pi\beta(at^2 + bt) + \psi), \tag{13.21}$$

where ψ is a uniform $[0, 2\pi]$ random variable. To see that $X(t)$ in (13.21) is a GLC process, let $u = at^2 + bt = g(t)$. Then $Y(u) = A \cos(2\pi\beta u + \psi)$ is a harmonic process as discussed in Example 1.3. Since $X(t) = X(g^{-1}(u)) = Y(u)$, it follows that

$$\begin{aligned}
&\text{(i) } E[X(t) = E[Y(u)] = 0 \\
&\text{(ii) Var}[X(t)] = \text{Var}[Y(u)] = \frac{1}{2}A^2. \\
&\text{(iii) } E[X(t)X(g^{-1}(g(t) + g(\tau))) = E[Y(u)Y(u + g(\tau))] \\
&\qquad = \frac{1}{2}A^2 \cos(2\pi\beta g(\tau)) \\
&\qquad = C_u(g(\tau)) \\
&\qquad = R_X(\tau).
\end{aligned}$$

Therefore, $X(t)$ is a GLC process.

Figure 13.14a shows a realization of $X(t)$ with $A = 1, \beta = 1, a = 1, b = 1$, and $\psi = 0$ over the time interval $[0, 4]$, while Figure 13.14b shows the corresponding realization of $Y(u)$.

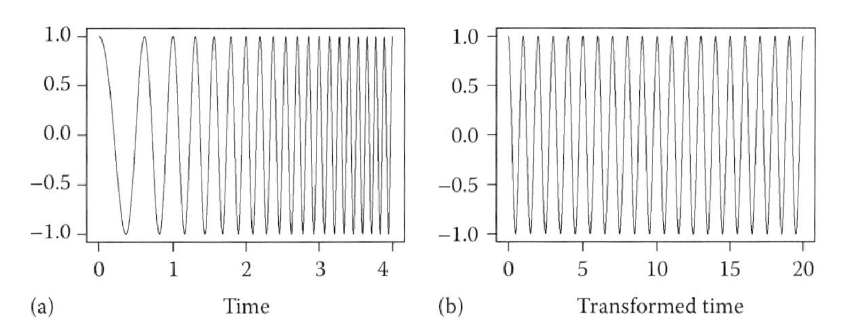

FIGURE 13.14
(a) Linear chirp realization from (13.21) with $A = 1, \beta = 1, a = 1, b = 1$, and $\psi = 0$, and (b) the dual realization based on the transformation $g(t) = at^2 + bt$.

It should be clear that $X(t)$ in Example 13.12 is easily generalized. That is, it follows that

$$X(t) = \sum_{k=1}^{m} A_k \cos\left[2\pi\beta_k(at^2 + bt) + \psi_k\right] + N(t), \tag{13.22}$$

is a GLC process, where the ψ_k are independent uniform $[0, 2\pi]$ random variables, and $N(t)$ is uncorrelated normal noise independent of the ψ_k's. We will refer to the GLC processes defined by (13.22) as *stochastic linear chirps (SLC)*.

Example 13.13 A Linear Chirp Signal with Two Frequencies

Figure 13.15a shows a realization from (13.22) with $m = 2$. Specifically, it is a realization from the model

$$X(t) = A_1 \cos\left[2\pi\beta_1(at^2 + bt) + \psi_1\right] + A_2 \cos\left[2\pi\beta_2(at^2 + bt) + \psi_2\right] + N(t), \tag{13.23}$$

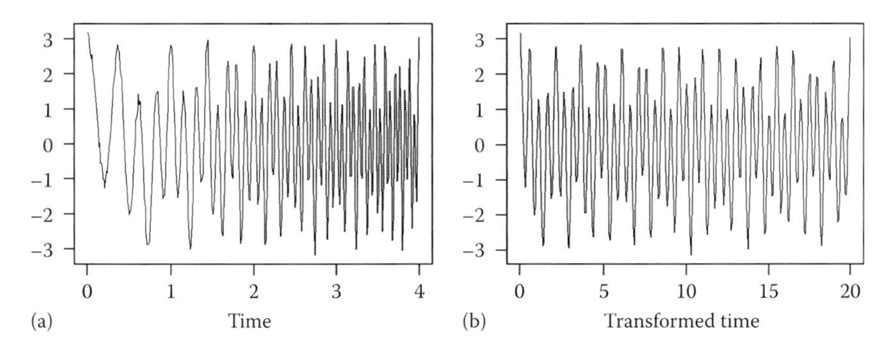

FIGURE 13.15
(a) Linear chirp realization from (13.23) with $A_1 = 1, \beta_1 = 1, \psi_1 = 0$, $A_2 = 2, \beta_2 = 2$, $\psi_2 = 1.5, a = 1, b = 1$, and $\sigma_N^2 = 0.04$, and (b) the dual realization.

where $A_1 = 1, \beta_1 = 1, \psi_1 = 0, A_2 = 2, \beta_2 = 2, \psi_2 = 1.5, a = 1, b = 1$, and $\sigma_N^2 = 0.04$. In Figure 13.15a, it can be seen that there is more than one frequency component that is increasing with time, and Figure 13.15b shows the dual realization in which the TVF behavior has been removed.

13.4.1 Models for Generalized Linear Chirps

In this section, we discuss autoregressive-type models for GLC processes that are similar to the Euler(p) and G($p;\lambda$) models. To begin the discussion, we define GLC white noise and then define the GLC(p) model. As before, we proceed formally assuming standard properties of the differential operator, D, since our purpose is to motivate the sampling scheme.

Definition 13.10 Let $z(u)$ be zero mean white noise and let $u = at^2 + bt, t > 0$. Then $\varepsilon(t) = z(u)$ is defined as *GLC white noise.*

Definition 13.11 Consider a process $\left\{X(t), t < \frac{-b}{2a} \text{ or } t > \frac{-b}{2a}\right\}$ defined by

$$\prod_{i=1}^{p}\left(\frac{1}{2at+b}D - \alpha_i\right)(X(t) - \mu) = \varepsilon(t), \tag{13.24}$$

where D is a differential operator, $a, b \in (-\infty, \infty), a \neq 0$ and the α_i are real or complex constants with $\text{Re}(\alpha_i) < 0, p > 0$, and $\varepsilon(t)$ is GLC (zero mean) white noise. Then, $X(t)$ is called a *GLC(p) process*. Note, if $a = 0$ and $t \in (-\infty, \infty)$, then $X(t)$ is an AR(p) process.

The GLC(p) process extends to a GLC(p, q) process. (See Robertson et al., 2010.) Just as SLC processes such as those in (13.23) are GLC processes, it can be shown that the GLC(p) and GLC(p, q) processes are also GLC processes and consequently are G-stationary as well. See Robertson (2008).

Note that if $u = at^2 + bt$, then $\frac{du}{dt} = 2at + b$. Consequently,

$$\begin{aligned}\left[\frac{1}{2at+b}D - \alpha\right]X(t) &= \frac{1}{2at+b}\frac{du}{dt}\frac{dX}{du} - \alpha X(t) \\ &= \frac{dX}{du} - \alpha X. \end{aligned}\tag{13.25}$$

Proceeding formally one can show

$$\prod_{i=1}^{p}\left(\frac{1}{2at+b}D_t - \alpha_i\right)(X(t) - \mu) = \prod_{i=1}^{p}(D_u - \alpha_i)(Y(u) - \mu),$$

where D_t and D_u are derivatives with respect to t and u, respectively, and where $Y(u) = Y(at^2 + bt) = X(t)$. It follows then that the dual process of $X(t)$ in (13.24) is an AR(p). For a further discussion of this, see Liu (2004), Robertson (2008), and Robertson et al. (2010).

As mentioned earlier, the time-transformation technique for G-stationary processes is essentially based on the concept of sampling at the appropriate points. If $X(t)$ is a GLC(p) process, we define $Y_k = Y(u_k) = X(t_k)$, where the sample points, $t_k \in (t_{(0)}, t_{(1)}]$, are given by

$$
\begin{aligned}
t_k &= \frac{-b - \sqrt{b^2 + 4au_k}}{2a}, \quad t_k \in \left(t_{(0)}, \frac{-b}{2a}\right] \\
&= \frac{-b + \sqrt{b^2 + 4au_k}}{2a}, \quad t_k \in \left(-\frac{b}{2a}, t_{(1)}\right],
\end{aligned} \tag{13.26}
$$

where $u_k = |k|\Delta_u$ and the Δ_u are equally spaced increments in U-space. The point $-b/2a$ is referred to as the *reflection point*. If Δ_u is sufficiently small, it can be shown that Y_k is a unique discrete ARMA(p, q) process with $q \leq p - 1$. See Jiang et al. (2006) and Robertson et al. (2010). Thus, again by sampling properly, one can convert the TVF (nonstationary) data to stationary data, perform the required analyses, and then transform back to the original time index.

Example 13.14 Sampling Scheme for an SLC Process

Let $X(t) = \cos(2\pi t^2 + \psi)$ where $\psi \sim \text{Uniform}[0, 2\pi]$, and let $t \in [-2, 2]$ with $\Delta_t = 0.05$. Then, as defined earlier, $t_{(0)} = -2$, $t_{(1)} = 2$, $a = 1$, and $b = 0$, and consequently $\dfrac{-b}{2a} = 0$. Taking $u = g(t) = t^2$, then from (13.26), if $t_k \in [t_{(0)}, -b/2a] = [-2, 0]$ then $u \in [0, 4]$. For this example we let $t_k = -\sqrt{u_k}$ where $u_k = |k|\Delta_u$, $k = -40, -39, \ldots, -1, -1, 0$, and $\Delta_u = 0.1$. Also, if $t_k \in (-b/2a, t_{(1)}) = (0, 2]$, then $u \in (0, 4]$ and $t_k = \sqrt{u_k}$, where $u_k = |k|\Delta_u$, $k = 1, 2, \ldots, 39, 40$, and $\Delta_u = 0.1$. We note that $\Delta_u = 0.1$ was selected to ensure that the number of data values in the dual realization is the same as the number of data values in the original realization.

Figure 13.16 is similar to Figure 13.1 for the case of a linear chirp. A realization of $X(t)$, where $\psi = 0$, is shown in Figure 13.16a which shows the plot of 81 equally spaced sampling points t between -2 and 2. Examining the plot shows that the sampling rate per cycle is lower in the regions of the data where the frequency is high. Conversely, the sampling rate per cycle is higher where the frequency is low, which is clearly undesirable. Figure 13.16b displays the same realization seen in Figure 13.16a sampled at time points t_k, where $t_k = \pm\sqrt{|k|\Delta_u}$. As evidenced by the tick marks in Figure 13.16b, this sampling strategy strategically defines t_k so that there are the

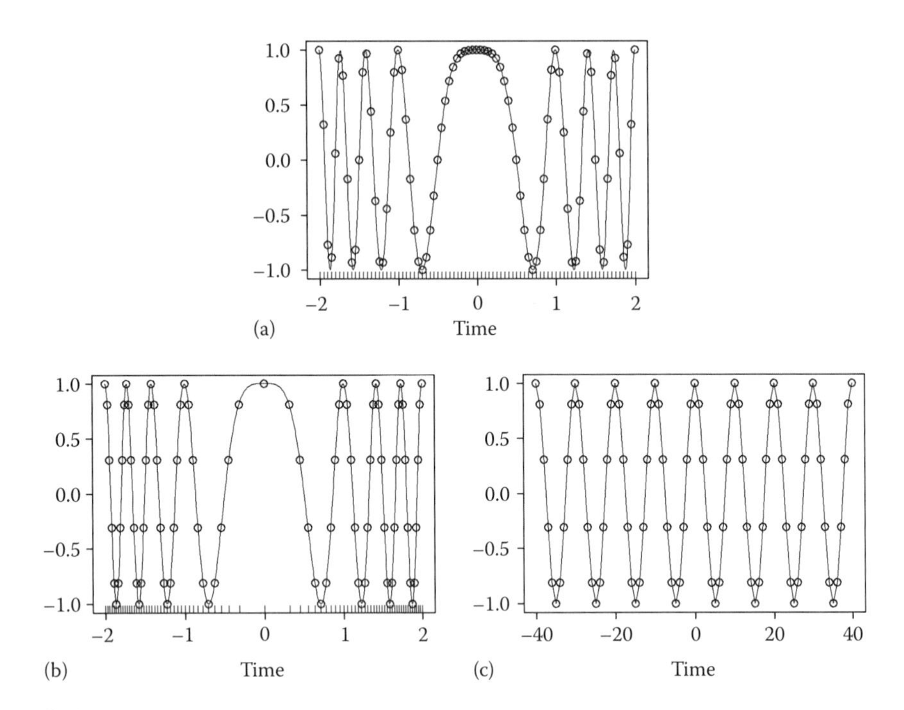

FIGURE 13.16
(a) Linear chirp signal showing equally-spaced time points, (b) same realization as in (a) showing the t_k time points given in (13.26), and (c) realization in (b) indexed on k.

same number of observations per cycle, and proceeds across the reflection point without interruption. Finally, Figure 13.16c shows a plot of the data in Figure 13.16b indexed on k. This is the dual process, and it can be seen that the frequencies are fixed.

The analysis steps for analyzing the GLC(p) processes are very similar to those for the G($p;\lambda$) processes, the only difference being the specific time transformation. As in the G($p;\lambda$) case, the t_k's are not equally spaced and are obtained by interpolation. Wang et al. (2009) give a Kalman filtering-based approach to avoid the need for interpolation. However, as stated previously, in most cases, we find the linear interpolation to be satisfactory.

Example 13.15 GLC(p) Analysis of Simulated Chirp Data

In this example, we consider the GLC(2) process associated with the discrete AR(2) dual process $(1 - 1.927B + 0.99B^2)Y_k = a_k$. A realization of 400 evenly spaced time points from this GLC(2) process is plotted in Figure 13.17a, in which it is clear that the frequency content is time varying, and in particular

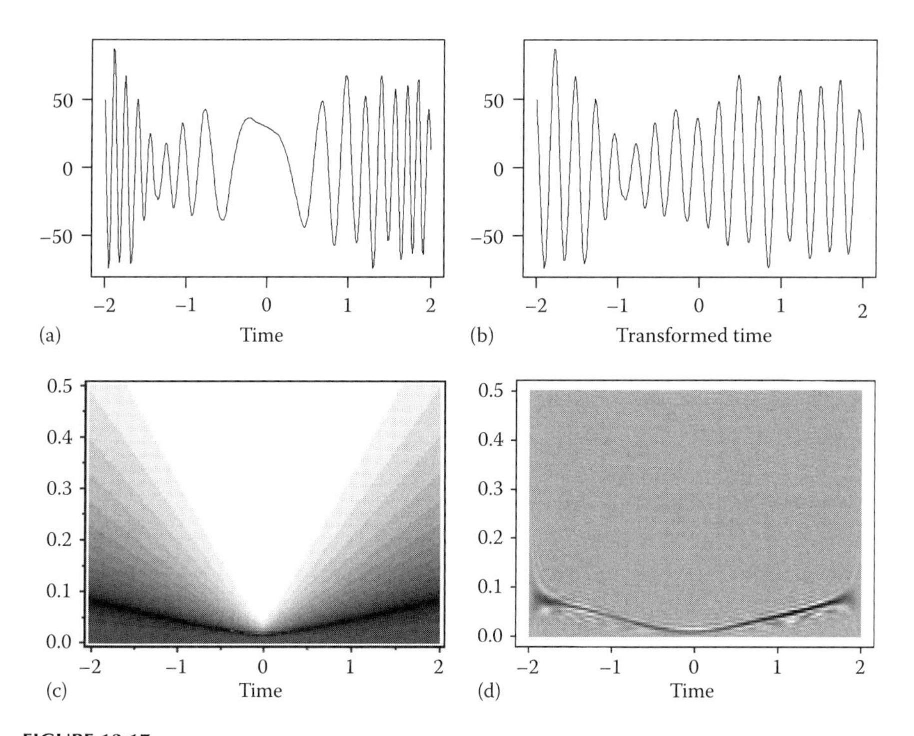

FIGURE 13.17
(a) GLC(2) data with $a = 1, b = 0$, (b) dual data based on a fitted GLC(5) model to the data in (a), (c) instantaneous spectrum base on GLC(5) fit, and (d) Wigner–Ville plot for data in (a).

is nonmonotonic in nature. The Wigner–Ville plot is shown in Figure 13.17d, where the linearity of the frequency change on each side of the reflection point can be seen. As in other cases, in order to analyze such a time-varying data set, we must find a stationary dual based on a transformation of the time index. The Wigner–Ville plot is used for estimating a and b in (13.21) based on the fact that the instantaneous frequencies change linearly on each side of the reflection point, and in fact change like $g'(t) = 2at + b$. See Robertson et al. (2010). Interpolation is used to approximate the values of the time series at the t_k's defined in (13.26). The resulting dual is found by indexing on k to yield $X(t_k) = Y_k$. A plot of Y_k is shown in Figure 13.17b, where it clearly looks stationary, especially compared to the data in Figure 13.17a. AIC selects an AR(5) model for the dual data. See Robertson et al. (2010). The instantaneous spectrum is shown in Figure 13.17c where it is seen that the linear frequency on each side of the reflection point is similar to that in the Wigner–Ville plot in Figure 13.17d. See Appendix 13.A.3 for details concerning the generalized instantaneous frequency and instantaneous spectrum in the linear chirp case.

13.5 Concluding Remarks

In this chapter, we have introduced important and relatively new methods for analyzing nonstationary time series with TVF behavior. The general approach is to transform the time axis to a dual space in order to obtain a stationary dual realization that can then be analyzed using standard techniques based on stationarity. A surprisingly large number of applications (chirp signals, Doppler signals, bat echolocation signals, seismic data, etc.) can be advantageously analyzed using these techniques. For more details concerning analysis and theoretical details, see the references listed throughout this chapter.

13.A Appendix

13.A.1 Generalized Instantaneous Period and Instantaneous Frequency of a G-Stationary Process

For nonstationary signals, whose spectral content vary with time, the frequency at a particular time is described by the concept of *instantaneous frequency* (IF). We consider first a simple sinusoidal signal of the form

$$X(t) = A\cos(\phi(t)), \tag{13.A.1}$$

where A is constant and ϕ is called the cumulative phase of the signal.

A common definition (Boashash, 1992) of the instantaneous frequency (IF) of $X(t)$ in (13.A.1) is

$$\text{IF} = \phi'(t)/2\pi.$$

Note that this definition shows the rate of change of $\phi(t)$ per 2π units. However, for some processes, this definition may not have a physical interpretation. Therefore, in Definition 13.A.2 we will give a new definition of instantaneous frequency for G-stationary processes. First, in Definition 13.A.1 we define the concepts of the G-period and generalized instantaneous period of a G-periodic function. These definitions were first given by Jiang et al. (2006). Consider the function $H(t)$ and let $u = g(t)$, so $t = g^{-1}(u)$ and $H(t) = H(g^{-1}(u)) = h(u)$. Now suppose $h(u + \delta) = h(u)$ for all u. Therefore $h(u)$ is a periodic function of u with period δ, and $h(u + \delta) = H(g^{-1}(u + \delta)) = H(g^{-1}(g(t) + \delta)) = H(t) = h(u)$. This leads to the following definition which generalizes the concept of periodicity to *periodicity on a given index set*. That is, although $H(t)$ is not a periodic function of t, $h(u)(=H(t))$ is a periodic function of u.

Definition 13.A.1 Let H be a function such that $H(t) = H(g^{-1}(g(t) + \delta_m))$, $m = 1, 2, \ldots$ where g is any monotonic function and $-\infty < \delta_m < \infty$. Then H is said to be a *G-periodic function* with the *G-period* δ which is equal to the δ_m for which $|\delta_m|$ is a minimum over m. The *generalized instantaneous period* (GIP) of the function H at time t, denoted $\mathcal{P}(t; g, \delta)$, is defined by

$$\mathcal{P}(t; g, \delta) = g^{-1}(g(t) + \delta) - t$$

Remark: If GIP = 0, then H is "a G-periodic", that is, not G-periodic. Note that $\delta = g(\mathcal{P} + t) - g(t)$.

Definition 13.A.2 The generalized instantaneous frequency (GIF), f, of a function $g(t)$ is a function of time defined as

$$f(t; g, \delta) = \frac{1}{\mathcal{P}(t; g, \delta)}.$$

If $\mathcal{P}(t; g, \delta) > 0$, then $\mathcal{P}$ is the period starting at t and measured in the positive direction. If $\mathcal{P}(t; g, \delta) < 0$, then $\mathcal{P}$ is the period starting at t but in the negative direction.

For a G(λ)-stationary process, such as the one given in Example 13.A.1 that follows with a single "periodic component", it is easy to show that the GIP of the G(λ)-autocovariance is

$$\mathcal{P}(t; g, \delta) = \begin{cases} (t^{\lambda} + \lambda\delta)^{1/\lambda} - t, & \lambda \neq 0 \\ t(e^{\delta} - 1), & \lambda = 0, \end{cases} \tag{13.A.2}$$

where $\lambda\delta > 0$.

The quantity δ can also be interpreted as the period of the autocovariance of the dual process. In the following, the GIP and GIF of the G(λ)-stationary processes are denoted by $\mathcal{P}(t; \lambda, \delta)$ and $f(t; \lambda, \delta)$ since λ is the only parameter of the function g. If $\lambda = 1$ (i.e., the stationary case), then $\mathcal{P}(t; 1, \delta) = \delta$ and $f(t; 1, \delta) = 1/\delta$. Consequently, as would be expected, in this case the GIP is the standard period, δ, for any $t > 0$, while $1/\delta$ is the usual frequency.

As already described in Section 13.3, differing values of λ produce a large variety of periodic behaviors of the process, or more specifically, of the G(λ)-autocovariance. Jiang et al. (2006) show that for $0 < \lambda < 1$, the GIP is a monotonically increasing concave function, while for $\lambda < 0$, it is monotonically increasing and convex. If $\lambda > 1$, the GIP is monotonically decreasing and convex.

From Example 13.A.1 which follows, it is clear that the GIF and IF are not the same. However, for G-stationary processes, the IF is a first-order approximation of the GIF.

Example 13.A.1 Illustration of the Difference between GIF and IF

Let $H(t) = \cos g(t)$, where $u = g(t) = (t^\lambda - 1)/\lambda$. Therefore, $t = g^{-1}(u) = (u\lambda + 1)^{1/\lambda}$ and $g^{-1}(g(t) + \delta) = [(g(t) + \delta)\lambda + 1]^{1/\lambda} = [t^\lambda + \delta\lambda]^{1/\lambda}$, for $\lambda > 0$. Therefore, if δ is the G-period, then from Definition 13.A.1, it follows that $H(t) = H(g^{-1}(g(t) + \delta))$. So, if $\lambda > 0$, then $\delta > 0$ and it follows that

$$\cos\left(\frac{t^\lambda - 1}{\lambda}\right) = \cos\frac{[(t^\lambda + \delta\lambda)^{1/\lambda}]^\lambda - 1}{\lambda} = \cos\left(\frac{t^\lambda - 1}{\lambda} + \delta\right).$$

Consequently, $\delta = 2\pi$. Note also that if $\lambda < 0$, then $t^2 + \delta\lambda > 0$ for $t > 0$, which implies that $\delta < 0$. Thus, in this case, $\delta = -2\pi$.

Now, returning to the case $\lambda > 0$, we have

$$\mathcal{P}(t;\lambda,2\pi) = (t^\lambda + 2\pi\lambda)^{1/\lambda} - t = t[1 + 2\pi\lambda t^{-\lambda}]^{1/\lambda} - t.$$

Then, for $t^\lambda > 2\pi\lambda$, it follows that the GIP is given by

$$\mathcal{P}(t;\lambda,2\pi) = \left\{t\left[1 + \frac{1}{\lambda}(2\pi\lambda t^{-\lambda}) + \frac{1}{2\lambda}\left(\frac{1}{\lambda} - 1\right)(2\pi\lambda t^{-\lambda})^2 + \cdots\right]\right\} - t.$$

So, the GIF is the reciprocal, which is

$$\text{GIF} = \frac{1}{2\pi t^{-\lambda+1} + \frac{1}{2}(1-\lambda)(2\pi)^2 t^{-2\lambda+1} + \cdots} = \frac{t^{\lambda-1}}{2\pi}\left[\frac{1}{1 + \pi(1-\lambda)t^{-\lambda} + \cdots}\right].$$

Thus, for large t, $\text{GIF} \simeq \dfrac{t^{\lambda-1}}{2\pi} = \dfrac{g'(t)}{2\pi} = \text{IF}$.

Jiang et al. (2006) show that the results of Example 13.A.1 can be generalized to any analytic monotonically increasing function $g(t)$. That is, if $H(t) = A\cos[g(t)]$, and $g(t)$ is monotonically increasing, then $\delta = 2\pi$ and $\mathcal{P}(t;g,2\pi) = g^{-1}(g(t) + 2\pi) - t$. They also show that the IF is a first-term Taylor series approximation to the GIF.

13.A.2 Instantaneous Spectrum of a G(λ) Process

Since in many signals, the entire frequency structure is evolving with time, the concept of an instantaneous spectrum of a G(λ) process is needed. In this regard, let $f^* = 1/\delta$, and let $G_X(f^*;\lambda)$ denote the G(λ) spectral density.

For example, in the case of a $G(p;\lambda)$ model, the $G(\lambda)$ spectral density is the $AR(p)$ spectral density associated with the $AR(p)$ model fit to the dual. From Definition 13.A.2 and (13.A.2), it is clear that f^* is a function of f (=GIF) and t. That is from (13.A.2)

$$
\begin{aligned}
f^* &= 0, \quad \text{if } f = 0, \\
&= \left[\ln\left(1 + \frac{1}{tf}\right)\right]^{-1}, \quad \text{if } f \neq 0, \lambda = 0, \\
&= \frac{\lambda}{\left(t + \frac{1}{f}\right)^{\lambda} - t^{\lambda}}, \quad \text{if } f \neq 0, \lambda \neq 0.
\end{aligned} \tag{13.A.3}
$$

We can, therefore, capture the instantaneous frequencies through f^* defined in (13.A.3). Note that since $G_X(f^*;\lambda) = G_X(-f^*;\lambda)$, we will always take $f^* \geq 0$ and $f \geq 0$.

Definition 13.A.3 If $X(t)$ is a $G(\lambda)$ process and $G_X(f^*;\lambda)$ is its $G(\lambda)$-spectrum, then the instantaneous spectrum of $X(t)$ at time t, denoted $S(f,t;\lambda)$, is defined as

$$S(f,t;\lambda) = G_X(f^*;\lambda). \tag{13.A.4}$$

Theorem 13.A.1

If $X(t)$ is a $G(p,q;\lambda)$ process, then the instantaneous spectrum of $X(t)$ at time t is given by

$$S(f,t;\lambda) = \sigma_a^2 \frac{\prod_{j=1}^{q} |\beta_j - i2\pi f^*|^2}{\prod_{k=1}^{p} |\alpha_k - i2\pi f^*|^2}, \tag{13.A.5}$$

where f^* is given by (13.A.3).

Proof

Follows from (13.A.4).

13.A.3 Generalized Instantaneous Frequency and Instantaneous Spectrum of a GLC(p) Process

In the case of the GLC(p) process, the generalized instantaneous frequency is given by

$$f_t = \frac{2a}{(2at + b)\left(-1 \pm \sqrt{\frac{1 + 8a\pi}{(2at + b)^2}}\right)}, \tag{13.A.6}$$

where the + and − relate to frequencies to the right and left of the reflection point, respectively. Corresponding to (13.A.3) and (13.A.4), it follows that the instantaneous spectrum, $S(f_t, t; a, b)$, is given by $S(f_t, t; a, b) = G(f^*; a, b)$, where $G(f^*; a, b)$ is the spectrum of the dual and

$$f^* = \frac{f_t^2}{(2at + b)f_t + a}. \quad (13.A.7)$$

See Robertson et al. (2010) and Jiang et al. (2006).

Exercises

The following problems depend on the availability of the `GWS S-Plus` software or other code that performs these functions. A `GWS` Users Guide is available on the ATSA website. The problems include step-by-step instructions.

13.1 Create a realization of length $n = 256$ from the Euler(4) model

$$X(h^k) - 1.75X(h^{k-1}) + 1.95X(h^{k-2}) - 1.70X(h^{k-3}) + 0.95(h^{k-4}) = a(h^k)$$

with offset 60. In S+ with the GWS software loaded, select `GWS` on the top menu, then select `Euler Process`, and then select `Generate Data` in order to see the dialog box in which the model information is entered.

(a) What system frequencies should the dual data contain?

(b) Fit an Euler model to the generated data set by selecting `GWS` on the top menu, then `Euler Process`, and then `Fit Model` in order to see the model fitting dialog box. Choose [0:200] as the offset range and 12 as the maximum AR order. (These are the defaults.) On the plots section of the dialog box select `Estimated Instant Euler Spectrum`.

(c) Using the output printed by the program, show and discuss plots of

(i) The data.

(ii) The dual.

(iii) The spectral density estimates of the original and dual data. Is there evidence that the expected system frequencies (from (a)) are present in the dual data?

(iv) The forecast performance of the Euler model versus the AR model fit to the original data.

(v) The estimated instantaneous Euler spectrum.

(d) Compute and plot the Wigner–Ville spectrum and compare it with the instantaneous spectrum in (c).

13.2 File PROB13.2.XLS contains a realization of length $n = 256$ from the M-stationary cosine plus noise process

$$X(t) = 5\cos[2\pi(0.2)\ln(t) + \psi] + a(t),$$

where $\psi = 1.2$, and where $\sigma_a^2 = 1$.

(a) Fit an Euler model to the data set by selecting GWS on the top menu, then Euler Process, and then Fit Model in order to see the model fitting dialog box. Choose 0:200 as the offset range and 12 as the maximum AR order. (These are the defaults.)

NOTE: You will need to import the data in PROB13.2.XLS into S+.

(b) Using the output printed by the program, show and discuss plots of the items listed in Problem 13.1(c).

13.3 Rerun the analyses in Problem 13.1 by generating a realization of length $n = 256$ from the G(4; 2) model with dual

$$X_t - 1.75X_{t-1} + 1.95X_{t-2} - 1.70X_{t-3} + 0.95X_{t-4} = a_t$$

with offset 60. In S+ with the GWS software loaded, select GWS on the top menu, then select G-lambda Process, and then select Generate Data in order to see the dialog box in which the model information is entered. Discuss the difference between the TVF behavior of an Euler(p) model and a G(p; 2) model.

Fit a G(p; λ) model to the data. This involves two steps:

(a) Identify offset and λ.
(b) Fit a G(p; λ) model to the data using the offset and λ obtained in (a).

These two steps are accomplished as follows:

(a) Select GWS on the top menu, then G-lambda Process, and then Data Explorer. Enter the λ-range 1:3 (to include the true value of 2) and offset range 0:100. This step will take a few minutes. The output file displayed on the screen shows the best choices of lambda and offset as measured by techniques discussed by Jiang et al. (2006) and Jiang (2003).
(b) Fit a G(p; λ) model by selecting GWS on the top menu, then G-lambda Process, then Fit Model in order to see the model fitting dialog box. Enter the values of offset and λ found in (a) and choose defaults for all other parameters.
(c) Using the output printed out by the program, show and discuss plots of the items listed in Problem 13.1(c).

13.4 Use steps (a)–(c) in Problem 13.3 to fit G(p; λ) models to the following data sets:

(A) DOPPLER.XLS—a Doppler signal
(B) PROB12.6C.XLS—a simulated data set
(C) NOCTULUS.XLS

References

Akaike, H. (1969). Fitting autoregressive models for regression, *Annals of the Institute of Statistical Mathematics* 21, 243–247.

Akaike, H. (1973). Information theory and an extension of the maximum likelihood principle. *2nd International Symposium on Information Theory*, B.N. Petrov and F. Csaki (eds.), Akademiai Kiado, Budapest, 267–281.

Akaike, H. (1976). Canonical correlations analysis of time series and the use of an information criterion, in *Advances and Case Studies in System Identification*, eds. R. Mehra and D.G. Lainiotis, Academic Press: New York.

Alder, A.C., Fomby, T.B., Woodward, W.A., Haley, R.W., Sarosi, G., and Livingston, E.H. (2010). Association of viral infection and appendicitis, *Archives of Surgery* 145, 63–71.

Andel, J. (1986). Long memory time series models, *Kybernetika* 22, 105–123.

Anderson, T.W. and Darling, D.A. (1952). Asymptotic theory of certain "goodness-of-fit" criteria based on stochastic processes, *Annals of Mathematical Statistics* 23, 193–212.

Ansley, C.F. (1979). AN algorithm for the exact likelihood of a mixed autoregressive-moving average process, *Biometrika* 66, 59–65.

Bartlett, M.S. (1946). On the theoretical specification and sampling properties of autocorrelated time series, *Journal of the Royal Statistical Society* B8, 27–41.

Beach, C.M. and MacKinnon, J.G. (1978). A maximum likelihood procedure for regression with autocorrelated errors, *Econometrica* 46, 51–58.

Beguin, J.M., Gourieroux, C., and Montfort, A. (1980). Identification of a mixed autoregressive-moving average process: The corner method, in *Time Series* (Ed. O.D. Anderson), North-Holland: Amsterdam, the Netherlands, pp. 423–436.

Bera, A.K. and Higgins, M.L. (1993). ARCH models: Properties, estimation and testing, *Journal of Economic Surveys* 7, 307–366.

Beran, J. (1994). *Statistics for Long-Memory Processes*, Chapman & Hall: New York.

Bhansali, R.J. (1979). A mixed spectrum analysis of the lynx data, *Journal of the Royal Statistical Society* A142, 199–209.

Billingsley, P. (1995). *Probability and Measure*, 3rd edn., John Wiley & Sons: New York.

Bloomfield, P. and Nychka, D.W. (1991). Climate spectra and detecting climate change, *Climatic Change* 21, 275–287.

Boashash, B. (1992a). Estimating and interpreting the instantaneous frequency—Part 1: Fundamentals, *Proceedings of the IEEE* 80, 520–538.

Boashash, B. (1992b). Estimating and interpreting the instantaneous frequency—Part 2: Algorithms and applications, *Proceedings of the IEEE* 80, 540–568.

Boashash, B. (2003). *Time Frequency Analysis*, Elsevier: Oxford, U.K.

Bolerslev, T. (1986). Generalized autoregressive conditional heteroskedasticity, *Journal of Econometrics* 31, 307–327.

Box, G.E.P. and Cox, D.R. (1964). An analysis of transformations, *Journal of the Royal Statistical Society, Series B* 26, 211–252.

Box, G.E.P, Jenkins, G.M., and Reinsel, G.C. (2008). *Time Series Analysis: Forecasting and Control*, Wiley: Hoboken, NJ.

Box, G.E.P. and Pierce, D.A. (1970). Distribution of the autocorrelations in autoregressive moving average time series models, *Journal of American Statistical Association* 65, 1509–1526.

Breakspear, M., Brammer, M., Bullmore, E., Das, P., and Williams, L. (2004). Spatiotemporal wavelet resampling for functional neuroimaging data, *Human Brain Mapping* 23, 1–25.

Brillinger, D.R. (1989). Consistent detection of a monotonic trend superimposed on a stationary time series, *Biometrika* 76, 23–30.

Brockwell, P.J. and Davis, R.A. (1991). *Time Series: Theory and Methods*, 2nd edn., Springer-Verlag: New York.

Brockwell, P.J. and Davis, R.A. (2002). *Introduction to Time Series and Forecasting*, 2nd edition, Springer Verlag: New York.

Bruce, A. and Gao, H. (1996). *Applied Wavelet Analysis with S+*, Springer Verlag: New York.

Bullmore, E., Long, C., Suckling, J., Fadili, J., Calvert, G., Zelaya, F., Carpenter, T., and Brammer, M. (2001). Colored noise and computational inference in neurophysiological time series analysis: Resampling methods in time and wavelet domains, *Human Brain Mapping* 12, 61–78.

Burg, J.P. (1975). Maximum entropy spectral analysis, PhD dissertation, Department of Geophysics, Stanford University, Stanford, CA.

Butterworth, S. (1930). On the theory of filter amplifiers, *Experimental Wireless and the Wireless Engineer* 7, 536–541.

Campbell, M.J. and Walker, A.M. (1977). A survey of statistical work in the Makenzic river series. If annual canadian lynx trapping for years 1821–1934 and a new analysis. *Journal of the Royal Statistical Society* A140, 411–431.

Canjels, E. and Watson, M.W. (1997). Estimating deterministic trends in the presence of serially correlated errors, *The Review of Economics and Statistics* 79, 184–200.

Carmona, R., Hwang, W., and Torrésani, B. (1998). *Practical Time-Frequency Analysis*, Academic Press: San Diego, CA.

Chan, W.S. (1994). On portmanteau goodness-of-fit tests in robust time series modeling, *Computational Statistics* 9, 301–310.

Cheng, Q.C. (1993). Transfer function model and GARMA II model, PhD dissertation, Department of Statistical Science, Southern Methodist University, Dallas, TX.

Cheng, Q.S. (1999). On time-reversibility of linear processes, *Biometrika* 86, 483–486.

Choi, B. (1986). An algorithm for solving the extended Yule-Walker equations of an autoregressive moving-average time series, *IEEE Transactions on Information Theory* 32, 417–419.

Choi, E., Woodward, W.A., and Gray, H.L. (2006). Euler (p,q) processes and their application to nonstationary time series with time-varying frequencies, *Communications in Statistics A* 35, 2245–2262.

Chui, C.K. (1992). *An Introduction to Wavelets*, Vol. 1, Academic Press: San Diego, CA.

Chung, C.F. (1996a). Estimating a generalized long memory process, *Journal of Econometrics* 73, 237–259.

Chung, C.F. (1996b). A generalized fractionally integrated autoregressive moving-average process, *Journal of Time Series Analysis* 17, 111–140.

Chung, C.L. (2001). *A Course in Probability Theory*, Academic Press: San Diego, CA.

Cochrane, D. and Orcutt, G.H. (1949). Application of least squares to relationships containing autocorrelated error terms, *Journal of the American Statistical Association* 44, 32–61.
Coifman, R. and Wickerhauser, V. (1992). Entropy-based algorithms for best basis selection, *IEEE Transactions on Information Control* 38, 713–718.
Cryer, J.D. and Chan, K.S. (2008). *Time Series Analysis with Applications in R*. Second Edition. Springer Verlag: New York.
Daubechies, I. (1988). Orthonormal bases of compactly supported wavelets, *Communications in Pure and Applied Mathematics* 41, 909–996.
Daubechies, I. (1992). *Ten Lectures on Wavelets*, Society for Industrial and Applied Mathematics, SIAM: Philadelphia, PA.
Davis, H.T. and Jones, R.H. (1968), Estimation of the innovation variance of a stationary time series, *Journal of the American Statistical Association* 63, 141–149.
Dickey, D.A. (1976). Estimation and hypothesis testing in Nonstationary Time Series, Ph.D. Dissertation, Iowa State University, Ames, Iowa.
Dickey, D.A. and Fuller, W.A. (1979). Distribution of the estimators for autoregressive time series with a unit root, *Journal of the American Statistical Association* 74, 427–431.
Diongue, A.K. and Guégan, D. (2008). Estimation of k-factor GIGARCH process: A Monte Carlo study, *Communications in Statistics-Simulation and Computation* 37, 2037–2049.
Donoho, D.L. (1993). Nonlinear wavelet methods for recovery of signals, densities, and spectra from indirect and noisy data, *Proceedings of Symposia in Applied Mathematics* 47, 173–205.
Donoho, D.L. and Johnstone, I.M. (1994). Ideal spatial adaptation via wavelet shrinkage, *Biometrika* 81, 425–455.
Donoho, D.L. and Johnstone, I.M. (1995). Adapting to unknown smoothness via wavelet shrinkage, *Journal of the American Statistical Association* 90, 1200–1224.
Durbin, J. (1960). Estimation of parameters in time series regression models, *Journal of the Royal Statistical Society B* 22, 139–153.
Engle, R. (1982). Autoregressive conditional heteroskedasticity with estimates of the variance of United Kingdom inflation, *Econometrica* 50, 987–1007.
Engle, R.F. and Granger, C.W.J. (1987). Co-integration and error-correction: Representation, estimation and testing, *Econometrica* 55, 251–276.
Engle, R.F. and Granger, C.W.J. (eds) (1991). *Long-Run Economic Relationships. Readings in Cointegration*, Oxford University Press: Oxford.
Eubank, R.L. (1999). *Nonparametric Regression and Spline Smoothing*, Marcel Dekker: New York.
Ferrara, L. and Guégan, D. (2001). Forecasting with k-factor Gegenbauer processes: Theory and applications, *Journal of Forecasting* 20, 581–601.
Findley, D.F. (1978). Limiting autocorrelations and their uses in the identification of nonstationary ARMA models, *Bulletin of the Institute of Mathematical Statistics* 7, 293–294.
Flandrin, P. (1992). Wavelet analysis and synthesis of fractional Brownian motion, *IEEE Transactions on Information Theory* 38, 910–917.
Ford, C.R. (1991). The Gegenbauer and Gegenbauer autoregressive moving average long-memory time series models, PhD dissertation, Department of Statistical Science, Southern Methodist University, Dallas, TX.

Fu, L. (1995). On the long-memory time series models, PhD dissertation, Department of Statistical Science, Southern Methodist University, Dallas, TX.
Geweke, J. and Porter-Hudak, S. (1983). The estimation and application of long memory time series models, *Journal of Time Series Analysis* 4, 221–237.
Giraitis, L. and Leipus, R. (1995). A generalized fractionally differencing approach in long-memory modeling, *Matematikos ir Informatikos Institutas* 35, 65–81.
Granger, C.W.J. and Joyeux, R. (1980). An introduction to long-memory time series models and fractional differencing, *Journal of Time Series Analysis* 1, 15–30.
Gray, H.L., Kelley, G.D., and McIntire, D.D. (1978). A new approach to ARMA modeling, *Communications in Statistics—Simulation and Computation* B7, 1–77.
Gray, H.L. and Woodward, W.A. (1981). Application of S-arrays to seasonal data, in *Applied Time Series Analysis II* (Ed. D. Findley), Academic Press: New York, pp. 379–413.
Gray, H.L. and Woodward, W.A. (1986). A new ARMA spectral estimator, *Journal of the American Statistical Association* 81, 1100–1108.
Gray, H.L. and Woodward, W.A. (1987). Improved ARMA spectral estimation, *Annals of Operations Research* 9, 385–398.
Gray, H.L. and Zhang, N.F. (1988). On a class of nonstationary processes, *Journal of Time Series Analysis* 9, 133–154.
Gray, H.L., Vijverberg, C.P., and Woodward, W.A. (2005). Nonstationary data analysis by time deformation, *Communications in Statistics* A34, 163–192.
Gray, H.L., Zhang, N.F., and Woodward, W.A. (1989). On generalized fractional processes, *Journal of Time Series Analysis* 10, 233–257.
Gray, H.L., Zhang, N.F., and Woodward, W.A. (1994). Correction to on generalized fractional processes, *Journal of Time Series Analysis* 15, 561–562.
Graybill, F.A. (1983). *Matrices with Applications in Statistics*, 2nd edn., Wadsworth: Belmont, CA.
Green, P.J. and Silverman, B.W. (1993). *Nonparametric Regression and Generalized Linear Models*, Chapman & Hall: New York.
Gupta, I.N., Chan, W.W., and Wagner, R.A. (2005). Regional source discrimination of small events based on the use of Lg wavetrain, *Bulletin of the Seismological Society of America* 95, 341–346.
Haney, J.R. (2011). Analyzing time series with time-varying frequency behavior and conditional heteroskedasticity. Ph.D. Dissertation, Department of Satistical Science, Southern Methodist University, Dallas, TX.
Hart, J.D. (1997). *Nonparametric Smoothing and Lack-of-Fit Tests*, Springer-Verlag: New York.
Hastie, T.J. and Tibshirani, R.J. (1990). *Generalized Additive Model*, Chapman & Hall: Boca Raton, FL.
Haugh, L.D. (1976). Checking the independence of two covariance-stationary time series: a univariate residual cross correlation approach, *Journal of the American Statistical Association* 71, 378–385.
Hosking, J.R.M. (1981). Fractional differencing, *Biometrika* 68, 165–176.
Hosking, J.R.M. (1984). Modeling persistence in hydrological time series using fractional differencing, *Water Resources Research* 20, 1898–1908.
Hurst, H.E. (1951). Long-term storage capacity of reservoirs, *Transactions of the American Society of Civil Engineers* 116, 770–799.
Hurst, H.E. (1955). Methods of using long-term storage in reservoirs, *Proceedings of the Institution of Civil Engineers* Part I, 519–577.

Hurvich, C.M. and Tsai, C.-L. (1989). Regression and time series model selection in small samples, *Biometrika* 76, 297–307.

James, J.F. (2011). *A Student's Guide to Fourier Transforms: With Applications in Physics and Engineering*. Cambridge University Press: Cambridge.

Jenkins, G.M. and Watts, D.G. (1968). *Spectral Analysis and Its Applications*, Holden-Day: San Francisco, CA.

Jensen, M.J. (1999a). An approximate wavelet MLE of short and long memory parameters, *Studies in Nonlinear Dynamics and Economics* 3, 239–253.

Jensen, M.J. (1999b). Using wavelets to obtain a consistent ordinary least squares estimator of the long-memory parameter, *Journal of Forecasting* 18, 17–32.

Jiang, H. (2003). Time frequency analysis: G(a)-Stationary processes. Ph.D. Dissertation, Department of Statistical Science, Southern Methodist University, Dallas, TX.

Jiang, H., Gray, H.L., and Woodward, W.A. (2006). Time frequency analysis—G(λ)-stationary process, *Computational Statistics and Data Analysis* 51, 1997–2028.

Johansen, S. (1988). Statistical analysis of cointegration vectors, *Journal of Economic Dynamics and Control* 12, 231–254.

Johansen, S. (1991). Estimation and hypothesis testing of cointegration vectors in Gaussian vector autoregressive models, *Econometrica* 59, 1551–1580.

Johansen, S. and Juselius, K. (1990). Maximum likelihood estimation and inferences on cointegration—with applications to the demand for money, *Oxford Bulletin of Economics and Statistics* 52, 169–210.

Jones, R.H. (1978). Multivariate autoregression estimation using residuals, in *Applied Time Series Analysis* (Ed. D. Findley), Academic Press: New York, pp. 139–162.

Kalman, R.E. (1960). A new approach to linear filtering and prediction problems, *Transactions of the ASME-Journal of Basic Engineering* 82, 35–45.

Kalman, R.E. and Bucy, R.S. (1961). New results in linear filtering and prediction theory, *Journal of Basic Engineering* 83, 95–108.

Kang, H. (1992). Bias correction of autoregressive parameters and unit root tests, *Proceedings of the Section on Business and Economic Statistics, American Statistical Association*, 45–50.

Kanwal, R.P. (1998). *Generalized Functions: Theory and Technique*, 2nd edn., Birkhäuser: Boston, MA.

Keeling, C.D., Bacastow, R.B., Carter, A.F., Piper, S.C., and Whore, T.P. (1989). American Geophysical Union, *Geophysical Monograph* 55, 165–236.

Keshner, M.S. (1982). 1/f noise, *Proceedings of the IEEE* 70, 212–218.

Koopmans, L.H. (1974). *The Spectral Analysis of Time Series*, Academic Press: New York.

Kouamé, E.F. and Hili, O. (2008). Minimum distance estimation of k-factors GARMA processes, *Statistics and Probability Letters* 78, 3254–3261.

Kumar, P. and Foufoula-Georgiou, E. (1994). Wavelet analysis in geophysics: An introduction, in *Wavelets in Geophysics* (Eds. E. Foufoula-Georgiou and P. Kumar), Academic Press: San Diego, CA.

Levin, D. and Sidi, A. (1981). Two new classes of nonlinear transformations for accelerating the convergence of infinite integrals and series, *Applied Mathematics and Computation* 9, 175–215.

Levinson, N. (1947). The Wiener (root mean square) error criterion in filter design and prediction, *Journal of Mathematical Physics* 25, 262–278.

Li, W.K. (1988). A goodness-of-fit test in robust time series modeling, *Biometrika* 75, 355–361.

Lighthill, M.J. (1959). *Introduction to Fourier Analysis and Generalised Functions*, The University Press: Cambridge, U.K.

Liu, L. (2004). Spectral analysis with time-varying frequency, Ph.D. Dissertation, Department of Statistical Science, Southern Methodist University, Dallas, TX.

Ljung, G.M. (1986). Diagnostic testing of univariate time series models, *Biometrika* 73, 725–730.

Ljung, G.M. and Box, G.E.B. (1978). On a measure of a lack of fit in time series models, *Biometrika* 65, 297–303.

Ljung, G.M. and Box, G.E.B. (1979). The likelihood function of stationary autoregressive-moving average models, *Biometrika* 66, 265–270.

Magnus, W., Oberhettinger, F., and Soni, R.P. (1966). *Formulas and Theorems for the Special Functions of Mathematical Physics*, Springer: Berlin.

Mallat, S.G. (1989). A theory for multiresolution signal decomposition: The wavelet representation, *IEEE Transactions on Pattern Analysis and Machine Intelligence* 11, 674–693.

Mandelbrot, B. (1983). *The Fractal Geometry of Nature*, W.H. Freeman and Company: New York.

Mandlebrot, B.B. and Van Ness, J.W. (1968). Fractional Brownian motions, fractional noises, and applications, *SIAM Review* 10, 422–437.

Mandelbrot, B.B. and Wallis, J.R. (1969). Computer experiments with fractional Gaussian noises, *Water Resources Research* 7, 1460–1468.

Mann, H.B. and Wald, A. (1943). On the statistical treatment of linear stochastic difference equations, *Econometrica* 11, 173–200.

Marple, S.L. and Nuttall, A.H. (1983). Experimental comparison of three multichannel linear prediction spectral estimators, *IEE Proceedings* Part F 130, 218–229.

Martin, W. and Flandrin, P. (1985). Wigner-Ville spectral analysis of nonstationary processes, *IEEE Transactions on Acoustics, Speech, and Signal Processing, ASSP* 33, 1461–1470.

McCoy, E.J. and Walden, A.T. (1996). Wavelet analysis and synthesis of stationary long-memory processes, *Journal of Computational and Graphical Statistics* 5, 26–56.

McQuarrie, A.D. and Tsai, C. (1998). *Regression and Time Series Model Selection*, World Scientific Publishing Co. Pte. Ltd., River Edge, NJ.

Miller, J.W. (1994). Forecasting with fractionally differenced time series models, PhD dissertation, Department of Statistical Science, Southern Methodist University, Dallas, TX.

Nelson, D.B. and Cao, C.Q. (1992). Inequality constraints in the univariate GARCH model, *Journal of Business and Economic Statistics* 10, 229–235.

Newbold, P. (1974). The exact likelihood function for a mixed autoregressive-moving average process, *Biometrika* 61, 423–426.

Nuttall, A.H. (1976a). FORTRAN program for multivariate linear predictive spectral analysis, employing forward and backward averaging, Naval Underwater Systems Center, Technical Report 5419, New London, CT.

Nuttall, A.H. (1976b). Multivariate linear predictive spectral analysis employing weighted forward and backward averaging: A generalization of Burg's method, Naval Underwater Systems Center, Technical Report 5501, New London, CT.

Ogden, R.T. (1997). *Essential Wavelets for Statistical Application and Data Analysis*, Birkhauser: Boston, MA.

O'Hair, J. (2010). Multidimensional signal detection in the presence of correlated noise with application to brain imaging, PhD dissertation, Department of Statistical Science, Southern Methodist University, Dallas, TX.
Park, R.E. & Mitchell, B.M. (1980). Estimating the autocorrelated error model with trended data, *Journal of Econometrics* 13, 185–201.
Parzen, E. (1962). *Stochastic Processes*, Holden-Day, San Francisco, CA.
Parzen, E. (1974). Some recent advances in time-series modeling, *IEEE Transactions on Automatic Control* AC-19, 723–730.
Parzen, E. (1980). Time series forecasting by ARARMA models, Technical Report, Texas A&M Institute of Statistics, College Station, TX.
Pavlicova, M., Santner, T.J., and Cressie, N. (2008). Detecting signals in fMRI data using powerful FDR procedures, *Statistics and Its Interface* 1, 23–32.
Peña, D. and Rodriguez, J. (2002). A powerful portmanteau test of lack of fit for time series, *Journal of the American Statistical Association* 97, 601–610.
Percival, D.B. and Walden, A.T. (1993). *Spectral Analysis for Physical Applications: Multitaper and Conventional Techniques*, Cambridge University Press: Cambridge, U.K.
Percival, D.B. and Walden, A.T. (2000). *Wavelet Methods for Time Series Data*, Cambridge University Press: Cambridge, U.K.
Porat, B. (1997). *A Course in Digital Signal Processing*, John Wiley & Sons: Hoboken, NJ.
Priestley, M.B. (1981). *Spectral Analysis and Time Series*, Academic Press, New York.
Quenouille, M.H. (1949). Approximate tests of correlation in time series, *Journal of the Royal Statistical Society, Series B* 11, 68–84.
Quinn, B.G. (1980). Limiting behaviour of autocorrelation function of ARMA process as several roots of characteristic equation approach unit circle, *Communications in Statistics—Simulation and Computation* 9, 195–198.
Rainville, E.D. (1960). *Special Functions*, Macmillan: New York.
Ramsey, F. (1974). Characterization of the partial autocorrelation function, *Annals of Statistics* 2, 1296–1303.
Reinsel, G.C. (1997). *Elements of Multivariate Time Series Analysis* (second edition). Springer-Verlag: New York.
Robertson, S.D. (2008). Generalizations and applications of the linear chirp, Ph.D. Dissertation, Department of Statistical Science, Southern Methodist University, Dallas, TX.
Robertson, S.D., Gray, H.L., and Woodward, W.A. (2010). The generalized linear chirp process, *Journal of Statistical Planning and Inference* 140, 3676–3687.
Sanders, J. (2009). A smoothing-based procedure for time series trend testing, PhD dissertation, Department of Statistical Science, Southern Methodist University, Dallas, TX.
Schwarz, G. (1978). Estimating the dimension of a model, *The Annals of Statistics* 6, 461–464.
Sendur, L. and Selesnick, I. (2002). Bivariate shrinkage functions for wavelet-based denoising exploiting interscale dependency, *IEEE Transactions on Signal Processing* 50, 2744–2756.
Sendur, L., Suckling, J., Whitcher, B., and Bullmore, E. (2007). Resampling methods for improved wavelet-based multiple hypothesis testing of parametric maps in functional MRI, *NeuroImage* 37, 1186–1194.
Shapiro, S.S. and Wilk, M.B. (1965). An analysis of variance test for normality (complete samples), *Biometrika* 52, 591–611.

Shen, X., Huang, H., and Cressie, N. (2002). Nonparametric hypothesis testing for a spatial event, *Journal of the American Statistical Association* 97, 1122–1140.

Shumway, R.H. and Stoffer, D.S. (1982). An approach to time series smoothing and forecasting using the EM algorithm, *Journal of Time Series Analysis* 3, 253–264.

Shumway, R.H. and Stoffer, D.S. (2006). *Time Series Analysis and Its Applications—With R Examples*, 2nd edn., Springer-Verlag: New York.

Strand, O.N. (1977). Multichannel complex maximum entropy (autoregressive) spectral analysis, *IEEE Transactions on Automatic Control* 22, 634–640.

Strang, G. and Nguyen, T. (1997). *Wavelets and Filter Banks*, Wellesley-Cambridge Press: Wellesley, MA.

Sun, H. and Pantula, S.G. (1999). Testing for trends in correlated data, *Statistics and Probability Letters* 41, 87–95.

Tang, L., Woodward, W., and Schucany, W. (2008). Undercoverage of wavelet-based resampling confidence intervals, *Communications in Statistics—Simulation and Computation* 37, 1307–1315.

Taqqu, M.S. (1986). A bibliographical guide to self-similar processes and long-range dependence, in *Dependence in Probability and Statistics* (Eds. E. Eberlein and M.S. Taqqu), Birkhäuser: Boston, MA, pp. 137–162.

Taylor, S. (1986). *Modelling Financial Time Series*, Wiley: Chichester, U.K.

Tewfik, A.H. and Kim, M. (1992). Correlation structure of the discrete wavelet coefficients of fractional Brownian motion, *IEEE Transactions on Information Theory* 38, 904–909.

Thomson, D.J. (1982). Spectrum estimation and harmonic analysis, *Proceedings of the IEEE* 70, 1055–1096.

Tiao, G.C. and Tsay, R.S. (1983). Consistency properties of least squares estimates of autoregressive parameters in ARMA models, *Annals of Statistics* 11, 856–871.

Tong, H. (1990) *Non-linear Time Series*, Clarendon Press, Oxford, U.K.

Tsay, R.S. (1992). Model checking via parametric bootstraps in time series analysis, *Journal of Applied Statistics* 41, 1–15.

Tsay, R.S. (2002). *Analysis of Financial Time Series Analysis,* John Wiley & Sons: Hoboken, NJ.

Tsay, R.S. and Tiao, G.C. (1984). Consistent estimates of autoregressive parameters and extended sample autocorrelation function for stationary and nonstationary ARMA models, *Journal of the American Statistical Association* 79, 84–96.

Tukey, J.W. (1961). Discussion emphasizing the connection between analysis of variance and spectrum analysis, *Technometrics* 3, 191.

Ulrych, T.J. and Clayton, R.W. (1976). Time series modelling and maximum entropy, *Physics of the Earth and Planetary Interiors* 12, 180–200.

van der Ziel, A. (1988). Unified presentation of 1/f noise in electronic devices: fundamental 1/f noise sources, *Proceedings of the IEEE* 76, 233–258.

Vijverberg, C.-P.C. and Gray, H.L. (2009). Discrete Euler processes and their applications, *Journal of Forecasting* 28, 293–315.

Vogelsang, T.J. (1998). Trend function hypothesis testing in the presence of serial correlation, *Econometrica* 66, 123–148.

Wahba, G. (1990). *Spline Models for Observational Data*, SIAM: Philadelphia, PA.

Waldmeier, M. (1961). *The Sunspot-Activity in the Years 1610–1960*, Schulthess and Co.: Zurich, Switzerland.

Wang, Z., Woodward, W.A., and Gray, H.L. (2009). The application of the Kalman filter to nonstationary time series through time deformation, *Journal of Time Series Analysis* 30, 559–574.

West, M. (1997). Time series decomposition, *Biometrika* 84, 489–494.

Whitcher, B. (2004). Wavelet-based estimation for seasonal long-memory processes, *Technometrics* 46, 225–238.

Whittle, P. (1953). Estimation and information in stationary time series, *Arkiv för Matematik* 2, 423–434.

Wiechecki, S.V. (1998). Semiparametric estimation for long memory parameters via wavelet packets, PhD dissertation, Department of Statistical Science, Southern Methodist University, Dallas, TX.

Woodward, W.A. (2006). Trend detecting. Encyclopedia of Environmetrics.

Woodward, W.A., Bottone, S., and Gray, H.L. (1997). Improved tests for trend in time series data, *Journal of Agricultural, Biological and Environmental Statistics* 2, 403–416.

Woodward, W.A., Cheng, Q.C., and Gray, H.L. (1998). A k-factor GARMA long-memory model, *Journal of Time Series Analysis* 19, 485–504.

Woodward, W.A. and Gray, H.L. (1978). A new model for Wölfer's sunspot numbers, *Communications in Statistics* 7B, 97–115.

Woodward, W.A. and Gray, H.L. (1981). On the relationship between the S Array and the Box-Jenkins method of ARMA model identification, *Journal of the American Statistical Association* 76, 579–587.

Woodward, W.A. and Gray, H.L. (1993). Global warming and the problem of testing for trend in time series data, *Journal of Climate* 6, 953–962.

Woodward, W.A. and Gray, H.L. (1995). Selecting a model for detecting the presence of a trend, *Journal of Climate* 8, 1929–1937.

Woodward, W.A., Gray, H.L., Gupta, I., and Haney, J.R. (2007). Use of time deformation methods to discriminate between earthquakes and explosions on the basis of Lg alone, *Bulletin of the Seismological Society of America* 97, 1196–1203.

Woodward, W.A., Gray, H.L., Haney, J.W., and Elliott, A.C. (2009). Examining factors to better understand autoregressive models, *American Statistician* 63, 335–342.

Wornell, G.W. (1993). Wavelet-based representation for the $1/f$ family of fractal processes, *Proceedings of the IEEE* 81, 1428–1450.

Wornell, G.W. (1996). *Signal Processing with Fractals: A Wavelet-Based Approach*, Prentice-Hall, Inc.: Upper Saddle River, NJ.

Yaglom, A.M. (1962). *An Introduction to the Theory of Stationary Random Functions*, Prentice-Hall, Inc.: Upper Saddle River, NJ.

Yule, G.U. (1927). On a method of investigating periodicities in disturbed series, with special reference to Wölfer's sunspot numbers, *Philosophical Transactions of the Royal Society* A226, 267–298.

Index

B

Q

R

S